THERMOMECHANICAL COUPLINGS IN SOLIDS

INTERNATIONAL UNION OF THEORETICAL AND APPLIED MECHANICS

THERMOMECHANICAL COUPLINGS IN SOLIDS

Jean Mandel Memorial Symposium
Paris, France, 1–5 September, 1986

Edited by

H. D. BUI

and

Q. S. NGUYEN

Laboratoire de Mécanique des Solides
Ecole Polytechnique
Palaiseau, France

1987

NORTH-HOLLAND
AMSTERDAM • NEW YORK • OXFORD • TOKYO

ISBN: 0 444 70236 9

Published by:
ELSEVIER SCIENCE PUBLISHERS B.V.
P.O. Box 1991
1000 BZ Amsterdam
The Netherlands

Sole distributors for the U.S.A. and Canada:
ELSEVIER SCIENCE PUBLISHING COMPANY, INC.
52 Vanderbilt Avenue
New York, N.Y. 10017
U.S.A.

PRINTED IN THE NETHERLANDS

PREFACE

The I.U.T.A.M. Symposium on "Thermomechanical Couplings in Solids" was held in Paris, in the first week of September 1986. The meeting offered an opportunity for Scientists, Mechanicians and Engineers to discuss various coupled problems in Solids Mechanics.

The volume contains the invited lectures presented at the Symposium. The contributions covered the following topics : thermodynamics of irreversible processes ; dissipative phenomena in plasticity and viscoplasticity, in fracture and damage ; behaviours of heteregeneous or complex media such as composites, woods and geomaterials ; phase transformation ; experimental techniques based on infrared thermography ; cyclic loads ; large deformation ; numerical methods for coupled problems, etc...

The organizers dedicated this Symposium to the memory of Jean Mandel who was Professor at the Ecole Polytechnique and the Ecole Nationale Supérieure des Mines, of Paris, and who was the President of the Laboratoire de Mécanique des Solides, Ecole Polytechnique, Palaiseau (France). His works covered a wide range of subjects, some of which were discussed in the present Symposium.

The Organizing Committee is indebted to many institutions for their financial supports of the Symposium : Centre National de la Recherche Scientifique, Electricité de France, I.U.T.A.M.

Enormous efforts in organizing the meeting were contributed by Geneviève Inglebert, Liliane Quéru, M. Amestoy, K. Dang Van and P. Navidi. The editors are indebted to Prof. I. Müller for many valuable discussions on the preparations of the Symposium project while he was a Visiting Professor at the Ecole Polytechnique.

Finally, we express our thanks to North-Holland for publishing the Symposium proceedings.

December, 1986 The Editors

SCIENTIFIC COMMITTEE

H.D. BUI	*(Chairman)* - Ecole Polytechnique, Palaiseau, France
J. CARLSSON	Royal Inst. of Technology, Stockholm, Sweden
P. GERMAIN	Académie des Sciences, Paris, France
K. HENNIG	Institut für Mechanik, Berlin, D.D.R.
F.A. LECKIE	University of Illinois, Urbana, U.S.A.
Th. LEHMANN	Lehrstuhl für Mechanik 1, Bochum, F.R.G.
P. PERZYNA	Polish Academy of Sciences, Varsovie, Pologne
A.R.S. PONTER	University of Leicester, Leicester, England

LIST OF PARTICIPANTS

ANTHOINE Armelle	L.M.S. - Ecole Polytechnique, Palaiseau, France
AMESTOY, M.	L.M.S. - Ecole Polytechnique, Palaiseau, France
ANDRIEUX, S.	Electricité de France, Clamart, France
ATLURI, S.N.	Georgia Inst. of Technology, Atlanta, U.S.A.
BAMBERGER, Y.	Electricité de France, Clamart, France
BENALLAL, A.	L.M.T. - E.N.S.E.T., Cachan, France
BEN CHEIKH, A.	L.M.T. - E.N.S.E.T., Cachan, France
BEREST, P.	L.M.S. - Ecole Polytechnique, Palaiseau, France
BLANC, R.H.	Labo. Mécanique & d'Acoustique, Marseille, France
BLANCHARD, D.	Sce Mathématiques - L.C.P.C., Paris, France
BRUN, L.	C.E.A., Villeneuve-Saint-Georges, France
BUISSON, M.	L.P.M.M. - Faculté des Sciences, Metz, France
CHADWICK, P.	University of East Anglia, Norwich, England
CHERET, R.	C.E.A., Villeneuve-Saint-Georges, France
COCKS, A.C.F.	Leicester University, Leicester, England
COMNINOU Maria	University of Michigan, Ann Arbor, U.S.A.
COUSIN, M.	I.N.S.A., Villeurbanne, France
CHRYSOCHOOS, A.	Université de Montpellier II, Montpellier, France
DANG VAN, K.	L.M.S. - Ecole Polytechnique, Palaiseau, France
DENIS Sabine	Ecole des Mines de Nancy, Nancy, France
DINH, C.M.	Institut Polytechnique de Hanoï, Hanoï, Viet-Nam
DUBOIS, Ph.	Université Paris Nord, Villetaneuse, France
DVORAK, G.	Rensselaer Polytechnic Institut, Troy, U.S.A.
EHRLACHER, A.	E.N.P.C., Noisy le Grand, France
FAN, J.	Chongqing University, Chongqing, Chine
FISCHER, F.D.	Institute of Mechanics, Leoben, Austria
FRANCFORT, G.	Sce Mathématiques, L.C.P.C., Paris, France
FREMOND, M.	Sce Mathématiques, L.C.P.C., Paris, France
FRELAT, J.	L.M.S. - Ecole Polytechnique, Palaiseau, France
GARY, G.	L.M.S. - Ecole Polytechnique, Palaiseau, France
GERALD, J.	S.N.E.A., Paris la Défense, France
GRIL, J.	L.M.S. - Ecole Polytechnique, Palaiseau, France
GUELIN, P.	I.M.G. - Saint-Martin d'Hères, France
HABIB, P.	L.M.S. - Ecole Polytechnique, Palaiseau, France
HALPHEN, B.	E.N.P.C., Noisy le Grand, France
HERRMANN, G.	Stanford University, Stanford, U.S.A.
HSIEH, R.K.T.	Royal Institute of Technology, Stockholm, Suède
HUET, Ch.	CERAM, E.N.P.C., Noisy le Grand, France
INGLEBERT Geneviève	L.M.S. - Ecole Polytechnique, Palaiseau, France
INOUE, T.	Kyoto University, Kyoto, Japon
JAMET, Ph.	DEMT - C.E.A. - Saclay, France
JULLIEN, J.F.	I.N.S.A. - Villeurbanne, France
KESTIN, J.	Brown University, Providence, U.S.A.
KLEPACZKO, J.R.	L.P.M.M. - Faculté des Sciences, Metz, France
KRUCH, S.	L.M.S. - Ecole Polytechnique, Palaiseau, France

LABORDE, P.	Université de Bordeaux I and L.M.S. - Ecole Polytechnique, Palaiseau, France
LAHOUD, A.	University of Sherbrooke, Sherbrooke, Canada
LATAILLADE, J.L.	Université de Bordeaux I, Talence, France
LEBLOND, J.B.	L.M.S. - Ecole Polytechnique, Palaiseau, France
LEE, E.H.	Lehrstuhl für Mechanik 1 and Rensselaer Polytechnic Institute, Troy, U.S.A.
LEHNER, K.	Shell E & P Laboratorium, Rijswijk, The Netherlands
LEMAITRE, J.	L.M.T. - ENSET - Cachan, France
LEXCELLENT, C.	L.M.A. - Faculté des Sciences, Besançon, France
LUBLINER, J.	University of California, Berkeley, U.S.A.
LUONG, M.P.	L.M.S. - Ecole Polytechnique, Palaiseau, France
MARIGO, J.J.	E.D.F. - Sce IMA - Clamart, France
MEZIERE, Y.	L.M.S. - Ecole Polytechnique, Palaiseau, France
MULLER, I.	Technische Universitat Berlin, Berlin, R.F.A.
NAVIDI, P.	L.M.S. - Ecole Polytechnique, Palaiseau, France
NGUYEN, Q.S.	L.M.S. - Ecole Polytechnique, Palaiseau, France
NOUILLANT, M.	Université de Bordeaux I, Talence, France
NOWACKI, W.K.	Polish Academy of Sciences, Varsovie, Pologne
ONAT, E.T.	Yale University, New Haven, U.S.A.
PAPADOPOULOS, Y.	L.M.S. - Ecole Polytechnique, Palaiseau, France
PARNES, R.	University of Tel Aviv, Tel Aviv, Israël
PREDELEANU, M.	L.M.T. - ENSET, Cachan, France
RADENKOVIC, D.	L.M.S., Ecole Polytechnique, Palaiseau, France
RANIECKI, B.	Polska Academia Nauf - Varsovie, Pologne
ROUSSELIER, G.	E.D.F. - Clamart, France
ROUSSET, G.	L.M.S. - Ecole Polytechnique, Palaiseau, France
SALENÇON, J.	L.M.S. - Ecole Polytechnique, Palaiseau, France
SCHMITT, N.	Lycée Technique M. Perret - Vincennes, France
STOLZ, C.	L.M.S. - Ecole Polytechnique, Palaiseau, France
SERRA, J.J.	E.T.C.A. - C.E.O. - Odeillo, France
TAYA, M.	University of Oxford, Oxford, England
TEODOSIU, C.	I.N.P.G. - GPMM - St-Martin d'Hères, France
ZAOUI, A.	Université Paris-Nord - LPMTM, Villetaneuse, France
ZARKA, J.	L.M.S. - Ecole Polytechnique, Palaiseau, France

CONTENTS

Preface v

Scientific Committee vii

List of Participants ix

Jean Mandel's Works
P. Habib 1

Simulation of Thermomechanical Properties of Materials with Shape Memory
I. Müller 5

Metal Plasticity as a Problem in Thermodynamics
J. Kestin 23

A New Approach for the Thermomechanics of Materials with Delayed Response
C. Huet 37

Thermomechanical Deformation and Coupling in Elastic-Plastic Composite Materials
G.J. Dvorak 43

On Thermoelastic Damping in Heterogenous Material
M. Buisson and A. Molinari 55

Infrared Technique in Solid Polymers Testing at High Strain Rates: Application to the Viscoplasticity of Polycarbonate
J.-L. Lataillade 71

The Heat Evolved During an Elastic-Plastic Transformation at Finite Strain
A. Chrysochoos 79

Infrared Observations of Damage and Fracture on Concrete
M.P. Luong 85

Thermoelastic Effects in Fracture
J.R. Barber and M. Comninou 95

Influence of Thermomechanical Coupling on Dynamic Fracture of Ductile Solids
P. Perzyna 105

Non-Isothermal Generalized Plasticity
J. Lubliner 121

Source of Heat in an Elastoviscoplastic Medium
P. Berest and G. Rousset 135

Modeling of Dimensional Change in Metal Matrix Composite Subjected to Thermal Cycling
M. Taya and T. Mori 147

Thermo-Hygro-Mechanical Couplings in Wood Technology and Rheological Behaviours
C. Huet 163

Unexpected Phenomena in Cyclic High-Temperature Plasticity
F.D. Fischer, F.G. Rammerstorfer, F.B. Bauer, and H.J. Boehm 183

Metallic Structure Under Mechanical and Cyclic Thermal Loading
J.F. Jullien, M. Cousin, and S. Ignaccolo 195

Dynamic Behaviour of a Carbon-Resin Composite
M. Nouillant, F. Joubert, and J.M. Delas 199

The Thermodynamics of Creep Damage
A.C.F. Cocks and F.A. Leckie 207

On a Thermomechanical Constitutive Theory and Its Application to CDM, Fatigue, Fracture and Composites
Jinghong Fan 223

Plastic Behaviour of Steels During Phase Transformations
J.B. LeBlond, J. Devaux, G. Mottet, and J.C. Devaux 239

Simulation of a Quenching Process for Carburized Steel Gear Under Metallo-Thermo-Mechanical Coupling
T. Inoue, Z.-G. Wang, and K. Miyao 257

A Uniaxial Model of Wood for Hygro-Thermo-Mechanical Loadings
J. Gril 263

Thermo-Mechanical Coupling in Large Deformations Particularly in Bifurcation Problems
Th. Lehmann 277

Asymptotic Transient Thermoelastic Behaviour
G.A. Francfort 291

Analysis of Thermomechanical Coupling by the Boundary Element Method
M. Predeleanu 305

The Two-Dimensional Problem of Thermoelasticity for an Infinite Region Bounded by a Circular Cavity
T. Honein and G. Herrmann 319

An Analysis of Thermomechanical Fields near Fast-Running Crack-Tips and in Welding
S.N. Atluri 325

Thermomechanical Coupling in Fracture Mechanics
H.D. Bui, A. Ehrlacher, and Q.S. Nguyen 327

Thermomechanical Coupling During Hot Working Processes
C. Teodosiu 343

Defect Models in Anisotropic Thermoelastic Materials
R.K.T. Hsieh 359

The Relationship Between Bounding Theory and the Thermodynamics of Metals with Special Emphasis on High Temperature Behaviour
A.R.S. Ponter and J.A. Scaife 369

Theoretical Schemes of Thermomechanical Coupling
D. Favier, P. Guelin, W.K. Nowacki, and P. Pegon 383

Modelling of Anisothermal Effects in Elasto-Viscoplasticity
A. Benallal and A. Ben Cheikh 403

Phase Change with Dissipation
D. Blanchard, M. Fremond, and A. Visintin 411

Thermoviscoelastic Behaviour of High Polymers: An Infrared Radiometry Study
R.H. Blanc and E. Giacometti 419

Thermomechanical Couplings in Solids
H.D. Bui and Q.S. Nguyen (Editors)
Elsevier Science Publishers B.V. (North-Holland)

JEAN MANDEL'S WORKS

It is now 4 years since Prof. Jean Mandel passed away and to day at the end of this meeting we shall designate the fifth prizewinner[1] of the Jean Mandel award. Jean Mandel founded this prize just before his death and he knew the first laureate but he was not with us when the first prize was given.

I am happy that this conference of the International Union of Theoretical and Applied Mechanics stands in Ecole Nationale Supérieure des Mines de Paris, one of the schools where he teached, since 1948 as a full Professor. At that time he was also Maître de Conference in Ecole Polytechnique, since 1942, and I remind he was known as a teacher who gave the students a taste for Mechanics. He became Professor of Mechanics in Ecole Polytechnique 1951 and remained till 1973.

Jean Mandel was Ingénieur des Mines, then Ingénieur en Chef, then Ingénieur Général. Some of our foreign guests know that it means that he remained in the French Administration all his life long.

In 1958, he asked me to come back to Ecole Polytechnique as a Preparateur of the course of Mechanics and to give some help for the Practical Works of the students. At this time, we started some activities in a very small laboratory about plexiglass viscoelasticity, about dust production during fluidization and about buckling of beams ; as you see it was very different subjects.

In 1961 Jean Mandel found some financial help and we were able to start with a full time laboratory, the name of which was soon Laboratoire de Mécanique des Solides,(Laboratory of Solid Mechanics). He was the Head. I was Assistant-Director. But that was all, and after that, we were completely alone.

---

(1) MM. Halphen - Mudry - Agassant - Coussy - Abouaf.

At that time H.D. Bui, the Scientific Committee Chairman of this Conference, was a very young engineer and he was among the very first to join the Laboratory of Solid Mechanics, with the financial support of Electricité de France.

Jean Mandel scientific activity was mainly devoted to continuum mechanics and more precisely to plasticity, viscoelasticity and viscoplasticity. At the beginning of his career, he found applications of these theories in Soil Mechanics and it is quite natural because soils exhibit viscoelastic deformations and because soil failure is something like a plastic flow. Jean Mandel solved practical problems and give fundamental contributions in each International Conference on Soil Mechanics and Foundation Engineering until 1965. But, when we started the Laboratory of Solid Mechanics, he thought to the applications of plasticity theory to metals and it was one of the first works of H.D. Bui ; Jean Mandel asked him to make the experimental determination of the subsequent yield surfaces of metals and to look at the effects of work hardening on yield surfaces.

In October 1964, Jean Mandel founded the "Groupe Français de Rhéologie" and he was chosen to be the first President of the Group. In 1980, he received the distinction of "Honorary Member" of the Group. In 1967, he became President of the "Comité Français de Mécanique des Roches" the french body of the International Society for Rock Mechanics.

With these different kinds of activities it was natural that the Laboratory of Solid Mechanics became a Common Laboratory of Ecole Polytechnique, Ecole Nationale Supérieure des Mines de Paris, Ecole Nationale des Ponts et Chaussées and was associated to C.N.R.S. (National Center for Scientific Research) and I think it is a very unusual situation to have groups of research in the fields of civil engineering, mining engineering, petroleum engineering, structural engineering in the same institution. I don't know many other examples of that, usually civil engineers don't know mining people and reciprocally.

This connection was in the ideas of Jean Mandel and appears clearly in his scientific work which cover a very large field with a bibliography listing five books and one hundred and fifty papers, most of them as a single author.

He presented original ideas about a lot of subjects, for instance the buckling of shells and beams, the finite deformation of solids, the laminar flow in porous media, the bearing capacity of shallow foundations, the punch resistance and the settlement of a two-layer medium, the plastic flow of metal, the propagation of plastic waves, the equilibrium of underground cavities, the effect of cyclical loading on structures, the large deformation of solids, about thermodynamics, about similarity, about rolling friction, about homogenization, about the distinction between rheological solids and liquids and so on.

For those who had the privilege to know him he will remain as an example of moral rigour and of exclusive scientific passion.

P. HABIB

Directeur du Laboratoire
de Mécanique des Solides

Thermomechanical Couplings in Solids
H.D. Bui and Q.S. Nguyen (Editors)
Elsevier Science Publishers B.V. (North-Holland)

SIMULATION OF THERMOMECHANICAL PROPERTIES OF MATERIALS WITH SHAPE MEMORY

INGO MÜLLER

FB 9 - Hermann-Föttinger-Institut für Thermo- und Fluiddynamik
Technische Universität Berlin, Straße des 17. Juni 135, 1000 Berlin 12
West Germany

1. PHENOMENOLOGY

1.1. Load - Deformation - Temperature Behaviour

Shape memory materials are characterized by a strong dependence of their load-deformation curves on temperature. This fact is demonstrated by the schematic load-deformation curves of Figure 1a through 1d, which correspond to increasing temperatures T_1 through T_4.

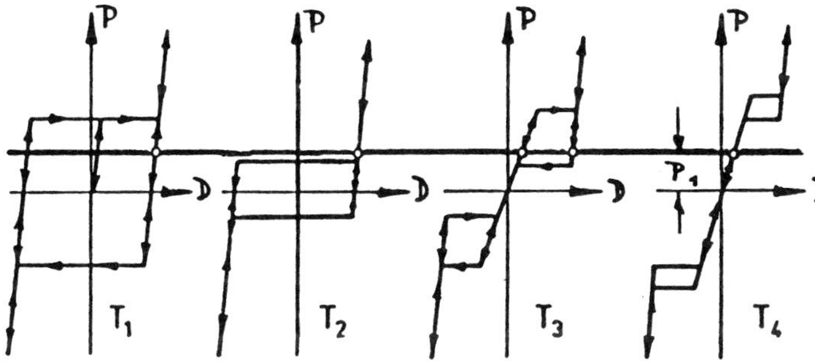

Figure 1: Schematized load-deformation diagrams

At low temperatures the behaviour is much like that of a plastic body with an initial elastic branch, and yield and creep upon loading and a residual deformation upon unloading. What is different from a plastic body is the existence of lateral elastic branches along which the body can be loaded far beyond the yield limit. At a slightly higher temperature this type of behaviour persists, but the yield limit is lower.

At high temperatures the load-deformation curves are entirely different: To be sure, there is still an initial elastic branch and a yield limit and a lateral elastic branch, but no residual deformation occurs upon unloading. Instead, as the load drops below a recovery load, the body regains the deformation that it yielded and eventually returns to the origin along the initial elastic line. This behaviour is called pseudo-elastic, because the body is elastic in that it returns to the origin in a loading-unloading cycle, but is only pseudo-elastic, because there is a hysteresis. At yet higher temperatures the pseudo-elastic behaviour persists but the yield limit and the recovery limit grow and the two limits grow closer together. It is clear that the load-deformation diagrams of Figure 1 imply shape memory. Indeed, if a body has started at low temperatures in the origin and then been given a residual deformation, heating to a high temperature will obviously restore it to the origin.

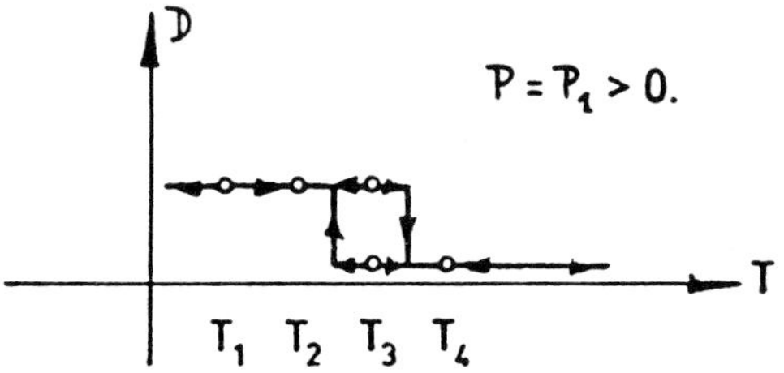

Figure 2: Schematic deformation-temperature diagram

The hystereses in the load-deformation diagrams of Figure 1 imply a hysteresis in the deformation-temperature diagram; see Figure 2, which is drawn for the constant load P_1 that is also marked in Figure 1. The dot in Figures 1 and 2 at temperatures T_1 through T_4 correspond to each other. We conclude that a loaded sample will contract upon heating and expand upon cooling. The hysteresis in a heating-cooling cycle reflects the hystereses in the loading - unloading cycles. This type of behaviour is at the basis of most of the applications of shape-memory alloys. Thermal actuators make use of the different deformations at different temperatures and medical applications for splints and braces make use of the hysteresis.

Typically the range of temperatures between T_1 and T_4 is 50^0C around room temperature and the recoverable deformations are about 6%.

Another aspect of the load-deformation-temperature behaviour of shape memory alloys is described by the Figure 3 which represents plots taken directly from a tensile test machine. In this figure the "input" consists of a frequently repeated triangular tensile load and of a temperature that first increases and then decreases. The "output" is the resulting deformation. In Figure 3 we see that the mean value of the deformation drops upon heating and then grows again upon cooling.

If the time t is eliminated between D(t) and T(t), we obtain the deformation temperature diagram at the bottom of Figure 3 which shows all the qualitative features of Figure 2.

The purpose of this paper is the description of a model that is capable of simulating the above described behaviour.

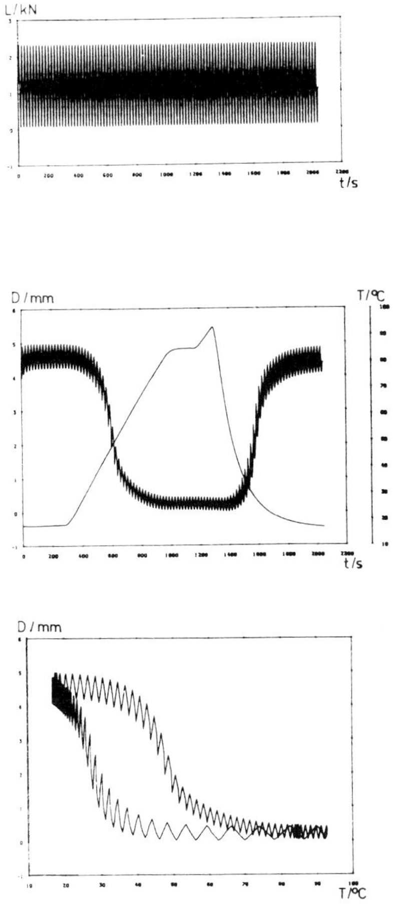

Figure 3: Deformation in response to time-dependent load and temperature

1.2. Phase Transition

The basis for the formulation of the model and for the qualitative understanding of the load-deformation-temperature behaviour of shape memory materials is the observation that between T_1 and T_4 the body undergoes a martensitic-austenitic phase transition and that the martensite is capable of twin formation. Thus in the origin of the (P,D)-diagram at low temperature the body consists of equal fractions of martensitic twins while on the lateral elastic branches one twin prevails. At high temperature the body is austenitic in the origin but it can be forced into one or the other martensitic twin by pulling or pushing it onto the lateral elastic branches.

2. THE MODEL

2.1. The Basic Element

The basic element of the model is what we may call a lattice particle, a small piece of the metallic lattice which is shown in the lower part of Figure 4 in three equilibrium configurations. The central highly symmetric configuration corresponds to t he austenitic phase A, while the lateral ones represent two martensitic twins $M_{\pm}$. These twins may clearly be seen as sheared versions of the austenitic particle. All intermediate shear length Δ are also possible and to each one corresponds a potential energy whose postulated form is shown in the upper part of Figure 4; it is characterized by two lateral stable minima for the martensitic twins and a central metastable minimum for the austenite. In between these minima there are potential barriers.

If the lattice particle is subject to a shear load P, the potential energy of that load must be taken into account. Since that potential energy is equal to - $P\Delta$, the lattice particle sees a total potential energy of the form $\Phi(\Delta) - P\Delta$, which is shown in Figure 5. We see that the heights of the barriers as well as the positions of the minima change with the load.

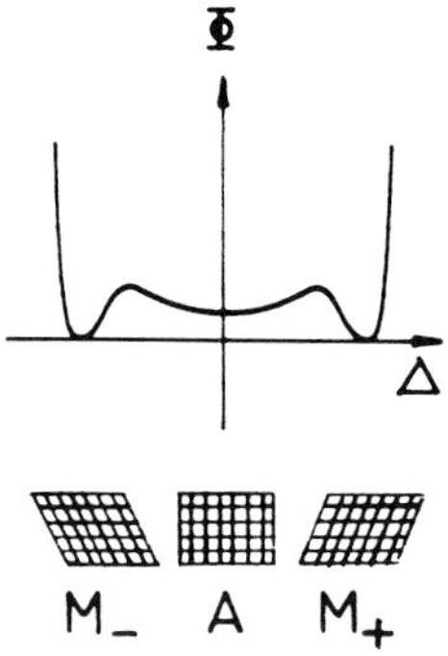

Figure 4: Lattice particles and their potential energy

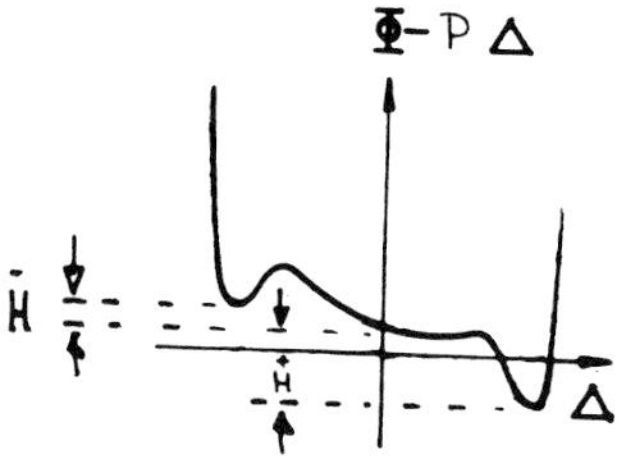

Figure 5: Potential energy under a shear load

2.2. The Body as a Whole

To construct the model for the body as a whole we arrange the lattice particles in layers and stack the layers on top of each other in the manner shown in Figure 6. The first four pictures in that figure correspond to the body at low temperature where the martensitic phase prevails. In particular in Figure 6 we have alternating layers of M_+ and M_-.

As the body is loaded by a small load the layers are sheared, i.e. the M_+ layers become flatter and the M_- layers become steeper. The vertical component of the shear length of each layer contributes to the overall deformation $D - D_0$ for which we thus obtain the formula

$$D - D_0 = \frac{1}{\sqrt{2}} \sum_{i=1}^{N} \Delta_i \,, \tag{2.1}$$

where N is the number of layers. When the load is removed the layers fall back to their initial positions, i.e. the deformation achieved by small loads is elastic.

When the load is increased to a critical value, the M_- layers flip into the M_+ position and thus a great change of deformation is achieved, see Figure 6c. Upon unloading all layers settle down in the equilibrium position of M_+, as shown in Figure 6d so that there is a residual deformation. We conclude that the properties of the model body do reflect the observed properties of shape memory materials at low temperature.

If indeed the body becomes austenitic at high temperature, it will contract to the shape of Figure 6e which macroscopically is the same shape as that of Figure 6a. Upon cooling the layers will become martensitic again and in the average we expect every other one to be of the M_+ and the M_- configuration so that we return to t he situation of Figure 6a. Thus in a qualitative manner we understand from the model the sequence of deformations involved in the shape memory experiment.

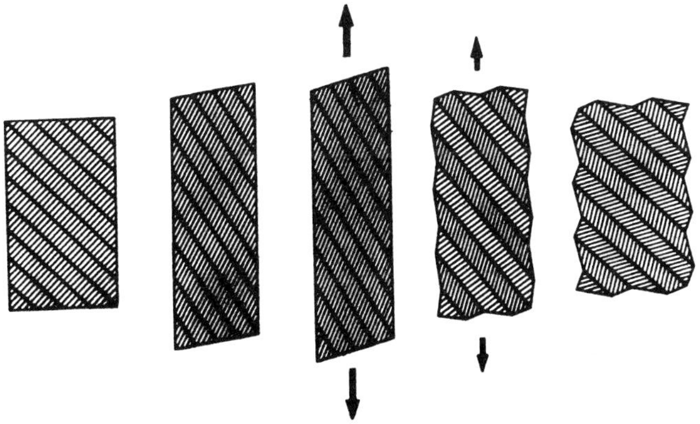

Figure 6: Model body and its deformation

The previous discussion may have some heuristic value, it is supposed to familiarize the reader with the model. In Chapter 3 we shall present more formal treatments that will allow us to predict deformations at higher temperatures and under time-dependent loads and temperature.

Figure 7 shows an etching of the martensitic body on which we clearly see the alternating layers of martensitic twins; pictures like these have motivated the construction of the model in the form described above.

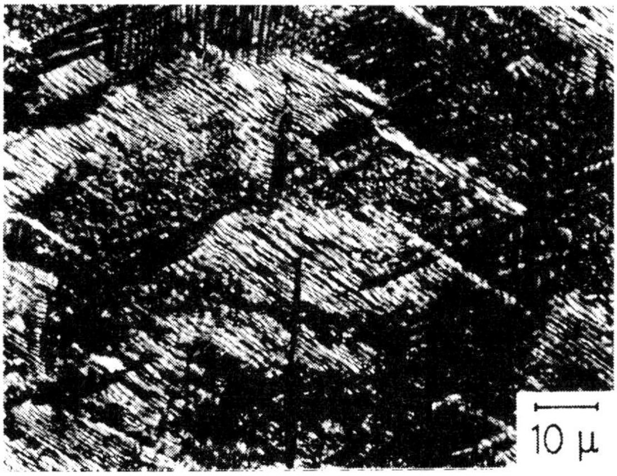

Figure 7: NiTi alloy in martensitic phase

2.3. Interfacial Energies

When two layers of martensitic and austenitc phase are in contact, we assume that there is an energy e associated with the interface. If there are K austenitic-martensitic interfaces, the energy is thus

$$E_I = K\, e\ .$$

(In principle,there should also be an energy associated with a M_+-M_- interface, but we ignore this, because the lattice distortion is minimal where two twins are in contact).

In the sequel we shall work with the martensitic and austenitic phase functions x_{M_+}, x_{M_-} and x_A which represent the fractions of lattice layers in the range of the right, left and central potential well of Figure 5 respectively. These phase fractions will serve as internal variables and it will be necessary to relate K -- the number of A-M interfaces -- to these fractions. This is done as follows: If there are Nx_A austenitic layers and Nx_M martensitic ones, it can be shown that there are

$$W_K = \binom{Nx_A - 1}{K/2 - 1} \binom{Nx_M - 1}{K/2 - 1}$$

possibilities to realize K interfaces. If W_K is small for $K = K_1$, the number K_1 of interfaces is improbable, but if W_K is big for $K = K_2$ the number K_2 is probable. We assume that we shall always see that number $\overline{K}$ for which W_K is maximal. This number turns out to be

$$\overline{K} = 2\,N\,x_A\;\;x_M$$

as can be shown by a simple condiseration.

Thus for given phase fractions x_M and x_A the interfacial energy is equal to

$$E_I = 2\,N\,e\,x_A\;\;x_M\,. \tag{2.2}$$

3. STATISTICAL MECHANICS OF THE MODEL

3.1. Thermal Fluctuations, Minimum of Free Energy

The lattice particles do not lie still in the minima of their potential wells, rather they participate in the thermal motion and therefore they fluctuate about those minima. Under these circumstances it is not practical to ascribe a shear length Δ_i to each layer, as was done in equation (2.1), rather we characterize the state of the model by giving the distribution function N_Δ whose value gives the number of layers with the shear length Δ.

At higher temperatures it is not easy to estimate the deformation of the body under a given load. Indeed, there are two conflicting tendencies to be considered: The tendency of energy E to become minimal by settling all layers into the deepest potential well and the tendency of entropy H to distribute all layers evenly over the whole range of shear lengths. In that competition it is the free energy

$$\Psi = E - TH \tag{3.1}$$

that actually assumes a minimum.

The free energy can be written in the form

$$\Psi = \sum_\Delta \Phi(\Delta) N_\Delta + 2\,Ne x_A x_M - kT \ln \frac{N!}{\prod_\Delta N_\Delta!}$$

The first two terms represent the energy while the last one is the entropic term in (1). The entropy is calculated from the number of possibilities to realize a given state $\{N_1 N_2 \ldots\}$ which is $N!/\prod_\Delta N_\Delta!$.

We write

$$x_A = \sum_{[\Delta]} \frac{N_\Delta}{N} \quad \text{and} \quad x_M = \sum_{]\Delta[} \frac{N_\Delta}{N} \tag{3.3}$$

where the brackets $[\Delta]$ and $]\Delta[$ indicate summation over the austenitic and the martensitic range respectively. Thus (2) becomes

$$\Psi = \sum_\Delta \Phi(\Delta) \; N_\Delta + \frac{2e}{N} \sum_{[\Delta]} N_\Delta \sum_{]\Delta[} N_\Delta - kT \ln \frac{N!}{\prod_\Delta N_\Delta!} \tag{3.4}$$

3.2. Thermodynamic State Functions

We minimize Ψ with respect to N_Δ under the constraints

$$\sum_\Delta N_\Delta = N \qquad \sum_\Delta \Delta N_\Delta = \sqrt{2}\,(D - D_0) \tag{3.5}$$

of which the latter is equivalent to (2.1). Thus we find the distribution $N_\Delta|_E$ that minimizes Ψ, viz.

$$N_\Delta = N e^{\frac{\alpha+P\Delta}{kT}-1} e^{-\frac{\phi(\Delta)+2ex_A}{kT}} \quad \text{for} \quad \Delta\in\,]\,[$$

$$N_\Delta = N e^{\frac{\alpha+P\Delta}{kT}-1} e^{-\frac{\phi(\Delta)+2ex_M}{kT}} \quad \text{for} \quad \Delta\in[\,] \tag{3.6}$$

α and P are the Lagrange multipliers that take care of the constraints (5), α can easily be calculated by insertion of (6) into (5), and P will later be interpreted as the load on the body.

Insertion of (6) into (2), $(5)_2$ and (3) gives

$$\Psi = -NkT\ln\left\{e^{-\frac{2ex_A}{kT}}\sum_{]\,[} e^{\frac{P\Delta-\phi(\Delta)}{kT}} + e^{-\frac{2ex_M}{kT}}\sum_{[\,]} e^{\frac{P\Delta-\phi(\Delta)}{kT}}\right\} +$$

$$+\sqrt{2}\,P(D-D_0) - 2Ne\,x_A x_M, \tag{3.7}$$

$$D-D_0 = \frac{N}{\sqrt{2}}\,\frac{e^{-\frac{2ex_A}{kT}}\sum_{]\,[}\Delta e^{\frac{P\Delta-\phi(\Delta)}{kT}} + e^{-\frac{2ex_M}{kT}}\sum_{[\,]}\Delta e^{\frac{P\Delta-\phi(\Delta)}{kT}}}{e^{-\frac{2ex_A}{kT}}\sum_{]\,[} e^{\frac{P\Delta-\phi(\Delta)}{kT}} + e^{-\frac{2ex_M}{kT}}\sum_{[\,]} e^{\frac{P\Delta-\phi(\Delta)}{kT}}}, \tag{3.8}$$

$$x_M = \frac{e^{-\frac{2ex_A}{kT}}\sum_{]\,[} e^{\frac{P\Delta-\phi(\Delta)}{kT}}}{e^{-\frac{2ex_A}{kT}}\sum_{]\,[} e^{\frac{P\Delta-\phi(\Delta)}{kT}} + e^{-\frac{2ex_M}{kT}}\sum_{[\,]} e^{\frac{P\Delta-\phi(\Delta)}{kT}}} \tag{3.9}$$

Note that $x_M = 1-x_A$ so that (9) is an implicit relation for the determination of the function $x_A = x_A\,(P,T)$, i.e. P and T determine the phase fractions x_A and x_M. Once this function is known, we may eliminate x_A and x_M from (8), thus getting $D = D\,(P,T)$ or, by inversion $P = P\,(D,T)$, the load as a function of deformation and temperature. Also by use of these results (7) determines the function $\psi = \psi(D,T)$. Differentiation of ψ with respect to D gives

$$P = \frac{\partial\Psi}{\partial D} \tag{3.10}$$

which confirms the above remark that the Lagrange multiplier P is equal to the load.

The actual calculation of the functions $P = P\ (D,T)$ and $\Psi = \Psi\ (D,P)$ must use numerical or graphical methods. For a typical potential $\Phi(\Delta)$ of the general form shown in Figure 4 we obtain functions of the form shown in Figures 8 and 9. The different curves in both figures refer to different temperatures. The temperature increases from left to right. The curves on the left refer to such a low temperature that statistical arguments can no longer be valid and therefore these curves must be ignored. On the other hand, at high temperature the curves on the right indicate that the body is non-linearly elastic. This is not observed in shape memory alloys because at high temperatures true plastic behaviour sets in, which disguises the shape memory phenomena and which is not accounted for in the model.

The load-deformation curves relevant to pseudo-elasticity are the two non-monotone ones at intermediate temperatures. The corresponding free energy curves are shown in Figure 8; they are those with a loop and a section of negative curvature.

3.3. Hystereses

The two non-monotone curves in Figure 9 imply pseudo-elastic hysteresis. Indeed, if we begin at $P = 0$ and $D = 0$ to load a body, the deformation grows smoothly until we reach the maximum of the curve $P = P\ (D,P)$. Upon further loading there is a break-through to the right branch of that curve. On that branch we may move up and down until the minimum is reached. If the load falls below the minimum there is a break-through back to the initial ascending branch. A break-through occurs along the dashed horizontal lines in the diagrams of Figure 9 and we conclude that in a load controlled experiment the model predicts the pseudo-elastic hysteresis loops that were reported in Fig. 1. Indeed the yield limit and the recovery limit in Figure 9 grow with increasing temperature and the two grow closer together.

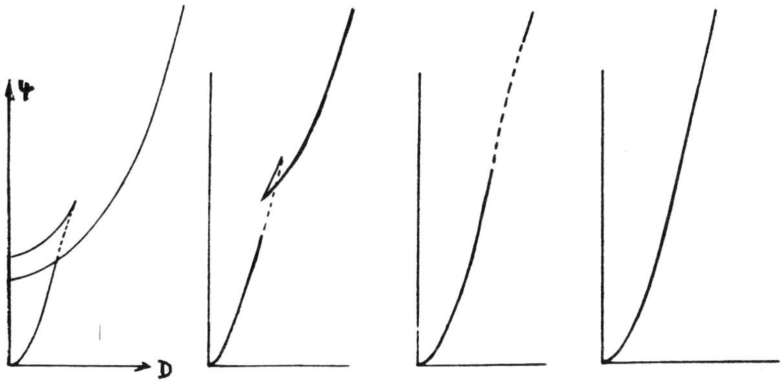

Figure 8: Free energy-deformation diagrams in their dependence on temperature

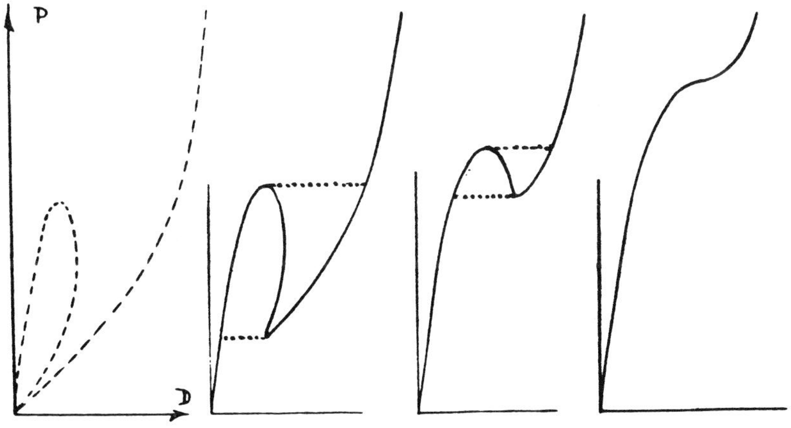

Figure 9: Load-deformation diagrams in their dependence on temperature

4. SHAPE MEMORY AS AN ACTIVATED PROCESS

4.1. The Deformation in Frozen Equilibrium

In this chapter we shall assume that, while t here is equilibrium in each potential well, equilibrium is not generally established between the wells. In such a situation we speak of "frozen equilibrium". The phase fractions are determined by initial conditions and they satisfy rate laws to be discussed below. In frozen equilibrium the distribution of lattice particles in the M_- -potential well (say) is given by

$$N_\Delta = \frac{e^{-\frac{\phi^{\wedge}(\Delta)-P\Delta}{kT}}}{\sum_{\Delta=-\infty}^{m_L} e^{-\frac{\phi^{\wedge}(\Delta)-P\Delta}{kT}}} \qquad -\infty < \Delta < m_L \tag{4.1}$$

and similar expressions hold for N_Δ in the ranges [A] and [M_+.

The expectation value for the shear length of the M_--particles is then

$$\overline{\Delta}^{M_-}(P,T) = \frac{\sum_{\Delta=-\infty}^{m_L} \Delta e^{-\frac{\phi^{\wedge}(\Delta)-P\Delta}{kT}}}{\sum_{\Delta=-\infty}^{m_L} e^{-\frac{\phi^{\wedge}(\Delta)-P\Delta}{kT}}} \tag{4.2}$$

and similarly for $\overline{\Delta}^{A}$ and $\overline{\Delta}^{M_+}$. Thus the formula (2.1) for the deformation can now be written in the form

$$D - D_0 = \frac{N}{\sqrt{2}}\left\{x_{M_-}\overline{\Delta}^{M_-}(P,T) + x_A\overline{\Delta}^{A}(P,T) + x_{M_+}\overline{\Delta}^{M_+}(P,T)\right\}. \tag{4.3}$$

This is a function of P, T and $x_{M_\pm}$.

4.2. Rate Laws in Activated Processes

Although the potential wells are separated by barriers that prevent the easy establishment of equilibrium, the layers on either side are fluctuating and every so often the barrier will be overcome by a layer thus leading to a change in the phase fractions. These changes are supposed to obey rate laws of the form

$$\begin{aligned}
\dot{x}_{M_-} &= -\overset{-o}{p}\, x_{M_-} + \overset{o-}{p}\, x_A \\
\dot{x}_A &= \overset{-o}{p}\, x_{M_-} - \overset{o-}{p}\, x_A - \overset{o+}{p}\, x_A + \overset{+o}{p}\, x_{M_+} \\
\dot{x}_{M_+} &= + \overset{o+}{p}\, x_A - \overset{+o}{p}\, x_{M_+}
\end{aligned} \tag{4.4}$$

Thus the rate of change of x_{M_-} (say) consists of two parts, a loss and a gain. The loss is due to particles that jump out of the left well into the central well and it is proportional to x_{M_-} with the transition probability $\overset{-o}{p}$ as a factor of proportionality. The gain is due to particles that jump from the middle to the left, it is proportional to x_A and the transition probability here is denoted by $\overset{o-}{p}$. The other rate laws are similarly constructed. Of course, the one for x_A has more terms, since the central minimum can exchange particles with both sides. We shall discuss the form of the transition probabilities in the next section

The temperature also satisfies a rate law which is a simplified version of the energy balance, viz.

$$C\,\dot{T} = \alpha(T-T_E) - (\dot{x}_{M_-}\,\overset{-}{H}(P) + \dot{x}_{M_+}\,\overset{+}{H}(P)). \tag{4.5}$$

C is the heat capacity of the body. Thus the temperature changes due to an energy efflux that is proportional to the difference of the body temperature and the external temperature T_E. Also there is a temperature change when layers jump from the left well to the central one, or vice versa. In the first case the potential energy $\overset{-}{H}(P)$ (see Figure 5) is converted into kinetic energy, i.e. thermal energy so that the body heats up; in the second case the body cools as kinetic energy is converted into potential energy.

4.3. The Transition Probabilities

The transition probability $\overset{-o}{p}$ consists of three factors:
The first factor is the number N_Δ for $\Delta = m_L$ from (1), i.e. the member of layers from the left well on the top of the left barrier.

The second factor is equal to $\frac{kT}{2\pi m}$ which determines the speed with which the layers of mass m fluctuate and therefore the frequency with which they run into the barrier.

The third factor is due to the fact that, by (2.2) the interfacial energy changes when a layer jumps from the left well into the middle. That change is equal to $\overset{-o}{\Delta E_I} = 2\,e\,(1 - 2\,x_M)$ and the probability that this energy is available is equal to

$$\exp\left(-\frac{2\,e\,(1 - 1\,x_M)}{kT}\right)$$

Thus the transition probability is equal to

$$\overset{-o}{p}(P,T,x_{M_\pm}) = \sqrt{\frac{kT}{2\pi m}}\;\frac{e^{-\frac{\Phi(m_L)-Pm_L}{kT}}}{\sum\limits_{\Delta=-\infty}^{m_L} e^{-\frac{\Phi(\Delta)-P\Delta}{kT}}}\; e^{-\frac{2e(1-2x_M)}{kT}} \qquad (4.6)$$

it is a function of P, T and $x_{M_\pm}$. The other transition probabilities are similarly constructed.

Whenever a transition probability is composed of Boltzmann factors like those in (6) we speak of the corresponding transition as an activated process. We may say that the process can be activated by a higher temperature or by a greater force, because both increase the transition probability.

4.4. The Determination of D(t). Examples

Given a potential $\Phi(\Delta)$ the equations (4) and (5) with (6) lend themselves for a determination of the phase fractions $x_{M_\pm}(t)$. $x_A(t)$ and of the temperature T(t) once the external load P(t) and the external temperature $T_E(t)$ are given as function of time. Insertion of $x_{M_\pm}(t)$, T(t) thus calculated and of P(t) into the formula (3) for D gives the deformation as a function of time.

Of course, since the set of equations (4), (5) are highly non-linear, this program cannot be carried out analytically. It must use numerical methods and some results - for special choices of $T_E(t)$ and P(t) are shown in the Figures 10 and 11.

The Figure 10 represents load deformation curves that are calculated for fixed external temperatures and for a triangular tensile and compressive loading cycle. The curves $x_{M_\pm}(t)$, $x_A(t)$ and D(t) are plotted, but most instructive are the (P,D)-curves on the right hand sides which must be compared to the schematic curves of Figure 1. We recognize great similarity between those curves and, in particular, we see that, as the temperature goes up, the hysteresis moves away from the origin and there appear hysteresis loops instead in the first and third quadrant which reflect the pseudo-elastic behaviour discussed in Chapter I.

Figure 11 shows the result of calculations that simulated the observed

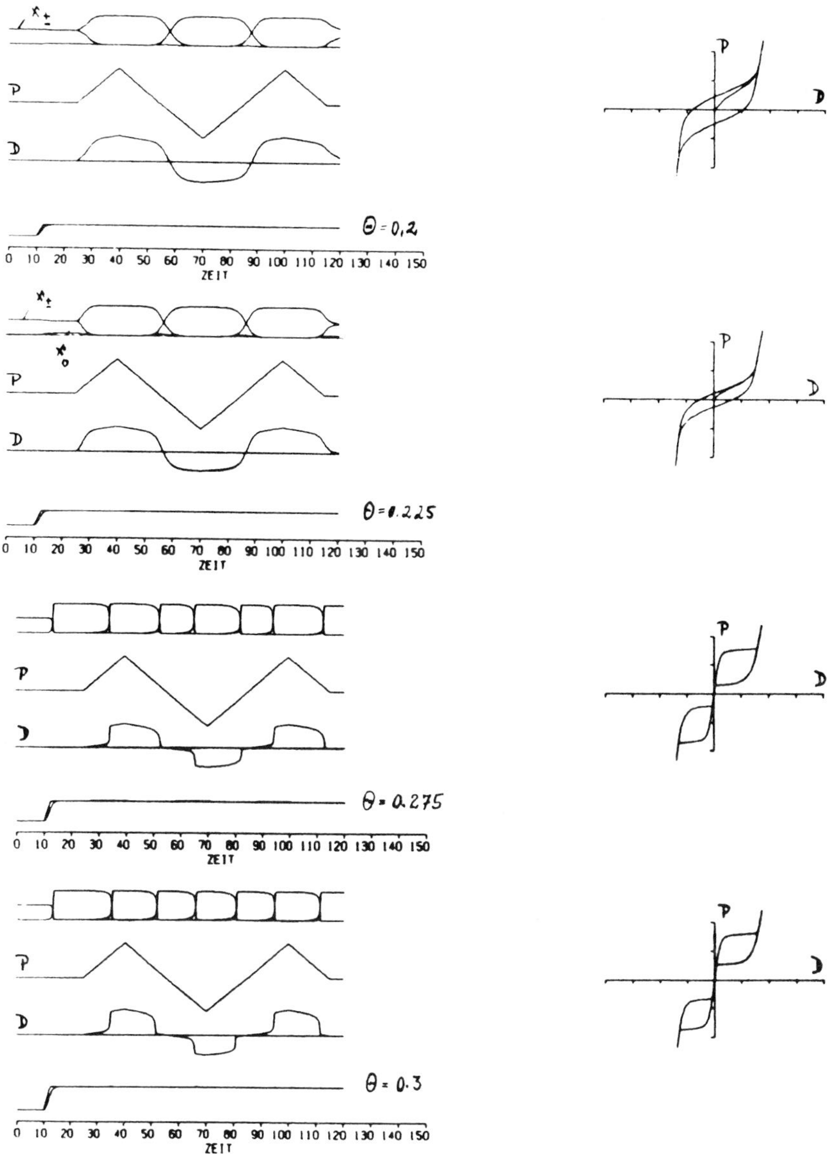

Figure 10: Response of the model to constant external temperatures and a tensile and compressive loading cycle.

behaviour reported in Figure 3. Here a triangular tensile load and the temperature is first increased and then decreased. Figure 11 obviously simulates all the qualitative features of Figure 3 and the phase factors reported in that part of Figure 11 show that the decrease in D is due to the increase of x_A at high temperature. The initial values of x_M are $x_{M_+} = 1$ and $x_{M_-} = 0$.

The great similarity of the simulated curves of Figures 10 and 11 with the observed curves of Figures 1 and 3 leads us to the conclusion that the model simulates the actual behaviour of shape memory alloys well.

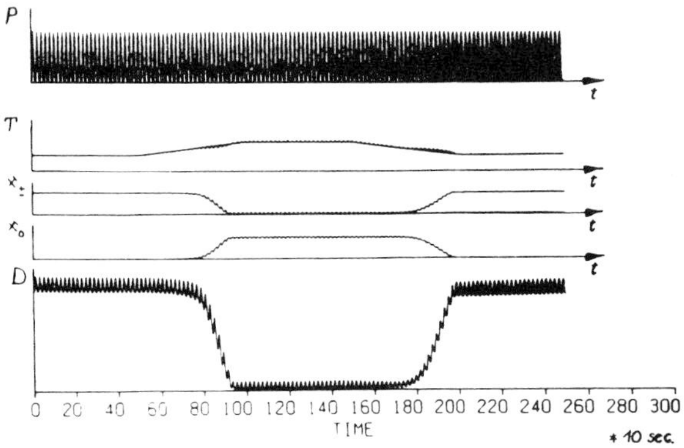

Figure 11: Deformation of the model in response to changing load and temperature

ACKNOWLEDGEMENTS

The present paper is abstracted from several research papers that can be found in the literature. For some of the material I am indebted to my co-workers M. Achenbach and H. Ehrenstein. Mr. Achenbach gives a full description of the model and its numerical exploitation. The results reported in Figures 10 and 11 are taken from that work. Dr. Ehrenstein has conceived a standard test of shape memory materials based on the plots reported in Figure 3. The etching of Figure 7 is taken from his thesis. Some literature is quoted below.

REFERENCES

On phenomenology

Perkins, J., (ed.) Shape Memory Effects in Alloys. (Plenum Press, New York, London, 1976)

Delaey, L., Chandrasekharan, L., (eds) Proc. int. Conf. on Martensitic Transformation, Leuven (1982), J. de Physique 43, (1982)

Ehrenstein, H., Formerinnerungsvermögen in NiTi, Dissertation, Technische Universität Berlin (1985)

On the model

Müller, I. Pseudoelasticity in Shape Memory Alloys - An Extreme Case of Thermoelasticity. IMA preprint No. 169, July 1986
Also: Proc. Convegno Termoelasticità, Rome, May 1985
Achenbach, M., Ein Modell zur Simulation des Last-Verformungs-Temperatur Verhaltens von Legierungen mit Formerinnerungsvermögen. Dissertation, Technische Universität Berlin (1986)

Thermomechanical Couplings in Solids
H.D. Bui and Q.S. Nguyen (Editors)
Elsevier Science Publishers B.V. (North-Holland)

METAL PLASTICITY AS A PROBLEM IN THERMODYNAMICS

Joseph Kestin, Brown University
Providence, RI 02912 USA

This lecture presents in broad outline what is claimed to be a complete thermodynamic theory of plastic deformation. The idea of Volterra cuts and Somigliana dislocations leads to the concept of a fictitious reversible process of changing the internal state created by plastic deformation. In turn, this makes it possible to calculate entropy, and to formulate Gibbs equations which differ from those valid in elasticity by the addition of internal deformation variables with their conjugate affinities. The virtual reversible work of the affinities along the internal deformations is exactly equal to the work dissipated, and is proportional to the entropy produced during the irreversible process as well as to the resulting decrease in the Helmholtz free energy. The relation between this dissipated internal work and the Helmholtz free energy (stored strain energy) makes available a large body of results (due in large measure to Eshelby) which can be utilized in the further, detailed development of a more complete theory. It is shown that Eshelby's, supposedly mysterious, "configurational force" is identical with the affinity conjugate to the internal variable. As a consequence, it becomes possible to clarify "the mystery" and to formulate an exact physical interpretation of this force.

The connection made to Volterra and Somigliana cuts enables us to recognize that plastic deformation introduces into the system a noncompatible strain field which is not derivable from an irrotational displacement field. This is accompanied by the emergence of a self-equilibrating system of internal stresses.

A sketchy example of a rod deformed plastically, drawn from earlier work, illustrates the theory and demonstrates that it does faithfully reproduce observed behavior from first principles, in addition to providing an equation for the yield surface.

A concluding digression indicates that this methodology can be extended to include fracture and damage.

1. INTRODUCTION

There exists a well-developed, empirical theory of plastic deformation in metals which allows engineers successfully to predict the behavior of a variety of structures and machine elements loaded beyond the elastic limit for purposes of design. However, this theory is not entirely satisfactory; it has failed to develop a method of calculating the yield surface directly from the properties of the material, to predict that surface's movement in stress space, to calculate the accumulation of strain energy after repeated loading cycles, to include temperature effects or to determine the temperature fields which result from dissipation. Furthermore, it relies on a variety of more-or-less *ad hoc* "constitutive laws" related to the "history of loading"; they are seldom derived from broader physical principles.

Some authors (e.g. Bridgman [1,2] or Lehmann [3]) and the present lecturer, [4-7], have become convinced that the shortcomings of the empirical theory can be eliminated by embedding metal plasticity in the wider discipline of thermodynamics, due attention being paid to the results provided by crystallography and materials science (see, e.g., Nabarro [8]). It is hoped that the present paper has succeeded in presenting a framework within which such a goal will eventually be achieved.*

The empirical theory has failed to merge with thermodynamics because it has not incorporated in its formalism the distinction between the indispensable *fundamental equation of state* and the result of an analysis in the form of a *process*. Thermodynamicists took a long time in identifying the independent variables needed for success, [4,7], and were deterred by Bridgman's "sea of irreversibility" from deriving expressions for the entropy of a plastically deformed metal. Some asserted that such a state, being one "far from equilibrium", cannot be handled by classical thermodynamics even "in principle". This led authors to propose a variety of more-or-less radical extensions and re-formulations of thermodynamics, without, however, meeting with conspicuous success. It has been the author's conviction for many years (4-6) that classical thermostatics, with some not very radical adjustments, is adequate for the description of inelastic deformations and, even, fracture and damage. A summary of such a conservative formalism was given recently in [9].

2. ELASTICITY

We model a polycrystalline metal as a continuum** and consider that all states during a process of elastic deformation are states of stable equilibrium. Consequently, elastic deformations are modelled as reversible processes [9]. The sequence of states in such a process is described by some suitable form of the fundamental equation. In the Helmholtz free-energy representation this is

$$f = f(T,\epsilon_{ij}) , \quad \epsilon_{ij} = \epsilon_{ji} , \tag{2.1}$$

in the usual notation of the subject. The state-space contains seven independent variables, and the surface (2.1) must be convex for stability. Neighboring equilibrium states are linked by the Gibbs equation

$$df = - sdT + \rho^{-1}\sigma_{ij}d\epsilon_{ij} , \tag{2.2}$$

or an equivalent Legendre transform. When analysis is confined to isothermal processes ***, we conveniently use

$$df = \rho^{-1}\sigma_{ij}d\epsilon_{ij} , \quad (T=\text{const}) \tag{2.3}$$

and recall that the Helmholtz free energy is identical with the conventional *stored strain energy* of elasticity theory. Evidently

$$- s = (\partial f/\partial T)_{\epsilon_{ij}} , \quad \rho^{-1}\sigma_{ij} = (\partial f/\partial \epsilon_{ij})_T \quad \text{etc.} \tag{2.4}$$

as well as

$$- (\partial s/\partial \epsilon_{ij})_T = \rho^{-1}(\partial \sigma_{ij}/\partial T)_{\epsilon_{ij}} \quad \text{etc.} \tag{2.5}$$

With the aid of fundamental equations, we create a complete thermodynamic version of conventional elasticity theory. Examples of simple, explicit fundamental equations can be found in [4] and [5]. On this, all research workers, of whatever persuasion, agree.

In our present paper, we demonstrate that the preceding equations retain their validity for plastic (generally, inelastic) deformations with the simple addition of *internal deformation variables.*

3. PLASTIC DEFORMATION

Plastic deformation# occurs when the internal state of the metal is modified by the movement and multiplication of dislocations. This process is irreversible and, therefore, accompanied by dissipation which manifests itself in the appearance of temperature fields and acoustic emissions. Such internal changes of state are modelled with the aid of internal variables. Their nature is best described by quoting Bridgman [2]: "I believe that in general the analysis of such systems will be furthered by the recognition of a new type of ... thermodynamic parameter of state, namely the parameter of state which can be measured but not controlled. Examples are ... dislocations in a solid. These parameters are measureable, but they are not controllable which means that they are coupled to no external force variable which might provide the means of control. And not being coupled to a force variable, they cannot take part in mechanical work. Such a parameter ... can take part only in irreversible changes." Similar ideas were expressed by Sommerfeld [11] as well as Leontovich and Meixner.

With each dissipative mechanism there is associated virtual work which is dissipated. Consequently, a typical internal variable ξ_k must be of the nature of an extensity or internal deformation, and the work it dissipates must involve a related force-like affinity A_k. The virtual-work term is

$$dW_k^{\circ} = A_k d\xi_k \quad \text{(sum over n mechanisms)} , \tag{3.1}$$

and the mental picture formed is that the irreversible process is akin to an unresisted expansion in a gas which dissipates work at the rate $P\dot{V}$. The extent of a chemical reaction is an example of an internal variable.

It is our thesis that each such irreversible step as well as its reverse can be performed reversibly with the aid of a suitable *Gedankenexperiment*. This is performed by imagining that the internal site displaced by $d\xi_k$, is connected to the surroundings by conceptual forces A_k, which then produce the work dW° of eqn. (3.1). We shall examine such a thought-experiment for dislocations in more detail in Section 4.

A comparison with standard forms shows that the work of a reversible process is then

$$dW^\circ = dW^\circ_{ext} + dW^\circ_k = -\rho^{-1}\sigma_{ij}d\epsilon_{ij} + A_k d\xi_k \ , \tag{3.2}$$

and eqn. (2.2) becomes

$$df = - sdT + \rho^{-1}\sigma_{ij}d\epsilon_{ij} - A_k d\xi_k \ . \tag{3.3}$$

The form of the Gibbs equation (3.3) allows us to recognize that the state-space of a polycrystalline metal must be increased by adding to the seven independent variables the internal variables ξ_k, say n in number. In the Helmholtz-function representation

$$f = f(T,\epsilon_{ij},\xi_k) \ , \tag{3.4}$$

we must have

$$\sigma_{ij}(T,\epsilon_{ij},\xi_k) = \rho(\partial f/\partial\epsilon_{ij})_{T,\xi_k} \quad \text{and} \quad - A_k = (\partial f/\partial\xi_k)_{T,\epsilon_{ij}} \ , \tag{3.5,6}$$

with

$$\rho^{-1}(\partial\sigma_{ij}/\partial\xi_k)_{...} = - (\partial A_k/\partial\epsilon_{ij})_{...} \ , \quad \text{etc.} \tag{3.7}$$

An elastic process occurs when either the internal state is one of *constrained equilibrium* (ξ_k=const) or when the system traverses a sequence of *unconstrained equilibrium* states ($A_k=\dot{\xi}_k=0$), as we shall explain in Secs. 5 and 9. A process of even very slow plastic deformation consists of a sequence of very closely spaced *equilibrium* states, constrained or unconstrained, the transition from one such state to its closest neighbor occurring by an irreversible jump, unresisted by the forces A_k. Whether the intermediate states are constrained ones or unconstrained ones depends on the characteristic time τ_m of loading implied in $\dot{\epsilon}_{ij}$. Fast loading involves the former, slow loading traverses the latter. The irreversible unresisted transition in the absence of an equilibrating force connected to some external mechanism dissipates work at a rate $\dot{W}^\circ_k$ and is a source primarily of heating and, to a much lesser extent, of acoustic radiation.

4. REVERSIBLE CREATION OF A DISLOCATION

So far, the symbol s for entropy in the preceding equations is restricted to states traversed by elastic processes. To extend its applicability to plastic deformations, we must be able, at least conceptually, to show that two neighboring states which differ in their dislocation structure, can be reached from each other, by a *reversible process.##*

The fictitious process of creating two dislocation lines is illustrated in figure 1. First, a fissure is made along line AB by applying fictitious forces +K,-K to overcome the interatomic attractions and to create a narrow gap of atomic dimensions. Secondly, we apply fictitious forces +F and -F and supply work Fy to displace the two sides of the elastic body, pinned at A and B, by a distance y which is a multiple of the Burgers vector b. This creates two edge dislocations as suggested by figure 1c. On releasing force ±K we recover the work done against the interatomic forces because the fields acting between them are conservative. This allows the cut to "heal". The

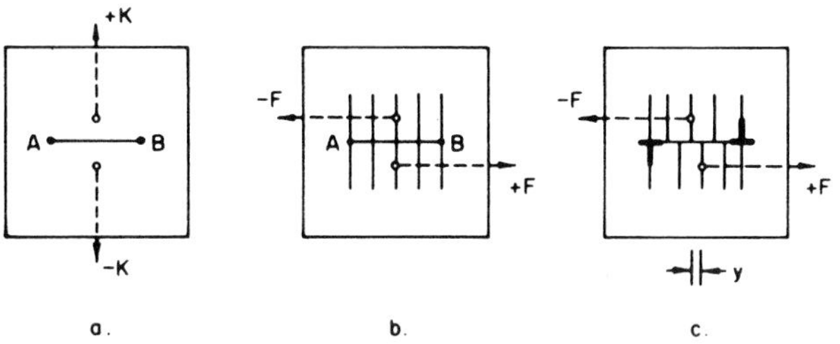

Figure 1. Reversible creation of dislocations

total work performed reversibly to create the dislocations is

$$dW^{\circ} = Fy \quad , \tag{4.1}$$

assuming that the strain ϵ_{ij} has been kept fixed. In this example A_i=F and $d\xi_i$=y. This fictitious operation is obviously reversible.

The procedure just outlined is seen to be a special case of a Volterra cut or of the creation of a Somigliana dislocation, [7]. Furthermore, it shows that an incompatible and self-equilibrating stress field is introduced and added to the one which existed before.

Depending on the direction of displacement of forces ±F in figure 1, the fictitious external mechanism may perform work on the system, increasing its Helmholtz function or decreasing it by producing external work. When the equilibrating forces are absent, only a decrease in the Helmholtz function (at T=const, ϵ_{ij}=const) is compatible with the Second Law of thermodynamics.

If at any step in plastic deformation the process is arrested, the resulting equilibrium state does not differ in principle from one which sets in after an elastic deformation and both are described by the same fundamental equations. The difference consists in the fact that the strains produced during an elastic deformation are in constant, compatible equilibrium with the changing external loads which drive the process. During plastic deformation the phenomenon of internal slip caused by the creation of new dislocations introduces an additional self-equilibrating field of stresses and strain fields linked to them (in our case) by Hooke's law. Thus portions which fit together compatibly in elastic deformation cease to do so after a step of plastic deformation. In particular, the strain field $\epsilon_{ij}(x_k)$ cannot be calculated with the aid of the field equations of elasticity without additional information concerning the nature of the singularity introduced by the cut AB. The classical compatibility conditions no longer apply and the strain is not derivable from an irrotational displacement field. Naturally, the entropy and strain in eqn. (2.1), or the strain in eqn. (2.4), have acquired different values, having become dependent on ξ_k.

The phenomenon of strain hardening and the existence of the Bauschinger effect provide experimental evidence of the creation of such additional strains by internal slip. The fact that in the early stages of unloading the system undergoes a reversible process of elastic deformation confirms that the instantaneous equilibrium state attained after plastic deformation satisfies the elastic fundamental equation of state. This is certainly true as long as the process of unloading does not cause the self-equilibrating stresses to change sign anywhere.

The interpretation of experimental facts subscribed to in this section and the resulting definition of the state of a plastically deformed metal are intrinsically incompatible with the ideas of "history" and "constitutive law". The latter is simply the result of finding a class of solutions of the field equations -- a class of processes. Equally incompatible with this picture is the suggestion, advanced by some, that the internal variable can be identified with plastic strain. Finally, it should have become clear that it is not really necessary to insist on measuring strains with reference to a "virgin state" free of internal stresses.

5. THREE INTERPRETATIONS OF THE GIBBS EQUATION

We now continue the discussion with reference to the isothermal form of the Gibbs equation

$$df = \rho^{-1}\sigma_{ij}d\epsilon_{ij} - A_k d\xi_k \quad \text{(T-const)} \quad . \tag{5.1}$$

Since the force A_k is not connected to the outside, the external work performed by an element in a continuum during an irreversible process is

$$dW = -\rho^{-1}\sigma_{ij}d\epsilon_{ij} \quad , \tag{5.2}$$

whereas the expression for reversible work is that given by eqn. (3.2). It follows that the dissipated work is

$$dW^\circ - dW = A_k d\xi_k = Td\theta \geqslant 0 \quad , \quad \text{[Second Law]} \tag{5.3}$$

where θ is the volumetric entropy production. In accordance with the Prigogine-Meixner elementary theory of irreversible processes [14], we interpret the affinities A_k as the generalized forces, and the time-rates $\dot{\xi}_k$ as the generalized fluxes.

If only one dissipative mechanism is active, we can adopt the linear phenomenological relation###

$$\dot{\xi}_k = KA_k$$

which is homogeneous, as it should. The rates $\dot{\xi}_i$ are descriptive of the velocities of propagation and formation of dislocations. Experiments show that such processes are comparatively fast [13], so that the relaxation times

$$\tau_\xi = \xi/\dot{\xi} \quad \text{are short and} \quad \tau_\xi << \tau_m \ . \qquad (5.5,6)$$

With these remarks in mind we revert to eqn. (5.1) and record that it can be given three interpretations [15,16]:

(a) When

$$A_k = \dot{\xi}_k = 0 \qquad (\tau_\xi << \tau_m) \ , \qquad (5.7a)$$

the Gibbs equation describes a state of *unconstrained* equilibrium. If such a state is maintained during a process, the process can be considered reversible because of the vanishing entropy production, eqn. (5.3).

(b) When

$$A_k \neq 0 \quad \text{with} \quad \dot{\xi}_k = 0 \ (\tau_m < \tau_\xi, \text{ or very fast loading}) \ , \qquad (5.7b)$$

the Gibbs equation describes a state of *constrained* equilibrium. It follows that points in the phase-space ϵ_{ij},ξ_k which do not lie in the hyperplane $\{\epsilon_{ij},(\xi_k=0)\}$ represent states of such constrained equilibrium.

(c) When

$$A_k \neq 0 \quad \text{and} \quad \dot{\xi}_k \neq 0 \ , \qquad (\tau_m \sim \tau_\xi) \qquad (5.7c)$$

the Gibbs equation describes accompanying equilibrium projections of *nonequilibrium* states, and their sequence constitutes an irreversible process which produces entropy at a rate

$$\dot{\theta} = A_k\dot{\xi}_k/T \qquad (5.7d)$$

as a result of creating dislocations with the unopposed forces A_k. This is the process which occurs when one unconstrained equilibrium state is transformed into a neighboring unconstrained equilibrium state while dislocations are being moved and created under a slowly varying external load.

In all cases, the fact that $A_k d\xi_k=0$ implies that f reaches an extremum with respect to ξ_k in the appropriate state-space. Such an extremum corresponds to equilibrium which is stable if the Helmholtz function is convex there. Otherwise, the equilibrium point is unstable.

6. DISSIPATION

Equation (5.3) allows us to reach the very important conclusion that the energy dissipated is measured by

$$T\theta \;=\; T\int_{t_1}^{t_2}\dot{\theta}dt \;=\; \int_{t_1}^{t_2}A_k\dot{\xi}_k dt \quad , \qquad (T=\text{const},\ \epsilon_{ij}=\text{const})$$

or

$$T\theta \;=\; f\left[T,\epsilon_{ij},\xi_k(t_1)\right] - f\left[T,\epsilon_{ij}\xi_k(t_2)\right] \;, \tag{6.1}$$

i.e., *either by an integral over the rate processes* $\dot{\xi}_k$ *or by the difference in the Helmholtz potential* between two equilibrium states attained at times t_1 and t_2, respectively. This quantity of dissipated energy predominantly goes toward the creation of a temperature field; it must be removed by cooling if the temperature is to remain constant.

Equation (6.1) provides a basis for the calculation of the temperature field which accompanies plastic deformation, and incidentally, also in fracture and damage (see Sec. 10).

7. IDENTIFICATION OF THE NATURE OF THE INTERNAL VARIABLE. ESHELBY'S CONFIGURATIONAL FORCE

From the preceding it follows that a calculation of the change in Helmholtz function (stored strain energy) yields an explicit expression for the unrecovered work of the internal variables. Furthermore, in view of eqn. (3.6), a knowledge of the algebraic structure of $f(T,\epsilon_{ij},\xi_k)$ provides us with *a method of identifying the physical character of the internal variable* associated with a particular dissipation mechanism, as well as of the conjugate affinity.

There exists a large number of mathematical studies which contain derivations and explicit expressions for the stored strain energy exactly under the conditions specified for eqn. (6.1) namely at ϵ_{ij}=const. and T=const. These refer to a variety of cuts, dislocations and inclusions and describe the change in the Helmholtz free energy upon the introduction of a specified defect. Notable among these studies is the work of J.D. Eshelby [12], among others.

In the literature of the subject, $-A_k$ calculated from eqn. (3.6), is known under a variety of names, such as the force on a dislocation or Peach-Koehler force in a slip plane [17]. They are generically identical with Eshelby's *configurational forces.* In the context in which they have been introduced, these forces appear mysterious, because their physical interpretation could not be made clear. For example, in a recent paper, Nabarro [18] opened his study with the statement: "It is generally held that the forces acting on and between elastic defects are *configurational forces* analogous to, but entirely different in kind from, the Newtonian mechanical forces that may act on the material in which the elastic forces are present." Thermodynamics can be invoked to resolve this "mystery." The difficulties of interpretation arise from a failure to seek an answer in terms of a clearly specified *process.*

The validity of the Gibbs equation is limited to reversible processes. If it were possible to balance the internal forces by introducing an external force A_k, as discussed in Sec. 4, the result would be a *reversible* modification of the internal state. By Newton's third law, as is the case, for example, during a reversible compression or expansion of a gas behind a piston, the applied forces would be "Newtonian mechanical forces," as in figure 1. Thus, quite simply, the configurational force is that "Newtonian mechanical force" which would permit us to perform a reversible change of the internal state, such as, e.g., reversing an irreversible step in a process.

During an irreversible step, such as that considered in conjunction with eqn. (5.7c), the system departs from equilibrium and the forces acting on it no longer conform to the Gibbs equation (3.3). Even more, its state requires more than the 7+n variables for its description, (e.g., see [9]). Nevertheless, the configurational force, calculable from the fundamental equation (3.4) serves as a precise basis for the calculation of dissipation via eqn. (5.7d) when the rate $\dot{\xi}$ is known or via eqn. (6.1) if the Helmholtz function is known. This very important interpretation from the point of view of future developments failed to be recognized without the present thermodynamic interpretation.

8. THE DIRECT METHOD OF IDENTIFICATION OF THE NATURE OF THE INTERNAL VARIABLE

The progress achieved by many experimental and analytic studies of plasticity in the fields of crystallography and materials science makes it possible to establish the geometric structure of the Helmholtz free-energy surface from first principles. A successful example of the application of such a method is contained in a paper by Ponter, Bataille and Kestin [19]. In that paper, the method is illustrated with the aid of a simple model of a system undergoing plastic deformation; it operates with a single internal variable ξ which is a measure of the area swept by deforming dislocation lines of N Frank-Read sources [20].

The model is concerned with metals at low rates of strain at which thermally activated processes are negligible and the initial dislocation density is low. If a rod of such a material is subjected to a slowly increasing homogeneous strain E, its initial response is elastic until a critical stress level is reached. This is the initial yield stress when the dislocation density increases rapidly and subsequent irreversible deformation occurs through the operation of Frank-Read sources.

We devote the remainder of this Section to a brief statement of results, because more detailed calculations can be found in References [7] and [19].

The theory of Frank-Read sources allows us to write down an expression for the Helmholtz free energy of the model in the form

$$F = Nel/\rho L^3 \tag{8.1}$$

where e is the so-called dislocation line tension and l is its length. Due to the appearance of a dislocation line, the external surface of the system suffers a displacement, as a result of which the average strain increases by

$$E_p = SNb/L^3 = \xi \tag{8.2}$$

Here S is the area of an equivalent Volterra cut, ξ is a measure of the internal deformation, L^3 is a measure of the size of the system and b is the Burgers vector length. To restore the prescribed value E of strain, it is necessary to compensate E_p by the imposition of suitably distributed surface tractions which amounts to the superposition of strain $(E-E_p)$. Hence the total Helmholtz free energy of the system becomes

$$F(E,\xi) = \left[\frac{Nel}{L^3} + \frac{1}{2} C(E-\xi)^2\right]\rho^{-1} , \tag{8.3}$$

where C is the elastic compliance of the material and $l=l(\xi)$. The stress is then

$$\Sigma(E,\xi) = \rho\left(\frac{\partial F}{\partial E}\right)_\xi = C(E-\xi) \tag{8.4}$$

and the affinity can be shown to be

$$A(E,\xi) = -\left(\frac{\partial F}{\partial \xi}\right)_E = -\left(\frac{eN}{L^3 R(\xi)} + \tau\right)\rho^{-1} . \tag{8.5}$$

Here τ is the resolved shear component of Σ on the slip plane, and $R(\xi)$ is the varying radius of curvature of the deforming, near-circular dislocation loop of a Frank-Read source.

The unexpected result is that the expression for surface $F(E,\xi)$ contains an essentially convex second term $\frac{1}{2}C(E-\xi)^2$ added to a term which is not essentially convex. This is due to the fact that the total length l of all dislocation loops of a Frank-Read source *is a periodic function* of the radius of curvature R of the loops; it passes through a minimum when $R=l_o/2$, where l_o is the length over which the source is pinned.

Remembering that the relaxation time τ_ξ of dislocation formation is much shorter than the relaxation time $\tau_m=E/\dot{E}$ of external intervention, we reduce the study of the stability of the configuration to the plane F,ξ at the constant, current value of E, figure 2. Hence we examine the sign of

$$\left(\frac{\partial^2 F}{\partial \xi^2}\right)_E = -\left(\frac{\partial A}{\partial \xi}\right)_E = \left[\frac{eN}{L^3 R^2}\frac{dR}{d\xi} + C\right]\rho^{-1} . \tag{8.6}$$

Applying the conventional theory of stability we conclude that

$(\partial^2 F/\partial \xi^2)_E < 0$ corresponds to unstable equilibrium;

$(\partial^2 F/\partial \xi^2)_E = 0$ corresponds to neutral equilibrium;

$(\partial^2 F/\partial \xi^2)_E > 0$ corresponds to stable equilibrium.

Owing to the properties of the function $R(\xi)$, its derivative can pass through zero and become negative. Whereas the first term in eqn. (8.3), where l is a function of $R(\xi)$, contributes little to F, the contribution of the derivative $dR/d\xi$ is decisive, because it affects the sign of the second derivative of F in eqn. (8.6) due to the change in the sign of $dR/d\xi$ when $R=l_o/2$; its periodic nature renders the function $F(\xi)$ also periodic at E=const.

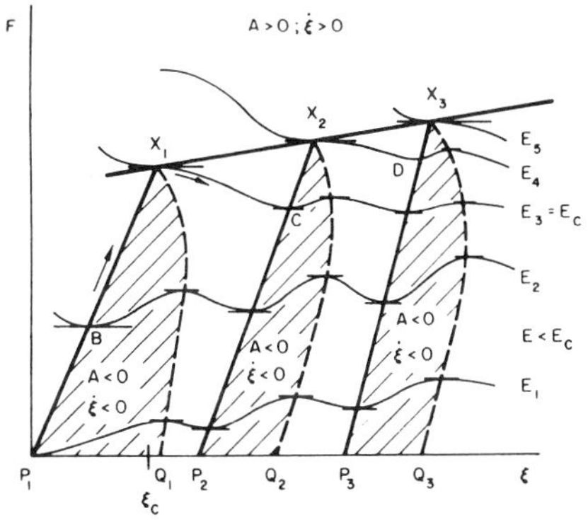

Figure 2. Level lines of the Gibbsian surface $F(E,\xi)$; $E=E_C$ and $\xi=\xi_C$ determine the first yield point; along P_iX_i and Q_iX_i we have A=0.

9. ELASTIC AND PLASTIC DEFORMATION

The explicit form of the Helmholtz free energy function presented in its simple, not to say schematic, form in eqn. (8.3), accounts fully for the behavior of the model system under all conditions of slow isothermal loading or unloading. The diagram in figure 2 depicts the projection of the intersection of planes E=const ($E_2>E_1$, etc.) with the Gibbsian surface $F(E,\xi)$ into the plane F,ξ. Each such intersection exhibits a succession of stable minima and unstable maxima whose loci correspond to solutions of eqn. (8.3) in which now the strain E plays the role of a parameter. The set of solutions is multivalued.

The loci of stable minima, P_1X_1, P_2X_2, P_3X_3, etc. represent reversible paths of elastic deformation with zero entropy production because along them

$$A = -(\partial F/\partial \xi)_E = 0 . \tag{9.1}$$

The loci of stable minima, P_iX_i, and unstable maxima, Q_iX_i, intersect at points X_1, X_2, ... etc. where

$$(\partial F/\partial \xi)_E = (\partial^2 F/\partial \xi^2)_E = 0 . \tag{9.2}$$

The shaded areas between P_iX_i and Q_iX_i enclose the only sets of points of stable equilibrium. Outside these and above X_1X_2 ... the points correspond to states of constrained equilibrium which, however, are all unstable in the absence of the fictitious balancing forces which served to illustrate the Volterra cut in figure 1. This means that if ξ is disturbed by an infinitesimal amount when the state is at X_1, X_2, etc., the affinity A becomes positive and the state-point moves away at a jump to the nearest stable equilibrium state, such as C,D etc. The transition from X_1 to C etc. is irreversible and involves a net entropy production θ (dissipation $T\theta$) in accordance with eqn. (6.1), giving

$$T\theta = [F(X_1) - F(C)] > 0 . \tag{9.3}$$

It follows that to the present order of approximation, the amount of entropy produced in a jump is independent of the exact form of the rate equation (5.4) as explained earlier.

The locus X_1, X_2, ... separates the F,ξ plane into a lower region where a large number of reversible paths exists and an upper region where *no* such paths exist. This thermodynamic result is interpreted as the conclusion that plastic yielding with subsequent strain hardening corresponds to the state point moving along the locus X_1X_2 ... . Consequently, *the yield surface turns out to be a solution of the set of seven-dimensional analogs of equations* (9.2).

Incidentally, it should by now be quite clear that the process of unloading from the yield locus X_1X_2... is reversible and elastic, as is well-known in plasticity theory. The usual Σ,E stress-strain curve, which is implied by the F(E,ξ) surface of figure 2, is depicted in figure 3. The pairs of states X_1C, X_2D, etc. lie so close together that they cannot be easily observed on the usual scale of a stress-strain diagram. Their reality is confirmed by acoustic observations and by the measurable heat of dissipation which occurs along the irreversible "jumps" X_1C, X_2D, etc. These are the observable manifestations of many thousands (millions?) of instabilities on the microscale of crystallites. The strain-hardening curve appears as a zig-zag line confined between the two broken lines h_1,h_2, and the reversible character of the process of unloading emerges quite clearly.

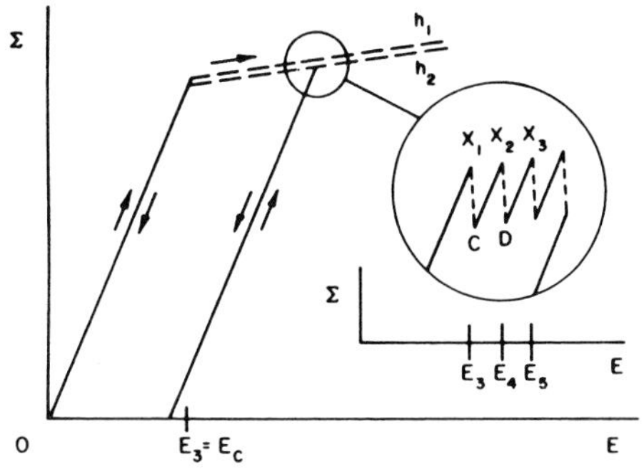

Figure 3. Stress-strain diagram.

10. REMARK ON FRACTURE AND DAMAGE

The two sides of a crack in a metallic sample fractured under a high vacuum grow together upon restored contact, unless prevented from doing so by allowing them to acquire adsorbed layers of gas. This suggests that crack formation in a metal is a special case of a Somigliana dislocation and can be *imagined* performed reversibly in a manner resembling the Volterra cut of figure 1. This identifies the "force on a crack" as the affinity A and the internal variable ξ as the area of the crack. It follows that the thermodynamic theory presented here does not endorse the view that the energy made available in a small extension of a crack must be interpreted as an increase in the surface energy of the specimen. Rather, it represents the energy dissipated in the form of a production of entropy which creates the observable temperature field around the crack. A similar point of view is represented in Reference [21].

Professor G. Herrmann has extended the same ideas to the study of damage [22].

ACKNOWLEDGEMENTS

The research done for this paper was not supported by a formal grant or contract.

I owe a great debt of gratitude to several professional colleagues with whom I collaborated over the years along the path which led me to the present results. Here I wish to mention J. Bataille, the late A. Golebiewska-Herrmann, G. Herrmann, J. H. Lambermont, J. Meixner, A. R. S. Ponter, J. R. Rice and R. S. Rivlin, as well as many discussions with O. P. Manley.

Thanks are also due to Mrs. Susan Whitney for her skillful composition of this camera-ready copy.

Last, but not least, I wish to thank the organizers of this Symposium for their invitation to attend and for partial financial assistance. Further financial assistance was received from R. A. Dobbins, Chairman, Division of Engineering of Brown University and from the Exxon Educational Foundation.

FOOTNOTES

* Owing to space limitations, only the barest outline can be given here. The outline assumes the simplest metal properties, and argues on the basis of infinitestimal deformations and Hooke's law. Nevertheless, it is implied that such restrictions do not invalidate the underlying formalism.

** The obvious limitations of such a theory have been made explicit in [9].

*** Restricting the argument to isothermal processes has no serious, practical consequences because the relaxation time τ_T for temperature equalization is very short in a metal compared with the characteristic times τ_M of loading. This restriction must be lifted when it is desired to determine the temperature field produced by dissipation.

\# The succeeding sections summarize, but also extend, the contents of a recent paper on a closely related subject [10].

\## We do not believe in introducing entropy by fiat, as a "primitive concept" about whose calculation we are forbidden to inquire.

\### The adoption of the linear relations (5.4) does not linearize the main problem because, generally speaking, the field equation remain nonlinear.

REFERENCES

Owing to space limitations, only the most relevant papers can be mentioned. Apologies are extended to authors whose contributions have been passed over.

[1] Bridgman, P. W., Rev. Mod. Phys. 22 (1950) 56-63.

[2] Bridgman, P. W., The Nature of Thermodynamics (Harper, 1961).

[3] Lehmann, Th. The Constitutive Law in Thermoplasticity (Springer-Verlag, 1984), esp. pp. 1-11.

[4] Kestin, J. On the Application of The Principles of Thermodynamics to Strained Solid Materials in: Parkus, H. and Sedov, L. I. (Eds.). Irreversible Aspects of Continuum Mechanics (Springer-Verlag, 1968), pp. 177-212.

[5] Kestin, J., A Course in Thermodynamics, Vols. 1 & 2 (Hemisphere, 1979).

[6] Kestin, J. and Bataille, J., Thermodynamics of Solids in: Kröner, E. and Anthony, K. H. (Eds.), Continuum Models of Discrete Systems (U. of Waterloo Press, 1980), pp. 99-147.

[7] Kestin, J. and Rice, J. R., Paradoxes in the Application of Thermodynamics to Strained Solids in: Stuart, E. B., Gal-Or, B. and Brainard, A. J. (Eds.) A Critical Review of Thermodynamics (Mono Book Corp., 1970), pp. 275-298.

[8] Nabarro, R. F. N., Dislocations in Solids. (North-Holland, 1979).

[9] Kestin, J., Irreversible Thermodynamics in: Wapnewski, P. (Ed.) Jahrbuch 1984/85, Wissenschaftskolleg zu Berlin (Siedler-Verlag 1986), in print.

[10] Kestin, J., Thermodynamics of Plastic Deformation in: Prigogine, I. (Ed.) Patterns, Defects and Microstructures in Nonequilibrium Systems, in print.

[11] Sommerfeld, A., Thermodynamics and Statistical Mechanics (Academic Press, 1956).

[12] Eshelby, J. D., Boundary Problems in: Nabarro, F. R. N. (Ed.) Dislocations in Solids. (North-Holland, 1979) pp. 168-221.

[13] Vreeland, I., Dislocation Velocity in Copper and Zink in: A. R. Rosenfield (Ed.) (McGraw-Hill, 1968) pp. 529-549.

[14] DeGroot, S. R. and Mazur, P., Non-Equilibrium Thermodynamics (North-Holland, 1962).

[15] Bataille, J. and Kestin, J., Jour. de Mec. 14 (1975) 365-384.

[16] Bataille, J. and Kestin, J., Jour. Non-Eq. Thermo. 4 91979) 229-258.

[17] Peach, M. and Koehler, J. S., Phys. Rev. 80 (1960) 436-439.

[18] Nabarro, F. R. N., Proc. Roy. Soc. A398 (1985) 209-222.

[19] Ponter, A. R. S., Bataille, J. and Kestin, J., Jour. de Mec. 18 (1979) 511-539.

[20] Lambermont, J. H., Int. J. Engng. Sci. 12 (1974) 937-965.

[21] Rivlin, R. S. and Thomas, A. G., Jour. Polymer Sci. 10 (1959) 291-318.

[22] Herrmann, G. in Prigogine, I. (Ed.) Patterns, Defects and Microstructures in Nonequilibrium Systems, in print.

Thermomechanical Couplings in Solids
H.D. Bui and Q.S. Nguyen (Editors)
Elsevier Science Publishers B.V. (North-Holland)

A NEW APPROACH FOR THE THERMOMECHANICS OF MATERIALS WITH DELAYED RESPONSE

Christian HUET

Ecole Nationale des Ponts et Chaussées, Centre d'Enseignement et de Recherches en Analyse des Matériaux
ENPC - CERAM - La Courtine - B.P. 105 - F 93194 NOISY-LE-GRAND Cedex

1. INTRODUCTION

The thermomechanics of bodies made of materials exhibiting memory effects in form of delayed response is considered. The problem is studied in a framework involving macroscopic variables only, with no use of internal variables. Based on the analysis of the response of the material to the adiabatic relaxation experiment, a new approach proposed by the author in a recent past, [1] to [7], for the statement of constitutive equations and the handling of thermomechanical problems is described. It leads to constitutive equations of the differential type for materials with delayed response in the classical experiments. A few new developments are presented.

2. THE FULL SET OF UNIVERSAL RELATIONS

The proposed approach lies on the consideration of an extended set of local universal balance equations (Table 1). This set involves the balance equation for the entropy in addition to the classical set of conservation equations for mass, momentum, moment of momentum and energy generally used in the calculations. This contrasts with the current use for which the entropy balance equation and CLAUSIUS inequality are considered as providing only restrictions on the constitutive equations after elimination of the heat supply density and heat current with the energy balance equation, this process resulting in the so-called CLAUSIUS-DUHEM inequality. This contrasts also with an other method sometimes used, for instance in fluid mechanics, which consider an entropy balance equation in place of the energy balance equation (and not, as here, in addition to). Thus, in our approach, the set of the universal balance equations involves one more scalar equation than the two classical approaches quoted above.

This new approach can be justified by the fact that, for materials with delayed response in the adiabatic relaxation experiment, the fixation of the mass, of the displacements and of the energy is not sufficient to fix the value of the entropy, because of the production of entropy due to the delayed internal relaxation mechanisms.

Table 1

THE FULL SET OF LOCAL UNIVERSAL BALANCE EQUATIONS

(1.1)	$\rho\dot{\tau} = \text{div } v$	(mass)
(1.2)	$\rho\dot{v} = F + \text{div } \sigma$	(momentum)
(1.3)	$\sigma^T - \sigma = 0$	(moment of momentum)
(1.4)	$\rho\dot{u} = \rho P + r - \text{div } q$	(energy)
(1.5)	$\rho\dot{s} = \rho\xi + \frac{r}{T} - \text{div } \frac{q}{T}$	(entropy)
(1.6)	$\xi \geqslant 0$	(CLAUSIUS inequality)

Notations

$\tau = \rho^{-1}$	volume density per unit mass
v	particle velocity
σ	CAUCHY stress tensor
F	volumic density of the external force
u	internal energy density per unit mass
s	entropy density per unit mass
T	local absolute temperature
q	eulerian heat current vector
ξ	density of the entropy production rate per unit mass

$\dot{\psi} = \frac{\partial\psi}{\partial t} + v \cdot \text{grad } \psi$: material derivative of the physical quantity $\psi(x,t)$

$P = \frac{1}{\rho}\, \sigma..d = \frac{1}{\rho}\, \sigma_{ij}\, d_{ij}$: density of the strain power per unit mass

$d = \text{Sym grad } v$: stretch tensor.

3. A NATURAL FULL SET OF INDEPENDENT VARIABLES

The availability of one more scalar equation leads then to the idea that it can be taken as independent thermodynamic variables, in the set of constitutive equations, one more variable than for the classical approaches with macroscopic variables. Thus, corresponding to the case of the classical approach where the strain ε and the internal energy density u are taken as the set $\{\varepsilon,u\}$ of independent variables, it is proposed in our new formalism to take $\{\varepsilon, u, s\}$ as such a set, where s is the entropy density. Further, it is proposed to take the entropy production rate density ξ as an additional dependent field variable, for which an explicit constitutive equation must be provided. The other dependent variables are the stress tensor σ, the temperature T and the heat current vector q (Table 2).

Of course the CLAUSIUS statement that ξ cannot be negative, and is strictly positive in every non-reversible processes, must be added to the balance equations in order to form a full set of universal relationships between the macroscopic variables of the problem.

Table 2

SETS OF VARIABLES INVOLVED IN THE CONSTITUTIVE EQUATIONS

The natural full set of independent variables

(2.1) $\Lambda = \{\rho_o , \varepsilon, u, s\}$

The associated dependent variables

(2.2) $\Phi(t) = \{\sigma(t), T(t), q(t), \xi(t)\}$

with

ρ_o volumic mass density in the reference configuration.

---

4. THE FUNDAMENTAL PROPERTY

From CLAUSIUS statement follows a fundamental property of the new formalism : the continuation functional, in the sense of COLEMAN [8], which appears in the time derivative of the dependent variables must vanish (Table 3). This contrasts with the classical approaches in macroscopic variables for which the continuation functional does not vanish for materials with delayed response. On the contrary, this classical continuation functional expresses, in the time derivative of the stress for instance, the very fact that the material exhibits delayed response to classical experiments.

Table 3

CONSTITUTIVE EQUATION FOR THE STRESS

Classical formalism

(3.1) $\sigma = \underset{\sim}{\sigma}(\varepsilon, u, \dot{\varepsilon}, \dot{u}, \ldots, \varepsilon^{(m)}, u^{(m)} ; H^{-}(\varepsilon,u))$

with

(3.2) $H^{-}(\varepsilon,u) = \{\varepsilon(\tau), u(\tau) \mid -\infty < \tau < t\}$ (previous history)

$$\Rightarrow \quad \dot{\sigma} = \frac{\partial \underset{\sim}{\sigma}}{\partial \varepsilon}\dot{\varepsilon} + \frac{\partial \underset{\sim}{\sigma}}{\partial u}\dot{u} + \ldots + \frac{\partial \underset{\sim}{\sigma}}{\partial \varepsilon^{(n)}}\varepsilon^{(m+1)} + \frac{\partial \underset{\sim}{\sigma}}{\partial u^{(n)}}u^{(m+1)} + \dot{\sigma}_{\varepsilon u}$$

$\dot{\sigma}_{\varepsilon u}$ continuation functional (in COLEMAN sense) of σ at constant ε and u (adiabatic relaxation).

Natural variables formalism

(3.3) $\sigma = \underset{\sim}{\sigma}(\varepsilon, u, s, \dot{\varepsilon}, \dot{u}, \dot{s}, \ldots, \varepsilon^{(n)}, u^{(n)}, s^{(n)} ; H^{-}(\varepsilon, u, s))$

$\Rightarrow$

(3.4) $$\dot{\sigma} = \frac{\partial \underset{\sim}{\sigma}}{\partial \varepsilon}\dot{\varepsilon} + \frac{\partial \underset{\sim}{\sigma}}{\partial u}\dot{u} + \frac{\partial \underset{\sim}{\sigma}}{\partial s}\dot{s} + \ldots + \frac{\partial \underset{\sim}{\sigma}}{\partial \varepsilon^{(n)}}\varepsilon^{(n+1)} + \frac{\partial \underset{\sim}{\sigma}}{\partial u^{(n)}}u^{(n+1)} + \frac{\partial \underset{\sim}{\sigma}}{\partial s^{(n)}}s^{(n+1)} + \dot{\sigma}_{u\varepsilon s}$$

Fundamental property

(3.5) $\dot{u} = \dot{\varepsilon} = \dot{s} = q = 0 \quad \Rightarrow \quad \xi = 0$

$\Rightarrow$

(3.6) $\dot{\underset{\sim}{\sigma}}_{u\varepsilon s} = 0$

The proof of the fundamental property comes from the classical CLAUSIUS-DUHEM inequality, which applies still to this formalism. If the continuation functional, which denotes the time variation rate of a dependent variable at constant values of the independent ones, were not zero, one would have an irreversible process (delayed adiabatic relaxation) for an isolated system without growth of the entropy. This circumstance is explicitly forbidden by CLAUSIUS statement of the second principle of thermodynamics. Thus, one must conclude that the continuation functionals of the formalism must vanish. For instance, if one states that σ must be a function of the present values of $\{\varepsilon, u, s\}$ and of the present values of their time derivatives, and must be also a functional of the past history of $\{\varepsilon, u, s\}$, the continuation functional $\dot{\sigma}_{u\varepsilon s}$, denoting the rate of variation of σ at constant values of the dependent variables u, ε, s, must be zero (Table 3). In other words, delayed isentropic adiabatic relaxation cannot exist. Of course, from a physical point of view, this comes from the fact that fixation of the entropy in addition to the strain and to the energy implies the fixation of all the internal variables, and thus the relaxation must stop. On another hand, as we have shown elsewhere [1], instantaneous isentropic adiabatic relaxation can exist, but is not involved in the continuation functional.

5. OTHER FULL SETS OF MACROSCOPIC VARIABLES

From the various forms taken by the CLAUSIUS-DUHEM inequality through the use of the various classical thermodynamic functions, such as the free energy density f, the enthalpy density h and the free enthalpy density g, and also the three corresponding MASSIEU functions, it can be shown that the fundamental property quoted above can be transfered to other full sets of independent variables, such as $\{\varepsilon, s, u\}$, $\{\varepsilon, T, f\}$, $\{\sigma, T, g\}$ for instances, where T denotes the temperature (Table 4). The corresponding continuation functionals must all vanish (in Table 4, ρ is taken equal to one, in small strains).

Table 4

VARIOUS UNIVERSAL FORMS OF THE DISSIPATION AND CONSEQUENCES

(4.1) $D = -(\dot{u} - T\dot{s} - \sigma..\dot{\varepsilon}) \Rightarrow \Lambda_u = \{\varepsilon, s, u\}$

(4.2) $D = -(\dot{f} + s\dot{T} - \sigma..\dot{\varepsilon}) \Rightarrow \Lambda_f = \{\varepsilon, T, f\}$

(4.3) $D = -(\dot{h} - T\dot{s} + \varepsilon..\dot{\sigma}) \Rightarrow \Lambda_h = \{\sigma, s, h\}$

(4.4) $D = -(\dot{g} + s\dot{T} + \varepsilon..\dot{\sigma}) \Rightarrow \Lambda_g = \{\sigma, T, g\}$

$\Rightarrow$

(4.5) $\dot{\sigma}_{\varepsilon su} = \dot{\sigma}_{\varepsilon Tf} = \dot{\varepsilon}_{\sigma sh} = \dot{\varepsilon}_{\sigma Tg} = 0$.

With these various forms of the fundamental property, it becomes possible to formulate, in the framework of our new formalism, constitutive equations of the (extended) differential type for materials that exhibit delayed response, not only in the adiabatic relaxation experiment, but also in the isothermal relaxation experiment, or in the isothermal creep experiment as well.

6. EXAMPLES

As an illustration, applications are made on Table 5 to the cases of two unidimensional models in isothermal conditions. For each model, two forms must be considered for the stress, according to the fact that $\dot{\varepsilon}$ is zero (relaxation) or not. This because when $\dot{\varepsilon}$ is zero, the first form becomes undeterminate (ratio of two zeros).

<u>Table 5</u>

EXAMPLES IN ISOTHERMAL CONDITIONS

<u>Unidimensional MAXWELL model</u> :

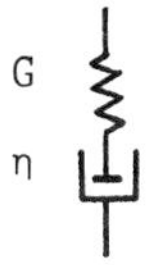

(5.1) $\sigma = \dfrac{1}{1 + \dfrac{\tau}{2}\dfrac{\dot{f}}{f}}\, \eta\dot{\varepsilon}$ $\quad \forall \dot{\varepsilon} \neq 0 \quad$ with $\quad \tau = \dfrac{\eta}{G}$

(5.2) $\sigma = [\,\mathrm{sign}\, \sigma(t^-)\,]\, \sqrt{2Gf}$ $\quad \forall \dot{\varepsilon} = 0$ (relaxation)

(5.3) $D_T = \dfrac{2}{\tau} f$.

<u>Unidimensional model made of KELVIN and MAXWELL models in parallel</u> :

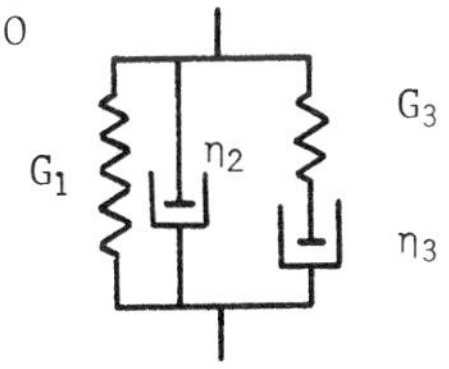

(5.4) $\sigma = G_1\varepsilon + \left(\eta_2 + \dfrac{1}{1 + \dfrac{\tau_3}{2}\dfrac{\dot{f} - G_1\,\varepsilon\,\dot{\varepsilon}}{f - \frac{1}{2}G_1\,\varepsilon^2}}\,\eta_3\right)\dot{\varepsilon}$ $\quad \forall \dot{\varepsilon} \neq 0$

with $\quad \tau_3 = \dfrac{\eta_3}{G_3}$

(5.5) $\sigma = G_1\varepsilon + [\,\mathrm{sign}\,\{\sigma(t^-) - G_1\varepsilon\}\,]\,\sqrt{2G_3(f - \frac{1}{2}G_1\varepsilon^2)}$ $\quad \forall \dot{\varepsilon} = 0$

(5.6) $D = \eta_2\dot{\varepsilon}^2 + \dfrac{2}{\tau_3}(f - \dfrac{1}{2}G_1\varepsilon^2)$.

Nevertheless, it turns out that the fundamental property predicted by the general formalism is indeed fulfilled for these two particular examples. Furtherly, we have recently shown [9] that a unidimensional model made of N MAXWELL elements grouped in parallel are of the differential type in this formalism with time derivatives of ε and f up to the order $n = 2N - 1$ involved in the constitutive equations.

REFERENCES

[1] Huet, C., Sc. et Tech. de l'Armement 53, 210 (1979) 611-652.
[2] Huet, C., Rheol. Acta 21 (1982) 360-365.
[3] Huet, C., Rheol. Acta 22 (1983) 245-259.
[4] Huet, C., Macroscopic rheology without functionals : the natural variables formalism, in : Mena B. et Alii, Proceedings of the IXth Int. Congress on Rheology. Mexico University Press (1984) pp. 497-507.
[5] Huet, C., A new formalism for the thermodynamics of rheological behaviours in macroscopic variables : the natural variables formalism (submitted for publication).
[6] Huet, C., Changing the set of variables in the natural variables formalism for the thermodynamics of rheological behaviour (submitted for publication).
[7] Huet, C., Une nouvelle approche de thermodynamique macroscopique pour la rhéologie des matériaux à réponse différée : le formalisme par variables naturelles, in : D. Bourgoin (Ed.). Proc. of the 20th Annual Colloquium of the French Group of Rheology, Paris, in print.
[8] Coleman, B.N., Archives Rat. Mech. Anal., 17 (1964) 1-46.
[9] Huet, C., unpublished result (1986).

Thermomechanical Couplings in Solids
H.D. Bui and Q.S. Nguyen (Editors)
Elsevier Science Publishers B.V. (North-Holland)

THERMOMECHANICAL DEFORMATION AND COUPLING IN ELASTIC-PLASTIC COMPOSITE MATERIALS

George J. Dvorak

Rensselaer Polytechnic Institute, Troy, NY 12180, USA

Abstract. It is shown that spatially uniform stress and strain fields can be caused in certain heterogeneous media by simultaneous application of a uniform thermal change and uniform overall stress. These results are found for a binary composite consisting of isotropic phases of any shape, and for binary and three-phase fibrous composites made of transversely isotropic phases of any transverse plane geometry. For these systems, the results make it possible to derive exact relations which describe macroscopic response to combined thermal and mechanical loads in terms of equivalent mechanical loads. If one of the phases deforms plastically, then the thermal and mechanical loading responses are found to be coupled.

1. INTRODUCTION

This paper is concerned with the macroscopic response of certain composite materials to combined incremental loading by a uniform thermal change and macroscopically uniform stresses or strains. Both elastic and elastic-plastic behavior at small strains are included; phase properties are independent of temperature. The limitation is to a recently identified class of composites [1,2] that in their response to uniform thermal change exhibit a striking similarity to that of unreinforced, isotropic solids. In particular, it is shown that uniform isotropic strain fields can be created in the heterogeneous microstructure of these composites if the thermal change is applied simultaneously with prescribed surface tractions. This property is used to derive exact relations which convert the elastic or elastic-plastic response under combined thermomechanical loading into an equivalent response under purely mechanical loading. These relations also clarify the nature of thermomechanical coupling in this class of composite materials.

The results are found to be valid for two binary composite systems which represent most materials of practical interest: one which consists of two distinct elastically isotropic phases of any geometry; and one in which a transversely isotropic or isotropic matrix is reinforced by aligned isotropic or transversely isotropic fibers. A similar, albeit less general result is derived for a three-phase fibrous composite where the matrix is reinforced by two different kinds of fiber, and each of the phases is either isotropic or transversely isotropic.

Throughout the paper it is assumed that the response of the composite aggregate can be evaluated under incremental mechanical loading. Specifically, in the notation introduced by Hill [3], the overall constitutive relations are

$$d\bar{\varepsilon} = M\, d\bar{\sigma} \quad , \qquad d\bar{\sigma} = L\, d\bar{\varepsilon} \tag{1}$$

where M, L are (6x6) instantaneous overall compliance and stiffness matrices, and $d\bar{\varepsilon}$, $d\bar{\sigma}$ are overall strain and stress increments. These relations are assumed to be valid for a certain volume which is representative of macroscopic behavior of the aggregate, and on that scale, $d\bar{\sigma}$, and $d\bar{\varepsilon}$ are both uniform. Also, the averages of local fields in each phase r can be written as:

$$d\varepsilon_r = A_r\, d\bar{\varepsilon} \quad , \qquad d\sigma_r = B_r\, d\bar{\sigma} \quad , \tag{2}$$

where A_r, B_r are known strain and stress concentration factors.

The thermomechanical response of each phase r is also known, in the form:

$$d\varepsilon_r = M_r\, d\sigma_r + m_r\, d\theta \quad , \quad d\sigma_r = L_r\, d\varepsilon_r - \ell_r\, d\theta \tag{3}$$

where all quantities are phase volume averages, M_r, L_r are again instantaneous phase compliance and stiffness, m_r, ℓ_r are local thermal strain and stress vectors, and $d\theta$ is the uniform thermal change. In the case of plastic straining A_r, B_r, M_r, L_r, in (2) and (3) are stress dependent and therefore variable in the phase; they may need to be evaluated locally at each point, or in a uniformly strained subelement of the inelastic phase, and the phase averages obtained as shown by Dvorak [2]. While ℓ_r may also become stress-dependent in the plastically deforming phase, only elastic values of ℓ_r will be needed in the sequel.

2. ARBITRARY PHASE GEOMETRY

Consider a binary composite material consisting of two perfectly bonded phases of any geometry such that it is statistically homogeneous in its representative volume, and free of voids or cracks. In the elastic range, each phase is isotropic. One or both phases may be elastic-plastic and if so, it is required that their response be plastically incompressible.

Assume that the composite has been loaded to current state by a certain uniform overall stress $\bar{\sigma}^o$, or strain $\bar{\varepsilon}^o$, and uniform thermal change θ_o. Next, simultaneous increments $d\bar{\sigma}$ and $d\theta$, or $d\bar{\varepsilon}$ and $d\theta$, are applied, and the macroscopic response is to be determined. To solve this problem, the phases are initially separated and loaded by the prescribed $d\theta$, and by certain unknown tractions which correspond to an isotropic stress change $d\hat{\sigma}_r$ in each phase r.

Let the symbols r = f and r = m denote the two phases, f may denote the reinforcement and m the matrix. Then, $d\theta$ and $d\hat{\sigma}_r$ create the following fields in the separated phases:

$$
\begin{aligned}
d\hat{\sigma}^{f}_{11} &= d\hat{\sigma}^{f}_{22} = d\hat{\sigma}^{f}_{33} = dS_f \\
d\hat{\sigma}^{m}_{11} &= d\hat{\sigma}^{m}_{22} = d\hat{\sigma}^{m}_{33} = dS_m \\
d\hat{\varepsilon}^{f}_{11} &= d\hat{\varepsilon}^{f}_{22} = d\hat{\varepsilon}^{f}_{33} = dS_f/(3K_f) + \alpha_f \, d\theta \\
d\hat{\varepsilon}^{m}_{11} &= d\hat{\varepsilon}^{m}_{22} = d\hat{\varepsilon}^{m}_{33} = dS_m/(3K_m) + \alpha_m \, d\theta
\end{aligned}
\qquad (4)
$$

where α_f, α_m are linear thermal expansion coefficients and K_f, K_m are elastic bulk moduli. Note that the fields (4) are all uniform. If the phases are to be reassembled, it is necessary to make them compatible, and assure internal equilibrium. Since the fields (4) are uniform, these requirements can be satisfield if

$$dS_f = dS_m = dS \quad , \qquad d\hat{\varepsilon}^{f}_{ij} = d\hat{\varepsilon}^{m}_{ij} \ , \qquad (5)$$

and

$$dS = s \, d\theta \quad , \qquad s = -3(\alpha_f - \alpha_m)/(1/K_f - 1/K_m) \ . \qquad (6)$$

The composite can now be reassembled. It is loaded by the thermal change $d\theta$, and by surface tractions which correspond to an overall hydrostatic stress dS. Internal stress and strain increments satisfy (5), hence they are uniform in the entire volume, without regard to the actual internal geometry of the heterogeneous medium.

To complete the solution of the problem, we now consider only the case when $d\theta$ and $d\bar{\underset{\sim}{\sigma}}$ are applied. The final overall strain increment $d\bar{\underset{\sim}{\varepsilon}}$ consists of the strains (5), and of such strains caused in the aggregate by application of surface tractions which cancel the overall stress dS. The result is:

$$d\bar{\underset{\sim}{\varepsilon}} = \underset{\sim}{q} \, d\theta + M(d\bar{\underset{\sim}{\sigma}} - \underset{\sim}{s} \, d\theta) \qquad (7)$$

where

$$
\begin{aligned}
\underset{\sim}{q} &= q \, [1 \ 1 \ 1 \ 0 \ 0 \ 0]^T \\
\underset{\sim}{s} &= s \, [1 \ 1 \ 1 \ 0 \ 0 \ 0]^T \\
q &= s/(3K_f) + \alpha_f = s/(3K_m) + \alpha_m
\end{aligned}
$$

The first term in (7) is the isotropic strain introduced while the phases were separated. If one or both phases f, m are elastic-plastic, but plastically incompressible, then $\underset{\sim}{q}\, d\theta$ is elastic. However, the second term in (7) is not generally isotropic, even under pure thermal change, and it may cause plastic straining in the aggregate. Of course, this part of the overall $d\bar{\underset{\sim}{\varepsilon}}$ is due to the mechanical loading change $(d\bar{\underset{\sim}{\sigma}} - \underset{\sim}{s}\, d\theta)$, and can be evaluated from (1), providing that M is taken for the new loading path in the case of plastic loading.

In analogy with (2) one can write the stresses in the phases as:

$$d\underset{\sim}{\sigma}_f = \underset{\sim}{s}\, d\theta + B_f(d\bar{\underset{\sim}{\sigma}} - \underset{\sim}{s}\, d\theta)$$
$$d\underset{\sim}{\sigma}_m = \underset{\sim}{s}\, d\theta + B_m(d\bar{\underset{\sim}{\sigma}} - \underset{\sim}{s}\, d\theta) \tag{8}$$

where B_f, B_m are instantaneous stress concentration factors for the overall mechanical load increment $(d\bar{\underset{\sim}{\sigma}} - \underset{\sim}{s}\, d\theta)$.

Next, consider loading by $d\theta$ and $d\bar{\underset{\sim}{\varepsilon}}$. The decomposition sequence (4) to (6) is again employed, and that creates the strain $\underset{\sim}{q}\, d\theta$, and stress $dS = s\, d\theta$, in the aggregate. However, since $d\bar{\underset{\sim}{\varepsilon}}$ is now prescribed, the overall stress $d\bar{\underset{\sim}{\sigma}}$ needed to create $(d\bar{\underset{\sim}{\varepsilon}} - \underset{\sim}{q}\, d\theta)$ must be found. It is easy to verify that the result is:

$$d\bar{\underset{\sim}{\sigma}} = \underset{\sim}{s}\, d\theta + L(d\bar{\underset{\sim}{\varepsilon}} - \underset{\sim}{q}\, d\theta) \tag{9}$$

where L is the instantaneous overall stiffness (1) for the overall strain increment $(d\bar{\underset{\sim}{\varepsilon}} - \underset{\sim}{q}\, d\theta)$.

Again, the local strains are:

$$d\underset{\sim}{\varepsilon}_f = \underset{\sim}{q}\, d\theta + A_f(d\bar{\underset{\sim}{\varepsilon}} - \underset{\sim}{q}\, d\theta)$$
$$d\underset{\sim}{\varepsilon}_m = \underset{\sim}{q}\, d\theta + A_m(d\bar{\underset{\sim}{\varepsilon}} - \underset{\sim}{q}\, d\theta) \tag{10}$$

where A_f, A_m are instantaneous strain concentration factors for an overall mechanical strain increment equal to $(d\bar{\underset{\sim}{\varepsilon}} - \underset{\sim}{q}\, d\theta)$.

Equations (7), or (9), convert in an exact way the thermomechanical loading problem into a mechanical one along a loading path $(d\bar{\underset{\sim}{\sigma}} - \underset{\sim}{s}\, d\theta)$, or a strain path $(d\bar{\underset{\sim}{\varepsilon}} - \underset{\sim}{q}\, d\theta)$, respectively. It is seen that the response to combined thermomechanical loading is coupled whenever the mechanical loading response depends on the loading path. This follows from the dependence, if it exists, of M in (7), B_f, B_m in (8), L in (9), and A_f, A_m in (10), on the respective loading terms.

As a simple example of application of the above results, (7) may be used to derive the expression for the overall linear thermal expansion coefficient of a binary composite which is macroscopically isotropic. The result is

$$\alpha = \alpha_m + \frac{(\alpha_f - \alpha_m)}{\left(\frac{1}{K_f} - \frac{1}{K_m}\right)} \left(\frac{1}{K} - \frac{1}{K_m}\right)$$

where K is the overall bulk modulus. This equation was derived in a different way by Levin [4].

3. FIBROUS COMPOSITES

The decomposition procedure can also be applied to two and three-phase fibrous composites. For simplicity of the derivation, assume that, in the elastic range, all phases are transversely isotropic about the fiber axis, which will be taken as parallel to the Cartesian coordinate x_3; x_1 and x_2 are in the transverse plane. The elastic composite is, of course, also transversely isotropic. The first two strain and stress invariants for a transversely isotropic medium can be written as:

$$d\varepsilon_1 = d\varepsilon_{11} + d\varepsilon_{22} \qquad d\varepsilon_2 = d\varepsilon_{33}$$
$$d\sigma_1 = \frac{1}{2}\left(d\sigma_{11} + d\sigma_{22}\right) \qquad d\sigma_2 = d\sigma_{33} \tag{11}$$

and connected by elastic constitutive relations:

$$\begin{Bmatrix} d\varepsilon_1 \\ d\varepsilon_2 \end{Bmatrix} = \frac{1}{kE}\begin{bmatrix} n & -\ell \\ -\ell & k \end{bmatrix}\begin{Bmatrix} d\sigma_1 \\ d\sigma_2 \end{Bmatrix} + \begin{Bmatrix} \alpha \\ \beta \end{Bmatrix} d\theta$$
$$\begin{Bmatrix} d\sigma_1 \\ d\sigma_2 \end{Bmatrix} = \begin{bmatrix} k & \ell \\ \ell & n \end{bmatrix}\begin{Bmatrix} d\varepsilon_1 \\ d\varepsilon_2 \end{Bmatrix} - \begin{Bmatrix} k\alpha + \ell\beta \\ \ell\alpha + n\beta \end{Bmatrix} d\theta \tag{12}$$

where k, ℓ, n are Hill's [5] elastic moduli, $E = n - \ell^2/k$, $\alpha = 2\alpha_T$, $\beta = \alpha_L$, and α_T and α_L are linear coefficients of thermal expansion in the transverse plane and longitudinal direction, respectively.

With appropriate values of elastic moduli and coefficients α, β, equations (11) and (12) can be applied either to the composite medium or to each of the phases.

In the first step of the procedure the fiber and matrix phases are separated and surface tractions which preserve the current local stresses $\underset{\sim}{\sigma}_r$ and strains $\underset{\sim}{\varepsilon}_r$ are applied to each phase r = f,m. Alternatively, surface displacements corresponding to $\underset{\sim}{\varepsilon}_r$ may be prescribed to preserve $\underset{\sim}{\sigma}_r$. In addition, a uniform thermal change $d\theta$ is applied to both phases. The local strains caused by $d\theta$ would make the phases incompatible if the composite was to be reassembled. Therefore, uniform tractions $d\hat{\sigma}_1^r$, $d\hat{\sigma}_2^r$ of as yet unknown magnitude are applied to the phases simultaneously with $d\theta$. (The top hats indicate auxilliary uniform fields used in the decomposition and reassembly of the composite.) This leads to the following uniform strain increments in the separated phases:

$$
\begin{aligned}
d\hat{\varepsilon}_1^f &= (n_f\, d\hat{\sigma}_1^f - \ell_f\, d\hat{\sigma}_2^f)/k_f E_f + \alpha_f\, d\theta \\
d\hat{\varepsilon}_2^f &= (-\ell_f\, d\hat{\sigma}_1^f + k_f\, d\hat{\sigma}_2^f)/k_f E_f + \beta_f\, d\theta \\
d\hat{\varepsilon}_1^m &= (n_m\, d\hat{\sigma}_1^m - \ell_m\, d\hat{\sigma}_2^m)/k_m E_m + \alpha_m\, d\theta \\
d\hat{\varepsilon}_2^m &= (-\ell_m\, d\hat{\sigma}_1^m + k_m\, d\hat{\sigma}_2^m)/k_m E_m + \beta_m\, d\theta
\end{aligned}
\tag{13}
$$

In the second step of the procedure, the tractions $d\hat{\sigma}_1^r$ and $d\hat{\sigma}_2^r$ must be adjusted to assure compatibility of the phases and equilibrium of these tractions at phase interfaces and on the surface S of the representative volume V. The strain and stress increments in (3) obviously satisfy the equations $d\hat{\varepsilon}_1^r = 2d\hat{\varepsilon}_{11}^r = 2d\hat{\varepsilon}_{22}^r$ and $d\hat{\sigma}_1^r = d\hat{\sigma}_{11}^r = d\hat{\sigma}_{22}^r$ in each phase r = f,m. Shear components vanish except in the immediate vicinity of fiber ends. Since the magnitude of average fiber diameter is assumed to be very small, the shear components may be neglected. Therefore, compatibility and equilibrium conditions for the increments can be written in terms of the invariants as follows:

$$
d\hat{\varepsilon}_1^m = d\hat{\varepsilon}_1^f \quad , \quad d\hat{\varepsilon}_2^m = d\hat{\varepsilon}_2^f \tag{14}
$$

$$
d\hat{\sigma}_1^f = d\hat{\sigma}_1^m = dS_T \tag{15}
$$

$$
c_f\, d\hat{\sigma}_2^f + c_m\, d\hat{\sigma}_2^m = dS_A \; . \tag{16}
$$

The dS_T and dS_A are surface stresses which need to be added at S to preserve overall equilibrium of V while $d\hat{\sigma}^r_1$ and $d\hat{\sigma}^r_2$ are applied to the phases. The magnitudes of phase volume fractions $c_f + c_m = 1$ need also be known at this point; $c_f > 0$.

All strain and stress increments in (13) are uniform and transversely isotropic, hence equations (14) to (16) are exact for any transverse plane geometry. These relations suggest that spatially uniform strain fields can be created in the fibrous composite by superposition of uniform eigenstrains $\alpha_r d\theta$, $\beta_r d\theta$ in the phases, with local strains caused by piecewise uniform stress fields which are in equilibrium with surface stresses dS_A, dS_T.

Internal equilibrium and compatibility of the phases in V depend only on the eight unknown strains and stresses $d\hat{\varepsilon}^r_1$, $d\hat{\varepsilon}^r_2$, $d\hat{\sigma}^r_2$, $d\hat{\sigma}^r_1$, and not on dS_A, dS_T. These unknowns can be determined from (13), (14), and (15), when an additional constraint is imposed.

A particular choice which will be useful in the sequel is:

$$d\hat{\sigma}^m_2 = \rho \, d\hat{\sigma}^m_1 , \tag{17}$$

where $\rho \neq 0$ is a constant.

These equations give magnitudes of dS_A, dS_T as:

$$\begin{aligned} dS_T &= s_T \, d\theta \quad , \quad dS_A = s_A \, d\theta \\ s_T &= (a_2 b_3 - a_3 b_2)/(a_1 b_2 - a_2 b_1) \\ s_A &= (a_3 b_1 - a_1 b_3)/(a_1 b_2 - a_2 b_1) \end{aligned} \tag{18}$$

where:

$$\begin{aligned} a_1 &= \frac{n_f}{k_f E_f} - \frac{n_m}{k_m E_m} + \rho \left[\frac{c_m \ell_f}{c_f k_f E_f} + \frac{\ell_m}{k_m E_m}\right] \\ a_2 &= -\frac{1}{c_f} \frac{\ell_f}{k_f E_f} \quad , \quad a_3 = \alpha_f - \alpha_m \\ b_1 &= \frac{\ell_f}{k_f E_f} - \frac{\ell_m}{k_m E_m} + \rho \left[\frac{c_m}{c_f E_f} + \frac{1}{E_m}\right] \\ b_2 &= -\frac{1}{c_f E_f} \quad , \quad b_3 = -\beta_f + \beta_m \end{aligned} \tag{19}$$

It can be verified that S_T, S_A do not depend on c_f.

Now, the local strains can be written as (6x1) column vectors, rather than the invariants (11):

$$d\hat{\varepsilon}^f_{11} = d\hat{\varepsilon}^f_{22} = \frac{1}{2}\, d\hat{\varepsilon}^f_1 = g_1\, d\theta \quad , \quad d\hat{\varepsilon}^f_{33} = d\hat{\varepsilon}^f_2 = g_2\, d\theta \tag{20}$$

$$d\hat{\varepsilon}^m_{11} = d\hat{\varepsilon}^m_{22} = \frac{1}{2}\, d\hat{\varepsilon}^m_1 = h_1\, d\theta \quad , \quad d\hat{\varepsilon}^m_{33} = d\hat{\varepsilon}^m_2 = h_2\, d\theta \tag{21}$$

From (13), with (14) to (16), and (18):

$$\begin{aligned} g_1 &= [\tfrac{1}{2}\,(n_f - \gamma\, \ell_f)/(k_f\, E_f)]\; s_T + \tfrac{1}{2}\,\alpha_f \\ g_2 &= [(-\ell_f + \gamma\, k_f)/(k_f\, E_f)]\; s_T + \beta_f \\ h_1 &= [\tfrac{1}{2}\,(n_m - \rho\, \ell_m)/(k_m\, E_m)]\; s_T + \tfrac{1}{2}\,\alpha_m \\ h_2 &= [(-\ell_m + \rho\, k_m)/(k_m\, E_m)]\; s_T + \beta_m \end{aligned} \tag{22}$$

where

$$\gamma = (s_A - \rho\, c_m\, s_T)/(c_f\, s_T) \; , \tag{23}$$

and, according to (14): $g_1 = h_1 \quad , \quad g_2 = h_2$.

The stresses are:

$$\begin{aligned} d\hat{\sigma}^f_{11} &= d\hat{\sigma}^f_{22} = d\hat{\sigma}^f_1 = s_T\, d\theta \quad , \quad & d\hat{\sigma}^f_{33} &= d\hat{\sigma}^f_2 = \gamma\, s_T\, d\theta \\ d\hat{\sigma}^m_{11} &= d\hat{\sigma}^m_{22} = d\hat{\sigma}^m_1 = s_T\, d\theta \quad , \quad & d\hat{\sigma}^m_{33} &= d\hat{\sigma}^m_2 = \rho\, s_T\, d\theta \end{aligned} \tag{24}$$

The decomposition procedure is completed when the composite is reassembled and the overall stresses (18) are removed. The final results assume a concise form with the definitions

$$\underset{\sim}{h} = [h_1 \;\; h_1 \;\; h_2 \;\; 0 \;\; 0 \;\; 0]^T \quad , \quad \underset{\sim}{s}_a = [s_T \;\; s_T \;\; s_A \;\; 0 \;\; 0 \;\; 0]^T$$
$$\underset{\sim}{\gamma} = [1 \;\; 1 \;\; \gamma \;\; 0 \;\; 0 \;\; 0]^T \quad , \quad \underset{\sim}{\rho} = [1 \;\; 1 \;\; \rho \;\; 0 \;\; 0 \;\; 0]^T \tag{25}$$

<u>For loading by $d\underset{\sim}{\bar{\sigma}}$ and</u> $d\theta$, the overall strain increment $d\underset{\sim}{\bar{\varepsilon}}$ is:

$$d\underset{\sim}{\bar{\varepsilon}} = \underset{\sim}{h}\, d\theta + M(d\underset{\sim}{\bar{\sigma}} - \underset{\sim}{s}_a\, d\theta). \tag{26}$$

<u>For loading by $d\underset{\sim}{\bar{\varepsilon}}$ and</u> $d\theta$, the overall strain increment $d\underset{\sim}{\bar{\sigma}}$ is:

$$d\underset{\sim}{\bar{\sigma}} = \underset{\sim}{s}_a\, d\theta + L(d\underset{\sim}{\bar{\varepsilon}} - \underset{\sim}{h}\, d\theta). \tag{27}$$

Figure 1 shows schematically the steps in the decomposition and reassembly of a fibrous composite. The circular fiber cross section is shown for convenience, in reality the procedure holds for any cylindrical fiber.

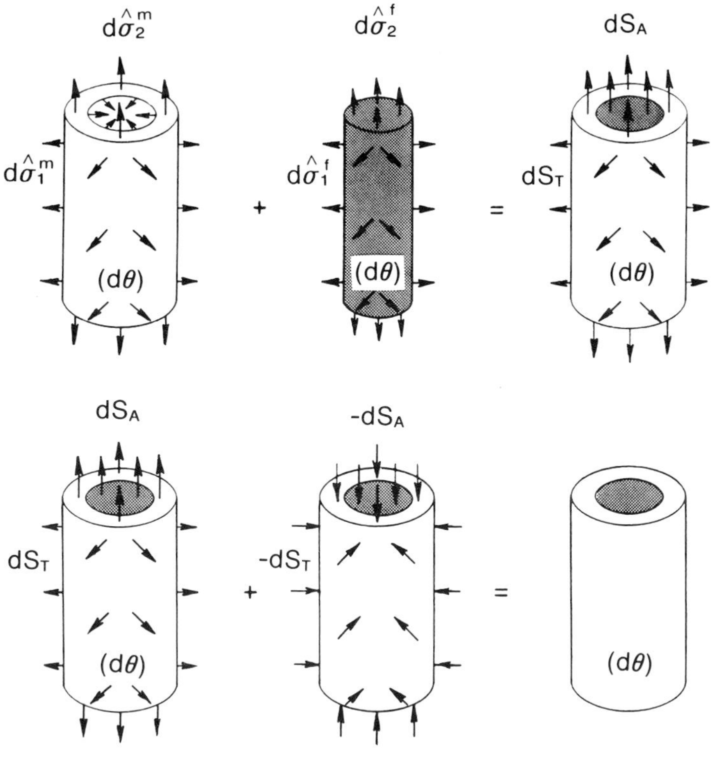

Fig. 1 Decomposition Procedure

These results can be applied to elastic-plastic fibrous composites in which the fiber is elastic, but the matrix may deform plastically while remaining plastically incompressible. This suggests a metal matrix, which is usually elastically isotropic. Thus the matrix elastic moduli in (12) become related as follows:

$$\ell_m = k_m - m_m, \qquad n_m = k_m + m_m, \qquad E_m = n_m - \ell_m^2/k_m \tag{28}$$

$$\frac{n_m}{k_m E_m} = \frac{2(1-\nu_m)}{E_m}, \qquad \frac{\ell_m}{k_m E_m} = \frac{2\nu_m}{E_m}, \qquad \alpha_m = 2\beta_m$$

where E_m, ν_m are the isotropic constants, and β_m is the linear thermal expansion coefficient of the matrix. Also, to assure that the composite remains elastic while loaded by $d\theta$, dS_A, and dS_T, it is necessary to select $\rho = 1$ in (17) and thus make $d\underset{\sim}{\sigma}_m$ isotropic. This simplifies the terms in (25), in particular

$$h_1 = h_2 = s_T/(3K_m) + \beta_m \,. \tag{29}$$

Equations (26) and (27) again connect in an exact way the thermomechanical loading problem into a mechanical problem along a modified loading path. Local fields can be obtained in terms of mechanical stress and strain concentration factors (2). Coupling of thermal and mechanical effects is again evident, as it was in (7) and (9).

The decomposition procedure can also be applied to three-phase elastic composites. The system consists of a matrix reinforced by two different types of aligned fiber arranged in such a way that the aggregate is again statistically homogeneous. Each of the phases can be either isotropic or transversely isotropic. Matrix properties are denoted by r = m, while the fiber properties are identified by letters f and g. Phase constitutive relations are given by (11) and (12). The procedure leading to (13) is repeated to yield:

$$\begin{aligned}
d\hat{\varepsilon}_1^f &= (n_f\, d\hat{\sigma}_1^f - \ell_f\, d\hat{\sigma}_2^f)/k_f E_f + \alpha_f\, d\theta \\
d\hat{\varepsilon}_2^f &= (-\ell_f\, d\hat{\sigma}_1^f + k_f\, d\hat{\sigma}_2^f)/k_f E_f + \beta_f\, d\theta \\
d\hat{\varepsilon}_1^g &= (n_g\, d\hat{\sigma}_1^g - \ell_g\, d\hat{\sigma}_2^g)/k_g E_g + \alpha_g\, d\theta \\
d\hat{\varepsilon}_2^g &= (-\ell_g\, d\hat{\sigma}_1^g + k_g\, d\hat{\sigma}_2^g)/k_g E_g + \beta_g\, d\theta \\
d\hat{\varepsilon}_1^m &= (n_m\, d\hat{\sigma}_1^m - \ell_m\, d\hat{\sigma}_2^m)/k_m E_m + \alpha_m\, d\theta \\
d\hat{\varepsilon}_2^m &= (-\ell_m\, d\hat{\sigma}_1^m + k_m\, d\hat{\sigma}_2^m)/k_m E_m + \beta_m\, d\theta
\end{aligned} \tag{30}$$

And, in analogy with (14) to (16), internal equilibrium and compatibility of the phases is assured in terms of spatially uniform fields when

$$d\hat{\varepsilon}^f_1 = d\hat{\varepsilon}^g_1 = d\hat{\varepsilon}^m_1 \quad ; \quad d\hat{\varepsilon}^f_2 = d\hat{\varepsilon}^g_2 = d\hat{\varepsilon}^m_2 \tag{31}$$

$$d\hat{\sigma}^f_1 = d\hat{\sigma}^g_1 = d\hat{\sigma}^m_1 = dQ_T \tag{32}$$

$$c_f \, d\hat{\sigma}^f_2 + c_g \, d\hat{\sigma}^g_2 + c_m \, d\hat{\sigma}^m_2 = dQ_A \tag{33}$$

The system of equations (30) to (32) is determinate, hence additional constraints such as (17) cannot be prescribed, and that limits the applicability of the results to elastic three-phase systems. If the solution exists, then it leads to equations for overall strain and stress increments, which are similar to (26) and (27). These can be utilized to find the overall thermal strain vector $\underset{\sim}{m}$, or the thermal stress vector $\underset{\sim}{\ell}$ of the composite medium in terms of thermoelastic properties of the phases and the overall mechanical stiffness L, or compliance M.

4. CONCLUDING REMARKS

The decomposition procedure reveals that spatially uniform fields can be created in certain heterogeneous media by application of a uniform thermal change and certain surface tractions which correspond to uniform overall stress fields. This property finds a natural application in solution of the thermomechanical problem for the three composite systems considered herein. Even with the limitations involved, the results apply to most composite materials of practical interest. The results are exact. They make it possible to convert a plasticity theory, or a corresponding computer program, developed for mechanical loading to a theory applicable to combined thermal and mechanical loading. Of particular interest is the explicit derivation of thermomechanical coupling effects which are caused by mutual constraints of the bonded phases.

Additional applications to similar differential dilatation problems in composite media are possible. Phase transformations and uniform moisture absorption are obvious examples.

We note in passing that the evaluation of thermoelastic properties of composite media in terms of thermoelastic properties of the phases and overall mechanical properties has been explored rather extensively by several authors, e.g., [4,6]. While the present approach is entirely different, the results obtained can be shown to coincide in the elastic range with those found in the earlier work, as in the example given at the conclusion of Section 2.

ACKNOWLEDGEMENT

This work was supported by a grant from the Office of Naval Research, Dr. Yapa Rajapakse served as contract monitor.

REFERENCES

[1] Dvorak, G.J., 1983, "Metal Matrix Composites: Plasticity and Fatigue," Mechanics of Composite Materials: Recent Advances, Hasin, Z., and Herakovich, C.T., eds., Pergamon Press, pp. 73-91.
[2] Dvorak, G.J., 1986, "Thermal Expansion of Elastic-Plastic Composite Materials," Journal of Applied Mechanics, to appear.
[3] Hill, R., 1963, "Elastic Properties of Reinforced Solids: Some Theoretical Principles," Journal of the Mechanics and Physics of Solids, Vol. 11, pp. 357-372.
[4] Levin, V.M., 1967, "Thermal Expansion Coefficients of Heterogeneous Materials," Mekhanika Tverdogo Tela, Vol. 2, pp. 88-94.
[5] Hill, R., 1964, "Theory of Mechanical Properties of Fiber-Strengthened Materials: I. Elastic Behaviour," Journal of the Mechanics and Physics of Solids, Vol. 12, pp. 199-212.
[6] Rosen, B.W., and Hashin, Z., 1970, "Effective Thermal Expansion Coefficients and Specific Heats of Composite Materials," International Journal of Engineering Science, Vol. 8, pp. 157-173.

Thermomechanical Couplings in Solids
H.D. Bui and Q.S. Nguyen (Editors)
Elsevier Science Publishers B.V. (North-Holland)

ON THERMOELASTIC DAMPING IN HETEROGENOUS MATERIAL

M. BUISSON, A. MOLINARI

Laboratoire de Physique et Mécanique des Matériaux - C.N.R.S. -
Faculté des Sciences de Metz, Ile du Saulcy - 57045 METZ Cedex 1

We investigate thermoelastic internal friction in an isotropic cylindrical rod subjected to an uniaxial quasistatic stress. Young's modulus is assumed to have longitudinal sinusoidal space-variations characterized by a small parameter of heterogeneity. Using a direct perturbation technique relative to this parameter, approximate analytical expressions of macroscopic behaviour and thermoelastic dissipation are given. Other arguments (Fourier-series, numerical computation, exact results from a thermodynamical analysis) check the accuracy of this technique. Further, on the basis of references with Zener's work, the macroscopic anelastic behaviour and the role of the length-scale of the inhomogeneities are exhibited.

1. INTRODUCTION

An example of thermomechanical coupling is examined here. It concerns thermoelastic damping arising in heterogeneous media. Such a matter is basically established in the writing of Zener [1] who investigated thermoelastic internal friction due to various causes like elastic anisotropy of polycrystals, macrohomogeneous media with internal cavities or stress inhomogeneities in a vibrating body. It is observed in each case that an external mechanical (or thermal) loading produces stress-and strain-inhomogeneities in the body coupled with heat-gradients and hence leads to irreversible entropy production.

In thermoelastic materials, the phenomenon of dissipation appears closely linked to the presence of temperature gradients. This remark is the key argument of our work. Using a quantitative method, we discuss the role of the length-scale of the inhomogeneities. We consider two types of thermal gradients :

- first, at the level of the body, is a macroscopic thermal non-uniformity. This case may be related, for example, to boundary conditions and has been quantitatively treated by Zener with a homogeneous material. The specimen can undergo thermal relaxation due to heat flow in relation to its surroundings or to another macroscopic-part.
- The second type of thermal gradients, at microscopic level, is linked to non-uniform thermomechanical properties (for example microheterogeneities in polycrystals).

The calculation of dissipation due to thermoelastic effect generally requires the complex resolution of a boundary value problem linked to the three dimensional set of partial differential equations of coupled thermoelasticity.

For the sake of simplicity, our work is dealing with a one dimensional problem where the amplitude of variations in the thermomechanical properties (e. g. Young's modulus and the adiabatic index) are assumed small. Using a direct perturbation technique for the governing equations given in paragraph (3), we establish an analogy with the behaviour of a standard viscoelastic Zener solid. This fact is not surprising when looking at the work carried out by Zener where thermoelastic coupling appears as a source of anelasticity. This anelastic behaviour is exhibited in paragraph (3) and (4) where a creep-test and internal friction take place. Comparative quantitative results with Zener's work are given. The accuracy of the perturbation technique is checked with other considerations and the role of the length-scale is discussed in paragraph (5).

2. GOVERNING EQUATIONS

We look at an isotropic cylindrical rod subjected to an uniaxial (axis-x) quasistatic stress (e. g. inertial effects are neglected). Microscopic thermal currents associated to microheterogeneities are present by assuming only a periodic variation of the isothermal Young's modulus E(x)

$$E(x) = E^\circ + H\cos(\pi x l^{-1}) \tag{2.1}$$

where E° is the average value of E(x) over a period of length 2 l and 2 H is the total amplitude of these variations.

In case of an uniaxial quasistatic stress loading, the Duhamel Neumann law and the equilibrium equation read respectively :

$$\begin{cases} \varepsilon = \varepsilon_{11} = \sigma E^{-1} + \alpha\theta \\ \varepsilon_{22} = -\nu E^{-1}\sigma + \alpha\theta \\ \varepsilon_{33} = -\nu E^{-1}\sigma + \alpha\theta \\ \varepsilon_{ij} = 0 \qquad (i \neq j) \end{cases} \tag{2.2}$$

$$\sigma_{,x} = 0 \qquad (\text{partial derivative}) \tag{2.3}$$

where ε_{ij} (i, j = 1 to 3), ν (assumed here to be uniform), α (assumed here to be uniform) and θ are the components of the strain-tensor, Poisson's coefficient, the free linear thermal expansion and the variation of temperature from some reference temperature $T_\circ$.

The evolution of the temperature distribution is governed by the uniaxial heat conduction equation

$$c_\sigma \theta_{,t} = k\theta_{,xx} - \alpha T_\circ \sigma_{,t} \tag{2.4}$$

where k is the uniform coefficient of conduction and c_σ the specific heat per unit volume at constant stress. We assume here that the temperature field is uniform in each section perpendicular to the x-axis (e. g. transverse thermal currents are neglected).

In order to calculate the thermoelastic damping due to thermal currents acting on the scale of the heterogeneity, we set as boundary conditions that the bar is infinitely extended. This assumption annihilates effectively macroscopic thermal currents due to some macroscopic boundary-thermal difference. By symmetry, the preceding boundary-conditions may be replaced by

$$\theta_{,x} = 0 \quad \text{at} \quad x=0 \quad \text{and} \quad x=l \tag{2.5}$$

without changing the temperature profile. We report on this research about the temperature distribution on the elementary cell constituted by a half period *.

<u>Heterogeneity parameter</u> :

c and γ (x) being respectively the specific heat at constant strain (assumed here to be uniform) and the adiabatic index, it is well known that [3] :

$$c_\sigma = \gamma c \quad \text{and} \quad \gamma(x) = 1 + 3E(x)T_\circ\alpha^2(1-2\nu)^{-1}c^{-1} \tag{2.6}$$

Taking (2.1) into account, we present (see appendix 1) γ (x) in the form

$$\gamma(x) = \gamma_\circ[1 + h\cos(\pi x l^{-1})] \tag{2.7}$$

where h is a parameter specific of the material's heterogeneity varying around the mean value $\gamma_\circ$.

In the following, we deal with the analysis of the macroscopic behaviour of this heterogeneous rod. Two methods are applied, one being based on the application of a stress-increment whereas in the second we observe the strain-response to a cyclical stress.

---

* Observing that the positions of maximum temperature are separated by l [2], it will be verified that l is the caractheristic length linked to the relaxation time.

3. STRESS INCREMENT LOADING

The bar is subjected to the application of a stress increment σ_o . We observe a deformation creep which may be assimilated, with a small heterogeneity-parameter h to the creep of a linear Zener-type solid. J (t) being the relaxation function, the strain response is written :

$$\varepsilon(t) = J(t)\sigma_o \quad \text{with} \quad J(t) = J_u + (J_r - J_u)[1-\exp(-t\tau^{-1})]$$

where τ , J_r and J_u are respectively the characteristic relaxation time at constant stress, the relaxed and unrelaxed compliances.

3.1. Unrelaxed and relaxed compliances

Let us decompose the incremental test OAB (figure 1)

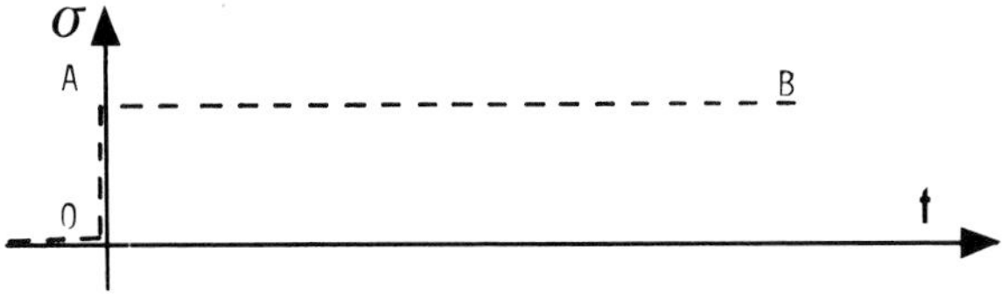

figure 1 : test OAB

- phase OA is a very rapid increase of the stress so that heat conduction is not acting. Assuming uniformity of initial-temperature and adiabaticity, integration of the heat equation (2.4) in relation to time leads to the thermal distribution corresponding to point A :

$$\theta_A(x) = -\alpha T_o \sigma_o c^{-1} \gamma^{-1}(x) \tag{3.1}$$

and the corresponding instantaneous strain : $\varepsilon_A(x) = E(x)\sigma_o^{-1} + \alpha\theta_A(x)$

If < > represents the spatial-average over the length of the cell, we have access to the unrelaxed compliance linked to the instantaneous macroscopic strain :

$$J_u = <\varepsilon_A>\sigma_o^{-1} = <E^{-1}> -\alpha^2 T_o c^{-1} <\gamma^{-1}> \tag{3.2}$$

- phase AB corresponds to the stage where stress is constant ($\sigma_{,t}$ = 0).

Breaking in mind that the cell is thermally-insulated, it is easy to show after space-integration and with k uniform that

$$<k\theta_{,xx}> = 0 \tag{3.3}$$

After a sufficiently long time, the final state corresponding to point B is attained when the temperature θ_B is uniform. Taking the average of (2.4) and using (3.3), time-integration from A to B provides : $<c_\sigma>\theta_B = <c_\sigma\theta_A>$

Knowing the final equilibrium temperature and consequently the final macroscopic-strain $<\varepsilon_B>$ gives us access to the relaxed compliance J_r . We obtain

$$J_r = <E^{-1}> - \alpha^2 T_o c^{-1} <\gamma>^{-1} \tag{3.4}$$

3.2 Relaxation function

We now seek information about evolution-temperature between point A and point B. It implies that the partial derivative equation (2.4) (with $\sigma_{,t}$ = 0) is solved with the space-variable coefficient $C_\sigma(x)$.

We use a direct perturbation technique based on the heterogeneity coefficient h giving an approximate analytical formulation (appendices 2-3). We obtain the perturbated compliance function $J^p(t)$:

$$J^p(t) = J_u^p + (J_r^p - J_u^p)[1 - \exp(-t\tau^{-1})] \tag{3.5}$$

where for the relaxation time τ , the unrelaxed and relaxed compliances

$$\tau = c\langle\gamma\rangle l^2 k^{-1} \pi^{-2} \tag{3.6}$$

$$J_u^p = \langle E^{-1}\rangle - \alpha^2 T_\circ c^{-1} \langle\gamma\rangle^{-1}(1 + 0.5h^2)$$

$$J_r^p = \langle E^{-1}\rangle - \alpha^2 T_\circ c^{-1} \langle\gamma\rangle^{-1}$$

It is shown in appendices (4.5) that J_u^p and J_r^p are respectively the values in the order of h of the exact values J_u^u and J_r^r . This comparison allows us to state that the perturbation technique is valid for the small heterogeneity-parameter h. The following paragraph will define the domain of validity of the results.

We conclude that, for a small parameter h, the thermoelastic rod has a macroscopic behaviour similar to the response of a Zener-type viscoelastic model characterised by the approximate relaxation time τ (3.6), the exact relaxed and unrelaxed compliances J_r (3.4) and J_u (3.2).

The role of length-scale is obvious when looking at the relaxation time obtained by Zener with thermoelastic homogeneous rod subjected to heat currents acting on the macroscopic scale (indice ° is used for the corresponding uniform values)

$$\tau_\circ = c\gamma_\circ L^2 k^{-1} \pi^{-2} \tag{3.7}$$

Results (3.6) and (3.7) have the same expression but the characteristic lengths are related to the microscopic scale l in the first case and the macroscopic scale L in the second.

In the following, the dissipation due to heat conduction is calculated when the bar is subjected to an elementary cyclic sollicitation.

3.3. Elementary cyclic stress

We describe a dissipative-cycle with the application of two successive opposite stress increments. The duration of each increment is long enough to allow a significant uniformisation of the temperature. The evolution of the stress and the corresponding temperature is represented in diagramm (1).

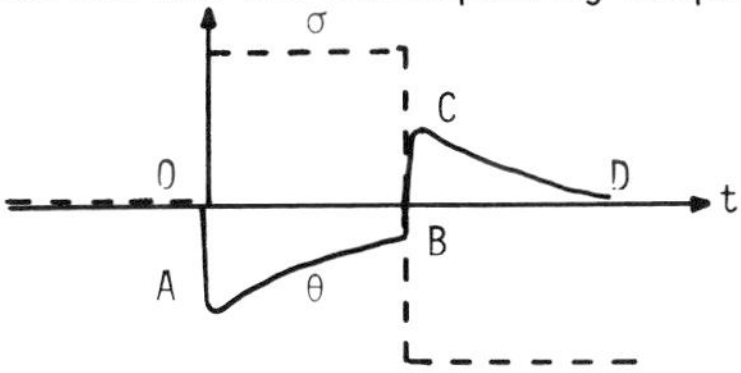

diag 1 : Evolution of the stress and of the corresponding temperature.

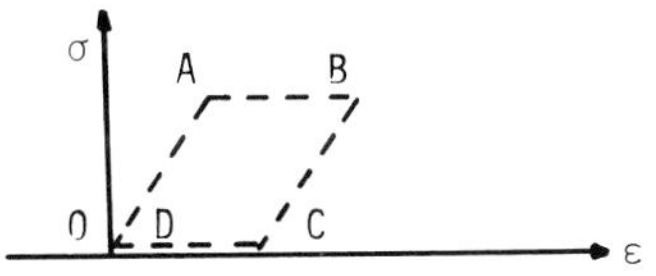

Figure 2 : Stress-strain cycle.

Using the similar calculation process as in paragraph 3.1, the successive temperatures are :

$$\theta_A(x) = -\alpha T_\circ \sigma_\circ c^{-1} \gamma^{-1}(x) \ ; \ \theta_B = -\alpha T_\circ \sigma_\circ c^{-1} \langle\gamma\rangle^{-1} \ ; \ \theta_C(x) = \theta_B - \theta_A(x) \ ; \ \theta_D = 0$$

The annulation of θ_D closes the cycle OABCD represented in figure (2). This is a consequence of the linearisation of coupled thermoelasticity.

According to the thermoelastic effect, we have access to the energy dissipated per cycle by two methods giving the same results :

- first, loss of energy W is usually given by the area of the cycle OABCD.

$$W^d = \sigma_\circ(\langle\varepsilon_B\rangle - \langle\varepsilon_A\rangle)$$

After substitution of $<\varepsilon_B>$ and $<\varepsilon_A>$ we obtain the exact value

$$W^d = \sigma_o{}^2 \alpha^2 T_o c^{-1} \, [<\gamma^{-1}> - <\gamma>^{-1}]$$

It appears from appendix (4) that the approximate value W^{dp} of W^d in the order of h^2 is :

$$W^{dp} = 0.5 \, \sigma_o{}^2 \, \alpha^2 T_o c^{-1} h^2 <\gamma>^{-1}$$

- the second method uses the expression of the production of irreversible entropy. The loss of energy is given by :

$$W^{d'} = T_o (\underset{0}{\overset{B}{\Delta}} \eta_I + \underset{B}{\overset{D}{\Delta}} \eta_I)$$

where $\underset{0}{\overset{B}{\Delta}} \eta_I$ and $\underset{B}{\overset{D}{\Delta}} \eta_I$ are respectively the variations of irreversible entropy from 0 to B and from B to D. In case of heat conduction, the rate of entropy production is written for the cell

$$\dot{\eta}_I = l^{-1} \int_0^l k(\theta_{,x})^2 T_o{}^{-2} dx$$

Using the approximate analytical expression in the order of h^2 of θ (x,t) (appendix 2), we integrate $\dot{\eta}_I$ from time 0 to time B (simular result is given from B to D when looking at the symmetry of loading and unloading). We obtain (appendix 6) :

$$W^{dp} = W^{d'}$$

It means that the method based on the entropy-production gives an expression of the approximate dissipation $W^{d'}$ equal to that obtained by the approximate dissipation-cycle area W^{dp}. Morever, it is possible to test the validity of the perturbation technique by comparing the exact value of energy losses W^d with the approximate one W^{dp}.

As an illustration, we give the ratios $(W^d - W^{dp}) / W^d$ obtained with various values of h :

h	$(W^d - W^{dp}) / W^d$
0.001	
0.25	4.7 %
0.3	6,8 %
0.99	99,7 %

The common values of (γ-1) with metals are about a few per cents [3] so that the application of the perturbation technique implies an error practically inferior to 5 %.

In the following, we use this perturbation technique to quantify dissipation occurring when the sollicitation is a cyclically varying stress.

4. CYCLIC STRESS LOADING

4.1. Complex compliance function

We apply a cyclic stress $\sigma(t)$ which is represented in a complex form

$$\sigma(t) = \sigma_o \exp(i\omega t)$$

The macroscopic strain observed has the same pulse ω but present a phase lag δ with the stress. Using two methods supplying equivalent results (perturbation technique or a decomposition in Fourier-series), the corresponding complex compliance function $J^*(\omega)$ is calculated in appendices (7,9). It appears that the thermal-insulated heterogeneous thermoelastic rod verifies the Debye-relations for small parameter h*.

It is then an easy task to calculate the internal thermoelastic damping.

4.2. Internal friction

The measure of internal friction is usually given by tgδ with

$$tg\ \delta = \Delta\omega\tau(1+\omega^2\tau^2)^{-1}$$

where the intensity of relaxation Δ is presented in the form (appendix 10)

$$\Delta = \alpha^2 T_o c^{-1} \langle E^{-1}\rangle^{-1} [\langle\gamma^{-1}\rangle\langle\gamma\rangle - 1]\ \langle\gamma\rangle^{-1}$$

In the following, these results are compared with those calculated by Zener :

$$tg\delta^o = \alpha^2 T_o c^{-1} \gamma_o^{-1} E^o\ \omega\tau(1+\omega^2\tau^2)^{-1} \quad \text{with} \quad \tau_o = c\gamma_o L^2 k^{-1} \pi^{-2}$$

5. DISCUSSION

5.1. Dissipation per cycle or per time unit

The two expressions tg δ and tg δ^o when considered as functions of the pulse ω, allow us to plot two typical Lorentz - curves :

- with low or high pulsations, dissipation is negligible. In each

---

* It should be noted that the Debye-relations should be directly used without the calculations of appendixes (7,9) in case of a rigorous proof of this analogy with a viscoelastic behaviour. A mathematical treatment of this question should be interesting.

case, the temperature-field is respectively uniform or adiabatic

- maximum dissipation per cycle occurs only at an intermediate pulse.

With the heterogeneous bar, the microscopic scale is involved whereas the macroscopic length L is used in case of macroscopic heat currents :

$$\omega_{max} = \tau^{-1} = k\pi^2 c^{-1} \langle\gamma\rangle^{-1} l^{-2} \quad \text{whereas} \quad \omega^{\circ}_{max} = k\pi^2 c^{-1} \gamma_{\circ}^{-1} L^{-2} \tag{5.1}$$

For the sake of comparing results, similar macroscopic characteristic values are set for the heterogeneous bar we compared to the homogeneous case studied by Zener. We propose * for Young's modulus $\langle E^{-1}\rangle = E_{\circ}^{-1}$ and we assume that a minor error is made when taking $\langle\gamma\rangle = \gamma_{\circ}$

Looking at the respective maximums of dissipation per cycle

$$tg\delta_{max} = 0.5\alpha^2 T_{\circ} c^{-1} \langle\gamma\rangle^{-1} \langle E^{-1}\rangle^{-1} [\langle\gamma^{-1}\rangle\langle\gamma\rangle - 1]$$

$$tg\delta^{\circ}_{max} = 0.5\alpha^2 T_{\circ} c^{-1} \gamma_{\circ}^{-1} E_{\circ}$$

it can be deduced that the measure of dissipation per cycle acting at the microscopic level is generally low when compared to the macroscopic level :

$$[tg\delta_{max}] \; [tg\delta^{\circ}_{max}]^{-1} = [\langle\gamma^{-1}\rangle\langle\gamma\rangle - 1] \ll 1$$

We should notice however that each of these maximum values occurs at different pulses depending on the characteristic length

$$\omega_{max} \gg \omega^{\circ}_{max} \quad \text{if} \quad L \gg l$$

This remark becomes crucial when looking at the values of thermoelastic damping per time-unit. It effectively appears that the value of dissipation per time-unit is proportionnal to the product of tg δ with the pulse ω . Then, looking at the ratio

$$[tg\delta_{max}\,\omega_{max}] \; [tg\delta^{\circ}_{max}\,\omega^{\circ}_{max}]^{-1} \cong [\langle\gamma^{-1}\rangle\langle\gamma\rangle - 1]\, L^2 l^{-2} \cong 0.5 L^2 l^{-2} h^2 \tag{5.2}$$

we obtain, with a sufficiently small microscopic scale l, the condition

$$tg\delta_{max}\,\omega_{max} > tg\delta^{\circ}_{max}\,\omega^{\circ}_{max}$$

---

* A longitudinal variation of E may be schematically represented by a juxtaposition of n springs assembled in series of alternating stiffnesses k_i . In such a case (Francfort, [4]) the overall stiffness k is : $k^{-1} = \Sigma k_i^{-1}$

It means that, with the well defined elevated pulse ω_{max}, thermal currents acting on a sufficiently small scale play a greater part per time unit in the production of irreversible entropy than some macroscopic heat-boundary fluxes. This result, which is not predictable when the bar is considered macroscopically as a homogeneous body, arises from the difference between microscopic and macroscopic length-scales. The role of the heterogeneity of the material described by the parameter h is obvious when looking at formula (5.2). To give an example, we have carried out a numerical resolution.

5.2. Numerical resolution.

A a numerical computation constitutes an other mean to test the validity of the perturbation technique .

An implicit sequence for the resolution of the heat conduction is used. Some typical Lorentz-curves are shown in the following figure

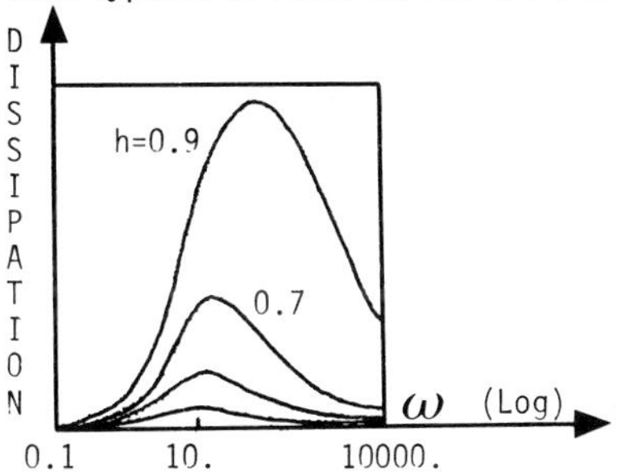

This example is based on the a-dimensional numerical parameters :
$\langle\gamma\rangle=1$, $l=1$, $\sigma_o=1$, $\alpha=1$, $k=1$, $c=1$

With these values and with small parameters h, we observe that maximum dissipation occurs at a pulse close to 10. and the value below-which is approximate-was supplied by the perturbation technique (cf formula 5.1)

$$\omega_{max} = \pi^2$$

6. CONCLUSION

Most of the above results have been obtained by using a direct perturbation technique whose domain of accuracy has been fully respected.Some practical cases may be investigated by this technique.

Two specific anelastic testes (incremental or vibratory) have been used in order to *show* that the thermoelatic heterogeneous rod adopts a macroscopic behaviour similar to a viscoelastic one. One relaxation-time was exhibited whereas due to the hypothesis of a small heterogeneity-parameter other smaller times were hidden in the development of the perturbation technique.

However, let us point out that the use of only one test should be theoretically sufficient to characterize the macroscopic behaviour. Proof of this argument, on the basis of Debye's relations, should be deduced from Molinari's and Ortiz's [5] work who *demonstrate* , in tridimensional case, assuming thermal insulated boundaries, that thermoelastic heterogeneous solids behave macroscopically like viscoelastic solids.

The results presented here in a one-dimensional problem lead to an analytical evaluation of the thermoelastic damping and clearly illustrate the role of the characteristic length of the microdefects.

REFERENCES

1 Zener, C., Elasticity and Anelasticity of Metals (The University of Chicago Press, 1952)
2 Nowick, A.S., Berry, B.S., Anelastic Relaxation in Crystalline Solids (Academic Press, New York and London, 1972)
3 Germain, P., Cours de Mécanique,III (Ecole Polytechnique, éd 1984)
4 Francfort, G.A., Two Variational Problems in Thermoelasticity, P.H.D. Stanford University, 1982.
5 Molinari,A., Ortiz, M., Global Viscoelastic Behavior of Heterogeneous Materials, in print.

APPENDIX 1: *ADIABATIC INDEX* $\gamma(X)$

$$\gamma(X) = c_G(X)c^{-1} = 1 + 3E(X)\alpha^2 T_0(1-2\nu)^{-1}c^{-1}$$

$$E(X) = E^0 + H\cos(\pi X l^{-1})$$

we note $\gamma_0 = 1 + 3E^0\alpha^2 T_0(1-2\nu)^{-1}c^{-1}$ *and* $h = 3H\alpha^2 T_0(1-2\nu)^{-1}c^{-1}\gamma_0^{-1}$

conclusion: $\gamma(X) = \gamma_0[\,1 + h\cos(\pi X l^{-1})\,]$

APPENDIX 2: *Relaxation function*

Resolution of $c_G(X)\theta^*_{,t^*} = k\theta^*_{,XX}$ (θ^* *and* t^* *are dimensional variables*) (A2-1)

boundary-conditions: $\theta^*_{,X} = 0$ *at* $X=0$ *and* $X=l$

initial-condition: $\theta^*(X,0) = T_0\Phi(X)$

A2a) *Adimensionalization*:

we note $\theta^*=T_0\theta$, $t^*=\gamma_0 c l^2 k^{-1} t$, $X=lx$ *and* $X_j=\cos(j\pi x)$ *for* $j=0,\ldots,n$ (A2a1)

(A2-1) *reads* $(1 + hX_1)\theta_{,t} = \theta_{,xx}$ *that we note* $(1 + hX_1)\dot\theta = \theta''$ (A2a2)

A2b) *Resolution scheme* :

we develop θ *in the form* $\theta(x,t)=\theta_0(x,t)+h\theta_1(x,t)+h^2\theta_2(x,t)+\ \ldots$ (A2b1)
(A2b1) *is replaced in* (A2a2) *and,after identification in the orders of* h, *we obtain successive partial differential equations*:

$\dot\theta_0=\theta_0''$ *with* $\theta_0'=0$ *at* $x=0$ *and* $x=1$; $\theta_0(x,0)=\Phi(x)$

and for $j \geqslant 1$

$\dot\theta_j=\theta_j''+ f^j$ *with* $\theta_j'=0$ *at* $x=0$ *and* $x=1$; $\theta_j(x,0)=0$

where $f^j=-X_1\dot\theta_{j-1}$ *is the heat generation term of the* j^{st} *equation*.
Developpment will be extended in order of h^2 .

A2c) *Resolution of* : $\dot\theta_0=\theta_0''$ *with* $\theta_0'=0$ *at* $x=0$ *and* $x=1$; $\theta_0(x,0)=\Phi(x)$

we develop $\Phi(x)$ *in Fourier-serie*: $\Phi(x)=\Phi_0+\Phi_1 X_1+\Phi_2 X_2+\ \ldots$ (A2c1)

we obtain θ_0 *in the form*

$$\theta_0 = T_0^0 + X_1 T_1^0 + X_2 T_2^0 + X_3 T_3^0 + \ldots$$

where for $i \geqslant 0$ $T_i^0=\Phi_i\exp(\alpha_i t)$ and $\alpha_i = -(i\pi)^2$

A2d) Resolution of $\theta_j = \theta_j'' + f^j$ *with* $\theta_j' = 0$ *at* $x = 0$ *and* $x = 1$; $\theta_j(x,0)=0$

we develop f^j *in Fourier's serie* : $f^j = f_0^j(t) + f_1^j(t)X_1 + f_2^j(t)X_2 + \ldots$

we obtain θ_j *in the form* : $\theta_j = T_0^j(t) + X_1 T_1^j(t) + X_2 T_2^j(t) + \ldots$

where $\dot{T}_0^j = f_0^j(t)$ and $\dot{T}_k^j = f_k^j(t) + \alpha_k T_k^j \quad (k \geqslant 1)$

time-integration gives the terms T_k^j *with initial-conditions* : $T_k^j\ (0)=0 \quad (k>0)$

$$T_0^j(t) = \int_0^t f_0^j(\tau)d\tau \ ; \ T_k^j(t) = \int_0^t exp[\alpha_k(t-\tau)]f_k^j(\tau)d\tau \ \textit{for} \ (k \geqslant 1)$$

a2e) Development in Fourier's series of $f^j(x,t)$ *and* $\Phi(x)$:

* *we have* $f^j(x,t) = -X_1\dot{\theta}_{j-1}$ *with* $\dot{\theta}_{j-1} = \dot{T}_0^{j-1} + X_1\dot{T}_1^{j-1} + X_2\dot{T}_2^{j-1} + \ldots$

f^j *reads* $f^j = f_0^j(t) + X_1 f_1^j(t) + X_2 f_2^j(t) + X_3 f_3^j(t) + \ldots$

where $f_0^j = -0.5\ \dot{T}_1^{j-1}$ *and* $f_k^j = -0.5\ [\dot{T}_{k-1}^{j-1} + \dot{T}_{k+1}^{j-1}]$; $k \geqslant 1$

* *we have* $\Phi(x) = \theta(x,0) = -\alpha\sigma_0 c^{-1}\gamma_0^{-1}\ (1+hX_1)^{-1}$

we note $\beta = \alpha\sigma_0 c^{-1}\gamma_0^{-1}$

we develop $(1+hX_1)^{-1}$ *in the form*: $(1 + hX_1)^{-1} = (1 - hX_1 + h^2X_1^2/2 - h^3X_1^3/3 + \ldots)$

neglecting terms in order of h^3 *or more, we obtain* :

$$\Phi(x) = \Phi_0 + \Phi_1 X_1 + \Phi_2 X_2 + O(h^3)$$

where

$$\Phi_0 = -\beta(1+0.5h^2) + O(h^3)$$
$$\Phi_1 = \beta h + O(h^3)$$
$$\Phi_2 = -0.5\beta h^2 + O(h^3)$$

A2f) Expression of $\theta_1(x,t)$

we have $\theta_1(x,t) = T_0^1(t) + X_1 T_1^1(t) + X_2 T_2^1(t) + \ldots$

Terms $T_i^1(t)$ *are given in (A2d). For example...*: $T_2^1 = \int_0^t exp\ [\alpha_2(t-\tau)]f_2^1(\tau)d\tau$

Terms $f_k^j(t)$ *are given in (A2e). For example...*: $f_2^1 = -0.5\ [\dot{T}_1^0 + \dot{T}_3^0]$

Terms $T_i^0(t)$ *are given in (A2c). For example...*: $T_1^0 = \Phi_1 exp\ [\alpha_1 t]$

Let us give for example the calculation of T_2^1 :

$$T_2^1(t) = \int_0^t exp\ [\alpha_2(t-\tau)]f_2^1(\tau)d\tau = exp(\alpha_2 t)\int_0^t -0.5\alpha_1\Phi_1 exp\ [(\alpha_1-\alpha_2)\tau]d\tau$$
$$+\ exp(\alpha_2 t)\int_0^t -0.5\alpha_3\Phi_3 exp\ [(\alpha_3-\alpha_2)\tau]d\tau$$

$$T_2^1(t) = 0.5\alpha_1\Phi_1(\alpha_2-\alpha_1)\ [exp(\alpha_1 t) - exp(\alpha_2 t)]$$
$$+\ 0.5\alpha_3\Phi_3(\alpha_2-\alpha_3)\ [exp(\alpha_3 t) - exp(\alpha_2 t)]$$

Conclusion : $\theta_1(x,t) = 0.5\ \Phi_1\ [1-exp(\alpha_1 t)]$

$$+0.5\ X_1\ [\alpha_2\Phi_2(\alpha_1-\alpha_2)^{-1}(exp(\alpha_2 t) - exp(\alpha_1 t))]$$

$$+0.5\ X_2\ \begin{bmatrix}\alpha_1\Phi_1(\alpha_2-\alpha_1)^{-1}(exp(\alpha_1 t) - exp(\alpha_2 t)) \\ +\alpha_3\Phi_3(\alpha_1-\alpha_3)^{-1}(exp(\alpha_3 t) - exp(\alpha_1 t))\end{bmatrix}$$

$$+\dots$$

$$+0.5\ X_n\ \begin{bmatrix}\alpha_{n-1}\Phi_{n-1}(\alpha_n-\alpha_{n-1})^{-1}(exp(\alpha_{n-1} t)-exp(\alpha_n t)) \\ +\alpha_{n+1}\Phi_{n+1}(\alpha_n-\alpha_{n+1})^{-1}(exp(\alpha_{n+1} t)-exp(\alpha_n t))\end{bmatrix}$$

A2g) Expression of $\theta_2(x,t)$

We have $\theta_2(x,t) = T_0^2 + X_1 T_1^2 + X_2 T_2^2 + \dots$

We *observe that* T_0^2, *proportional to* Φ_2, *is in order of* h^2. *In the same way,* T_1^2 *is linearly written with* Φ_1 *and* Φ_3 *which are of order* h *or more. For* n>2, *terms* T_n^2 *are linearly written with* Φ_{n-2}, Φ_n *and* Φ_{n+2} *which are in orders of* h *greater than* h^3.

Consequently, $\theta_2(x,t)$ *is in order of* h *so that the term* $h^2\theta_2(x,t)$ *in the development of* $\theta(x,t)$ *is in order of* h^3.

A2h) Expression of $\theta^*(X,t^*)$ *in order of* h^2

$$\theta^*(X,t^*) = T_0\ \theta(x,t) = T_0\ (\theta_0 + h\theta_1 + h^2\theta_2 + \dots)$$

but $\theta_0 + h\theta_1 + h^2\theta_2 + \dots = \Phi_0 + \Phi_1 X_1 exp(\alpha_1 t) + h[0.5\Phi_1(1-exp(\alpha_1 t))] + O(h^3)$

$$\theta_0 + h\theta_1 + h_2\theta_2 + \dots = -\beta(1+0.5h^2) + \beta h X_1 exp(\alpha_1 t) + 0.5h^2\beta[1-exp(\alpha_1 t)] + O(h^3)$$

Conclusion : *Replacing* β *from (A2e) and* α_1 *from (A2c), we have in order of* h^2

$$\theta^*(X,t^*) = -\alpha\sigma_0\gamma_0^{-1}T_0 c^{-1}[1+0.5h^2 exp(-t^*\tau^{-1}) - h\cos(\pi X l^{-1})exp(-t^*\tau^{-1})]$$

where $\tau = \gamma_0 c\ l^2\pi^{-2}k^{-1}$

APPENDIX 3 : Relaxation function deduced from the approximate solution $\theta(x,t)$

We have in dimensional form: $J^p(t) = \langle\varepsilon(x,t)\rangle\sigma_0^{-1}$

$$\text{with : } \varepsilon(x,t) = \sigma_0 E^{-1}(x) + \alpha\theta(x,t)$$

We obtain, from (A2h): $J^p(t) = \langle E^{-1}\rangle - \alpha^2 T_0\gamma_0^{-1}c^{-1}[1 + 0.5h^2 exp(-t\tau^{-1})]$

$$\text{which is written : } J^p(t) = J^p_U(t) + [J^p_R(t) - J^p_U(t)]\ [1 - exp(-t\tau^{-1})]$$

$$\text{with : } J^p_U(t) = \langle E^{-1}\rangle - \alpha^2 T_0\gamma_0^{-1}c^{-1}(1+0.5h^2)$$

$$J^p_R(t) = \langle E^{-1}\rangle - \alpha^2 T_0\gamma_0^{-1}c^{-1}$$

APPENDIX 4 : Average value $\langle\gamma^{-1}(X)\rangle$:

$$\gamma^{-1}(X) = \gamma_0^{-1}[1+h\cos(\pi X l^{-1})]^{-1}$$

$$\langle\gamma^{-1}\rangle = \gamma_0^{-1}l^{-1}\int_0^1[1+h\cos(\pi x l^{-1})]^{-1}dx = \gamma_0^{-1}\pi^{-1}\int_0^\pi[1+h\cos x]^{-1}dx$$

Let us recall the formula :

$$(a+b\cos x)^{-1}dx = 2(a^2-b^2)^{-1/2}artg[(a^2-b^2)^{1/2}(a+b)^{-1}tg(0.5x)]$$

Conclusion : (*with* $\gamma_0^{-1} = \langle\gamma\rangle^{-1}$)

$$\langle\gamma^{-1}\rangle = \langle\gamma\rangle^{-1}(1-h^2)^{-1/2} = \langle\gamma\rangle^{-1}(1+0.5h^2) + O(h^3).$$

APPENDIX 5 : Comparison of exact and approximate compliances.

we have $J_u = \langle E^{-1}\rangle - \alpha^2 T_0 c^{-1}\langle\gamma^{-1}\rangle \cong \langle E^{-1}\rangle - \alpha^2 T_0 c^{-1}\langle\gamma\rangle^{-1}(1+0.5h^2)$

and $J_r = \langle E^{-1}\rangle - \alpha^2 T_0 c^{-1}\langle\gamma\rangle^{-1}$

conclusion $J_u = J_u^p + O(h^3)$

and $J_r = J_r^p$

APPENDIX 6 : Irreversible entropy production :

we have $$\overset{B}{\underset{0}{\Delta}}\,\eta_I = \int_0^{+\infty} \dot{\eta}_I\, dt^* = \int_0^{+\infty} l^{-1}\int_0^{l} k[T_0^{-1}\theta^*_{,X}]^2 dX dt^*$$

Adimensionalization (A2a) leads to : $$\overset{B}{\underset{0}{\Delta}}\,\eta_I = \gamma_0 c\int_0^{+\infty}\int_0^1 (\theta_{,x})^2 dx dt$$

From (A2h) $\theta_{,x} = -\Phi_1\pi\, sin\,(\pi x)\, exp\,(\alpha_1 t) + O(h^3)$

After space-and time-integration we obtain :

$$\overset{B}{\underset{0}{\Delta}}\,\eta_I = -0.25\,\gamma_0 c\pi^2\Phi_1^2\alpha_1^{-1} + O(h^3) = 0.25\,\alpha^2\sigma_0^2 h^2\gamma_0^{-1}c^{-1} + O(h^3)$$

By symmetry of loading and unloading : $\overset{D}{\underset{B}{\Delta}}\,\eta_I = \overset{B}{\underset{0}{\Delta}}\,\eta_I$

Conclusion : Irreversible entropy production from O to D reads :

$$\overset{D}{\underset{0}{\Delta}}\,\eta_I = 0.5\alpha^2\sigma_0^2 h^2\langle\gamma\rangle^{-1}c^{-1} + O(h^3)$$

APPENDIX 7 : Complex Compliance Function – Resolution of

$$c\gamma(X)\theta^*_{,t^*} = k\theta^*_{,XX} - \alpha T_0 \frac{\partial(\sigma_0\, exp\,(i\omega t^*))}{\partial t^*}$$

boundary-conditions : $\theta^*_{,X} = 0$ *at* $X = 0$ *and* $X = 1$

initial-condition : we observe that at time t=0, the bar is abruptly subjected to the stress σ_0. *To be consistent with paragraph 3, we take as initial temperature :*
$\theta^*(X,0) = -\alpha T_0\sigma_0 c^{-1}\gamma_0^{-1}$
corresponding to the final temperature which would be attained if the stress were constant.

A7a) *<u>Adimensionalization</u> :*

we note $\theta^* = T_0\theta$, $X = xl$, $t^* = t_0 t$ *with* $t_0 = c\gamma_0 l^2 k^{-1}$

$\omega t_0 = a$, $\delta = -\alpha\sigma_0 c^{-1}\gamma_0^{-1}$, $\beta = ia\delta$

$X_j = \cos(j\pi x)$ *and* $\alpha_j = -(j\pi)^2$

the partial differential equation reads : $(1+hX_1)\dot{\theta} = \theta'' + \beta\, exp(iat)$
boundary-conditions : $\theta_{,x} = 0$ *at* $x = 0$ *and* $x = 1$
initial-condition : $\theta(x,0) = \delta$

A7b) Resolution Scheme :

Same development (A2b1) leads to the following successive equations :

$\dot{\theta}_0 = \theta_0'' + \beta\, exp(iat)$ *with* $\theta_0(x,0) = \delta$ *and* $\theta_{0,x} = 0$ *at boundaries.*

$i \geqslant 1 \quad \dot{\theta}_i = \theta_i'' - X_1 \dot{\theta}_{i-1}$ *with* $\theta_i(x,0) = 0$ *and* $\theta_{i,x} = 0$ *at boundaries.*

A7c) Successive resolutions :(Restricted in order of h^2*).*

$\theta_0 = \delta exp(iat)$

$\theta_1 = X_1 \beta(\alpha_1 - ia)^{-1} [exp\,(ia) - exp(\alpha_1 t)]$

$$\theta_2 = 0.5\beta[exp(iat) - exp(\alpha_1 t)](ia-\alpha_1)^{-1}$$
$$+ 0.5X_2\beta(ia-\alpha_1)^{-1} ia[exp(iat) - exp(\alpha_2 t)](ia-\alpha_2)^{-1}$$
$$+ 0.5X_2\beta(ia-\alpha_1)^{-1}\alpha_1[exp(\alpha_2 t) - exp(\alpha_1 t)](\alpha_1-\alpha_2)^{-1}$$

A7d) Average value $\langle\theta\rangle$ *in order of* h^2

It appears that the mean temperature $\langle\theta\rangle$ *of the cell fluctuates periodically after a transient regime around the zero value. We note* $\langle\theta\rangle = \tilde{\theta}(t)$ *to express the periodic aspect of the permanent temperature pattern.*

$$\langle\tilde{\theta}\rangle = \beta\, exp\,(iat)\,[(ia)^{-1} + 0.5h^2(ia - \alpha_1)^{-1}] + O(h^3)$$

or in dimensional form :

$$\langle\tilde{\theta}^*\rangle = -\alpha T_0 \sigma_0 c^{-1} \gamma_0^{-1} exp(i\omega t^*)[1 + 0.5h^2 \omega\tau(1 + \omega^2\tau^2)^{-1}(\omega\tau + i)]$$

APPENDIX 8 : COMPLEX COMPLIANCE

We have in dimensional form : $J^*(\omega) = \langle\varepsilon\rangle\sigma_0^{-1}\, exp(-i\omega t)$

with : $\langle\varepsilon\rangle = \langle E^{-1}\sigma_0\, exp(i\omega t) + \alpha\tilde{\theta}^*\rangle$

from A7d), $J^*(\omega)$ *is written in the form :* $J^*(\omega) = J_1(\omega) - iJ_2(\omega)$
with :
$J_1 = \langle E^{-1}\rangle - \alpha^2 T_0 c^{-1}\gamma_0^{-1}(1 + 0.5h^2) - 0.5\alpha^2 T_0 h^2 c^{-1}\gamma_0^{-1}(1 + \omega^2\tau^2)^{-1}$
and :
$J_2 = 0.5\,\alpha^2 T_0 h^2 c^{-1}\gamma_0^{-1}(1 + \omega^2\tau^2)^{-1}$

conclusion : Replacing the expressions of J_u^p *and* J_r^p *from A3, we obtain the : Debye-relations:*

$J_1 = J_u^p + (J_r^p - J_u^p)\,(1 + \omega^2\tau^2)^{-1}$

$J_2 = (J_r^p - J_u^p)\omega\tau(1 + \omega^2\tau^2)^{-1}$

APPENDIX 9 : RESOLUTION OF $(1 + hX_1)\dot{\theta} = \theta'' + \beta\, exp(iat)$ *(A9-1)*

using a decomposition in Fourier's serie.

We seek for the solution of the form: $\theta = \beta[b_0 + b_1X_1 + b_2X_2 + b_3X_3 + \ldots]exp\,(iat)$

After replacing θ *in (A9-1), we linearize products of the form* X_iX_j *.*

Bearing the respective coefficients of X_j *to zero provides the successive equations :*

$$0.5iab_1h = 1-iab_0$$

$$(b_0 + 0.5b_2)iah = (\alpha_1 - ia)b_1$$

$$0.5(b_1 + b_3)iah = (\alpha_2 - ia)b_2$$

Taking $b_3 = 0$, *we truncate the development. Using approximation in the order of* h^2 *provides the expression of* b_0

$$b_0 = [(ia)^{-1} + 0.5(ia - \alpha_1)^{-1}h^2] + O(h^3)$$

It follows that $\langle\theta\rangle$ *reads :*

$$\langle\theta\rangle = \beta\,[(ia)^{-1} + 0.5(ia - \alpha_1)^{-1}h^2]exp(iat)$$

which is the similar result given in (A7-d)

APPENDIX 10 : INTERNAL FRICTION. INTENSITY OF RELAXATION Δ.

It is well known that : $tg\delta = J_2J_1^{-1}$

Using the debye-relations, the phase lag is written in terms of J_u *and* J_r *(we replace the approximate values* J_u^p *and* J_r^p *by the exact ones* J_u *and* J_r)

$$tg\delta = \Delta\omega\tau(1 + \omega^2\tau^2)^{-1}$$

with : $$\Delta = (J_r - J_u)\,[J_u + (J_r - J_u)\,(1 + \omega^2\tau^2)^{-1}]^{-1}$$

A minor error is induced when we replace the pulse-dependent value $\Delta(\omega)$ *with the following :* $(J_r - J_u)J_u^{-1}$.

We obtain, from (3.2) and (3.4) :

$$\Delta = [\langle\gamma\rangle\langle\gamma^{-1}\rangle - 1]\,[1 - \alpha^2T_0c^{-1}\langle\gamma\rangle^{-1}\langle E^{-1}\rangle]^{-1}\,\alpha^2T_0c^{-1}\langle\gamma\rangle^{-1}\langle E^{-1}\rangle^{-1}$$

or approximately : $\Delta \cong [\langle\gamma\rangle\langle\gamma^{-1}\rangle - 1]\,\alpha^2T_0c^{-1}\langle\gamma\rangle^{-1}\langle E^{-1}\rangle^{-1}$

Thermomechanical Couplings in Solids
H.D. Bui and Q.S. Nguyen (Editors)
Elsevier Science Publishers B.V. (North-Holland)

INFRARED TECHNIQUE IN SOLID POLYMERS TESTING AT HIGH STRAIN RATES: APPLICATION TO THE VISCOPLASTICITY OF POLYCARBONATE

Jean-Luc LATAILLADE

Laboratoire de Mécanique Physique, Unité Associée au CNRS 867
Université de Bordeaux I, 351 cours de la Libération
33405 Talence Cedex - France -

The scope of this work is to explain how it is possible, with a standard infrared camera, to measure the temperature evolution of a polymeric material, during its dynamic plastic deformation at high strain rates. The present paper describes the testing stand based on a torsional Hopkinson pressure bar and on a BARNES infrared microscope.

1. INTRODUCTION

In rheological investigation of solid polymers at high strain rates ($10 - 10^4$ s^{-1}), it can be interesting - or necessary - to measure the temperature history of sample, in order to set up a thermomechanical approach; when such an attempt is made several difficulties arise : for example the sample can undergo a complex stress and strain field.

Even when the strain field may be supposed uniform the major difficulty lies in the very short time range of the experiments. Generally the time duration is of the order of ten or twenty microseconds. This fact requires a rather broad bandwidth of the associated electronic devices and transducers.

In the case of thermocouples, such a requirement is not well satisfied because of their relatively high response times, which is conversely proportional to the junction volume. Whatever the cares taken can be (for instance through an electrical welding) the minimum retardation time is of one millisecond order |1| |2|; in this condition, it is not possible to measure the current value of the temperature. Due to the above, the infrared technique including fast sensors, seems to us a convenient way for solving the problem. Moreover, it presents the great advantage of leading to non-contact measurements, which implies a sample whitout a special machining for the thermocouple insertion.

If many experimental works involving temperatures'measurements have been developped in the field of fatigue or slow fracture |3|, only a very few results, as far as we know, have been reported in the case of high strain rate testing |4|, especially for solid polymers |5|.

The aim of our study is the use of an infrared radiometer for determining the evolution of surface temperature when a polycarbonate tubular sample is twisted at high shear strain rates, by means of a torsional Hopkinson pressure bar apparatus. The increase of temperature T is due to the viscoplastic work at rates ranging from 10 s^{-1} to 10^3 s^{-1}.

2. MECHANICAL EQUIPMENT AND GOVERNING EQUATIONS

The apparatus consists of aligned circular elastic bars sandwiching a short tubular specimen (active length of about 2 mm) and is similar in its principle to that one described by Lewis and Campbell |6| : see figure 1. The length is short enough to ensure a quasi-static stress state.

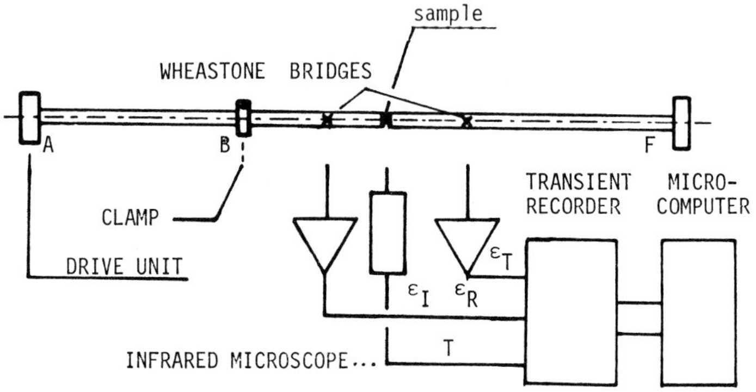

FIGURE 1

The active part of the input bar (AB) is previously twisted to a desired value of torque Γ_S, between the drive unit and the clamp. The notched bolt's fracture implies the sudden release of the stored energy. Incident, reflected and transmitted waves (resp Γ_I, Γ_R and Γ_T) are registered through semi-conductor strain gages, during the time t. The signals $\varepsilon_I(t)$, $\varepsilon_R(t)$ and $\varepsilon_T(t)$ are stored on a transient recorder using a sampling rate of 400 kHz. The governing equations are :

$$\dot{\gamma}(t) = \frac{2\ rm}{1 J_b \rho_b} C_b \quad \varepsilon_R(t) \qquad (1)$$

where C_b is the shear wave velocity in the bars ρ_b the specific mass, J_b the polar inertial moment ; rm and l are the geometrical characteristics of the specimen (i.e. the radius and the active length), and assuming a quasi-uniform stress distribution $\sigma_{z\theta}$ (or τ) - across the thickness e :

$$\sigma_{z\theta}(t) = \frac{\Gamma_T(t)}{2\pi e\ r^2 m} \qquad (2)$$

The energy momentum equation inside the active part is :

$$\rho \frac{dU}{dt} + \rho\vec{v}.\frac{d\vec{v}}{dt} = \vec{\rho g}\ \vec{v} + \mathrm{div}\ (\underset{\sim}{\sigma} \cdot \vec{v}) + \rho\alpha - \mathrm{div}\ \vec{\Phi} \qquad (3)$$

with U as the internal energy, ρ the specific mass of the material tested, g the gravitational acceleration, $\vec{\Phi}$ the heat exchange density, $\vec{v}$ the particle velocity, $\underset{\sim}{\sigma}$ the stress tensor and α the thermal production ; associating the local movement equation i.e. :

$$\rho \{ \frac{d\vec{v}}{dt} - \vec{g} \} = \mathrm{div}\ \underset{\sim}{\sigma} \qquad (4)$$

and assuming that the deformation in the viscoplastic range is isovolume (Ω = cte), i.e. :

$$dS = (\frac{\partial S}{\partial T})_{\Omega} \qquad (5)$$

and

$$dU = TS - P\ d\ \Omega = T\ dS \qquad (6)$$

or

$$dU = C_v\ dT \qquad (6bis)$$

it comes :

$$\rho\ C_v \frac{dT}{dt} = \mathrm{div}\ (\underset{\sim}{\sigma} \cdot \vec{v}) - \vec{v} \cdot (\mathrm{div}\ \underset{\sim}{\sigma}) + \rho\alpha - \mathrm{div}\ \vec{\Phi} \qquad (7)$$

where C_v denotes the heat capacity. The application of these equations can be easily done in the case of a tubular specimen.

Assuming a pure twisting and taking into account the linear Fourier law it is easily shown that :

$$\rho C_v \frac{dT}{dt} = \frac{\partial}{\partial z} (z\dot{\gamma}\ \sigma_{z\theta}) - \rho z^2 \dot{\gamma}\ \ddot{\gamma} - \lambda \frac{d^2T}{dz^2} \qquad (8)$$

This equation can be simplified :

- actually in the viscoplastic range the shear strain rate $\dot{\gamma}$ remains nearly constant,
- the loading time is short enough (1.9 ms) for assuming an adiabatic process,
- and the shear stress and the shear strain rate are averaged over the cross section (i.e. $\partial/\partial z\ (z\dot{\gamma}\ \sigma_{z\theta}) = \dot{\gamma}\sigma_{z\theta}$).

In these conditions one gets :

$$\rho C_V \frac{dT}{dt} = \dot{\gamma} \quad \sigma_{z\theta} \qquad (9)$$

It is clear that the checking of our assumptions needs the measurement of T(t) as well.

3. THE INFRA-RED TECHNIQUE - GENERALITY

For recording temperature signals, describing - in our experiments - the true phenomenon, at the outer surface of the specimen, we have chosen an infra-red technique in such a way to take into account some specific features of the experiment.

i) the input signals are rather low in the case of polymers : the maximum temperature increase is about 50°C.

ii) the signals are very fast ; the rise time is governed by that one of the input wave (the value of which varies from 40.10^{-6} s to 70.10^{-6} s).

iii) the surface emissivity factor in the case of polycarbonate is quite small. So it has been useful to paint the active surface with a very thin black film.

By doing this, all the artefacts due to the emissivity dependance onto the wavelength, or to the surface state, are eliminated.

The infra-red "microscope" set up is a BARNES Engineering Company's one (model RM-2A), covering the wavelengths interval from 1.8 10^{-6} m to 5.5 10^{-6} m. The sensor is a photovoltaïc Indium Antimonide detector, in which the photon beam modifies the potential barrier of a junction : it is cooled at 77 K by means of liquid nitrogen, in order to cancel the radiation of the detector itself, and by reducing the thermal agitation which causes noise.

This apparatus, operating from 15°C to 165°C, involves a spot diameter of 0.5 mm, for a focal length equal to 540 mm.

Because the response time of the detector associated to the preamplifier was unknown, it has been necessary to experimentally establish its transfer function.

Let us denote e(t) the input signal directly related to the surface temperature $T_s(t)$ and s(t) the voltage output delivered by the preamplifier ; if we assume a linear behaviour then the relation between e(t) and s(t) is based on a convolution product (*) as follows:

$$s(t) + \beta(t) = \{e(t) + \delta(t)\} * h(t) \qquad (10)$$

In this relation $\beta(t)$ and $\delta(t)$ express the noises. Because it is randomly distributed we may omit to take it into account, it comes then :

$$s(t) = e(t) * h(t) \qquad (11)$$

or $S(\nu) = E(\nu) \cdot H(\nu)$ (11 bis)

by means of the Fourier transform ($A(\nu)$ denotes the Fourier transform of $a(t)$, ν being the frequency).

Hence the input signal $e(t)$ can be easily derived from the output $s(t)$ provided the impulse response $h(t)$ - or its Fourier transform $H(\nu)$ - is established. In fact, the detector in association with the supplied preamplifier exhibits a non linear behaviour so that equation (11) has to be modified, in a simple way by introducing a weightening function g :

$$g\{s(t)\} = e(t) * h(t) \qquad (12)$$

(here $g(s(t))$ can be expressed in the degree Celsius scale).

4. CALIBRATION PROCEDURE

4.1. Input signal with gradient

By setting up a semi-circular shaped strip brass as shown on figure 2a, we managed to get a monotonous increasing profile of temperature ; the process consists of cooling (by means of liquid nitrogen) one end, while the other end is electrically heated by means of a resistive wire. The temperature field obtained is measured in a static mode with the infra-red camera and the mirror immobilized in various positions to aim regularly spaced points on the circular surface (figure 2a).

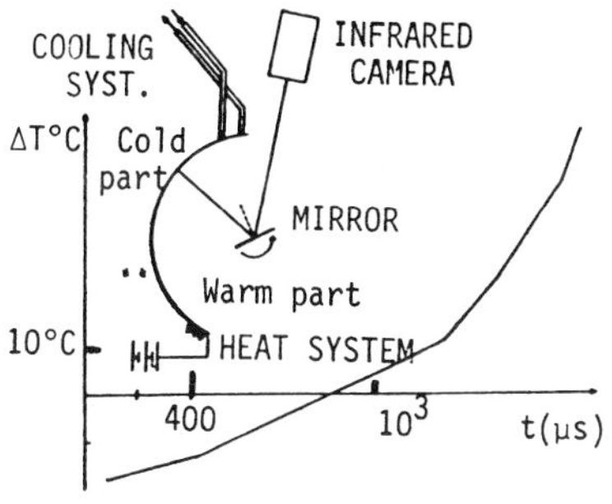

Figure 2a

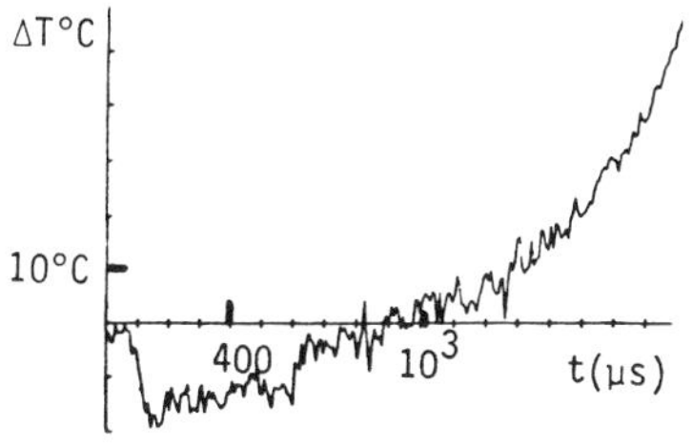

Figure 2b

In the dynamic operating mode, the mirror rotated at a 110 Hz frequency and the time duration for a total scanning of the strip was 1.76 mS, which is of the same magnitude order as the loading time (1.92 mS) : see figure 2b).

Excepting the noise it can be noticed that the dynamic record is nearly in agreement with the static one, which can easily be converted as the input signal $e(t)$. In these conditions, this arrangement is quite useful for determining the function $H(\nu)$ in the frequency range involved in our experiments.

Because on one hand it is the matter of a transient signal, and on the other one the noise spectral bandwidth partially covers the pure signals'one, the digital treatment techniques commonly used (digital filtering for instance) do not apply.

Since our work a new technique has been described for solving such a problem |7|; at that time several records of this type have been made in order to smooth the so averaged experimental curve $\bar{s}(t)$, from which the continuous component has been cut off, (step 2 in the flow chart). At this stage, the function g(s(t)) has to be determined in order to take into account the non-linearity of the equipment ; the way for obtaining this result is described below. The flow chart A indicates the next stage of the treatment either of $g(\bar{s}(t))$ or of e(t) and then shows how the Fourier transform H(ν) is obtained.

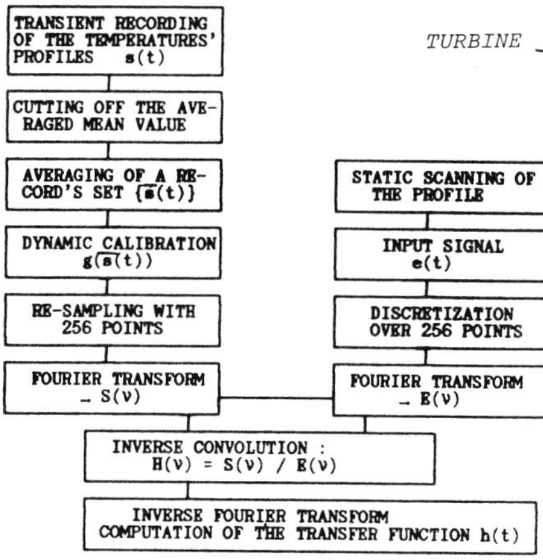

Flow chart A

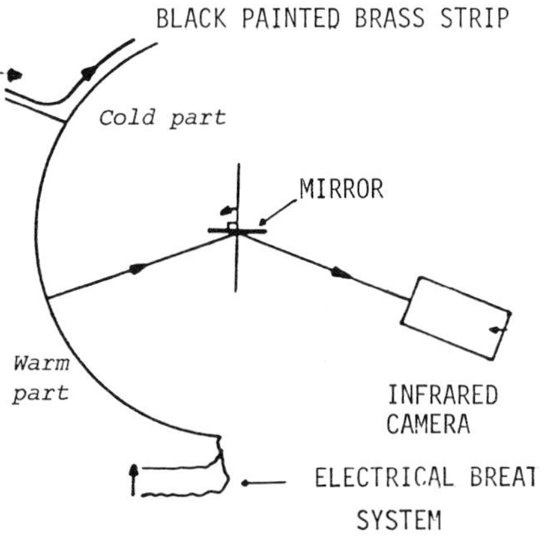

Figure 3

4.2. Evaluation of the non-linearity of the I.R. camera

In order to quantify the non-linearity of the I.R. camera which appears in the mathematical expressions through the function g(s(t)) (eq. 12), the experimental arrangement described in figure 3 was used.

The target consists of two black painted brass strips. The first is maintained at room temperature by means of a turbine blower (part AB), when the second is electrically heated. In this way a temperature step is created at the junction (in fact because of thermal exchanges there is a gradient). The mirror reflects the infra-red radiation from one point M towards the detector. The dynamic calibration was properly achieved with an angular frequency for the mirror of 325 Hz so that the arc AC was scanned within 1.9 mS.

As examples, four experimental records are reported in figure 4 corresponding to temperature deviations of 20°C, 35°C, 50°C and 65°C. With such results, it is then possible to draw the calibration curve s(mv) against ΔT(°C) = g(s) (fig. 5) because the last recorded values on figure 4 are not affected by hysteresis effects.

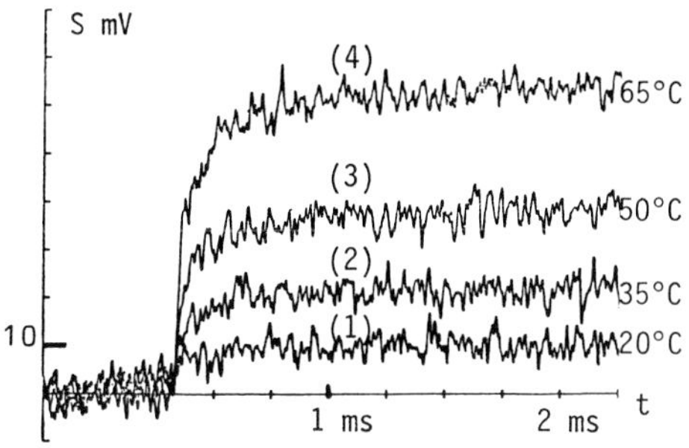

Figure 4

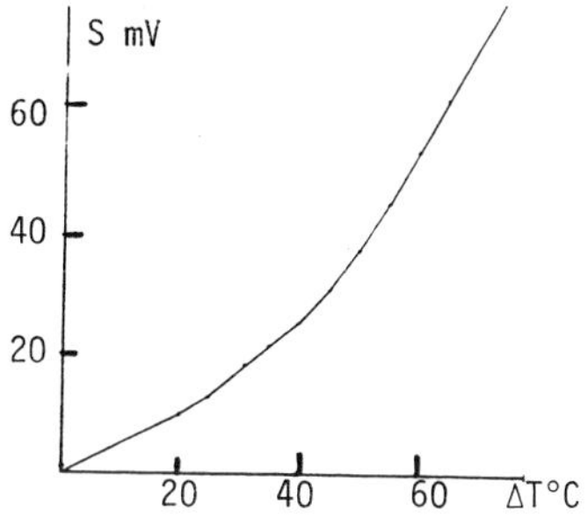

Figure 5

5. APPLICATION TO POLYCARBONATE

During the test, the camera does not aim at the same material point on the outer gage surface but the spot continuously covers - at the same geometrical point - many points representing successive various states of a single one point, owing to the pure twisting hypothesis.

The results shown in figure 6 concern a test performed at room temperature, and for which the mean viscoplastic strain rate value was 1136 s^{-1}.

The temperature is derived from the signal supplied by the preamplifier and the transfer function following the flow chart B.

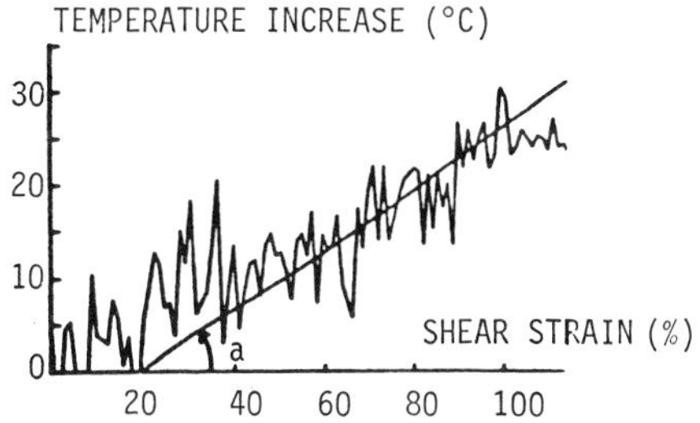

Figure 6

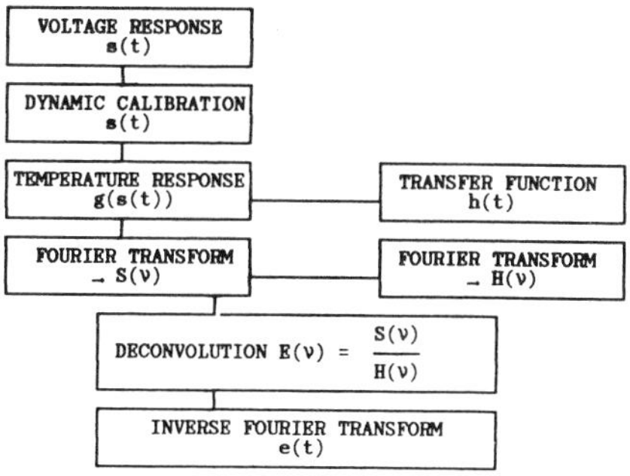

Flow chart B

According to equation (9) and assuming that the mechanical process is non dissipative up to the yield strain value γ_y (which is not strictly true because of the sligth viscoelastic character of the material) an integration leads to :

$$\rho C_v \; \Delta T \; (\gamma_o) = \int_{\gamma}^{\gamma_o} \tau(\gamma) \; \dot{\gamma} \; dt \qquad (13)$$

The right hand side of this relation can be determined through equations (1) and (2) and the corresponding measurements. Hence the temperature increase, derived from the mechanical parameters ($\tau(t)$ and $\gamma(t)$), can be compared to the directly measured increase. Figure 6 shows a good agreement and justifies the adiabatic hypothesis.

6. CONCLUSION

In the testing of solid polymers at impact loading rate, the surface temperature of the sample can be monitored by infra-red techniques. The response time of the radiometer being non zero it is necessary to describe the output signal as a convolution between the real signal - i.e. the surface temperature - and the transfer function of the electronic device. When this latter is determined it is then easy to obtain the temperature history of a point of the sample's surface.

REFERENCES

1. J.W. Philips - Brown University, Providence, Dept. of Appl. Math. Contr., D.A.31.K4, AROD 358, janv. 1969.
2. T. Vinh, M. Azfali and A. Roche - Mechanical behaviour of materials. ICM3, vol. 2, Edit. K.J. Miller, R.F. Smith, Pergamon Press, p.633, 1979.
3. B. Cassagne et G. Leroy - Mesure de température par contact en régime variable. Revue Phys. Appl., 17, p. 153, 1982.
4. M. Malatynski, W.K. Nowacki and W. Oliferuk - Temperature determination in dynamic plasticity by infra-red radiation detection. Arch. Mech. n°4, p.475, 1983.
5. P. Törmälä, E.J. Pääkönen and P. Kemppainen - Deformation and relaxation studies of polycarbonate using the split Hopkinson pressure bar method. J. of Mat. Sc., vol. 16, p. 275, 1981.
6. J. Lewis and J.D. Campbell - The development and use of a torsional Hopkinson bar apparatus. Exper. Mech., 16, n°11, p.520, 1972.
7. S.F. Lin and R.S. Brodkey - Time domain rheological properties of thin materials utilizing frequency analysis. Proc. of IX Int. Cong. on Rheology (Mexico 1984). Edit. B.MENA, vol. 3, p.147 (Univers. Nacional Autonoma de Mexico).

Thermomechanical Couplings in Solids
H.D. Bui and Q.S. Nguyen (Editors)
Elsevier Science Publishers B.V. (North-Holland)

THE HEAT EVOLVED DURING AN ELASTIC-PLASTIC TRANSFORMATION AT FINITE STRAIN

André CHRYSOCHOOS

L.M.G.M.C., Université Montpellier II, place Eugène Bataillon
34060 MONTPELLIER Cédex, France, et L.M.A. Marseille, pour la partie Infra-Rouge

In order to verify the thermomechanical validity of classical behaviour laws, the energy balance is studied in case of traction tests, for several materials, by two experimental approaches. Microcalorimetric and infra-red technics are used to determine respectively the heat globaly and locally evolved during the deformation processes.

1. INTRODUCTION

In spite of the important part played by the Thermodynamics, in the Solid Mechanics, for the determination of behaviour laws, the precise experimental results of thermal type are rare in comparison with the mechanical ones.

We try to have a better knowledge of the dissipative and non dissipative phenomena (energy storage) associated to quasi-static deformation process. The main objective is to measure the stored energy, in order to be able to propose new behaviour laws, suggested by energetical considerations.

The main difficulty, from an experimental point of view, is the precise measurement of the heat evolved during elastic-plastic transformation. Therefore, two independant approaches have been used :

- in the first case, a microcalorimeter, adapted to traction tests, has been realized. The heat, globaly evolved by the sample, deformed in a microcalorimetric cell, is determined.
- in the second case, the infra-red device developped by [1,2] , in the "Laboratoire de Mécanique et d'Acoustique", is adapted to the energy balance study. The infra-red camera gives temperature maps which are used to evaluate the heat locally evolved during the deformation process.

After the presentation of both apparatuses, the main energetical results are shown. The last discussed point is the use of such results in the determination of behaviour laws. As an example, two classical and simple models are considered, we verify they can't predict our observations on the stored energy evolution.

2. THEORETICAL BACKGROUND

2.1 Energy Balance

The deformation energy can be decomposed into an elastic and reversible part

W_e , and a complementary part W_a , that will call "anelastic" energy :

(1) $W_{ext} = W_e + W_a$

The "anelastic" energy can also be decomposed into the dissipated energy W_c and a complementary part W_b , which is the stored energy of cold work (index b stands for "bloquée", in French)

(2) $W_a = W_c + W_b$

The term W_{is} , which appears on figure 1 , during the load and the elastic unload, is the isentropic energy coming from the thermoelastic effects.

The load-unload cycle represents a thermomechanical process (0,A,B) . The complete energy balance associated shows that the stored energy variations can be considered as the internal energy variations ΔE of the material :

(3-a) $\Delta E|_0^A = W_{ext} + W_{is} - W_c$, (3-b) $\Delta E|_A^B = -W_e - W_{is}$,

(4) $\Delta E|_0^B = W_b$.

2.2 Thermodynamical background

The classical theory of Irreversible Processes Thermodynamics is used [3,4]. The use of classical results allows to define the intrinsic dissipation D_1 :

(5) $D_1 = -p_i - \rho \frac{\partial \Psi}{\partial \alpha_j} \cdot \dot{\alpha}_j$,

where p_i represents the internal power, and $\rho(\frac{\partial \Psi}{\partial \alpha_j} \dot{\alpha}_j)$, the power associated to the set of internal state variables (α_j), $j = 1,\ldots,n$, as a function of the free energy $\Psi(T,\alpha_j)$; ($T = \alpha_0$ being the absolute temperature, and ρ the mass density).

The dissipation appears as heat source in the second member of the heat conduction equation :

(6) $\rho C_\alpha \dot{T} + \mathrm{div}(q) = D_1 + r + \rho T \frac{\partial^2 \Psi}{\partial \alpha_j \partial T} \cdot \dot{\alpha}_j$,

where C_α is the heat capacity when (α_j) $j = 1,\ldots,n$ remain constant, q is the heat influx, r the external heat supply.

If we assume that in the vicinity of the thermal equilibrium the variations of the hardening parameters (thermodynamical forces related to the hardening), $\rho(\frac{\partial \Psi}{\partial \alpha_j})$, $j = 2,..,n$, with the temperature are negligible, then :

(7) $\rho C_\alpha \dot{\theta} - k\Delta\theta = D_1 + \rho T \frac{\partial^2 \Psi}{\partial \theta \partial \varepsilon_e} \cdot \dot{\varepsilon}_e = \dot{w}_{ch}$,

if $\theta = T - T_0$, (T_0 : equilibrium temperature), $r = cte$, $q = -k \,\mathrm{Grad}\, \theta$ (Fourier's law), and where $\varepsilon_e = \alpha_1$ is an elastic strain tensor. In the second member, the dissipation and the isentropic term can be recognized.

2.3 Dissipation forms at finite strain

The dissipation form has been studied for several choices of elastic plastic decomposition in the general case and in the case of traction test. Some of them [5] , are based on propositions made by Green and Naghdi [6] , Lee [7], or Nemat-Nasser [8] . We have verified that in case of traction test, the dissipation and the energy balance forms can be made independant of the most classical kinematical approaches, if the elastic strain remains small ; that allows to interprete our energetical measures, without formal ambiguity.

Classicaly, in case of homogeneous traction test, load and deformation signals give an evaluation of stress and strain states, which are used to calculate the anelastic energy :

$$(8) \qquad W_a = \int_0^t \int_{\Omega(t)} \sigma . \dot{\varepsilon}_p \, dx \, d\tau \quad ,$$

where $\Omega(t)$ is the geometrical domain occuped by the sample, σ the Cauchy's stress tensor, and $\dot{\varepsilon}_p$ the plastic strain rate tensor. If the elastic behaviour is assumed to be linear and isotropic, the isentropic energy can be deduced :

$$(9) \qquad - W_{is} = \int_0^t \int_{\Omega(t)} \rho \frac{\partial^2 \Psi}{\partial \varepsilon_e \partial T} \dot{\varepsilon}_e \, dx \, d\tau = - \int_0^t \int_{\Omega(t)} \lambda T \, tr(\dot{\sigma}) \, dx \, d\tau \quad ,$$

where $\dot{\varepsilon}_e$ is the elastic strain rate, λ the coefficient of linear thermal dilatation.

3. MICROCALORIMETRY

A microcalorimeter is placed on the crosshead of the testing machine. The sample passes through a cell composed by two cylinders, electrically insulated. Between them, take place 1200 thermocouples. The thermopile must be very sensitive because the thermal phenomena accompanying the quasi-static deformation processes are very weak. To get a correct signal against noise ratio and a good stability of the thermopile, the thermostatation device must be very efficacious. The unstability and the fluctuations of the signal do not exeed 10^{-3} °C .

If we assume that in the vicinity of thermal equilibrium, the heat flux by convection, radiation, conduction can be modelized by linear law in temperature, the thermal balances on the sample, on the internal cell cylinder and on the external one are :

$$(10) \qquad \left\{ \begin{array}{l} \mu_1 \dot{T}_m + H_1(T_m - T_i) + H_2(T_m - T_a) = \dot{W}_{ch} \\ \mu_2 \dot{T}_i + H_3(T_i - T_e) = H_1(T_m - T_i) \\ \mu_3 \dot{T}_e + H_4(T_e - T_a) = H_3(T_i - T_e) \end{array} \right. \quad ,$$

where respectively, T_m , T_i , T_e are the average temperature of the sample, the internal and the external cell cylinder ; T_a is the room temperature ;

(H_i) $i = 1,\dots,4$ are heat transfert coefficients, and (μ_k) $k = 1,2,3$ are respectively the heat capacity of the sample, the internal and external cell cylinder ; $\dot{W}_{ch}$ is the heat rate globaly evolved by the sample.

The signal s, given by the thermopile is proportional to the thermal desequilibrium through the cell :

$$s = G(T_i - T_e)\ , \quad \text{where} \quad G = C^{te}. \tag{11}$$

With equations (10) and (11), the relationship between the signal and the heat evolved can be deduced :

$$t_a\dot{s}(t) + s(t) + \int_0^t \left(\frac{1}{t_b} + \frac{1}{\tau_1}\left(1 - \frac{t_a}{\tau_2}\right).\exp\left(-\frac{t-x}{\tau_2}\right)\right)s(x)dx = K\,\dot{W}_{ch}(t)\ . \tag{12}$$

The five constants which appear in the equation can be determined by playing on thermoelastic effects. To illustrate the performances of such device, we mention that thermoelastic effects have been precisely observed during cyclic elastic sollicitations for cycle until 10^{-4}Hz of frequency.

4. INFRA-RED THERMOGRAPHY

The second experimental approach is based on I.R. technics. The I.R. device developped in the L.M.A. is adapted to the energy balance study. The camera gives temperature maps which are used to determine, by the interposition of the heat conduction equation, the heat locally evolved. To estimate the first member of eq. (7), the choosen method is the following :

$$\frac{\partial\theta}{\partial t} + u.\text{Grad}\ \theta - \frac{k}{\rho C_\alpha}\Delta\theta = \frac{1}{\rho C_\alpha}(D_1 - \dot{w}_{is})\ , \tag{13}$$

where u is the velocity vector.

In our experiments, the convection term is negligible. The local equation is integrated on a little surface S of the observation zone, in order to eliminate the Laplacian term, which is a big noise amplificator. The noise on the video signal comes from the parasite reflexions and from the limited performances of the different electronic devices

$$\frac{\partial\overline{\theta}}{\partial t} - \frac{k}{\rho C_\alpha}\int_{\partial S}\frac{\partial\theta}{\partial n}\,d(\partial S) = \frac{1}{\rho C_\alpha}(\overline{D}_1 - \overline{\dot{w}}_{is})\ , \tag{14}$$

the bar symbolizing the average on S.

Then, we assume, once more, that in the vicinity of the thermal equilibrium, the heat flux by radiation and conduction (no convection, experiments are made under a primary vacuum), can be modelized by linear law in temperature :

$$\frac{\partial\overline{\theta}}{\partial t} + \frac{\overline{\theta}}{\tau_{th}} = \frac{1}{\rho C_\alpha}(\overline{D}_1 - \overline{\dot{w}}_{is})\ ; \tag{15}$$

the constant τ_{th} is determined during the sample cooling (or reheating) after an elastic load (or unload).

5. EXPERIMENTAL RESULTS

The behaviour of three metals have been studied : one carbon steel, one stainless steel, and one duralumin. The dispertion of the measures between both approaches is sufficiently short to admit the following results :

- the evolution of the stored energy during a monotonic traction test seems to be limited (asymptotic saturation state) ;
- at the begining of the strain hardening the fraction of the stored energy reaches 50 to 60 % and the stored energy ratio is a decreasing function of the plastic strain.
- in first approximation, the evolution of the stored energy is linear with the applied stress.

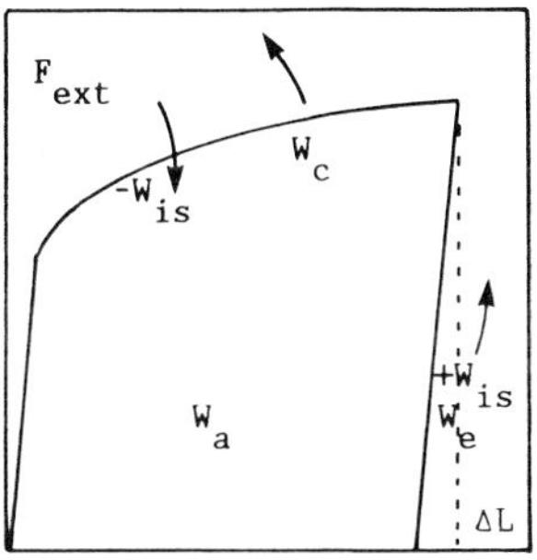

FIGURE 1

Energy balance for load-unload cycle.

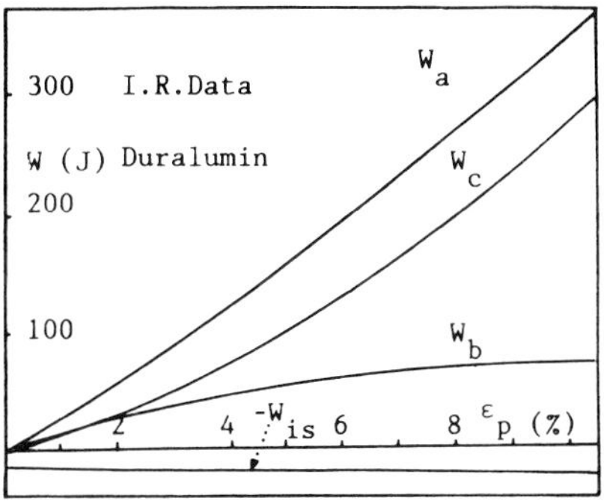

FIGURE 2

Evolution of the energy balance.

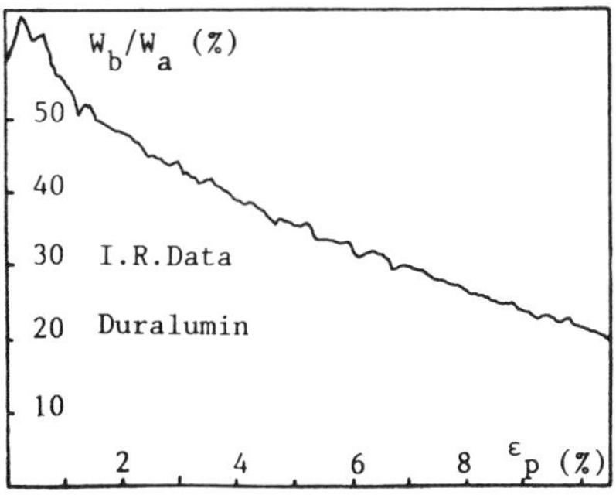

FIGURE 3

The decreasing evolution of the stored energy ratio.

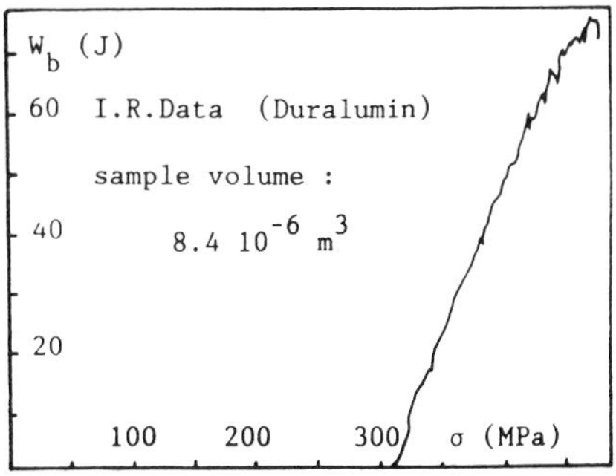

Figure 4

The stored energy as a function of stress.

6. MODELING AND CONCLUSIONS

The last point to be discussed is the utilization of such results, in the determination of behaviour laws. Two classical models, which can be included in the general theory of the generalized standard materials (Prandtl-Reuss, and Prager's law) have been studied. In each case, it is verified that such models cannot predict the observations we have made on the stored energy evolution, (the stored energy evolution is unlimited, the stored energy ratio is an increasing function of the plastic strain). This models have been transformed, taking into account our observations [9] . New hardening state variables have been defined, based on energetical data.

To conclude that first study, we would like to stress the importance of such new investigations. The incidence of the knowledge of energetical phenomena on the determination of behaviour laws seems to be fundamental even if the associated thermal effects are negligible. Experiments will be multiplied, playing on several sollicitations, and using several materials in order to increase the thermomechanical validity domain of the behaviour laws.

REFERENCES

[1] B. NAYROLES et All "Téléthermographie Infra-Rouge et Mécanique des structures", Int. J. Eng. Sci., 19, p. 929-947, 1981.

[2] R. BOUC, B. NAYROLES, "Méthodes et résultats en thermographie I.R. des Solides", J. de Mec. Théo. et Appl., vol. n° 4, 1985.

[3] J. BATAILLE, J. KESTIN, "L'interprétation physique de la Thermodynamique rationnelle", J. de Mec., Vol. 14, n° 2, 1975.

[4] G. LEBON, P. MATHIEU, "Etude comparée de diverses théories de Thermodynamique du non équilibre", Entropie n° 100, 1981.

[5] A. CHRYSOCHOOS, "Bilan énergétique en élastoplasticité grandes déformations", J. Mec. Théo. Appl. Vol. 14, n° 5, 1985.

[6] A.E. GREEN, P.M. NAGHDI, "A General Theory of an Elastic Plastic Continuum", Arch. Rat. Mec. Anal., 1965.

[7] E.H. LEE, "Elastic Plastic Deformation at Finite Strains", J. Appl. Mech., Vol. 36, 1969.

[8] S. NEMAT-NASSER, "Decomposition of Strain Measure...", Int. J. of Solids Structures, Vol. n° 36, 1969.

[9] A. CHRYSOCHOOS, "Deux propositions de Modèles Thermomécaniques...", Note Interne L.M.G.M.C. 85-4, 1985.

Thermomechanical Couplings in Solids
H.D. Bui and Q.S. Nguyen (Editors)
Elsevier Science Publishers B.V. (North-Holland)

INFRARED OBSERVATIONS OF DAMAGE AND FRACTURE ON CONCRETE

Minh Phong LUONG

Maître de Recherche CNRS, UA317 -
Laboratoire de Mécanique des Solides - Ecole Polytechnique -
91128 Palaiseau Cedex - France.

Infrared vibrothermography has been used as a nondestructive and noncontact technique to analyze the unstable crack propagation and to examine the mechanical response of concrete under a given unconfined compression and subjected to a vibratory excitation. The experimental investigation consists of the study of thermographic or heat patterns which are recorded or observed in real time during a specific vibratory loading applied to the specimen. The parameter investigated is the heat generation due to energy dissipation by the material which has been excited beyond its stable reversible limit.

The thermomechanical coupling offers a quantitative evaluation for the growth rate of thermal dissipation indicating the damage evolution of the material. This useful technique allows an accurate illustration of the onset of unstable crack propagation and/or flaw coalescence when increasing irreversible microcracking is generated by vibratory loading. In addition, it may provide a convenient method of field inspection and evaluation of stress concentration on loaded concrete structures, especially on areas where serious defects are known to be the most probable.

1. INTRODUCTION

Fatigue and damage behaviour of plain concrete in compressive loading is an important consideration in connection with design or regulations of bridges, offshore concrete structures, earthquake resistant buildings, subjected to high allowable working stresses or in cases where the dead load forms a smaller part of the total load capacity. In such situations, failure of the construction material may occur at stresses below its static strength. Accurate knowledge should therefore be obtained of the mechanical behaviour of concrete subjected to various loadings.

In response to these problems, various fatigue analysis methodologies have been developed in recent years which isolate the factors affecting crack initiation and growth, and enable the prediction of their cumulative effect on the fatigue performance of structural components. They are based on :

1- the formulation of analytical models for fatigue crack initiation and growth ;
and

2- the acquisition of supporting baseline data and validation of such models by means of a comprehensive testing procedure.

The existing partial damage hypothesis gives unsatisfactory results in evaluating the fatigue behaviour of plain concrete [1]. Volumetric, ultrasonic and acoustic emission measurements have not permitted a complete interpretation of the deterioration of concrete [2]. Research on absorption of energy in fatigue loading of plain concrete has been investigated [3] to aid understanding of the physical process of degradation when concrete is subjected to varying dynamic loads. The absorbed energy is believed to be used in the material in forming microcracks, crushing material, redistributing stress and causing a rise in temperature. Unhappily the difficulties involved with the measurement systems were such that the results obtained by these methods were not completely satisfactory.

The aim of the present paper is to illustrate the use of infrared vibrothermography as a nondestructive and noncontact technique to detect the physical process of fatigue, particularly the onset of unstable crack propagation and/or flaw coalescence, due to the thermomechanical coupling, when increasing irreversible microcracking is generated by vibratory loading. In addition, an application of this technique is presented as a nondestructive method for inspection and evaluation of stress concentration on a concrete structure subjected to an earthquake-type loading.

2. CHARACTERISTICS OF PLAIN CONCRETE

Concrete is a composite material consisting of coarse aggregates embedded in a continuous matrix of mortar which is a mixture of hydraulic binding materials, additives and admixtures distributed in a suitable homogeneous dosage. This construction material has been extensively used for structural applications because of its excellent processibility, low cost and versatility in manufacture together with desirable mechanical properties for engineering design.

Under applied loading, the material as a whole deforms in spite of significant incompatibilities between the aggregates and the matrix which promote further breakdown. At the macroscopic level, breakdown is accompanied by both loss in stiffness and the accumulation of irrecoverable deformation. At the structural level, breakdown appears as micro-cracking and possibly slippage at the aggregate-cement paste interfaces.

Failure in plain concrete may be viewed as a microstructural process through the activation and the growth of one preexisting flaw or site of weakness, or through the coalescence of a system of interacting small flaws and growing microcracks. The stress level corresponding to the activation of the flaws is related to the flaw size and connected with the encompassing microstructure. Flaws initiating concrete failure may be divided into two classes : the intrinsic flaws develop during the processes of hydration and curing of the current paste, the extrinsic flaws result from significant incompatibilities between the aggregate and the matrix when the material as a whole deforms under applied loading.

It may be said that fatigue of concrete is associated with the development of internal microcracks, probably both at the cement matrix/aggregate interface and in the matrix itself.

The occurrence of microcracking and slippage lead to non linearity and softening in the stress-strain response and a marked dependence on the mean normal stress [4].

The figure 1 describes a monotonic uniaxial compression test on a two years old concrete specimen.

The volume change detected by strain gages in transverse and longitudinal directions of the cylindrical specimen is seen to be highly relevant in characterizing the deformation of concrete under loads. It may be attributed to (1) elastic variations of minerals,(2) or to dilatancy due to growing microcracking.

Under repetitive stresses, fatigue mechanism is a progressive, permanent internal structural changes occurring in the concrete. These changes result in progressive growth of cracks and complete fracture.

The formation and propagation of microcracks have been detected by means of different measuring methods :

a- The ultrasonic pulse velocity technique involves measurement of the transit time of an ultrasonic pulse through a path of known length in a specimen. The velocity of the ultrasonic pulse in a solid material will depend on the density and elastic properties of the material and it will therefore be affected by the presence of cracks.

b- The acoustic emission method works on the principle that the formation and propagation of the microcracks are associated with the release of energy. When a crack forms or spreads, part of the original strain energy is dissipated in the form of heat, mechanical vibrations and in the creation of new surfaces. The mechanical vibration component can be detected by acoustic methods and recorded, hence microcracking may be detected by studying sounds emitted from the concrete.

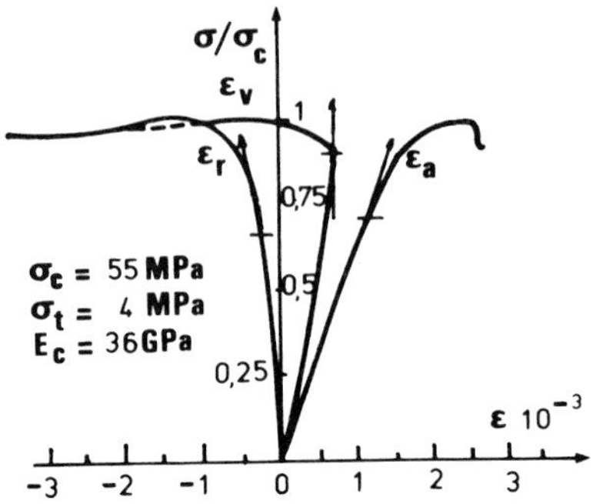

FIGURE 1 :
Axial (ε_a), radial (ε_r) and volumetric (ε_v) strains of concrete specimen under uniaxial compression.

It can be thought that fatigue mechanism of plain concrete consists primarily in the formation and propagation of microcracks. The formations of microcracks are often associated with points of stress concentration in the material. The stress concentrators may be flaws present in the material, existing cracks or notches. Fatigue cracks initiate quite early at a site of stress concentration, then propagate through the plastic zone and into the elastic zone. In other cases, flaws are inherent in the material owing to the process of fabrication. These defects exist prior to the application of any load. They may be some initiation period during which the material at the tip of the flaw undergoes dislocation pile-up, microvoid formation, and coalescence, etc..., prior to the onset of cycle-by-cycle growth.

3. THERMAL DISSIPATION

Infrared thermography has been successfully used as an experimental method for the detection of plastic deformation during crack propagation of a steel plate under monotonic loading [5] or as a laboratory technique for investigating damage, fatigue and creep mechanism [6].

This experimental tool is used here to detect the onset of unstable crack propagation and/or flaw coalescence due to the thermomechanical coupling, when increasing irreversible microcracking is induced by vibratory loading.

1- Thermoplasticity :

Under small perturbation hypothesis and taking into account the two principles of thermodynamics and the Fourier heat conduction law, the theory of internal variables leads to the coupled heat condition of thermoelastic plasticity [7] :

$$c\,\dot{T} = \text{div}\;k\;\text{grad}\;T - T\,\alpha\,\dot{\sigma} + D + \phi \quad ,$$

where c denotes the specific heat, T the absolute ambient temperature, k the heat conductivity, α the thermal dilatation, σ the stress tensor, D the intrinsic dissipation and ϕ the heat source per unit volume.

The first term on the right side governs the thermal diffusion which tends to make the temperature uniform in the specimen. The second term represents the thermoelastic effect that may be significant in cases of isentropic loading. The nature of factors k and α is tensorial in case of anisotropic material. The third term is the thermal dissipation generated by viscosity or plasticity. The last term show the existence of heat source in the specimen.

2- Electro-magnetic radiation :

Electromagnetic radiation is a form of energy characterized as waves or as particles called photons. Visible light is the most familiar form of electromagnetic energy. Other forms include radio waves, heat radiation, ultraviolet rays, X-rays and gamma rays. All this energy is similar and radiates in accordance with basic wave theory. Electromagnetic radiation is produced by the acceleration of charged particles. More rapid acceleration produces higher energy (shorter wavelength) waves. The electromagnetic spectrum is a categorization by wavelength of electromagnetic energy. The range from 2 to about 100 micrometres is called "thermal infrared".

All matter radiates energy because it contains charged particles being accelerated (changing speed or directions), the higher the temperature the greater the acceleration. The amount of energy radiated depends on the object's temperature and its ability to radiate.

3- Infrared vibrothermography :

Infrared thermography utilizes a photon-effect detector in a sophisticated electronics system in order to detect radiated energy and to convert it into a detailed real time thermal picture on a video system. Temperature differences in heat patterns as fine as 0.2°C. are discernible instantly and represented by several distinct hues.

This technique is sensitive,nondestructive and noncontact, thus ideally suited for records and observations in real time of heat patterns produced by the heat transformation of energy due to stress concentration and/or plastic strains. No interaction at all with the specimen is required to monitor the thermal gradient.

The quantity of energy W_r emitted by infrared radiation is a function of the temperature and the emissivity of the specimen. The higher the temperature, the more important is the emitted energy. Differences of radiated energy correspond to differences of temperature, since $W_r = h\ t^4\ \omega$, where h denotes a constant, T the absolute temperature and ω the emissivity.

Concrete presents a low thermomechanical conversion under monotonic loading. Plastic deformation -whereby microcracking and slips occur creating permanent changes globally or locally- is however one of the most efficient heat production mechanisms. Most of the energy which is required to cause such plastic deformations is dissipated as heat. Such heat development is more easily observed when it is produced in a fixed location by reversed applied loads. These considerations define the use of vibrothermography as a nondestructive method for observing concrete damage.

4- Experimental set-up and Results :

The high-frequency servo-hydraulic test machine Servotest, used at the Laboratoire de Mécanique des Solides, provides a means of vibration and dynamic testing of engineering materials. Control of the machine is provided by a sophisticated closed-loop electronic control system. This utilizes feedback signals from the force and displacement transducers. The programming section comprises a digital function generator and a frequency sweep controller which enables resonant phenomena testing.

The sample is observed in a nondestructive, noncontact manner by means of an infrared thermographic system AGA THERMOVISION 782. The thermal image is shown on the monitor screen.

The parameter investigated in this test in heat generation due to the energy dissipated by the concrete which has been excited beyond the stable reversible domain. A vibratory loading at 100 Hz on the specimen subjected to a given static compression (figure 2) exhibits in a non destructive manner the irreversible plastic strain concentrations around gaps or cracks generated by stresses exceeding locally the stability limit of the material. The contribution of the plasticity term is revealed by the rapid evolution of heat dissipation once the stable reversible domain has been exceeded.

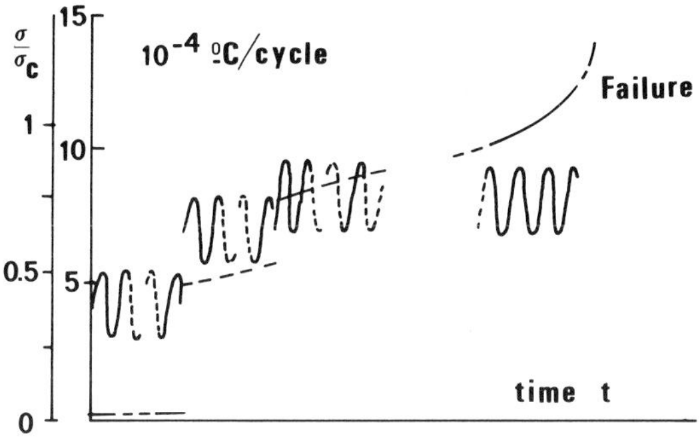

FIGURE 2 :
Vibratory excitation on a specimen under static uniaxial compression and evolution of the growth rate of heat on the warmest point.

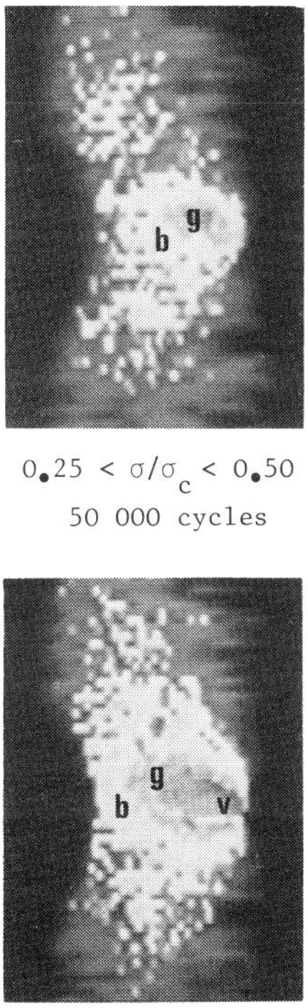

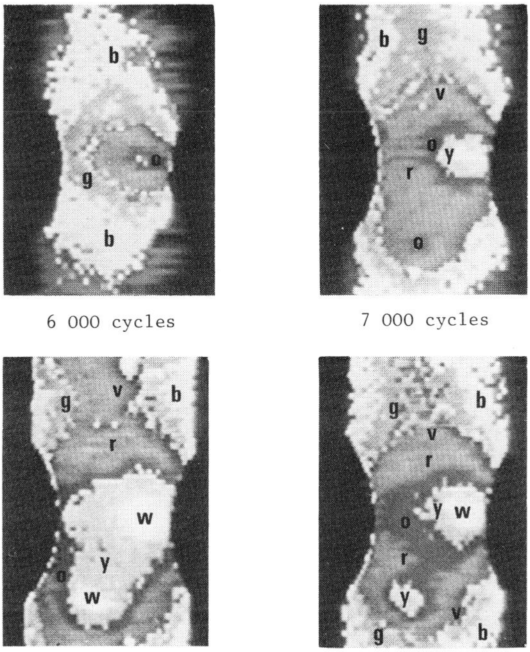

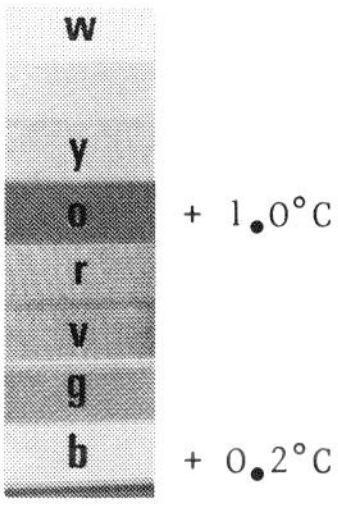

FIGURE 3 : Heat patterns on loaded concrete specimen.

FIGURE 4 : Experimental concrete structure under seismic type loading generated by a rotating mass excitator on its top.

	m	b	g	r	o	y
ΔT		+0.2	+0.4	+0.6	+0.8	+1.0

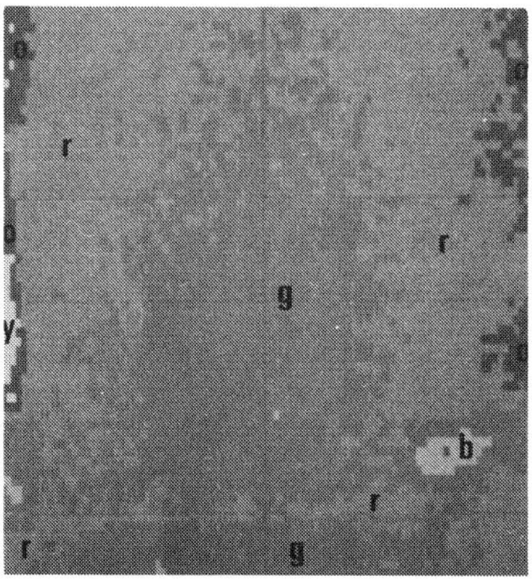

FIGURE 5 : Thermogram of the base of a column before the start of the excitator.

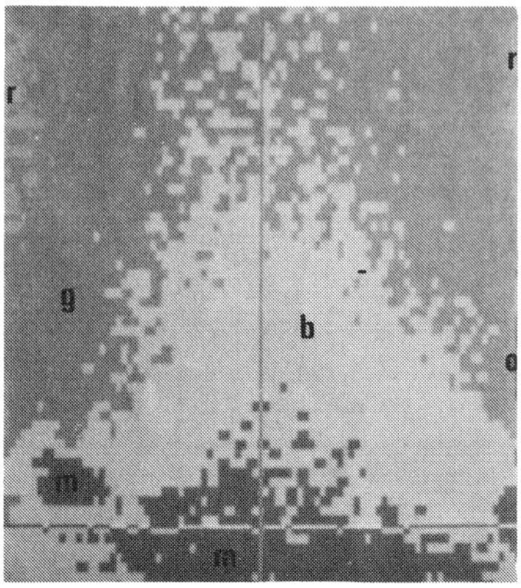

FIGURE 6 : Thermogram at the same location recorded during loading.

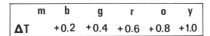

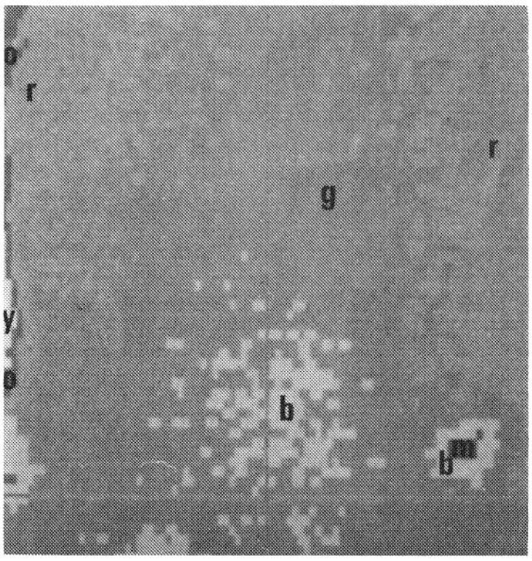

FIGURE 7 : Thermogram at the same location recorded before cracks are visible.

It can be seen on both figures 2 and 3 that :

a- for an sinusoïdal load excitation varying between 25 to 50 percent of the compression strength σ_c, heat dissipation is negligible even on the warmest point.

b- With $0.50 < \sigma/\sigma_c < 0.75$, stress concentrations around gaps or cracks are readily located on the surface of the specimen.

c- When the vibratory excitation is loaded between $0.63 < \sigma/\sigma_c < 0.88$, flaw coalescence can be observed progressively and the growth rate of heat on the warmest point enables the detection of the onset of specimen failure.

Thus the thermomechanical coupling offers a quantitative evaluation for the growth rate of thermal dissipation monitoring the damage evolution of the material. The damaged areas are located and highlighted by heat patterns [8].

4. FULL-SCALE EXPERIMENTAL INSPECTION OF A CONCRETE STRUCTURE

Information about the location and significance of structural defects needed as a basis for maintenance decisions, including the extreme case of removal from service, can be obtained through inspection and non destructive evaluation. The proposed infrared thermographic procedure involves careful examination of areas where defects are most likely to occur. The critical areas can be identified by analyzing the structure and the service histories of similar structures in similar environments.

The application of infrared scanning to inspection of concrete structure relies on the fact that energy is dissipated during the process of failure when internal cracks or flaws develop. It is indeed well known that the fatigue process involves crack initiation and/or nucleation, stable crack propagation, and final fracture or unstable crack propagation.

Figure 4 describes an experimental reinforced concrete building frame, intended for earthquake resistance studies. The most severe likely earthquake can be survived if the members are sufficiently ductile to absorb and dissipate seismic energy by inelastic deformations. It is recognized that the connections present adequate ductility in order to undergo large inelastic deformations with little decrease in strength. Under seismic loading, simulated by a rotating mass excitator placed on the top of the building, plastic hinges form progressively at the column bases where heat dissipation can be observed by infrared thermography.

Figures 5, 6 and 7 show the progressive evolution of heat dissipation at a column base before crack line is visible.

5. CONCLUDING REMARKS

Infrared vibrothermography thanks to thermomechanical coupling offers the possibility of a non destructive, non contact test of concrete degradation. It allows a mesure of the material damage and permits the detection of the limit of a progressive damaging process under load beyond which the material is destroyed.

This useful technique allows accurate illustration of the onset of unstable crack propagation and/or flaw coalescence when increasing irreversible microcracking is generated by vibratory loading.

REFERENCES

[1] Tepfers, R., Friden, C., Georgsson, L., A study of the applicability of the Palmgren-Miner partial damage hypothesis, Magazine of Concrete Research, vol. 29, n° 100, Sept. 1977, pp. 123-130.

[2] Bergues, J., Terrien, M., Study of concrete's cracking under multiaxial stresses, Advances in Fracture Research, 5th Int. Conf. on Fracture, Cannes, France, 29 March-3 April 1981.

[3] Tepfers, R., Hedberg, B., Szczekocki, G., Absorption of energy in fatigue loading of plain concrete, Matériaux & Construction, vol. 17, n° 97, pp. 59-64, Bordas-Dunod, 1984.

[4] Kotsovos, M.D., Newman, J.B., Generalized stress-strain relationsphips for concrete, Journal Eng. Mech. Div., ASCE, vol. 194, n°EM4, pp. 845-856, 1978.

[5] Bui, H.D., Ehrlacher, A., Nguyen, Q.S., Etude expérimentale de la dissipation dans la propagation de fissure par thermographie infrarouge, Comptes-Rendus Ac. Sci., 293, série II, pp. 1015-1018, 1981.

[6] Reifsnider, K.L., Henneke, E.G., Stinchcomb, W.W., The mechanics of vibrothermography, in Mechanics of Non destructive Testing, ed. by W.W. Stinchcomb, pp. 249-276, 1980.

[7] Nguyen Q.S., Thermodynamique des Milieux Continus, Cours D.E.A., E.N.P.C., 1984.

[8] Luong, M.P., Infrared Vibrothermography of plain concrete, magnetic sound 16mm film, video Umatic and VHS PAL-SECAM, Systems edited by IMAGICIEL, 1984, Ecole Polytechnique, 91128 Palaiseau, France.

Thermomechanical Couplings in Solids
H.D. Bui and Q.S. Nguyen (Editors)
Elsevier Science Publishers B.V. (North-Holland)

THERMOELASTIC EFFECTS IN FRACTURE

J.R. BARBER and Maria COMNINOU

Department of Mechanical Engineering and Applied Mechanics,
University of Michigan, Ann Arbor, MI 48109-2125 USA.

1. INTRODUCTION

A wide range of mechanical devices and structures are subject to thermal loading. However, practising designers do not usually attach great importance to analytical predictions of high thermal stresses, since experience shows that they are seldom catastrophic, at least for ductile materials. The reason for this behavior is that thermal stresses are induced by geometrical constraints and hence can be relieved by local yielding.

The situation is very different if the thermal loading is cyclic, since alternating yielding - even if very localized - will now contribute to the initiation and propagation of fatigue cracks. For many machines, the major alternating stresses arise from start-up and shut-down, and a comparatively small number of cycles is accumulated during the design life. However, the stress levels obtained can be high enough to cause low cycle fatigue failure.

Once a fatigue crack is initiated, it will interact with the thermoelastic field in two ways: the far-field thermal stresses will be locally perturbed by the inability of the crack to transmit tensile tractions, but also the crack will act as an obstruction to heat flow, causing an additional local perturbation in temperature and hence stress.

2. SOME CLASSICAL SOLUTIONS

2.1 The penny-shaped crack

In view of these considerations, it is surprising that relatively little attention is given to thermoelastic effects in the fracture mechanics literature. The earliest published solution is that of Olesiak and Sneddon [1] for the steady-state problem of the penny-shaped crack in an infinite homogeneous isotropic solid, with prescribed axisymmetric temperature or heat flux at the crack faces. The solution was obtained by representing the stress and temperature fields in terms of Hankel transforms and hence reducing the problem to the solution of a pair of dual integral equations. The same problem was considered by Williams [2], who showed that certain general features of the solution can be derived directly from a harmonic potential function solution of the steady-state equations of thermoelasticity.

These solutions were restricted to the case in which the temperature field is also symmetric about the crack plane, which implies the continuous generation or absorption of heat within the crack - a situation for which it is difficult to envisage an appropriate physical mechanism. The symmetry condition also implies that there are purely normal stresses on the crack plane and hence that only a mode I stress intensity factor is obtained at the crack tip. The relative displacement of the opposing crack faces is normal to the crack plane, corresponding to crack opening or closing.

We should note that crack closure - often signalled by the analytical prediction of a negative stress intensity factor - is physically unrealistic, unless the problem is re-solved using unilateral contact conditions between the crack faces. Very few authors in the fracture mechanics literature take the trouble to do this or indeed even note where the problem arises. This is unfortunate, since as we shall see, crack closure adds considerably to the mathematical richness of the problem, in addition to being more physically realistic, In symmetric problems, crack closure is likely to occur when the crack is heated, and crack opening when it is cooled.

2.2 The antisymmetric problem

The physically more interesting case is that in which the temperature field is antisymmetric, in which case there are no heat sources or sinks in the crack and the solution corresponds to the perturbation of a heat flux by an insulating or partially insulating crack. In this case the crack does not open or close, but there is a tendency for shear displacement between the crack faces, leading to a mode II stress intensity factor at the crack tip.

The case of a uniform heat flux obstructed by an insulating penny-shaped crack was first solved by Florence and Goodier [3,4], who first considered the crack as the limiting case of an insulating spheroidal cavity [3] and then gave a more numerically efficient solution, using Sneddon's Hankel transform method [4]. The corresponding problem for an arbitrary temperature field which is not necessarily axisymmetric was given by Barber [5], using Williams' potential function representation and a solution of the resulting boundary value problem due to Green [6].

Since the crack faces do not separate, it is perhaps more realistic to assume that some heat flow occurs across the crack, rather than that it is completely insulating. Barber [7,8] gave a solution of this problem for the penny-shaped crack, assuming a *radiation* boundary condition - i.e. the heat flux across the crack was assumed to be proportional to the local temperature difference. This paper also uncovers the interesting fact that the shear stresses on the crack plane (and hence the stress intensity factor) depend only on the average temperature difference between the crack faces and not on the exact distribution. For symmetric problems, a corresponding result relates the normal stresses on the crack plane to the total heat generated within the crack [1,2].

2.3 The external crack

A closely related problem concerns the axisymmetric external crack, which can be conceived as two similar elastic half-spaces bonded over a circular region of their common interface. The axisymmetric problem was first discussed by Kassir and Sih [9], who gave stress intensity factors for various symmetric and antisymmetric boundary conditions. Shail [10] gave a general solution to both the symmetric and the antisymmetric problems, with arbitrary non-axisymmetric boundary conditions. However, Rubenfeld [11] later showed that Shail's method breaks down except in the axisymmetric case. He gave an alternative solution to the non-axisymmetric problem, using a potential function representation due to Youngdahl [12].

2.4 Some other three-dimensional problems

Kassir and Bregman [13] treated the problem of two parallel coaxial penny-shaped cracks. Two cases were considered: (i) a uniform temperature was prescribed at all the crack faces, giving a solution with heat generation in the cracks which is symmetric about the plane midway between the two cracks. (ii) the cracks were assumed to be insulating and obstructing a uniform flow of heat, giving a solution which is antisymmetric about the mid-plane. Both mode I and mode II stress intensity factors are obtained in each case, since the cracks do not lie on the plane of symmetry.

For case (ii), we can argue by symmetry that if one crack opens at a given point, the corresponding point on the other crack must close by the same amount. Hence, for all applied heat fluxes there must be some closure in at least one of the cracks.

The only non-axisymmetric three-dimensional geometry which has received much attention is the elliptical crack in an infinite solid. Kassir [14,15] and Kassir and Sih [16] give solutions with prescribed or heat flux, using a series expansion in ellipsoidal harmonics. Also, Kassir [17,18] has obtained Green's functions for prescribed temperature or heat flux on the faces of a three-dimensional semi-infinite plane crack for the symmetric and the antisymmetric case.

2.5 Two-dimensional problems

The simpler geometry and the complex variable representation permit more varied geometries to be considered in two-dimensional thermoelasticity. The basic problems of a plane crack in an infinite medium were treated by Sih [19,20].

Sekine considered the problem of an inclined plane crack in a half-plane with an insulated free surface in the presence of a uniform heat flux. The boundary condition at the crack surface was one of prescribed temperature or zero heat flux [21,22]. The case where the crack is perpendicular to the free surface was also considered by Tweed and Lowe [23] though they used a constant temperature boundary condition at the free surface. They recorded negative stress intensity factors at the tip nearest the free surface when the crack was long in comparison with its depth. They concluded that partial closure would then occur, but did not extend the analysis to consider this case. Similar behavior can be observed in the results of [21], though it is not commented upon by the author. Sekine has also treated the problem of a cooled crack in the vicinity of an insulating hole in a two-dimensional infinite solid, [24].

2.6 Anisotropic materials

The two-dimensional problem of a plane crack with prescribed temperature or heat flux in a generally anisotropic infinite solid has been considered by Atkinson and Clements [25] using a particular solution of the anisotropic thermoelastic equations due to Clements [26] and the isothermal solution of Stroh [27]. Fourier transformation was used to reduce the problem to a pair of dual integral equations. The temperature distribution for the plane crack in an infinite slab was given by Clements and Tauchert [28]. When the crack is not on a plane of material symmetry, each problem gives rise to stress intensity factors in modes I, II and III. In particular, the insulating crack obstructing a uniform heat flux will tend to open or close. Clements and Atkinson give a solution for this problem, but they inadvertently used an incorrect boundary condition at the crack face, so that the results are difficult to evaluate. More recently, Sturla [29] has reworked this problem using a Green's function formulation [30] and shows that for either direction of heat flow, the crack will tend to close at one end or the other.

If the material anisotropy is symmetrical with respect to the crack plane, not such coupling occurs and the results are very similar to those of the isotropic problems already discussed. Tsai gives a solution of this kind for the penny-shaped crack in an infinite transversely isotropic solid for the symmetric [31] and antisymmetric case [32].

2.7 Composite materials

Various authors have considered problems involving layered materials with a crack parallel to the interfaces. Fourier or Hankel transform methods provide an appropriate method of solution. Srivastava [33] lists a number of solutions of this kind and also treats the problem of a penny-shaped crack symmetrically disposed in a layer, sandwiched between and bonded to two half-spaces. The material properties of the half-spaces are the same, but differ from those of the layer, so the geometry is symmetrical about the crack plane and exhibits similar behavior to that recorded for the simpler homogeneous case.

New features are observed when the crack occurs at the interface between two dissimilar materials. Bregman and Kassir treat the case of an insulating penny-shaped crack obstructing a uniform heat flux between two dissimilar half-spaces [34]. There is now no symmetry about the crack plane and the crack generally opens or closes. All quantities are linearly proportional to the imposed flux and hence closure must occur for at least one direction of heat flow. Also the solution exhibits the usual oscillatory behavior in modes I and II at the crack tip, suggesting the presence of local contact zones. The effect of closure on the penny-shaped interface crack is discussed in more detail in section 3.2 below.

3. CRACK CLOSURE AND THERMAL CONTACT RESISTANCE

As we have seen from the previous section, there are many situations in which the crack obstructing a uniform heat flux has a tendency to open or close, depending on the direction of heat flow. In general, this arises whenever the body is not symmetrical with respect to the crack plane, for example, if the material has general anisotropy, if the crack occurs at an interface between two dissimilar materials or if the crack is not plane.

When closure is predicted, it is important that the problem be re-solved as a contact problem - i.e. imposing unilateral constraints on the boundary conditions requiring the tractions to be non-tensile and the crack opening displacement to be non-negative. Furthermore, the boundary conditions on the heat conduction problem are influenced by crack closure. We anticipate that heat will be conducted between the crack faces in regions of mechanical contact. An important consequence is that the solution then depends in a non-linear fashion upon the magnitude and direction of the heat flux. The crack may only close partially, in which case we have to solve a thermoelastic contact problem to determine the initially unknown extent of the contact region.

3.1 Existence and Uniqueness

Further complication is introduced by the fact that the classical existence and uniqueness theorems do not apply to thermoelastic contact problems with unilateral constraints. Indeed, some problems are known to have no solution unless boundary conditions involving an interfacial thermal resistance are imposed, while others have multiple solutions, raising questions of stability.

The difficulties which arise when heat flows across a unilateral interface into the material of lower distortivity were first observed by Barber [35]. The distortivity δ is defined as $\delta = \alpha(1+\nu)/K$, where α is the coefficient of thermal expansion, ν is Poisson's ratio and K is thermal conductivity. Briefly stated, if continuity of temperature (no thermal contact resistance) is assumed in all contact regions and perfect insulation (no heat flux) in all separation regions, there are some conditions for which all solutions violate the physical requirements of non-tensile tractions and non-negative gap. Comninou and Dundurs [36] show by asymptotic analysis that a transition from perfect contact to separation cannot in fact occur if the heat flux is directed into the less distortive body.

The difficulty can be avoided by making the more realistic assumption of a pressure or gap dependent interface thermal resistance. Duvaut [37] has proved an existence theorem for the case where the contact resistance varies inversely with contact pressure. A limiting case of this is the state of *imperfect thermal contact* defined by Barber [38]. In this state, the contact pressure is vanishingly small, but there is some finite resistance to heat flow across the interface, resulting in a temperature discontinuity. When the heat flows into the material of lower distortivity, a zone of imperfect contact is interposed between the zone of perfect contact - in which the conventional boundary conditions apply - and the separation zone or gap. Comninou and Dundurs [36] show that the asymptotic behavior of these transitions is acceptable. The advantage of introducing the limiting case of imperfect contact, rather than using a more general pressure dependent thermal contact resistance, is that linearity is preserved in the resulting boundary value problem, although superposition does not hold in general, since the extents of the various zones depend on the applied conditions.

As remarked above, when the heat flows into the more distortive material, non-uniqueness of solution is often observed. The stability of the resulting steady-state solutions is explored in a one-dimensional model by Barber et al. [39]. This question is outside the scope of the present review, but we might note that recent work has uncovered the remarkable conclusion that a steady-state solution can be unstable **even when it is unique** [40]. In such cases, we must presume that the system settles into an oscillatory state, but as yet no solutions to such transient problems have been obtained.

3.2 The interface crack

To illustrate some of these phenomena in the context of fracture, we consider in more detail the penny-shaped interface crack obstructing a uniform heat flux, originally treated by Bregman and Kassir [34]. This problem, in the absence of heat flux and in a uniform tension field, was solved by Keer et al. [41], who showed that the crack opens except for a small annular region of contact adjacent to the crack tip. If the crack is first loaded in tension σ_o and then subjected to a heat flux q_o, it seems reasonable to expect a similar configuration with a possible change in the extent of the contact region. To facilitate the discussion we consider separately the case of perfect contact, i.e. heat flowing into the more distortive material solved by Martin-Moran et al. [42], and the case of imperfect contact, i.e. heat flowing into the less distortive material, Barber and Comninou [43]. We also need to introduce two dimensionless parameters:

$$\beta = 0.5[(1 - 2\nu_1)\mu_2 - (1 - 2\nu_2)\mu_1]/[(1 - \nu_1)\mu_2 + (1 - \nu_2)\mu_1];$$

is one of the Dundurs constants representing the mismatch in the mechanical properties and varying in the range $-0.5 < \beta < 0.5$, where μ is the shear modulus; also

$$\gamma = (\delta_1 - \delta_2)/ (\delta_1 + \delta_2)$$

is a parameter representing the mismatch in the distortivities and varies in the range $-1 < \gamma < 1$.

The sign of the product $\beta\gamma$ proves to have a major effect on the behavior of the interface crack with heat flow. It is positive if the material with the higher distortivity has also the higher value of $(1-2\nu)/\mu$. Figure 1 shows the values of these two material properties for various metals, alloys and ceramics. In this figure, the slope of a straight line joining any two points has the same sign as the product $\beta\gamma$ for the corresponding interface. For example, an interface between copper and zinc gives $\beta\gamma > 0$, while one between copper and mild steel gives $\beta\gamma < 0$. Many practical examples of each kind occur, as can be seen from Figure 1.

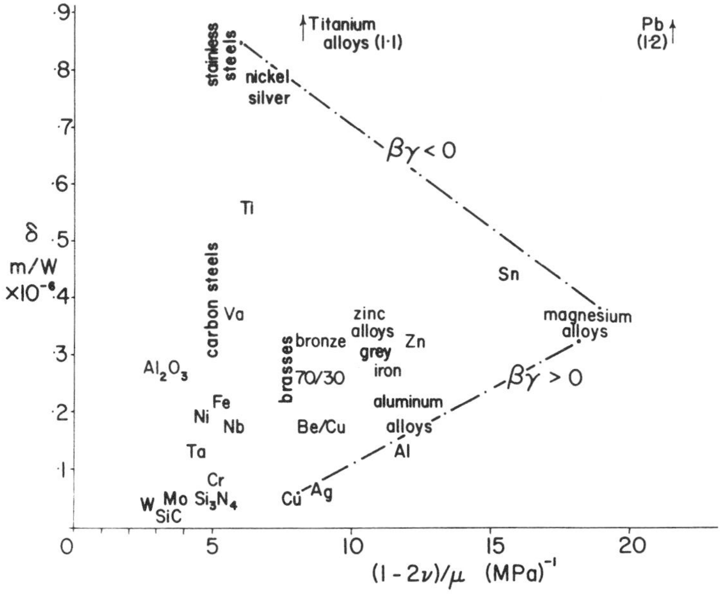

Figure 1

Distortivity δ and $(1-2\nu)/\mu$ for various structural materials

Perfect contact. Suppose we start with the Keer et al. solution (uniform tension) and gradually increase the heat flux from zero. The results of [42] show that the contact region will get larger if $\beta\gamma > 0$ and smaller if $\beta\gamma < 0$. The case with $\beta\gamma < 0$ corresponds to an infinitesimal contact region and will not be discussed further. For $\beta\gamma > 0$, the contact region increases as the normalized ratio of the applied tension to the applied heat flux, SQ decreases, as shown in Figure 2. Further increase in contact can be obtained by applying compression at infinity. Something interesting happens then: there is a range of compression to heat flux ratio for which three solutions are possible, corresponding to the points B, C and D in Figure 2. All three solutions satisfy the boundary conditions, including the inequalities, of the problem. The point D corresponds to the crack being fully closed. This solution can be reached by starting with compression and zero heat flux, and gradually increasing the heat flux. Note that the fully closed crack is always a possible solution if $\sigma_0 < 0$. Point B can possibly be reached by loading the crack in tension and then slowly changing σ_0 to the desired compressive value. If the compressive stress is increased further, point A would be reached, from which the solution would jump to the state with the fully closed crack. It is not clear what sequence of loading, if any, would establish solution C.

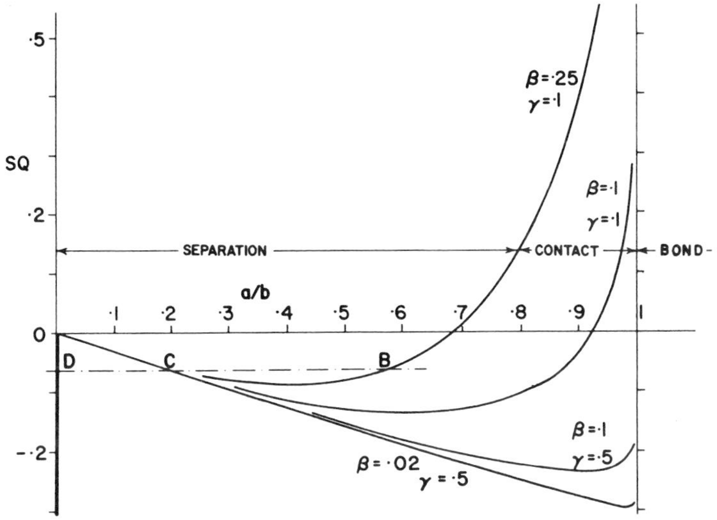

Figure 2
Variation of normalized contact radius a/b with the normalized applied traction to heat flux ratio SQ for various material combinations

Imperfect contact. In this case a large contact zone is obtained for $\beta\gamma < 0$. Note that the contact zone consists now of a zone of perfect contact and a zone of inperfect contact, introducing extra complexity in the problem. We find that there is a region with a unique solution, a region with two or three solutions (one without separation), and a region in which the only possible solution does not involve separation.

If we consider the range of heat flux for which multiple solutions were obtained for the steady-state problem and we slowly cycle the heat flux through this range, there must be a point where the crack moves from one branch to another. This transient process will probably be relatively rapid and involve higher thermal stresses than those in the steady-state. It could be an important factor in the propagation of interface cracks in thermally loaded structures. However, the problem of the interface crack in a thermal field varying slowly with time has not yet been solved.

There are a few solutions of transient thermal stress problems with cracks. Noda and Sumi [44] investigated the case of the Griffith crack, while the axisymmetric external crack was studied by Noda et al. [45]. There are also some approximate transient solutions in which it is assumed that the crack does not alter the temperature distribution in the body, [46,47,48]. A finite element solution for an internal crack in a semi-infinite body under thermal transient stress was obtained by Ting and Jacobs [49].

However, in all these cases, the solution for the temperature could be obtained independently of the thermoelastic solution. In problems involving transient thermoelastic contact, there is a continuous interaction in time between the temperature field and the elastic field, through the varying extent of the contact zone. Producing efficient solutions of such problems is currently one of the most challenging problems in the field of thermoelastic contact and fracture.

REFERENCES

[1] Olesiak, Z., and Sneddon, I.N., Archives of Rational Mechanics and Analysis 4 (1960) 238.
[2] Williams, W.E., Z. angewandte Mathematik und Phyzik. 12 (1961) 452.
[3] Florence, A.L. and Goodier, J.N., Proceedings of the 4th U.S. National Congress of Applied Mechanics (1962) 595.
[4] Florence, A.L., and Goodier, J.N., Int. J. Engineering Science 1 (1963) 533.
[5] Barber, J.R., J. Strain Analysis 10 (1975) 19.
[6] Green, A.E., Proceedings Cambridge Philosophical Society 45 (1949) 251.
[7] Barber, J.R., Int. J. Heat and Mass Transfer 19 (1976) 956.
[8] Barber, J.R., J. Thermal Stresses 3 (1979) 77.
[9] Kassir, M.K., and Sih, G.S., Int. J. of Solids and Structures 5 (1968) 351.
[10] Shail, R., Int. J. Engineering Science 6 (1968) 685.
[11] Rubenfeld, L., Int. J. Engineering Science 8 (1970) 499.
[12] Youngdahl, C.K., Int. J. Engineering Science 7 (1969) 61.
[13] Kassir, M.K., and Bregman, A.M., Applied Scientific Research 25 (1971) 262.
[14] Kassir, M.K., Int. J. Engineering Science 7 (1969) 769.
[15] Kassir, M.K., Engineering Fracture Mechanics 2 (1971) 373.
[16] Kassir, M.K., and Sih, G.S.,Three-dimensional thermoelastic problems of planes of discontinuities or cracks in solids, in: Shaw, W.A, (ed.), Developments in Theoretical and Applied Mechanics 3 (Pergamon Press, New York, 1967), pp.117-146.
[17] Kassir, M.K., Int. J. Engineering Science 13 (1975) 703.
[18] Kassir, M.K., J. Applied Mechanics 43 (1976) 107.
[19] Sih, G.C., J.Applied Mechanics 29 (1962) 587.
[20] Sih, G.C., J. Heat Transfer 87 (1965) 293.
[21] Sekine, H., Engineering Fracture Mechanics 7 (1975) 713.
[22] Sekine, H., Engineering Fracture Mechanics 9 (1977) 499.
[23] Tweed, J., and Lowe, S., Int. J. Engineering Science 17 (1979) 357.
[24] Sekine, H., Int. J. Fracture 13 (1977) 133.
[25] Atkinson, C., and Clements, D.L., Int. J. Solids and Structures 13 (1977) 855.

[26] Clements, D.L., SIAM J.Applied Math. 24 (1973) 332.
[27] Stroh, A.N., Philosophical Magazine 3 (1958) 625.
[28] Clements, D.L., and Tauchert,T.R., Int. J. Engineering Science 17 (1979) 1141.
[29] Sturla, F.A., Private communication.
[30] Sturla, F.A. and Barber, J.R., J.Applied Mechanics, under review.
[31] Tsai, Y.M., J. Applied Mechanics 50 (1983) 24.
[32] Tsai, Y.M., J. Thermal Stresses 6 (1983) 379.
[33] Srivastava, K.N., Palaiya, R.M., and Choudhary, A., Int. J. Fracture 13 (1977) 27.
[34] Bregman, A.M., and Kassir, M.K., Int. J. Fracture 10 (1974) 87.
[35] Barber, J.R., Int. J. Heat and Mass Transfer 14 (1971) 751.
[36] Comninou, M., and Dundurs, J.D., J. Applied Mechanics 46 (1979) 849.
[37] Duvaut, G., Free boundary value problems, 11 (Pavia, 1979)
[38] Barber, J.R., J. Elasticity 8 (1978) 409.
[39] Barber, J.R., Dundurs, J. and Comninou, M., J.Applied Mechanics 47 (1980) 871.
[40] Barber, J.R., Proceedings Institution of Mechanical Engineers Conference on Tribology, London (1987), submitted.
[41] Keer, L.M., Chen, S.H., and Comninou, M., Int. J. Engineering Science 16 (1978) 765.
[42] Martin-Moran, C.J., Barber, J.R., and Comninou. M., J. Applied Mechanics 50 (1983) 29.
[43] Barber, J.R., and Comninou, M., J. Applied Mechanics 50 (1983) 770.
[44] Noda, N., and Sumi, N., J. Thermal Stresses 8 (1985) 173.
[45] Noda, N., Matsunaga, Y., and Nyuko, H., J. Thermal Stresses 9 (1986) 119
[46] Koizumi, T., and Niwa, H., Transactions of the Society of Japanese Mechanical Engineers 43 (1977) 442.
[47] Nied, H. F., and Erdogan, F., J. Thermal Stresses 6 (1983) 1.
[48] Nied, H. F., J. Thermal Stresses 6 (1983) 217.
[49] Ting, V.C., and Jacobs, H.R., J. Thermal Stresses 2 (1979) 1.

Thermomechanical Couplings in Solids
H.D. Bui and Q.S. Nguyen (Editors)
Elsevier Science Publishers B.V. (North-Holland)

INFLUENCE OF THERMOMECHANICAL COUPLING ON DYNAMIC FRACTURE OF DUCTILE SOLIDS

Piotr PERZYNA

Institute of Fundamental Technological Research, Polish Academy of Sciences, Świętokrzyska 21, 00-049 Warsaw, Poland

The paper aims at the description of the influence of thermomechanical coupling on dynamic fracture of ductile solids. Basing on the available experimental results an analysis of thermal effects on fracture for particular materials has been given. Fracture phenomenon is described as a sequence of the nucleation, growth and coalescence of microvoids. A thermomechanical elastic-viscoplastic model of voided solids is proposed. Particular attention is given to the formulation of the evolution equation for porosity parameter. This equation describes the work-hardening viscoplastic response of solid as well as thermal effects and takes also account of the interactions of microvoids. A dynamical criterion of fracture of ductile metals is proposed. The criterion describes the dependence of fracture phenomenon upon the evolution of the constitutive structure and upon the thermomechanical coupling. The procedure of the determination of the material functions and material constants is presented.

1. INTRODUCTION

The main objective of the present paper is the analysis of the influence of thermomechanical coupling on dynamic fracture of ductile solids.

Discussion of the influence of thermal effects on fracture phenomenon is based on available experimental results. Fracture phenomenon is described as a sequence of the nucleation, growth and coalescence of microvoids. It has been proved that thermal effects might have dominated role in the mechanism of growth and nucleation of microvoids, and as a result in fracture mechanism.

For example, for elevated temperature the nucleation and growth of microvoids in ductile metals will be govern by different mechanisms than for room temperature. Generally the mechanisms of nucleation and growth of microvoids are very sensitive to temperature changes.

As it has been investigated by J.N.Johnson [5] even for room temperature in surroundings of a body considered for heterogeneous plastic deformations there exists substantial localized heating effects. Plastic shear strains of several thousand percent are developed at the expanding pore walls. The magnitude of the corresponding temperature caused by thermomechanical coupling and estimated by calculating the plastic shear strain at the pore wall is a substantial fraction of the melting temperature.of a material. These localized thermal effects can influence the growth mechanism by softening of the strength of a material as well as the nucleation mechanism by changing rate of the thermally-activated process.

In this paper a heurestic examination of the thermal influence on growth and nucleation of microvoids has been given. Discussion of the influence of thermal effects on the equilibrium state is presented. Phenomenological evolution equation for porosity parameter has been postulated. This equation describes the work-hardening viscoplastic response of solid as well as thermal effects and takes also account of the interactions of microvoids.

A thermo-elastic-viscoplastic model of voided solid is proposed. This model is formulated within the framework of the internal state variable structure. The porosity parameter is introduced as a fundamental internal state variable. Basing on the Clausius-Duhem inequality the thermodynamic restrictions have been investigated.

An alternative approach, when the number of microvoids per unit volume and the average size of microvoids are introduced as the internal state variables, is also discussed. Comparison of the results showed very important consequences for the determination of the growth and nucleation material functions in the evolution equation for the porosity parameter.

A dynamic criterion of fracture of ductile metals is proposed. The criterion describes the dependence of fracture phenomenon upon the evolution of the constitutive structure and upon the thermomechanical coupling.

The procedure of the determination of the material functions and material constants is also presented. This procedure is based on postshot photomicrographic observations for ductile metals.

2. INTERNAL STATE VARIABLES WITH THERMOMECHANICAL COUPLINGS

2.1. Local thermodynamic process

In previous papers of the author [10,11] the general framework of the thermodynamics of dissipative materials was developed. The main objective of the present paper is to apply the theory proposed to the description of the thermomechanical coupling in dynamic fracture.

The consideration of a global thermodynamic process for a body $\mathcal{B}$ gives the following equations describing a local thermodynamic process in a particle X

$$\begin{aligned} &\mathrm{Div}\,[\,\underset{\sim}{F}(t)\,\underset{\sim}{T}(t)\,] + \varrho_{\varkappa}\, b(t) = \varrho_{\varkappa}\,\ddot{\underset{\sim}{x}}(t),\\ &\underset{\sim}{T}(t) = \underset{\sim}{T}(t)^{T},\\ &\tfrac{1}{2}\,\mathrm{tr}\,[\underset{\sim}{T}(t)\dot{\underset{\sim}{C}}(t)] - \mathrm{Div}\,\underset{\sim}{q}_{\varkappa}(t) - \varrho_{\varkappa}[\,\dot{\psi}(t) + \vartheta(t)\dot{\eta}(t) + \dot{\vartheta}(t)\,\eta(t)\,]\\ &\qquad\qquad + \varrho_{\varkappa}\, r(t) = 0,\\ &\dot{\psi}(t) - \dot{\vartheta}(t)\eta(t) + \frac{1}{2\varrho_{\varkappa}}\,\mathrm{tr}\,[\underset{\sim}{T}(t)\,\dot{C}(t)] - \frac{1}{\varrho_{\varkappa}\vartheta(t)}\,\underset{\sim}{q}_{\varkappa}(t)\cdot\nabla\vartheta(t) \geqslant 0, \end{aligned} \qquad 2.1$$

where $\underset{\sim}{F}(t)$ denotes the value of the deformation gradient of a particle X at time t and is determined by the function of motion $\underset{\sim}{\chi}$ by the relation

$$\underset{\sim}{F}(X,t) = \nabla\underset{\sim}{\chi}\,(X,t) = \nabla\underset{\sim}{x} \qquad 2.2$$

if ∇ denotes gradient with respect to the material coordinates $\underset{\sim}{X}$, the particle X is identified with its position $\underset{\sim}{X}$ in a fixed reference configuration $\underset{\sim}{\varkappa}$, $\underset{\sim}{T}(t)$ is the value of the Piola-Kirchhoff stress tensor of X at t, $\underset{\sim}{b}(t)$ the value of the density of the body force of X at t, the operator Div is computed with respect to the material coordinates $\underset{\sim}{X}$, $\varrho_{\varkappa}$ is the mass density in the reference configuration $\underset{\sim}{\varkappa}$, the dot denotes the material differentiation with respect to time t, $\underset{\sim}{C}(t)$ is the value of the right Cauchy-Green deformation tensor in X at t, $\underset{\sim}{q}_{\varkappa}(t)$ is the value of the heat flux vector per unit surface in the reference configuration in X at t, ψ denotes the specific free energy per unit mass, η entropy, ϑ is the absolute temperature and r the heat supply per unit mass.

The equations $(2.1)_1$ - $(2.1)_2$ are called the Cauchy's laws of motion, Eq. $(2.1)_3$ represents the first local law of thermodynamics and the inequality $(2.1)_4$ represents the second law of thermodynamics and is called the Clausius-Duhem inequality.

The three values

$$g = (\underset{\sim}{C}(t), \vartheta(t), \nabla\vartheta(t)) \qquad 2.3$$

we shall call the local deformation temperature configuration of X at t.

A set of all possible local configuration of a particle X will be denoted by $\mathfrak{G}$ and will be called the configuration space.

Similarly the four values

$$s = (\psi(t), \eta(t), \underset{\sim}{T}(t), q_{\varkappa}(t)) \qquad 2.4$$

given in a particle X at time t we shall call the local response of X at t.

A set of all possible local responses of a particle X will be denoted by $\mathcal{S}$ and will be called the response space.

We shall consider processes in the configuration space $\mathfrak{S}$ and processes in the response space $\mathcal{S}$.

A process

$$P \equiv (\underset{\sim}{C}, \vartheta, \nabla\vartheta) : [0, d_P] \to \mathfrak{S} \qquad 2.5$$

will determine the change of the deformation-temperature configuration of a particle X in the interval of time $[0, d_P]$. A number d_P will be called the duration of the process P, and $P^i = P(0)$ and $P^f = P(d_P)$ the initial and final values of the process P, respectively.

A process

$$Z \equiv (\psi, \eta, \underset{\sim}{T}, q_{\varkappa}) : [0, d_Z] \to \mathcal{S} \qquad 2.6$$

will determine the change of the response of a particle X in the interval of time $[0, d_Z]$, i.e., the change of the free energy, the entropy, the Piola-Kirchhoff stress and the heat flux.

It is important to note that if the deformation-temperature configuration g and the response s of a particle X at time t are known and we have the function of motion $\underset{\sim}{\chi}$ then we can determine the value of the body force $b(t)$ from the first Cauchy's law of motion (2.1) and the value of the heat supply per unit mass and unit time $r(t)$ from the first local law of thermodynamics (2.1).

Let us denote by

$$\Pi \equiv \{ P \mid P : [0, d_P] \to \mathfrak{S} \} \qquad 2.7$$

a set of all deformation-temperature configuration processes, and by

$$\mathfrak{L} \equiv \{ Z \mid Z : [0, d_Z] \to \mathcal{S} \} \qquad 2.8$$

a set of all response processes for a particle X.

Definition 1. Every pair $(P, Z) \in \Pi \times \mathfrak{L}$ such that $\mathrm{Dom}\,P = \mathrm{Dom}\,Z$ and for every instant of time $t \in [0, d_P]$ the dissipation principle in the form of the Clausius-Duhem inequality $(2.1)_4$ is satisfied will be called a local thermodynamic process.

2.2. General principle of determinism

In a class of local thermodynamic processes we shall consider a subset which will be compatible with the intrinsic constitutive assumptions describing the internal physical constitution of a body $\mathcal{B}$, i.e. compatible with a material of a body $\mathcal{B}$. Such a subset of a local thermodynamic process space will be called admissible for the constitutive assumptions in question.

To discuss the general relation between processes $P \in \Pi$ and $z \in \mathfrak{L}$ which defines a material structure of a body $\mathcal{B}$, let us introduced a space $\mathcal{K}$ connected with configuration space $\mathcal{G}$ in such a way that elements of the space $\mathcal{K}$, which will be denoted by $k \in \mathcal{K}$, are the method of preparation of the corresponding configuration $g \in \mathcal{G}$. The space $\mathcal{K}$ will be called the method of preparation space*/.

A main objective of thermodynamics of continuous media is to predict the response of a particle X of a body $\mathcal{B}$ of which physical properties are known, at the end of a deformation-temperature process. We can give an answer to this question if and only if we have full information about particle X before the test, i.e. before a deformation-temperature process. This information which is needed for unique prediction of a future response of a particle X for every deformation-temperature process is called the method of preparation of the actual deformation-temperature configuration. In other words, the method of preparation should give the additional information required to define uniquely the intrinsic state of particle X of a body $\mathcal{B}$ during the local thermodynamic process.

It will be shown that a method of preparation of the deformation-temperature configuration of a particle X is needed to describe the internal dissipation of a material.

Having the method of preparation space $\mathcal{K}$ we can define the intrinsic state σ of a particle X at time t as a pair of the deformation-temperature configuration and the method of preparation

$$\sigma = (P(t), A(t)) = (g, k), \quad g \in \mathcal{G}, \quad k \in \mathcal{K}_g, \qquad 2.9$$

where by A we denote a process in the method of preparation space $\mathcal{K}$, i.e.

$$A : [0, d_P] \rightarrow \mathcal{K}. \qquad 2.10$$

A set of all intrinsic states of a particle X is called the intrinsic state space Σ.

*/ For a notion of the method of preparation see Ref.[9]. The precise definition of the method of preparation space for a pure mechanical process was first given in Ref.[16].Cf also comments presented in Ref.[4].

The notion of a method of preparation is connected with a general principle of determinism in mechanics of solids.

A principle of determinism can be stated as follows.

Postulate. Between an initial deformation-temperature configuration, its method of preparation, a deformation-temperature process beginning at this configuration and a response process of a particle X there exists a functional relationship.

This functional relationship will describe a unique material structure in a particle X of a body $\mathcal{B}$.

There exists a mapping

$$\mathcal{R} : (\Sigma \times \Pi)^* \to \mathcal{L} \tag{2.11}$$

where

$$(\Sigma \times \Pi)^* \equiv \{(\sigma, P) \in \Sigma \times \Pi \mid \sigma \in \{P^i\} \times \mathcal{K}_{P^i}\}. \tag{2.12}$$

The mapping $\mathcal{R}$ is called the constitutive mapping. So, we can write

$$z = \mathcal{R}(\sigma, P), \quad (\sigma, P) \in (\Sigma \times \Pi)^*. \tag{2.13}$$

Let us assume that a unique material structure is given. If we have the initial intrinsic state and the deformation-temperature process beginning at this intrinsic state we are interested then in the intrinsic state at the end of the process. The problem will be solved if the mapping between the intrinsic state at the end of the deformation-temperature process and the initial intrinsic state will be given.

Definition 2. It is said that the mapping

$$\hat{e} : (\Sigma \times \Pi)^* \to \Sigma \tag{2.14}$$

is the evolution function if for every pair $(\sigma, P) \in (\Sigma \times \Pi)^*$ the equation

$$\mathcal{R}(\hat{e}(\sigma, P), P^f_{(0)}) = [\mathcal{R}(\sigma, P)]^f \tag{2.15}$$

is satisfied, where $[\mathcal{R}(\sigma, P)]^f$ denotes the final value of the response process $z = \mathcal{R}(\sigma, P)$ and $P^f_{(0)}$ is the deformation-temperature process of duration zero.

In practical application it will be convenient to have mapping from the intrinsic state space Σ into the response space $\mathcal{S}$. So, it is useful to define a new mapping

$$\hat{S} : \Sigma \to \mathcal{S} \tag{2.16}$$

by the expression

$$\hat{S}(\sigma) = \mathcal{R}(\sigma, P^f_{(0)}). \tag{2.17}$$

The principle of determinism can be expressed by the relation

$$s = z(t) = \hat{S}(\sigma) = \hat{S}\left(\hat{e}\left(\sigma_0, P_{[0,t]}\right)\right) \quad 2.18$$

for every pair $(\sigma_0, P) \in (\Sigma \times \Pi)^*$.

A unique value of the response $s \in \mathcal{S}$ corresponds to every intrinsic state $\sigma \in \Sigma$.

The mapping $\hat{S}$ is called the response function.

The response function $\hat{S}$ represents the free energy response function $\hat{\Psi}$, the entropy response function $\hat{N}$, the stress response function $\hat{T}$ and the heat flux response function $\hat{Q}$, i.e.

$$\hat{S} \equiv \{\hat{\Psi}, \hat{N}, \hat{T}, \hat{Q}\}. \quad 2.19$$

Definition 3. A local thermodynamic process compatible with a unique material structure will be called an admissible process.

2.3. Internal state variable material structure

Let us assume that a class of deformation-temperature processes contains only piecewise continuously differentiable processes.

Proposition 1. The method of preparation space is a finite dimensional vector space $\mathcal{V}_n$, i.e.

$$\mathcal{K} \equiv \mathcal{V}_n \quad 2.20$$

the intrinsic state space Σ is the set

$$\Sigma \equiv \{(g, k) \mid g \in \mathcal{G}, \quad k \in \mathcal{V}_{n_g}\}.$$

Proposition 2. There exists a mapping

$$\hat{\alpha} : \Sigma \to \mathcal{V}_n \quad 2.21$$

such that for every $P \in \Pi$ and $k_0 \in \mathcal{V}_{nP^i}$ the initial-value problem

$$\dot{A}(t) = \hat{\alpha}(P(t), A(t)), \quad A(0) = k_0 \quad 2.22$$

has a unique solution $A : [0, d_p] \to \mathcal{V}_n$.

The evolution function $\hat{e}$ is of the form

$$\hat{e}(\sigma_0, P) = (P^f, \mathcal{F}(P, k_0)), \quad 2.23$$

where $\mathcal{F}$ denotes the solution functional of the initial-value problem (2.22).

The principle of determinism for the material structure with internal state variables is expressed by the constitutive equation

$$z(t) = \hat{S}(P(t), A(t)) \quad 2.24$$

By using the evolution function (2.23) the constitutive equation 2.24 can be written in the form

$$z(t) = \hat{S}\left(\hat{e}\left(\sigma_0, P_{[0,t]}\right)\right). \quad 2.25$$

2.4. Thermodynamic restrictions

According to the Definition 1 every local thermodynamic process has to satisfy the dissipation principle in the form of the Clausius-Duhem inequality (2.1).

By using the Definition 3 of an admissible process we may state now the main problem of the thermodynamics of materials: In an assigned class of processes and within an assigned class of response functions $\hat{S} \equiv \{\hat{\Psi}, \hat{N}, \hat{T}, \hat{Q}\}$ to determine those that satisfy the Clausius-Duhem inequality $(2.1)_4$.

Thus it can be said that the main problem of the thermodynamics of materials is to determine an admissible thermodynamic process.

It follows from the above assumptions that in each admissible local thermodynamic process the inequality $(2.1)_4$ requires the results /cf. Refs. [10,11]/

$$
\begin{aligned}
\partial_{\nabla\vartheta(t)}\hat{\Psi}(\cdot) &= 0, \\
\underset{\sim}{T}(t) &= 2\varrho_{\varkappa}\partial_{\underset{\sim}{C}(t)}\hat{\Psi}(\cdot), \\
\eta(t) &= -\partial_{\vartheta(t)}\hat{\Psi}(\cdot), \\
-\partial_{A(t)}\hat{\Psi}(\cdot)\cdot\dot{A}(t) &- \frac{1}{\varrho_{\varkappa}\vartheta(t)}\hat{Q}(\sigma)\cdot\nabla\vartheta(t) \geqslant 0
\end{aligned}
\qquad 2.26
$$

for every time $t \in [0, d_P]$.

Let us introduce the following notations

$$
\begin{aligned}
\hat{d}(\sigma) &= -\partial_{A(t)}\hat{\Psi}(\cdot)\cdot\dot{A}(t) - \frac{1}{\varrho_{\varkappa}\vartheta(t)}\hat{Q}(\sigma)\cdot\nabla\vartheta(t), \\
\hat{\iota}(\sigma) &= -\frac{1}{\vartheta(t)}\partial_{A(t)}\hat{\Psi}(\cdot)\cdot\dot{A}(t).
\end{aligned}
\qquad 2.27
$$

The mapping $\hat{d}: \Sigma \to \mathbb{R}^{+}$ /where $\mathbb{R}^{+}$ denotes the set of non--negative real numbers/ is called the general dissipation function and $\hat{d}(\sigma)$ denotes the value of the general dissipation function at the intrinsic state $\sigma \in \Sigma$.

The mapping $\hat{\iota}: \Sigma \to \mathbb{R}$ /where $\mathbb{R}$ denotes the set of real numbers/ is called the internal dissipation function, and $\hat{\iota}(\sigma)$ is its value at the intrinsic state $\sigma \in \Sigma$.

The inequality (2.26) is called the general dissipation inequality and using the Eqs. (2.27) it can be written in the form

$$
\hat{d}(\sigma) = \vartheta(t)\hat{\iota}(\sigma) - \frac{1}{\varrho_{\varkappa}\vartheta(t)}\hat{Q}(\sigma)\cdot\nabla\vartheta(t) \geqslant 0. \qquad 2.28
$$

The results (2.26) express the criterion of the selection of the response functions $\hat{S} \equiv \{\hat{\Psi}, \hat{N}, \hat{T}, \hat{Q}\}$.

3. MICRO-DAMAGE PROCESS

3.1. Fundamental internal state variables

We can return now to the notion of the method of preparation as such information which is required for the description of the internal dissipation of an inelastic material. The expression $(2.27)_2$ which defines the value of the internal dissipation at the intrinsic state σ shows that information given in the method of preparation by means of the internal state variables introduced, i.e. $A(t) = k$ and the evolution equation $\dot{A}(t) = \hat{\alpha}(\sigma)$ with $A(0) = k_0$, essentially determines the internal dissipation at this intrinsic state σ.

Fracture is a final stage of the inelastic flow process and is influenced by many cooperative phenomena such as localization of plastic deformations, internal imperfections generated by the micro-damage process /nucleation, growth and coalescence of microvoids/, strain rate sensitivity effect, thermal effects, etc.

Experimental observations of dynamic fracture [19-21] suggest that two cooperative phenomena, namely the visco-plastic flow process and the micro-damage process play the most important role in the proper explanation of the final stage of the dynamic deformation process.

To describe both phenomena simultaneously two groups of internal state variables are introduced

$$A(t) = (\alpha(t), \omega(t)) \tag{3.1}$$

and it is postulated

$$\begin{aligned} \alpha(t) &= \underset{\sim}{C}^P(t), \\ \omega(t) &= \xi(t), \end{aligned} \tag{3.2}$$

where $\underset{\sim}{C}^P(t)$ denotes the inelastic deformation tensor and $\xi(t)$ is the porosity.

It is our conjecture that the inelastic deformation tensor $\underset{\sim}{C}^P$ can describe viscoplastic flow process while the porosity parameter ξ can control micro-damage process.

For this interpretation of the internal state variables we have

$$\hat{i}(\sigma) = \frac{1}{\vartheta(t)} \left\{ tr\left[\partial_{\underset{\sim}{C}^P(t)} \hat{\Psi}(\cdot)\, \dot{\underset{\sim}{C}}^P(t)\right] + \partial_{\xi(t)} \hat{\Psi}(\cdot)\, \dot{\xi}(t) \right\}. \tag{3.3}$$

The result (3.3) suggests that the internal dissipation is splitting into two parts, the first is implied by the viscoplastic flow process and the second is generated by the micro-damage process.

3.2. Evolution equations

Let us focus our attention on the micro-damage process. During impact processes for ductile solids one observes nucleation, growth and coalescence of microvoids. These phenomena lead to very serious changes of porosity, [1-3,7,17-19].

In the case when the micro-damage process is influenced by thermomechanical coupling it is very difficult to describe changes of porosity. To overcome this difficulty let us introduce two internal state variables $N(X,t)$ and $R(X,t)$, where N is interpreted as the density of microvoids /the number of microvoids per unit volume/ and R as the average size of a microvoid.

It is noteworthy that these two internal state variables are suggested by experimental observations, by analysis of physical mechanisms as well as by previous theoretical descriptions of ductile and brittle fracture of metals /cf. Refs.[19-24]/.

The internal state variables N and R help to represent the following phenomena [19,20]:

/i/ Nucleation of microcracks or microvoids as a function of stress level, temperature and stress impulse duration.

/ii/ Growth of microcracks or microvoids during the dynamic process considered.

/iii/ A range of microcrack or microvoid sizes at all stages of fracture process.

/iv/ Coalescence of microcracks or microvoids to form fragments.

/v/ A transition from no damage through fracture to full separation.

We postulate the evolution equations for N and R in the form /cf. Ref.[15]/

$$\dot{N} = \dot{N}_0 \left\{ \exp \frac{[I_n - T_N(\vartheta)]m}{k\vartheta} - 1 \right\},$$

$$\dot{R} = \frac{1}{\bar{\eta}(\vartheta)} [I_g - T_{eq}(\vartheta)] R, \qquad 3.4$$

where $\dot{N}_0$ and m are the material constants, k is the Boltzmann constant, $\bar{\eta}(\vartheta)$ is temperature dependent viscosity coefficient, T_N is the threshold stress for nucleation, T_{eq} is the equilibrium stress, I_n and I_g are the stress intensity invariants for nucleation and growth, respectively, i.e.

$$I_n = a_1 J_1 + a_2 \sqrt{J_2'} + a_3 (J_3')^{\frac{1}{3}},$$
$$I_g = b_1 J_1 + b_2 \sqrt{J_2'} + b_3 (J_3')^{\frac{1}{3}}, \quad 3.5$$

a_i and b_i $(i = 1,2,3)$ are the material constants and by J_1 we denote the first invariant of the Piola-Kirchhoff stress tensor T, J_2' and J_3' are the second and third invariants of the stress deviator, respectively.

3.3. Description of porosity effect

To take advantage of the porosity parameter /the void volume fraction/ $\xi(X,t)$ in the constitutive modelling of damaged solids we postulate the fundamental equation of state /cf. Ref. [15]/

$$\xi = \Xi N R^3, \qquad \xi \in [0,1], \quad 3.6$$

where Ξ is the material constant.

The evolution of ξ during the dynamic process is determined by the differential equation

$$\dot{\xi} = \xi \left(\frac{\dot{N}}{N} + 3 \frac{\dot{R}}{R} \right) \quad \text{with} \quad \xi(0) = \xi_0 . \quad 3.7$$

The equation (3.7) together with (3.4) leads to the evolution equation for the porosity parameter ξ as follows

$$\dot{\xi} = h(\xi, \vartheta) \left\{ \exp \frac{m[I_n - T_N(\vartheta)]}{k\vartheta} - 1 \right\} + g(\xi,\vartheta)\left[I_g - T_{eq}(\vartheta)\right], \quad 3.8$$

where

$$h(\xi,\vartheta) = \xi \frac{\dot{N}_0}{N},$$
$$g(\xi,\vartheta) = \frac{3\xi}{\eta(\vartheta)} . \quad 3.9$$

The result (3.8) shows that there exists equivalence between descriptions within the framework of the internal state variable theory by using the porosity parameter ξ or N and R, i.e. $\xi \Leftrightarrow \{N, R\}$.

4. VISCO-PLASTIC FLOW PROCESS

4.1. Yield function and softening effects

We postulate the yield function for damaged solids in the form

$$f(\cdot) = J_2' \left[1 - (n_1 + \xi n_2) \frac{J_3'^2}{J_2'^3} + (n_3 + \xi n_4) \frac{J_1^2}{J_2'} \right], \qquad 4.1$$

where n_i $(i=1,2,3,4)$ denote the material constants.

For ductile solids during dynamic process we observe significant work-hardening effect generated by large plastic deformations, pronounced thermal softening due to increase of temperature and very important softening induced by the micro-damage process.

To describe all these effects we propose the work-hardening-softening function in the form

$$\varkappa = \varkappa_0^2 \left(1 + h_0 \bar{\epsilon}^{P \frac{1}{\mu_0}}\right)\left(1 - h_1 \vartheta^{\frac{1}{\mu_1}}\right)\left(1 - h_2 \xi^{\frac{1}{\mu_2}}\right) \qquad 4.2$$

where $\varkappa_0$ is the yield constant, h_0 denotes the work-hardening coefficient, h_1 the thermal softening coefficient, h_2 the micro-damage softening coefficient and μ_0, μ_1 and μ_2 are the material constants, and $\bar{\epsilon}^P$ denotes the equivalent inelastic deformation, i.e.

$$\bar{\epsilon}^P = \left(tr\, \underset{\sim}{C}^{P^2}\right)^{\frac{1}{2}}. \qquad 4.3$$

4.2. Evolution equation

It is postulated that the evolution equation for the inelastic deformation tensor C^P has the form

$$\dot{\underset{\sim}{C}}^P(t) = \frac{\gamma}{\varphi} \left\langle \Phi \left[\frac{f(\cdot)}{\varkappa} - 1 \right] \right\rangle \partial_{\underset{\sim}{T}(t)} f(\cdot), \qquad 4.4$$

where γ denotes the viscosity coefficient, φ is the control function and is assumed to depend on $(I_2/I_2^s) - 1$, where I_2 is the second invariant of the rate of deformation tensor $\dot{\underset{\sim}{C}}$, I_2^s is its static value, and Φ denotes the viscoplastic overstress function. The symbol $\langle [\ ] \rangle$ is understood according to the definition

$$\langle [\ ] \rangle = \begin{cases} 0 & \text{if } f \leqslant \varkappa, \\ [\ ] & \text{if } f > 0. \end{cases} \qquad 4.5$$

The internal dissipation (3.3) can be written in the form as follows

$$\hat{i}(\sigma) = -\frac{1}{\vartheta(t)}\left\{\frac{\gamma}{\varphi}\left\langle\Phi\left[\frac{f(\cdot)}{\varkappa}-1\right]\right\rangle \operatorname{tr}\left[\partial_{\underset{\sim}{C}^P(t)}\hat{\Psi}(\cdot)\,\partial_{\underset{\sim}{T}(t)}f(\cdot)\right]\right.$$

$$+\partial_{\xi(t)}\hat{\Psi}(\cdot)\Big(h(\xi,\vartheta)\Big[\exp\frac{m\left[I_n-I_N(\vartheta)\right]}{k\vartheta}-1\Big]$$

$$\left.+\,g(\xi,\vartheta)\left[I_g-T_{eq}(\vartheta)\right]\Big)\right\}. \tag{4.6}$$

5. PHENOMENON OF DYNAMIC FRACTURE

5.1. Fracture criterion

As it has been suggested by experimental observations the criterion of dynamic fracture can be expressed by means of the porosity parameter ξ.

It is postulated that during the dynamic inelastic flow process the fracture phenomenon occurs when

$$\xi = \xi^F \quad \Longrightarrow \quad \bar{\epsilon}^P = \bar{\epsilon}^P_F\left(\dot{\underset{\sim}{C}}^P, \vartheta\right), \tag{5.1}$$

what leads to the condition

$$\varkappa = \hat{\varkappa}\left(\bar{\epsilon}^P, \vartheta, \xi\right)\Big|_{\substack{\xi=\xi^F\\ \bar{\epsilon}^P=\bar{\epsilon}^P_F}} = 0. \tag{5.2}$$

The condition (5.2) expresses the fact that fracture means a catastrphe or intrinsic failure when the material loses its stress carrying capacity. It can be said that at $\xi=\xi^F$ one observes full separation of the material /cf. Refs.[12-15]/.

It seems useful to take advantage of the equivalence between descriptions by using ξ and $\{N, R\}$. The criterion of dynamic fracture (5.1) and the fundamental equation of state (3.6) lead to the condition

$$R^F = \left(\frac{\xi^F}{\Xi N}\right)^{\frac{1}{3}} \tag{5.3}$$

which defines the critical average size of microvoid at fracture.

5.2. Determination of the softening constant

The condition (5.2) together with the assumption (4.2) yields the relation

$$\varkappa_0^2\left(1+h_0\bar{\epsilon}_F^{P\,\frac{1}{\mu_0}}\right)\left(1-h_1\vartheta^{\frac{1}{\mu_1}}\right)\left(1-h_2\xi^{F\,\frac{1}{\mu_2}}\right)=0 \tag{5.4}$$

As a result of the relation (5.4) we have the micro-damage softening coefficient

$$h_2 = \left(\xi^F\right)^{-\frac{1}{\mu_2}}. \quad 5.5$$

The remaining material constants h_0 and h_1 can be determined by using experimental investigations of work-hardening and thermal softening effects.

6. FINAL COMMENTS

To complete the thermomechanical elastic-viscoplastic theory of voided solids we need to have particular representation for the free energy response function $\hat{\Psi}$. We can proceed here in the same way as it has been proposed by D.Krajcinovic [6].

When the free energy function is given then the internal dissipation in particle X at any stage of the dynamic process is determined by the expression (4.6).

The result (4.6) is of great importance to us. It shows that the internal dissipation is govern by two cooperative phenomena, namely by the visco-plastic flow process and by the micro-damage process generated by nucleation and growth mechanisms. Of course, there exists coupling between these processes as well as between thermal and mechanical effects.

The expression for the internal dissipation (4.6) is valid until the final stage of the dynamic process. At final stage of the process when fracture occurs in the expression for the internal dissipation (4.6) arise some singularities. This is perhaps the most important feature of the proposed theory.

REFERENCES

1 T.B.Cox and J.R.Low,Jr., Metall.Trans.,5,1457-1470,1974.
2 L.Davison, A.L.Stevens and M.E.Kipp, J.Mech.Phys.Solids, 25, 11-28,1977.
3 A.G.Evans, Acta Metall., 28,1155-1163,1980.
4 K.Frischmuth, W.Kosiński, P.Perzyna, Arch.Mech.,38,1986 /in print/.
5 J.N.Johnson, J.Appl.Phys., 52,2812-2825,1981.
6 D.Krajcinovic, J.Appl.Mech.,52,829-834,1985.
7 G.Leroy, J.D.Embury, G.Edward and M.F.Ashby, Acta Metall., 29,1509-1522,1981.
8 W.Noll, Arch.Rat.Mech.Anal.,48,1-50,1972.
9 P.Perzyna, Arch.Mech.,23,845-850,1971.
10 P.Perzyna, Arch.Mech., 27,759-806,1975.
11 P.Perzyna, CISM Courses and Lectures No 262, Springer 1980, pp.95-220.
12 P.Perzyna, ASME J.Eng.Materials and Technology, 106,410-419, 1984.
13 P.Perzyna, Arch.Mech.,37,485-501,1985.

14 P.Perzyna, Int.J.Solids Structures, 1986 /in print/.
15 P.Perzyna, Arch.Mech.,38,1986 /in print/.
16 P.Perzyna and W.Kosiński, Bull.Acad.Polon.Sci.,Ser.Sci.Techn., 21,647-662,1973.
17 R.Ray and M.F.Ashby, Acta Metall.,23,653-666,1975.
18 R.O.Ritchie, J.F.Knott and J.R.Rice, J.Mech.Phys.Solids,21, 395-410,1973.
19 L.Seaman, D.R.Curran and D.A.Shockey, J.Appl.Phys.,47,4814-4826,1976.
20 L.Seaman, D.R.Curran and W.J.Murri, J.Appl.Mech.,107,593-600,1985.
21 D.A.Shockey, L.Seamann and D.R.Curran, Int.J.Fracture, 27, 145-157,1985.
22 P.J.Wray, J.Appl.Phys.,40,4018-4029,1969.
23 P.J.Wray, Metal.Trans.,6A,1379-1391,1975.
24 P.J.Wray, Metal.Trans.,7A,1621-1627,1976.

Thermomechanical Couplings in Solids
H.D. Bui and Q.S. Nguyen (Editors)
Elsevier Science Publishers B.V. (North-Holland)

NON-ISOTHERMAL GENERALIZED PLASTICITY

Jacob LUBLINER

Department of Civil Engineering
University of California, Berkeley
Berkeley, California 94720, USA

A model of rate-independent inelasticity of continua, called generalized plasticity, is formulated for non-isothermal behavior. In this model the local state is determined by the deformation, the temperature, and by a family of internal variables which are governed by rate equation such that the derivative of each internal variable is a function of the state and of the deformation and temperature rates, the dependence on these rates being homogeneous of the first degree. A maximum-dissipation postulate that implies a normality principle is likewise extended to non-isothermal processes, with the result that the plastic entropy production is necessarily positive for materials whose elastic range contracts with increasing temperature.

1. INTRODUCTION

It is well known that the *yield criterion* is the central concept of the mathematical theory of plasticity, as applied originally to metals and more recently to soils, rocks, concrete and the like. Historically, the reason for this focus is probably that the primary material to which the theory was applied by its founders was mild steel, which is indeed characterized by a well-defined yield stress. Very few other materials exhibit this characteristic, however; in most cases one has to settle for a more or less arbitrarily defined "offset" yield stress. Another feature of the classical theory is that, if a material specimen is loaded to a plastic state and unloaded elastically along some path, then reloading along the same path must proceed elastically. This feature, too, is shared by very few real materials. In spite of its widespread use, therefore, the theory appears quite unsatisfying when applied to materials other than steel, and even for steel it agrees poorly with empirical data for cyclic loading.

For a number of years I have put some effort into finding a rate-independent plasticity model that does not depend on (but does not exclude) the existence of a yield criterion. My inspiration has come from three sources: the work of Pipkin and Rivlin [1], which shifted the emphasis from the yield criterion to the elastic range as the central concept of plasticity; the definition due to Perzyna [2] of plasticity as behavior that is different in loading and unloading; and the theory of Eisenberg and Phillips [3] that divorced the concept of loading from that of the yield criterion. These led me to formulate what I at first [4] called a "simple" theory of plasticity, and what I have since come to call *generalized* plasticity, because it includes classical or conventional plasticity as a special case; see also [5-7].

In the present study, I attempt to lay a foundation for a non-isothermal formulation of generalized plasticity. In regard to many aspects of the phenomenology and thermodynamics of plasticity, I have benefited greatly from a long friendship with the late Jean Mandel.

2. VISCOPLASTICITY AND CLASSICAL PLASTICITY

One way of motivating classical rate-independent plasticity is as the limit of slow processes in viscoplastic materials. Here "viscoplastic" is used in the sense of Prager [8] and most other workers in the field, describing a material characterized by *yield hypersurface* in the state space. The state space may be taken as the space whose points are given by $(\mathbf{C}, \theta, \mathbf{q})$, where $\mathbf{C}$ is the right Cauchy-Green deformation tensor, θ is the absolute temperature, and $\mathbf{q}$ is the vector whose components are the internal variables which are coupled to inelastic deformation. The yield hypersurface is defined by $\{(\mathbf{C}, \theta, \mathbf{q}) \mid f(\mathbf{C}, \theta, \mathbf{q}) = 0\}$, f being a continuous real-valued function of $(\mathbf{C}, \theta, \mathbf{q})$, with $f<0$ defining the elastic (or rigid) domain. The internal-variable vector $\mathbf{q}$ is assumed to be governed by rate equations of the form

$$\dot{\mathbf{q}} = \mathbf{h}(\mathbf{C}, \theta, \mathbf{q}), \tag{1}$$

where, for those internal variables q_i that determine inelastic deformation, the corresponding h_i vanish for $f<0$. (In general, some irreversible processes coupled to deformation may go on even when the material behaves elastically; strain-aging is an example.) In Equation (1) and elsewhere, the superposed dot denotes, as usual, Eulerian differentiation with respect to time.

A number of writers (e.g. Bodner [9]) have used the term "viscoplastic" to describe nonlinear rate-dependent behavior of metals without any yield criterion. In terms of constitutive theory, however, it is difficult to see how viscoplasticity in this sense differs from nonlinear viscoelasticity.

If the multiplicative decomposition $\mathbf{F}=\mathbf{F}_e\mathbf{F}_p$ of the "deformation-gradient" tensor $\mathbf{F}$ is used, with $\mathbf{F}_e$ the "elastic" and $\mathbf{F}_p$ the "plastic" factor, then $\mathbf{q}$ may be defined in turn as the pair $(\mathbf{F}_p, \boldsymbol{\alpha})$, where $\boldsymbol{\alpha}$ is the vector whose components are the "structural" internal variables. It is then usually assumed that the thermomechanical state of a material neighborhood is determined by the quadruple $(\mathbf{C}, \theta, \mathbf{F}_p, \boldsymbol{\alpha})$ only through the combination $(\mathbf{C}_e, \theta, \boldsymbol{\alpha})$, where $\mathbf{C}_e = \mathbf{F}_e^T\mathbf{F}_e = \mathbf{F}_p^{T-1}\mathbf{C}\mathbf{F}_p^{-1}$. In the analysis to follow, however, this restriction is not important.

The viscoplastic rate equations for $(\mathbf{F}_p, \boldsymbol{\alpha})$ reduce in the limit of slow processes to the flow equations

$$\mathbf{L}_p = \lambda\mathbf{N}<\mathring{f}>, \quad \dot{\boldsymbol{\alpha}} = \lambda\mathbf{r}<\mathring{f}>, \tag{2}$$

where $\lambda = 0$ when $f<0$, $\lambda>0$ when $f = 0$, and states with $f>0$ are inadmissible. Here $\mathbf{L}_p = \mathbf{F}_p^{-1}\dot{\mathbf{F}}_p$, $\mathring{f} = (\partial f/\partial \mathbf{C}):\dot{\mathbf{C}}+(\partial f/\partial\theta)\dot{\theta}$, and $<\cdot>$ denotes the *Macauley bracket* defined by

$$<x> = \begin{cases} x, & x\geq 0, \\ 0, & x<0. \end{cases}$$

$\mathbf{N}$ and $\mathbf{r}$ are functions of $(\mathbf{C}, \theta, \mathbf{q})$, whose values are respectively second-rank tensors and r-dimensional vectors, r being the dimensionality of $\boldsymbol{\alpha}$.

It should be noted that it is more conventional to take a stress (for example, the second Piola-Kirchhoff stress $\mathbf{S}$) rather than the deformation ($\mathbf{C}$) as a state variable for f. It has been recognized over the past decade, however, that the "strain-space formulation" of plasticity offers certain advantages in general theoretical discussions, in particular the non-necessity of differentiating between hardening and softening materials ([10], [11]).

3. GENERALIZED PLASTICITY

In the present study rate-independent plasticity is postulated *a priori*, rather than as the limit of viscoplasticity. With this approach it is possible to define plasticity without assuming the existence of a yield hypersurface. Borrowing a notation introduced by Coleman [12], we shall write $\mathbf{\Lambda}$ for the deformation-temperature pair $(\mathbf{C}, \theta)$ (Coleman used it for the pair $(\mathbf{F}, \theta)$). Accordingly, points in the space S of local states will be denoted by $(\mathbf{\Lambda}, \mathbf{q})$. The set of *realizable* states will be denoted $\mathscr{S}$, and we shall define $\mathscr{S}(\mathbf{q}) \overset{def}{=} \{\mathbf{\Lambda} \mid (\mathbf{\Lambda}, \mathbf{q}) \in \mathscr{S}\}$. The rate equations will be assumed as

$$\dot{\mathbf{q}} = \mathbf{h}(\mathbf{\Lambda}, \mathbf{q}, \dot{\mathbf{\Lambda}}), \tag{3}$$

which is more general than Equation (1), but necessary for rate-independence [13]. Rate-independence means that Equation (3) is invariant under a replacement of t by $\phi(t)$ where $\phi(\cdot)$ is any monotonically increasing, continuously differentiable function. It is a standard result that a necessary and sufficient condition for rate-independence is that $\mathbf{h}(\mathbf{\Lambda}, \mathbf{q}, \cdot)$ be homogeneous of the first degree, that is, for any positive number c,

$$\mathbf{h}(\mathbf{\Lambda}, \mathbf{q}, c\,\dot{\mathbf{\Lambda}}) = c\,\mathbf{h}(\mathbf{\Lambda}, \mathbf{q}, \dot{\mathbf{\Lambda}}). \tag{4}$$

The assumption of rate-independence is, of course, an idealization; it must be restricted to process times that are either long or short in comparison to the characteristic internal times (such as relaxation times) of the material. The range of such process times, and hence the range of validity of the assumption of rate-independence, varies from material to material: it is fairly narrow in steels, so that different parameters must be used when rate-independent theory is applied to processes whose rates have different orders of magnitude, and viscoplastic theory must be used for processes spanning a wide range of rates. Precipitation-hardened aluminum alloys, on the other hand, seem to be rate-independent over a range of rates from those found in quasi-static tests to those found in dynamic tests (e.g. Kolsky and Douch [14]).

If a material neighborhood of a rate-independent inelastic body is treated as a small thermodynamic system, with a local process defined as a mapping $t \to (\mathbf{\Lambda}(t), \mathbf{q}(t))$ of a time interval (say $[t_0, t_1]$)—the process will then be said to go *from* the state $(\mathbf{\Lambda}(t_0), \mathbf{q}(t_0))$ *to* the state $(\mathbf{\Lambda}(t_1), \mathbf{q}(t_1))$—then every state is an equilibrium state, Consequently every process is quasi-static but not, in general, reversible.

A function $t \to \phi(t)$ such as was mentioned above in defining rate-independence produces a process $t \to (\mathbf{\Lambda}(\phi(t)), \mathbf{q}(\phi(t)))$. The relation between this process and the original process $(\mathbf{\Lambda}(\cdot), \mathbf{q}(\cdot))$ is an equivalence relation on the set of processes (to prove this, we need only note that any such function ϕ is invertible); the corresponding equivalence classes will be called *paths*.

3.1. ELASTIC PROCESS, RANGE, STATE, DOMAIN.

Rate-independent plasticity is closely tied to the concept of *elastic range,* first formalized by Pipkin and Rivlin [1] and later expanded by Owen [15, 16]. These studies used the framework of the theory of materials with memory without reference to internal variables. In this section the theory will be reformulated with the use of internal variables.

A local process will be called *elastic* if $\mathbf{q}(\cdot)$ is a constant function. If the Clausius-Planck inequality is assumed to represent the second law of thermodynamics [17], then in an elastic process (with heat conduction neglected) the internal entropy production vanishes; the process is still not necessarily reversible, but may be called *quasi-reversible* [18]. It is obvious that a process is elastic if and only if all processes having the same path are elastic, and hence we can speak of elastic and inelastic paths.

As a result of rate-independence—embodied in the constraint (2) on the rate equation (1)—a process with $\mathbf{\Lambda}(\cdot)$ constant is also elastic. In a rate-dependent (e.g. viscoplastic) body, such a process would be a relaxation process and hence inelastic, unless $\mathbf{\Lambda}$ is in the elastic region. Here, such a process (whose path consists of one point) will be called *trivial.*

The *elastic range* of a state $(\mathbf{\Lambda}, \mathbf{q}) \in \mathscr{S}$ will be defined as

$$\mathscr{E}(\mathbf{\Lambda}, \mathbf{q}) = \{\mathbf{\Lambda}^* \mid \text{there exists an elastic process from } (\mathbf{\Lambda}, \mathbf{q}) \text{ to } (\mathbf{\Lambda}^*, \mathbf{q})\}.$$

Since a trivial process is elastic, we have $\mathbf{\Lambda} \in \mathscr{E}(\mathbf{\Lambda}, \mathbf{q})$ at every state $(\mathbf{\Lambda}, \mathbf{q})$, and therefore every state has a non-empty elastic range. The set $\mathscr{E}(\mathbf{\Lambda}, \mathbf{q}) \times \{\mathbf{q}\}$, which may be naturally identified with $\mathscr{E}(\mathbf{\Lambda}, \mathbf{q})$, is obviously the union of the ranges of all elastic processes (paths) from $(\mathbf{\Lambda}, \mathbf{q})$, and is therefore path-connected, and hence connected. The elastic range of every state will furthermore be assumed closed relative to $\mathscr{A}(\mathbf{q})$ (in what follows, all topological notions regarding subsets of $\mathscr{A}(\mathbf{q})$ will refer to the relative topology induced by $\mathscr{A}(\mathbf{q})$); a similar assumption was made by Pipkin and Rivlin [1], while in the work of Owen [15, 16] the elastic range, differently defined, is an open set.

A state $(\mathbf{\Lambda}, \mathbf{q}) \in \mathscr{S}$ will be called *elastic* if $\mathbf{\Lambda} \in \mathring{\mathscr{E}}(\mathbf{\Lambda}, \mathbf{q})$, and *plastic* if $\mathbf{\Lambda} \in \partial\mathscr{E}(\mathbf{\Lambda}, \mathbf{q})$. The set of all elastic states in $\mathscr{S}$ will be called the *elastic domain* and denoted $\mathscr{D}$; *the projection of* $\mathscr{D}$ *into* $\mathscr{A}(\mathbf{q})$—that is, the set $\{\mathbf{\Lambda} \mid (\mathbf{\Lambda}, \mathbf{q}) \in \mathscr{D}\}$ for a given $\mathbf{q}$—will be denoted $\mathscr{D}(\mathbf{q})$, and also called the elastic domain (at $\mathbf{q}$). It is easy to show that $\mathscr{D}(\mathbf{q}) \subset \mathscr{E}(\mathbf{\Lambda}, \mathbf{q})$ for every $\mathbf{\Lambda} \in \mathscr{A}(\mathbf{q})$.

Note that a process whose range is entirely in the elastic domain is reversible and not merely quasi-reversible.

Since it is the rate equations that determine elastic and inelastic processes, it is the nature of these equations that ultimately decides the structure of the elastic range and, therefore, of the elastic domain. In general it is quite difficult to deduce the structure of the elastic range and domain from the rate equations, and this will not be done here; instead, the sets will be assumed to be sufficiently regular in some sense, and possibly deduce some necessary properties of $\mathbf{h}$. For example, if $\mathbf{\Lambda} \in \mathscr{D}(\mathbf{q})$ then $\mathbf{h}(\mathbf{\Lambda}, \mathbf{q}, \dot{\mathbf{\Lambda}}) = 0$, since every $(\mathbf{\Lambda}^*, \mathbf{q})$ with $\mathbf{\Lambda}^*$ in a sufficiently small neighborhood of $\mathbf{\Lambda}$ is then attainable elastically, hence $\dot{\mathbf{q}} = 0$ in any possible process through $(\mathbf{\Lambda}, \mathbf{q})$. If the elastic range of a state $(\mathbf{\Lambda}, \mathbf{q})$ has a non-empty interior, then its boundary will be assumed to be a piecewise smooth surface in $\mathscr{A}(\mathbf{q})$, called the *loading surface* in deformation-temperature space. Thus, if $(\mathbf{\Lambda}, \mathbf{q})$ is a plastic state and $\partial\mathscr{E}(\mathbf{\Lambda}, \mathbf{q})$ is locally smooth at $\mathbf{\Lambda}$ (such a state will be called a *regular* plastic state), with a normal $\mathbf{\Omega}$ pointing away from $\mathring{\mathscr{E}}(\mathbf{\Lambda}, \mathbf{q})$, then $\mathbf{h}(\mathbf{\Lambda}, \mathbf{q}, \cdot)$ must have the property that $\mathbf{h}(\mathbf{\Lambda}, \mathbf{q}, \dot{\mathbf{\Lambda}}) = 0$ if and only if $\mathbf{\Omega} \cdot \dot{\mathbf{\Lambda}} \leq 0$. (*Proof*: For a sufficiently small positive number h, $\mathbf{\Lambda} + h\dot{\mathbf{\Lambda}}$ is in the interior or exterior of $\mathscr{E}(\mathbf{\Lambda}, \mathbf{q})$, respectively, as $\mathbf{\Omega} \cdot \dot{\mathbf{\Lambda}} < 0$ or $\mathbf{\Omega} \cdot \dot{\mathbf{\Lambda}} > 0$. In the former case $\mathbf{\Lambda} + h\dot{\mathbf{\Lambda}}$ is attainable elastically, hence

$\dot{\mathbf{q}} = 0$; in the latter case $\mathbf{\Lambda} + h\dot{\mathbf{\Lambda}}$ is not attainable elastically, and in particular not by a straight-line path, hence $\dot{\mathbf{q}} \neq 0$.) The simplest form of $\mathbf{h}$ having this property and obeying the homogeneity constraint (4) is

$$\mathbf{h}(\mathbf{\Lambda}, \mathbf{q}, \dot{\mathbf{\Lambda}}) = \mathbf{g}(\mathbf{\Lambda}, \mathbf{q})\langle \mathbf{\Omega} \cdot \dot{\mathbf{\Lambda}} \rangle. \tag{5}$$

It can be readily seen that Equation (2), representing rate-independent plasticity as the limit of viscoplasticity in slow processes, is a special case of (5), with g identified with $\lambda(\mathbf{N}, \mathbf{r})$ and $\mathbf{\Omega}$ with $(\partial f/\partial \mathbf{C}, \partial f/\partial \theta)$. More particularly, if $\partial \mathscr{E}(\mathbf{\Lambda}, \mathbf{q})$ is locally given by $\{\mathbf{\Lambda}^* \mid f(\mathbf{\Lambda}^*; \mathbf{\Lambda}, \mathbf{q}) = \text{constant}\}$, where f is a function defining the loading surface in the same way that it defines the yield hypersurface in Section 2, then Equation (5) is of the same form as Equation (2).

The boundary of $\mathscr{D}(\mathbf{q})$ will likewise be assumed to be a piecewise smooth surface, called the *yield surface* (also in deformation-temperature space). This usage of the terms "loading surface" and "yield surface" is consistent with that introduced by Eisenberg and Phillips [3] in their theory of plasticity with non-coincident yield and loading surfaces (in stress space), though not with later usage by Phillips and other collaborators. The theory in which the surfaces do coincide—that is, in which $\partial \mathscr{E}(\mathbf{\Lambda}, \mathbf{q}) = \partial \mathscr{D}(\mathbf{q})$ for every $\mathbf{\Lambda} \in \mathscr{A}(\mathbf{q})$ at a given $\mathbf{q}$—coincides with what is called conventional or classical plasticity theory; the more general model studied here is what I call generalized plasticity.

3.2. PROPERTIES OF THE ELASTIC RANGE AND DOMAIN

We shall state and prove some useful propositions, using elementary set theory and topology. When we have a mapping defined on $\mathscr{A}(\mathbf{q})$ whose values are *sets*, such as $\mathscr{E}(\cdot, \mathbf{q})$ or $\partial \mathscr{E}(\cdot, \mathbf{q})$, we may wish to specify that such a mapping is continuous, and for that purpose we need a topology on a space of sets, in particular, on the space of non-empty closed subsets of $\mathscr{A}(\mathbf{q})$; this space is often denoted $2^{\mathscr{A}(\mathbf{q})}$. Fortunately, such a topology exists; it is defined by a metric introduced by Hausdorff and named for him. If d denotes a metric on $\mathscr{A}(\mathbf{q})$ (e.g. the natural finite-dimensional metric), then the corresponding Hausdorff metric $\bar{d}$ on $2^{\mathscr{A}(\mathbf{q})}$ is defined by

$$\bar{d}(\mathscr{A}, \mathscr{B}) = \max\{\sup_{\mathbf{\Lambda} \in \mathscr{B}} d(\mathbf{\Lambda}, \mathscr{A}), \sup_{\mathbf{\Lambda} \in \mathscr{A}} d(\mathbf{\Lambda}, \mathscr{B})\},$$

for any $\mathscr{A}, \mathscr{B} \in 2^{\mathscr{A}(\mathbf{q})}$, where $d(\mathbf{\Lambda}, \mathscr{A}) \overset{def}{=} \inf_{\mathbf{\Lambda}^* \in \mathscr{A}} d(\mathbf{\Lambda}, \mathbf{\Lambda}^*)$.

We now prove the following:

Proposition 1. If $\mathbf{\Lambda} \in \mathscr{D}(\mathbf{q})$ and if the mapping $\partial \mathscr{E}(\cdot, \mathbf{q}): \mathscr{A}(\mathbf{q}) \to 2^{\mathscr{A}(\mathbf{q})}$ is continuous at $\mathbf{\Lambda}$, then $\mathbf{\Lambda}$ is an interior point of $\mathscr{D}(\mathbf{q})$.

Proof. Assume the contrary, that is, $\mathbf{\Lambda} \in \partial \mathscr{D}(\mathbf{q})$. Since $\mathbf{\Lambda} \in \mathscr{D}(\mathbf{q})$ by hypothesis, we have $\mathbf{\Lambda} \in \mathring{\mathscr{E}}(\mathbf{\Lambda}, \mathbf{q})$ by definition and therefore, if we define $\mathscr{B} = \partial \mathscr{E}(\mathbf{\Lambda}, \mathbf{q})$ to simplify the notation, we have $r = d(\mathbf{\Lambda}, \mathscr{B}) > 0$. But if $\mathbf{\Lambda} \in \partial \mathscr{D}(\mathbf{q})$ then any neighborhood of $\mathbf{\Lambda}$ (in $\mathscr{A}(\mathbf{q})$) contains a $\mathbf{\Lambda}^* \notin \mathscr{D}(\mathbf{q})$; that is, $(\mathbf{\Lambda}^*, \mathbf{q})$ is a plastic state, so that by definition $\mathbf{\Lambda}^* \in \mathscr{B}^* \overset{def}{=} \partial \mathscr{E}(\mathbf{\Lambda}^*, \mathbf{q})$. By the triangle inequality, $d(\mathbf{\Lambda}, \mathscr{B}) \leq d(\mathbf{\Lambda}^*, \mathscr{B}) + d(\mathbf{\Lambda}, \mathbf{\Lambda}^*)$; hence $d(\mathbf{\Lambda}^*, \mathscr{B}) \geq r - d(\mathbf{\Lambda}, \mathbf{\Lambda}^*)$. Furthermore, $\bar{d}(\mathscr{B}, \mathscr{B}^*) \geq d(\mathbf{\Lambda}^*, RB)$; consequently $\lim_{d(\mathbf{\Lambda}, \mathbf{\Lambda}^*) \to 0} \bar{d}(\mathscr{B}, \mathrm{B}^*) \geq r$. Thus the assumption that $\partial \mathscr{E}(\cdot, \mathbf{q})$ is continuous is violated.

Corollary. If, at each $\mathbf{q}$, $\partial\mathscr{E}(\cdot,\mathbf{q})$ is continuous in $\mathscr{A}(\mathbf{q})$, then $\mathscr{D}(\mathbf{q})$ is open in $\mathscr{A}(\mathbf{q})$.

Before stating the next proposition, we need the following purely topological result:

Lemma. Let X be a topological space, A a non-empty open set in X, and B a closed set in X whose interior is connected. If $A \subset B$ and $\partial A \subset \partial B$, then $A = \mathring{B}$.

Proof. Clearly $A \subset \mathring{B}$, because, if there were a point $x \in A \cap \partial B$, then any neighborhood of x would contain points that do not belong to B and therefore do not belong to A, so that A would not be an open set. Now $X-(\mathring{B}-A) = (X-\mathring{B}) \cup A = (X-\bar{B}) \cup \partial B \cup A$; but $\partial B = \partial B \cup \partial A$ by hypothesis, so that $X-(\mathring{B}-A) = (X-\bar{B}) \cup \partial B \cup \partial A \cup A = (X-\mathring{B}) \cup \bar{A}$, the union of two closed sets and hence a closed set. Therefore $\mathring{B}-A$ is open. If, however, $\mathring{B}-A$ is not empty, then $\mathring{B}$ is the union of two disjoint non-empty open sets and is therefore not connected. Consequently $A = \mathring{B}$.

Proposition 2. If $\mathscr{D}(\mathbf{q})$ is path-connected, then for every $\mathbf{\Lambda} \in \overline{\mathscr{D}}(\mathbf{q})$, $\mathring{\mathscr{E}}(\mathbf{\Lambda},\mathbf{q}) = \mathscr{D}(\mathbf{q})$.

Proof. For every $\mathbf{\Lambda}, \mathbf{\Lambda}^* \in \overline{\mathscr{D}}(\mathbf{q})$ there is, by the hypothesis of path-connectedness, a curve lying entirely (except possibly the end points) in $\mathscr{D}(\mathbf{q})$ and joining $\mathbf{\Lambda}$ and $\mathbf{\Lambda}^*$. Consequently $\mathbf{\Lambda}^* \in \mathscr{E}(\mathbf{\Lambda},\mathbf{q})$ and $\mathbf{\Lambda} \in \mathscr{E}(\mathbf{\Lambda}^*,\mathbf{q})$, so that $\mathscr{E}(\mathbf{\Lambda},\mathbf{q})$ is the same for all $\mathbf{\Lambda} \in \overline{\mathscr{D}}(\mathbf{q})$—call it $\mathscr{E}_0(\mathbf{q})$. Clearly $\overline{\mathscr{D}}(\mathbf{q}) \subset \mathscr{E}_0(\mathbf{q})$. Now consider $\mathbf{\Lambda} \in \partial\mathscr{D}(\mathbf{q})$; then $(\mathbf{\Lambda},\mathbf{q})$ is a plastic state, hence $\mathbf{\Lambda} \in \partial\mathscr{E}(\mathbf{\Lambda},\mathbf{q}) = \partial\mathscr{E}_0(\mathbf{q})$, so that $\partial\mathscr{D}(\mathbf{q}) \subset \partial\mathscr{E}_0(\mathbf{q})$. We deduce (with the help of the preceding lemma) that $\mathscr{D}(\mathbf{q}) = \mathring{\mathscr{E}}_0(\mathbf{q})$.

Corollary. If $\mathscr{E}(\mathbf{\Lambda},\mathbf{q})$ equals the closure of its interior, then for every $\mathbf{\Lambda} \in \mathscr{D}(\mathbf{q})$ we have $\mathscr{E}(\mathbf{\Lambda},\mathbf{q}) = \overline{\mathscr{D}}(\mathbf{q})$ and $\partial\mathscr{E}(\mathbf{\Lambda},\mathbf{q}) = \partial\mathscr{D}(\mathbf{q})$. The last result may be phrased as follows [3]: *the yield surface is the initial loading surface.*

The following two-sided inclusion result is useful for the transition to classical plasticity.

Proposition 3. At every possible $\mathbf{q}$,

$$\bigcap_{\mathbf{\Lambda} \in \mathscr{A}(\mathbf{q})} \mathring{\mathscr{E}}(\mathbf{\Lambda},\mathbf{q}) \subset \mathscr{D}(\mathbf{q}) \subset \bigcup_{\mathbf{\Lambda} \in \mathscr{A}(\mathbf{q})} \mathring{\mathscr{E}}(\mathbf{\Lambda},\mathbf{q}).$$

Proof. The intersection is the set $\{\mathbf{\Lambda} \mid \mathbf{\Lambda} \in \mathring{\mathscr{E}}(\mathbf{\Lambda}^*,\mathbf{q})$ for *all* $\mathbf{\Lambda}^* \in \mathscr{A}(\mathbf{q})\}$; the union is the set $\{\mathbf{\Lambda} \mid \mathbf{\Lambda} \in \mathring{\mathscr{E}}(\mathbf{\Lambda}^*,\mathbf{q})$ for *any* $\mathbf{\Lambda}^* \in \mathscr{A}(\mathbf{q})\}$; and $\mathscr{D}(\mathbf{q}) = \{\mathbf{\Lambda} \mid \mathbf{\Lambda} \in \mathring{\mathscr{E}}(\mathbf{\Lambda}^*,\mathbf{q})$ for a particular $\mathbf{\Lambda}^*$, namely $\mathbf{\Lambda}^* = \mathbf{\Lambda}\}$.

3.3. CLASSICAL PLASTICITY AS A SPECIAL CASE.

It follows immediately from Proposition 3 that if $\mathscr{E}(\mathbf{\Lambda},\mathbf{q})$ is independent of $\mathbf{\Lambda}$, then its interior equals $\mathscr{D}(\mathbf{q})$ at every $\mathbf{q}$. Moreover, since $\mathscr{E}(\mathbf{\Lambda},\mathbf{q})$ is connected, it contains no isolated points and is therefore equal to the closure of its interior; thus $\partial\mathscr{E}(\mathbf{\Lambda},\mathbf{q}) = \partial\mathscr{D}(\mathbf{q})$—precisely the definition of classical plasticity given above. It further follows that $\partial\mathscr{D}(\mathbf{q})$ is the set of all plastic states:

$$\partial\mathscr{D}(\mathbf{q}) = \{\mathbf{\Lambda} \mid \mathbf{\Lambda} \in \partial\mathscr{D}(\mathbf{q})\} = \{\mathbf{\Lambda} \mid \mathbf{\Lambda} \in \partial\mathscr{E}(\mathbf{\Lambda},\mathbf{q})\}.$$

Another criterion for classical plasticity is that *the elastic range is unaffected by elastic deformations* [1], that is, $\mathscr{E}(\mathbf{\Lambda}^*,\mathbf{q}) = \mathscr{E}(\mathbf{\Lambda},\mathbf{q})$ if $\mathbf{\Lambda}^* \in \mathscr{E}(\mathbf{\Lambda},\mathbf{q})$. It is easy to show that this criterion is implied by the one used here: If $\mathscr{E}(\mathbf{\Lambda},\mathbf{q}) = \mathscr{E}(\mathbf{q})$ for all $\mathbf{\Lambda} \in \mathscr{A}(\mathbf{q})$, then in particular $\mathscr{E}(\mathbf{\Lambda}^*,\mathbf{q}) = \mathscr{E}(\mathbf{q})$ for all $\mathbf{\Lambda}^* \in \mathscr{E}(\mathbf{\Lambda},\mathbf{q}) = \mathscr{E}(\mathbf{q})$.

4. THERMODYNAMICS

4.1. APPLICABILITY OF COLEMAN'S METHOD.

Since his pioneering work with Noll [19], Coleman and various collaborators have applied the second law of thermodynamics, embodied in the Clausius-Duhem inequality, in order to find restrictions on constitutive equations of materials, and in particular, when deformation and temperature are the independent variables, to show how the stress and the entropy density can be derived from the density of the Helmholtz free energy. The application to materials with internal variables is due to Coleman and Gurtin [20]. It is instructive to try to apply Coleman's method to the rate-independent inelastic continuum discussed here, with the rate equations taking in particular the form (5). Assuming no coupling between heat conduction and plastic deformation, we may replace the Clausius-Duhem inequality by the Clausius-Planck inequality [17], which may be written in the form

$$\boldsymbol{\Sigma}\cdot\dot{\boldsymbol{\Lambda}} - \dot{\psi} = (\boldsymbol{\Sigma} - \partial\psi/\partial\boldsymbol{\Lambda})\cdot\dot{\boldsymbol{\Lambda}} + \sigma(\boldsymbol{\Lambda},\mathbf{q},\dot{\mathbf{q}}) \geq 0;$$

here ψ is the specific Helmholtz free energy (per unit mass), $\boldsymbol{\Sigma}$ is the pair $(\frac{1}{2\rho_0}\mathbf{S}, -\eta)$, ρ_0 being the mass density in the reference state, $\mathbf{S}$ the second Piola-Kirchhoff stress and η the specific entropy, and

$$\sigma(\boldsymbol{\Lambda},\mathbf{q},\dot{\mathbf{q}}) \stackrel{def}{=} -\partial\psi(\boldsymbol{\Lambda},\mathbf{q})/\partial\mathbf{q}\cdot\dot{\mathbf{q}} \tag{6}$$

is the *specific dissipation*. With the rate equation (5) inserted, the inequality becomes

$$(\boldsymbol{\Sigma} - \partial\psi/\partial\boldsymbol{\Lambda})\cdot\dot{\boldsymbol{\Lambda}} + \sigma(\boldsymbol{\Lambda},\mathbf{q},\mathbf{g}(\boldsymbol{\Lambda},\mathbf{q}))<\boldsymbol{\Omega}\cdot\dot{\boldsymbol{\Lambda}}> \geq 0;$$

but this is satisfied at all $(\boldsymbol{\Lambda},\mathbf{q},\dot{\boldsymbol{\Lambda}})$ if

$$\boldsymbol{\Sigma} = \partial\psi/\partial\boldsymbol{\Lambda} - k\,\sigma(\boldsymbol{\Lambda},\mathbf{q},\mathbf{g}(\boldsymbol{\Lambda},\mathbf{q}))\boldsymbol{\Omega} \tag{7}$$

for *any* real number $k\in[0,1]$ (see [21]). Thus Coleman's method does not lead to a unique dependence of $\boldsymbol{\Sigma}$ on the state without a further assumption, namely, that an elastic process—one in which $\boldsymbol{\Omega}\cdot\dot{\boldsymbol{\Lambda}}\leq 0$ by hypothesis—is quasi-reversible, that is, in such a process the Clausius-Planck inequality holds as an equality; this assumption leads to $k = 0$ as the only possible value and hence yields the classical thermodynamic relation $\boldsymbol{\Sigma} = \partial\psi/\partial\boldsymbol{\Lambda}$, that is,

$$\mathbf{S} = 2\rho_0\partial\psi/\partial\mathbf{C}, \quad \eta = -\partial\psi/\partial\theta.$$

4.2. ALTERNATIVE STRESS TENSORS

In order to discuss yield surfaces and flow rules in large-deformation plasticity, Mandel [22] introduced the stress tensor $\mathbf{P}$ defined by

$$\mathbf{P} = J\mathbf{F}_e^T\mathbf{T}\mathbf{F}_e^{T-1},$$

where $\mathbf{T}$ is the Cauchy stress tensor, and $J = \det\mathbf{F}$. The significance of the stress tensor $\mathbf{P}$ is that, if the dependence of the free energy on the state is of the commonly assumed form

$$\psi(\mathbf{C},\theta,\mathbf{F}_p,\boldsymbol{\alpha}) = \bar{\psi}(\mathbf{C}_e,\theta,\boldsymbol{\alpha}), \tag{8}$$

then the dissipation σ takes the form

$$\sigma = \frac{1}{\rho_0}\mathbf{P}:\mathbf{L}_p - (\partial\bar{\psi}/\partial\boldsymbol{\alpha})\cdot\dot{\boldsymbol{\alpha}},$$

where ρ_0 is the mass density in the reference state. In other words, $\mathbf{P}$ is conjugate to the plastic distortion rate $\mathbf{L}_p$. Note that $\mathbf{P}$ is not in general symmetric.

A symmetric tensor $\mathbf{S}_e$ may be defined by

$$\mathbf{S}_e = \mathbf{F}_p\mathbf{S}\mathbf{F}_p^T;$$

then

$$\mathbf{S}_e = 2\rho_0\partial\bar{\psi}/\partial\mathbf{C}_e, \tag{9}$$

and

$$\mathbf{P} = \mathbf{C}_e\mathbf{S}_e. \tag{10}$$

Equation (9) establishes, at given $(\theta, \mathbf{q})$, a one-to-one correspondence between $\mathbf{C}_e$ and $\mathbf{S}_e$ and therefore between $\mathbf{C}_e$ and $\mathbf{P}$. Consequently Equation (10) imposes a nonlinear constraint on $\mathbf{P}$ (namely, that $\mathbf{P}\mathbf{C}_e$ or $\mathbf{C}_e^{-1}\mathbf{P}$ must be symmetric) which restricts $\mathbf{P}$ to a six-dimensional manifold (not a subspace, as would be the case with symmetric tensors) of the nine-dimensional space of second-rank tensors.

If the elastic strains are small (so that $\mathbf{C}_e$ differs only slightly from the identity tensor $\mathbf{1}$) then $\mathbf{P}$ differs only slightly from $\mathbf{S}_e$, and the manifold to which $\mathbf{P}$ is confined is "almost" flat.

5. NORMALITY AND MAXIMUM-DISSIPATION POSTULATE.

Many of the most useful results of plasticity theory, such as uniqueness theorems, variational and extremum principles, and computational algorithms, rely—as is well known—upon an assumption according to which the plastic deformation-rate "vector" is "normal" to the loading surface. A flow rule (that is, an equation giving the "direction" of the "vector") that obeys the normality principle is also said, in classical plasticity theory, to be *associated* with the yield criterion. An examination of the normality principle and its relation to a "maximum-dissipation postulate", in both isothermal and non-isothermal classical plasticity (for large deformations), and finally in generalized plasticity, forms the remainder of this study.

5.1. NORMALITY IN ISOTHERMAL CLASSICAL PLASTICITY

Let the yield hypersurface be defined, as in Section 2, by $f(\boldsymbol{\Lambda}, \mathbf{q}) = 0$, with $\mathbf{q} = (\mathbf{F}_p, \boldsymbol{\alpha})$. More particularly, let us assume, with Mandel [22], that f is further given by

$$f(\boldsymbol{\Lambda}, \mathbf{q}) = \bar{f}(\mathbf{P}(\boldsymbol{\Lambda}, \mathbf{q}), \theta, \mathbf{q}).$$

As proposed by Mandel, normality holds (in a rather strong sense) if the stress-space yield function $\bar{f}$ serves as a *plastic potential* for $\mathbf{L}_p$, that is, if

$$\mathbf{L}_p = \mu\partial\bar{f}/\partial\mathbf{P}, \quad \mu \geq 0, \tag{11}$$

at a point where the yield surface is locally smooth, while at a corner of the yield surface $\mathbf{L}_p$ is in the cone formed by the normal vectors meeting there.

Mandel [23] regarded the flow rule embodied in Equation (11) as a nine-dimensional one, in the sense that it generates not only the symmetric part of $\mathbf{L}_p$ (the "plastic strain rate") but also its antisymmetric part (the "plastic spin"). In this he was followed by Dafalias [24] and Loret [25]. It is clear, however that the right-hand side of (11) must lie in the local cotangent space of the $\dot{\mathbf{P}}$ manifold, which is a six-dimensional vector space, though not identical with the space of symmetric second-rank tensors.

For hardening materials, the "loading" factor $<\overset{\circ}{f}>$ in the rate equations (2) may be replaced by $<\frac{\partial \bar{f}}{\partial \theta}\dot{\theta} + (\partial \bar{f}/\partial \mathbf{P}):\dot{\mathbf{P}}>$, and consequently in isothermal processes ($\dot{\theta}=0$) Equation (11) takes the more specific form

$$\mathbf{L}_p = \lambda<(\partial \bar{f}/\partial \mathbf{P}):\dot{\mathbf{P}}>\partial \bar{f}/\partial \mathbf{P}. \tag{12}$$

It follows that

$$\dot{\mathbf{P}}:\mathbf{L}_p \geq 0, \tag{13}$$

a result known as **Drucker's inequality** (in Mandel's version for finite deformations). Equation (11) and inequality (13) in turn imply (a) the *convexity* of the yield surface, and (b) the **principle of maximum plastic dissipation** in the form

$$(\mathbf{P} - \mathbf{P}^*):\mathbf{L}_p \geq 0 \text{ for any } \mathbf{P}^* \text{ such that } \bar{f}(\mathbf{P}^*, \theta, \mathbf{q}) \leq 0, \tag{14}$$

$\mathbf{P}$ and $\mathbf{L}_p$ being the actual values. Equations (13) and (14) together may be said to constitute the finite-deformation version of **Drucker's postulate**.

Conversely, if the elastic region is convex then an inequality such as (14) implies normality, though not necessarily the apparently nine-dimensional one given by (11). For infinitesimal deformations, the relations can be elegantly formulated by means of convex analysis [26]. A derivation for finite deformations will now be given [27].

Unlike Drucker's inequality (13), the maximum-dissipation inequality (14) is valid not only for hardening or perfectly plastic but also for softening materials. Inequality (14) is closely related to the plasticity postulate of Il'iushin [28]. At this point we return to using the deformation-temperature pair $\mathbf{\Lambda}$ directly as a state variable. With the free energy having the specific form (8) and with σ as defined by Equation (6), it can be shown that (14) can be written as

$$\sigma(\mathbf{\Lambda}, \mathbf{q}, \dot{\mathbf{q}}) - \sigma(\mathbf{\Lambda}^*, \mathbf{q}, \dot{\mathbf{q}}) \geq 0 \tag{15}$$

for any $\mathbf{\Lambda}^* \overset{def}{=} (\mathbf{C}^*, \theta)$ such that $f(\mathbf{\Lambda}^*, \mathbf{q}) \leq 0$. Suppose, in particular, that the process from $(\mathbf{\Lambda}^*, \mathbf{q})$ to $(\mathbf{\Lambda}, \mathbf{q})$ is elastic, and that its continuation from $(\mathbf{\Lambda}, \mathbf{q})$ is inelastic, the process rate at $(\mathbf{\Lambda}, \mathbf{q})$ being $(\dot{\mathbf{\Lambda}}, \dot{\mathbf{q}})$, with $\dot{\theta} = 0$. If $\mathbf{\Lambda}^*$ is close to $\mathbf{\Lambda}$, then $\mathbf{\Lambda}^* = \mathbf{\Lambda} - h\dot{\mathbf{\Lambda}} + o(h)$ for some small positive h. For the left-hand side of (15) we therefore have

$$\sigma(\mathbf{\Lambda}, \mathbf{q}, \dot{\mathbf{q}}) - \sigma(\mathbf{\Lambda}^*, \mathbf{q}, \dot{\mathbf{q}}) = h(\partial\sigma/\partial\mathbf{C}):\dot{\mathbf{C}} + o(h);$$

hence

$$(\partial\sigma/\partial\mathbf{C}):\dot{\mathbf{C}} \geq 0. \tag{16}$$

Furthermore

$$2\rho_0\partial\sigma/\partial C = -2\rho_0(\partial^2\psi/\partial C\partial q)\cdot\dot{q} = -(\partial S/\partial q)\cdot\dot{q} = (\partial S/\partial C):\dot{C} - \dot{S} = \Phi\cdot\dot{E}^p,$$

the last equality defining the *Lagrangian plastic strain rate* $\dot{E}^p \overset{def}{=} \dot{E}|_{\dot{S}=0,\dot{\theta}=0} = \dot{E} - \Phi^{-1}\cdot\dot{S}$, with $\Phi = \partial S/\partial E = 2\,\partial S/\partial C$ being the "Lagrangian" elastic tangent stiffness tensor. $\dot{E}^p$ is essentially the plastic strain rate used by Rice [29], and is related to L_p by the somewhat complicated relation

$$\dot{E}^p = F_p^T[(C_e L_p)^S + 2\Phi_e^{-1}\cdot(L_p S_e)^S]F_p,$$

where $\Phi_e = 2\partial S_e/\partial C_e$ is the elastic tangent stiffness referred to the intermediate configuration, and is related to Φ by

$$\Phi_e\cdot A = F_p[\Phi\cdot(F_p^T A\, F_p)]F_p^T,$$

where A is any symmetric second-rank tensor. Inequality (16) may consequently be written as

$$\dot{E}\cdot\Phi\cdot\dot{E}^p \geq 0. \tag{17}$$

Inequality (17) may be interpreted as follows: ignoring temperature as a variable (that is, identifying Λ with C), we may regard $\mathscr{A}(q)$ as a six-dimensional Riemannian manifold, with $\Phi(C, q)$ as the metric tensor at C. $\dot{E} = \frac{1}{2}\dot{C}$ and $\dot{E}^p$ are then tangent vectors to $\mathscr{A}(q)$ at C, and the left-hand side of (17) is their inner product. The inequality says that this inner product is never negative, or that the two vectors never oppose each other—that is, *the plastic strain rate never opposes the total strain rate*, an assertion that may be interpreted as implying *stability under strain (displacement) control.* Drucker's inequality, often referred to as a stability postulate, implies only stability under stress (load) control, and it is invalid whenever stress decreases with strain, i.e. when the material is a softening one.

The rate equation (1), with h given by (5), implies that $\dot{q}\neq 0$ only if $\Omega\cdot\dot{\Lambda} > 0$, that is (in an isothermal process), only if $\partial f/\partial C):\dot{C} > 0$. This last inequality is compatible with (17) only if

$$\dot{E}^p = \mu\Phi^{-1}\cdot(\partial f/\partial C), \quad \mu \geq 0, \tag{18}$$

a result that may be interpreted as a normality principle in deformation space in terms of the aforementioned treatment of $\mathscr{A}(q)$ as a Riemannian manifold with Φ as metric tensor, since $\partial f/\partial C$ is then a covector and Φ^{-1} is the reciprocal metric tensor. Moreover, whenever stress (i.e. S, S_e or P) can be used as a state variable in place of deformation, Equation (18) is equivalent to normality in stress space, but in a weaker sense than proposed by Mandel.

Suppose, first off, that the yield surface is given in S space by $\hat{f}(S, \theta, q) = 0$. With $S = S(C, \theta, q)$, we have

$$f(C, \theta, q) = \hat{f}(S(C, \theta, q), \theta, q),$$

so that $\partial f/\partial C = \partial\hat{f}/\partial S{:}\partial S/\partial C = \Phi\cdot\partial\hat{f}/\partial S$, since $\Phi = 4\rho_0\partial^2\psi/\partial C\partial C$ is symmetric. Consequently (18) is equivalent to

$$\dot{E}^p = \mu\partial\hat{f}/\partial S, \quad \mu \geq 0. \tag{19}$$

It can be shown [27] that (19) is equivalent to

$$\mathbf{L}_p = \mu \partial \bar{f} / \partial \mathbf{P} + \mathbf{L}'_p, \tag{20}$$

where $\mathbf{L}'_p$ may be any tensor belonging to the three-dimensional subspace defined by $\boldsymbol{\Delta} \cdot \mathbf{L}'_p = 0$. In other words, the normality principle derived from inequality (14) does not determine $\mathbf{L}_p$ but only its projection into the six-dimensional complement of this subspace, which is precisely the cotangent space of the six-dimensional manifold to which all values of $\mathbf{P}$ belong.

5.2. NORMALITY IN NON-ISOTHERMAL CLASSICAL PLASTICITY

A normality principle for non-isothermal classical plasticity will be derived by assuming that the validity of Inequality (15) is not limited to $\boldsymbol{\Lambda}^*$ given by $(\mathbf{C}^*, \theta)$ but may be extended to $\boldsymbol{\Lambda}^* = (\mathbf{C}^*, \theta^*)$, with $\theta^* \neq \theta$ in general, and that $\dot{\boldsymbol{\Lambda}} = (\dot{\mathbf{C}}, \dot{\theta})$, with $\dot{\theta} \neq 0$. As a result, (16) is replaced by

$$(\partial \sigma / \partial \mathbf{C}) : \dot{\mathbf{C}} + (\partial \sigma / \partial \theta) \dot{\theta} \geq 0. \tag{21}$$

But

$$\partial \sigma / \partial \theta = -(\partial^2 \psi / \partial \theta \partial \mathbf{q}) \cdot \dot{\mathbf{q}} = (\partial \eta / \partial \mathbf{q}) \cdot \dot{\mathbf{q}} \overset{def}{=} \dot{\eta}^p,$$

$\dot{\eta}^p$ being, by definition, the plastic entropy production rate per unit mass. Inequality (21) is therefore equivalent to

$$\dot{\mathbf{E}} \cdot \boldsymbol{\Phi} \cdot \dot{\mathbf{E}}^p + \rho_0 \dot{\theta} \dot{\eta}^p \geq 0. \tag{22}$$

The deformation-space normality principle (18), and its stress-space corollaries (19) and (20), are therefore supplemented by

$$\dot{\eta}^p = \frac{1}{\rho_0} \mu \partial f / \partial \theta. \tag{23}$$

Now $\partial f / \partial \theta$ is given by $\partial \bar{f} / \partial \mathbf{S}_e : \partial \mathbf{S}_e / \partial \theta + \partial \bar{f} / \partial \theta$. The first term represents thermoelastic coupling, and the second the temperature-dependence of the yield stress. In a typical elastic-plastic material, however, the first term is likely to be much less significant than the second, because the dependence of the yield criterion on the stress is usually through the stress deviator, while thermoelastic coupling is usually confined to the spherical stress. Consequently it can be argued that, as a rule, $\partial f / \partial \theta > 0$. This result may be interpreted as the *contraction of the elastic range in strain space with increasing temperature.*

Independently it can be argued that the plastic entropy production rate is positive, since an increase in dislocation density corresponds to an increase in their configurational entropy, as well as that of other defects [30]. (This result is independent of the second law of thermodynamics.)

Another argument for the positiveness of $\dot{\eta}^p$ is furnished by the Bauschinger effect. Suppose that the specific Helmholtz free energy $\psi(\boldsymbol{\Lambda}, \mathbf{q})$ is not only given by Equation (8) but has the even more specific form

$$\psi(\mathbf{C}, \theta, \mathbf{F}_p, \boldsymbol{\alpha}) = \psi_e(\mathbf{C}_e, \theta) + \epsilon_p(\boldsymbol{\alpha}) - \theta \eta_p(\boldsymbol{\alpha}), \tag{24}$$

where ϵ_p represents stored plastic energy and η_p plastic entropy. Then $\dot{\eta}^p = (\partial\eta_p/\partial\boldsymbol{\alpha})\cdot\dot{\boldsymbol{\alpha}} = \dot{\eta}_p$, and the Clausius-Planck inequality takes the form

$$\frac{1}{\rho_0}\mathbf{P}{:}\mathbf{L}_p - \dot{\epsilon}_p + \theta\dot{\eta}_p \geq 0.$$

When the Bauschinger effect occurs, then after a certain amount of plastic deformation the yield surface in stress space no longer encloses the origin, and the first term on the left-hand side may become negative. Since a change in the stored plastic energy, if it is significant, is usually positive, satisfaction of the second law of thermodynamics requires a sufficiently large positive value of $\dot{\eta}^p$.

The preceding arguments form, at least qualitatively, a basis for the extension of the maximum-dissipation postulate, and hence of the normality principle, to non-isothermal processes.

5.3. NORMALITY IN GENERALIZED PLASTICITY

As first pointed out by Eisenberg and Phillips [3], in generalized plasticity, with the yield and loading surfaces distinct, if a normality principle is defined then it must be with respect to the latter surface. Thus, let $\partial\mathscr{E}(\mathbf{\Lambda},\mathbf{q})$ be locally given by $\{\mathbf{\Lambda}^* \mid f(\mathbf{\Lambda}^*;\mathbf{\Lambda},\mathbf{q}) = \text{constant}\}$, with f now defining the *loading* hypersurface. Note that f is now defined on the set $\{(\mathbf{\Lambda}^*;\mathbf{\Lambda},\mathbf{q}) \mid \mathbf{\Lambda}^* \in \mathscr{A}(\mathbf{q}), (\mathbf{\Lambda},\mathbf{q}) \in \mathscr{S}\}$. Equations (18) and (23) will therefore represent non-isothermal normality in generalized plasticity if $\partial f/\partial\mathbf{C}$ in the former equation and $\partial f/\partial\theta$ in the latter are replaced respectively by $(\partial f/\partial\mathbf{C}^*)\,|_{\mathbf{C}^*=\mathbf{C}}$ and $(\partial f/\partial\theta^*)\,|_{\theta^*=\theta}$. These results are valid if the maximum-dissipation postulate is taken as a *local* one [7], namely that *for any plastic state* $(\mathbf{\Lambda},\mathbf{q})$ *there exists a neighborhood N of* $\mathbf{\Lambda}$ *(in* $\mathscr{A}(\mathbf{q})$*) such that (15) holds for every* $\mathbf{\Lambda}^* \in N \cap \mathscr{E}(\mathbf{\Lambda},\mathbf{q})$.

The condition

$$\left.\frac{\partial f}{\partial\theta^*}\right|_{\theta^*=\theta} > 0$$

means that an outward normal corresponds to increasing temperature, and can likewise be interpreted as the contraction of the elastic range (in the generalized-plasticity sense) with increasing temperature. It appears reasonable that this property is not limited to those materials that are described by the classical plasticity model. Such a supposition must, however, in the end be supported by experimental evidence.

REFERENCES

[1] Pipkin, A.C., and Rivlin, R.S., *Z. angew. Math. Phys.* **16** (1965) 313.
[2] Perzyna, P., On thermodynamic foundations of viscoplasticity, in: Lindholm. U.S. (ed.), *Mechanical Behavior of Materials Under Dynamic Loads* (Springer-Verlag, Wien-New York, 1968) pp. 61-76.
[3] Eisenberg, M.A., and Phillips, A., *Acta Mech.* **11** (1971) 247.
[4] Lubliner, J., *Int. J. Solids & Structures* **10** (1974) 313.
[5] Lubliner, J., *Int. J. Solids & Structures* **11** (1975) 1011.
[6] Lubliner, J., *Int. J. Solids & Structures* **16** (1980) 709.

[7] Lubliner, J., *Acta Mech.* **52** (1984) 225.
[8] Prager, W., *Introduction to Mechanics of Continua* (Ginn & Co., Boston, 1961).
[9] Bodner, S.R., Constitutive equations for dynamic material behavior, in: Lindholm, U.S. (ed.), *Mechanical Behavior of Materials Under Dynamic Loads* (Springer-Verlag, Wien-New York, 1968).
[10] Nguyen, Q.S., and Bui, H.D., *J. de Méc.* **13** (1974) 321.
[11] Naghdi, P.M., and Trapp, J.A., *Int. J. Engrg. Sci.* **13** (1975) 785.
[12] Coleman, B.D., *Arch. Rational Mech. Anal.* **17** (1964) 1.
[13] Lubliner, J., *Int. J. Math. & Math. Sci.* **7** (1984) 409.
[14] Kolsky, H., and Douch, L.S., *J. Mech. Phys. Solids* **10** (1962) 195.
[15] Owen, D.R., *Arch. Rational Mech. Anal.* **31** (1968) 91.
[16] Owen, D.R., *Arch. Rational Mech. Anal.* **37** (1970) 85.
[17] Paglietti, A., *Ann. Inst. H. Poincaré* **27A** (1977) 85.
[18] Fosdick, R.L., and Serrin, J., *Arch. Rational Mech. Anal.* **57** (1975) 97.
[19] Coleman, B.D., and Noll, W., *Arch. Rational Mech. Anal.* **13** (1963) 167.
[20] Coleman, B.D., and Gurtin, M.E., *J. Chemical Phys.* **47** (1967) 597.
[21] Lubliner, J., *Int. J. Non-Linear Mech.* **7** (1972) 237.
[22] Mandel, J., *Plasticité Classique et Viscoplasticité* (Springer-Verlag, Wien-New York, 1972).
[23] Mandel, J., *Int. J. Solids & Structures* **9** (1973) 725.
[24] Dafalias, Y.F., *J. Appl. Mech.* **105** (1983) 561.
[25] Loret, B., *Mech. of Matls.* **2** (1983) 287.
[26] Moreau, J. J., Application of convex analysis to the treatment of elastoplastic systems, in: Germain, P., and Nayroles, B. (eds.), *Applications of Methods of Functional Analysis to Problems in Mechanics* (Springer-Verlag, Wien-New York, 1976).
[27] Lubliner, J., *Mech. of Matls.* **5** (1986) 29.
[28] Il'iushin, A.A., *Prik. Mat. Mekh.* **25** (1961) 503.
[29] Rice, J.R., *J. Mech. Phys. Solids* **19** (1971) 433.
[30] Cottrell, A.H., *Dislocations and Plastic Flow in Crystals* (Clarendon Press, Oxford, 1956).

Thermomechanical Couplings in Solids
H.D. Bui and Q.S. Nguyen (Editors)
Elsevier Science Publishers B.V. (North-Holland)

SOURCE OF HEAT IN AN ELASTOVISCOPLASTIC MEDIUM

Pierre BEREST, Gilles ROUSSET

Laboratoire de Mécanique des Solides, Ecole Polytechnique,
91128 Palaiseau Cedex, France

An infinite elastoviscoplastic medium (Tresca's yield function) is loaded with a spherically symmetrical time dependent temperature distribution. The thermal expansion induced by changes in temperature can lead to the generation of viscoplastic zones. It is shown that relatively simple formulas can be used to calculate the boundaries of the viscoplastic zones for any temperature distribution. The case of a plastic medium, previously studied, is included in the present paper.

1. INTRODUCTION

1.1- Initial state :

We consider an infinite medium, isotropic and homogeneous in its thermal and mechanical properties. Initially, this medium is in a natural state : strains and stresses are equal to zero, and the temperature is constant and can be taken to be zero. The medium is then subjected to a time-dependent temperature distribution.

1.2- Medium behavior :

The medium has elastoviscoplastic behavior with Tresca's criterion and the associated flow rule. Mechanical parameters are temperature-independent ; the effect of the temperature is due to isotropic thermal expansion proportional to the variation in temperature. Only the quasi-static problem and small displacements will be considered. Under the previous hypothesis, the constitutive equations can be written as follows :

$$E\,\dot{\varepsilon} = (1+\nu)\,\dot{\sigma} - \nu\,(\mathrm{tr}\,\dot{\sigma})\,1 + E\,\alpha\,\dot{T}\,1 + E\,\dot{\lambda}\,\partial_\sigma F$$

$$F = \sigma_1 - \sigma_2 - 2\,C \quad \text{(the principal stresses are : } \sigma_1 > \sigma_3 > \sigma_2\text{)}$$

$$\dot{\lambda} = 0 \quad \text{if} \quad F \leqslant 0$$

$$\dot{\lambda} = \frac{1}{\eta}\,F \quad \text{if} \quad F \geqslant 0 \quad .$$

With the following nomenclature :

C cohesion of the material
E Young's modulus
F yield function (Tresca)
α coefficient of thermal expansion
ε strain tensor
ν Poisson's ratio
σ stress tensor
1 unit tensor

1.3- The case of plastic behavior :

The case of plastic behavior has been previously studied by Bérest [1]. In a certain sense, this case can be obtained as a limit of the more general visco-plastic case when $\eta = 0$. Then

$$\dot{\lambda} > 0 \quad \text{if and only if} \quad F = 0 \quad \text{and} \quad \dot{\sigma}\,\partial_\sigma F > 0 \quad .$$

1.4- Temperature distribution :

We will consider in the following the case of a spherically symmetrical temperature distribution of the form

$$T = T(r,t) \quad .$$

Such unidimensional problems have already been considered in the litterature, when plastic behavior is considered [3-5]. More recently Gamer [6] considered the case of a centrally heated disk, assuming plane stresses and linear hardening of a plastic material.

2. STATEMENT OF THE PROBLEM

2.1- Spherical symmetry :

Since we have spherical symmetry, spherical coordinates r, θ, φ will be used, σ_r, σ_θ and σ_φ are the principal stresses ; moreover, $\sigma_\theta = \sigma_\varphi$. The displacement u(r,t) is purely radial ; therefore :

$$\varepsilon_r = \partial_r u \qquad \text{and} \qquad \varepsilon_\theta = \varepsilon_\varphi = \frac{u}{r} \quad .$$

Since two principal stresses are equal, two yield functions must be used :

$$F_\varphi = \omega\,(\sigma_\varphi - \sigma_r) - 2\,C \quad ,$$

$$F_\theta = \omega\,(\sigma_\theta - \sigma_r) - 2\,C \quad ,$$

where $\omega = \pm 1$, depending on the order of the principal stresses. The total visco-plastic strain rate is, in this particular case, the sum of the two associated viscoplastic strain rates :

$$\frac{1}{2\eta}\,\partial_\sigma \quad (F_\theta^2 + F_\varphi^2) \quad .$$

However, due to spherical symmetry, the constitutive relationship reduces to :

$$E\,\partial_r v = \dot{\sigma}_r - 2\nu\dot{\sigma}_\varphi + E\alpha\dot{T} - \omega E\dot{\lambda} \tag{1}$$

$$E\frac{v}{r} = (1-\nu)\,\dot{\sigma}_\varphi - \nu\,\dot{\sigma}_r + E\,\alpha\,\dot{T} + \frac{1}{2}\,\omega\,E\,\dot{\lambda} \tag{2}$$

where $v = \dot{u}$, $\dot{\lambda} = 2\,\dot{\lambda}_\theta = 2\,\dot{\lambda}_\varphi = \frac{2}{\eta} < \omega\,(\sigma_\varphi - \sigma_r) - 2\,C >$, $\omega = \text{sgn}\,(\sigma_\varphi - \sigma_r)$.

2.2- Basic equations of the problem :

Since at t = 0 the state is natural, Eqs. *(1)* and *(2)* can be integrated with respect to time between t = 0 and any other instant t. If we take :

$$\varepsilon^{vp}(r,t) = \int_0^t \omega\,(r,\tau)\,\dot{\lambda}\,(r,\tau)\,d\tau \quad ,$$

we obtain :

$$E\,\partial_r u = \sigma_r - 2\nu\sigma_\theta + E\alpha T - E\,\varepsilon^{vp} \tag{1'}$$

$$E\frac{u}{r} = (1-\nu)\,\sigma_\varphi - \nu\,\sigma_r + E\alpha T + \frac{1}{2}\,E\,\varepsilon^{vp} \tag{2'}$$

$$\dot{\varepsilon}^{vp} = \frac{2}{\eta}\,\omega < \omega\,(\sigma_\varphi - \sigma_r) - 2\,C > \tag{3}$$

Moreover, the equation of equilibrium reduces to :

$$\partial_r\,\sigma_r = 2\,(\sigma_\varphi - \sigma_r)\,/\,r \tag{4}$$

3. GENERAL RELATION

3.1- Eliminating σ_φ and ε^{vp} between Eqs. *(1')*, *(2')* and *(4)* yields :

$$\partial_r\,[\,E\,r^2\,u - (1-2\nu)\,r^3\,\sigma_r\,] = 3\,E\alpha\,r^2\,T(r,t) \quad .$$

At the origin $r = 0$, $\sigma_r(0,t) = \sigma_\varphi(0,t)$ by spherical symmetry. Hence the origin always remains within the yield limit.

Thus Eq. *(2')* can be written as follows at the origin :

$$E\frac{u}{r}\,(0,t) - (1-2\nu)\,\sigma_r(0,t) = E\alpha\,T(0,t) \quad .$$

Then for any r,t :

$$E\frac{u}{r} - (1-2\nu)\,\sigma_r = \frac{1}{r^3}\int_0^r 3\,E\alpha\,\xi^2\,T(\xi,t)\,d\xi \quad .$$

Furthermore, by eliminating σ_φ between Eqs. *(2')* and *(4)* we have :

$$E\frac{u}{r} - (1-2\nu)\,\sigma_r = \frac{1}{2}\,E\,\varepsilon^{vp} + E\alpha\,T + (1-\nu)\,\frac{r}{2}\,\partial_r\,\sigma_r \quad .$$

Comparing these last two equations and setting :

$$\theta(r,t) = \frac{2\,E\alpha}{3(1-\nu)}\,T(r,t) \qquad \text{and} \qquad \sigma_E\,(r,t) = -\frac{1}{r^3}\int_0^r 3\,\xi^2\,\theta(\xi,t)\,d\xi$$

we have :

$$\boxed{\frac{E\,\varepsilon^{vp}}{(1-\nu)\,r} + \partial_r\,\sigma_r - \partial_r\,\sigma_E = 0} \tag{5}$$

Equation (5) is true everywhere in the medium (the yield law (4) has not yet been used), whatever the location of any viscoplastic zones.

3.2 - Comment on the location of the viscoplastic zones :

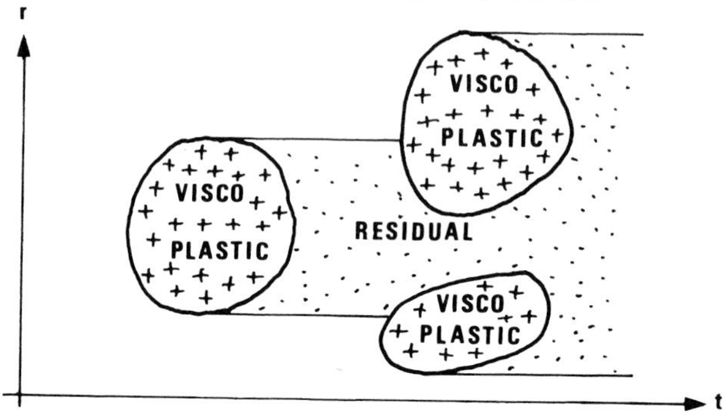

FIGURE 1 :
Location of the viscoplastic zones.

As the temperature T(r,t) can vary in any prescribed way, viscoplastic zones can appear, increase, decrease, and vanish. In some cases, a viscoplastic zone can appear in a purely elastic zone ; in other cases, a viscoplastic zone can appear in a previously viscoplastic zone (see Figure 1). The set of equations to be solved is different in each case. In the following we will focus on the evolution of the plastic boundaries.

4. VISCOPLASTIC BOUNDARIES

4.1- Solution in an elastic zone :

By *elastic zone* we mean a region where, at the instant t, $\varepsilon^{vp}(r,t) = 0$. Integration of Eq. (5) is then easy, provided σ_r is known at one point of the elastic zone. The following formula can be used for any point r in an elastic zone :

$$\sigma_r(r,t) = A(t) + \sigma_E(r,t) \quad ,$$

A(t) vanishes if the elastic zone under study extends to infinity.

4.2- Boundary between an elastic zone and a viscoplastic zone :

By viscoplastic zone, we mean a region where, except perhaps in a finite number of points,

$$\dot{\lambda}(r,t) = \omega(r,t)\ \dot{\varepsilon}^{vp}(r,t) > 0 \quad .$$

A boundary separating an elastic zone and a viscoplastic zone is referred to as an E - V P boundary, x = x(t) (see figure 2).

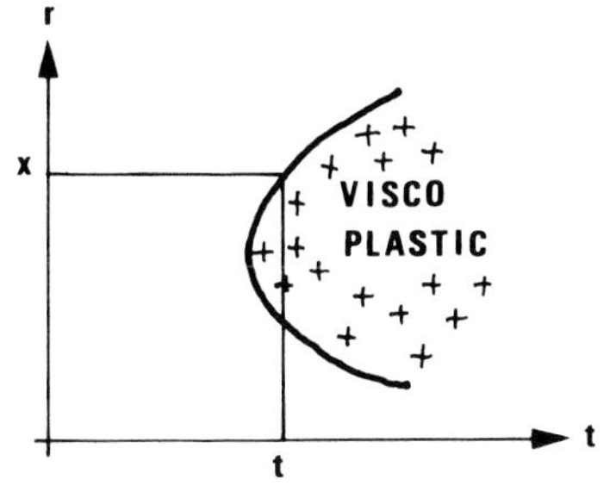

FIGURE 2
Elastic-viscoplastic boundary.

Attention, must be paid to the continuity through this boundary. T, ε^{vp}, u and σ_r are continuous ; then (2') proves that σ_φ is continuous too. Therefore $\partial_r\,\sigma_r$ is continuous and equal to 4C/x. As for ε^{vp} (x) = 0, Eq. (5) can be written :

$$\boxed{\varphi(x,t) = -\,x \quad \partial_r\,\sigma_E(x,t) + 4\,\omega\,C = 3\ \ (\theta + \sigma_E + 4\,\omega\,C/3)\ = 0} \qquad (6) \quad .$$

At time t, the E . VP boundaries are therefore given by the zero values of $\varphi(x,t) = 0$. For a simple zero, $\varphi(x,t)$ has the same sign as ω has on the elastic side. A double zero is obtained when a viscoplastic zone appears. In such a case $\partial_t\,\varphi$ is generally non zero, hence the curve $x = x(t)$ has a vertical tangent when a viscoplastic zone appears. (6) still holds in the special case of a plastic behavior, $\eta = 0$, [1].

4.3- Unloading viscoplastic boundary :

By *residual zone* we mean a region where, at the instant t, $\dot{\varepsilon}^{vp}(r,t) = 0$ and $\varepsilon^{vp}(r,t) \neq 0$. A boundary separating a residual zone and a viscoplastic *regressing* zone is referred to as an R - V P_{unload} boundary, $y = y(t_u)$ (see figure 3).

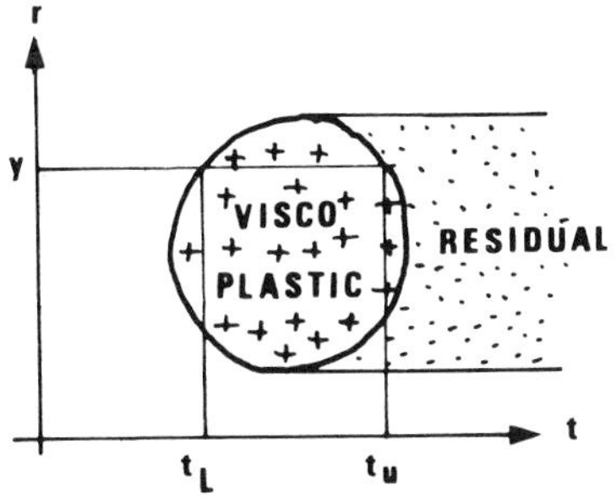

FIGURE 3 :
Unloading viscoplastic boundary.

t_L and t_u are defined as follows :

$$t_L < t < t_u \qquad \dot{\varepsilon}^{vp}(y,t) \neq 0 \quad ,$$

$$t \leqslant t_L \text{ and } t_u \leqslant t \qquad \dot{\varepsilon}^{vp}(y,t) = 0 \quad .$$

For $t_L \leqslant t \leqslant t_u$, *(4)* holds and can be written :

$$\dot{\varepsilon}^{vp} = \frac{r}{\eta} (\partial_r \sigma_r - \partial_r \sigma_p) \ , \quad \text{with} \quad \sigma_p = \frac{4\,\omega\,C}{3} \,\mathrm{Log}\, r^3 \qquad (4')$$

(4') and *(5)* can be combined in order to eliminate $\partial_r \sigma_r$:

$$\frac{1}{r} (\dot{\varepsilon} + \beta\,\varepsilon) + \frac{1}{\eta} (\partial_r \sigma_E - \partial_r \sigma_p) = 0 \qquad (7)$$

With $\beta = E/\eta(1-\nu)$. This relation can be derived with respect to time :

$$\frac{1}{r} (\ddot{\varepsilon} + \beta\,\dot{\varepsilon}) + \frac{1}{\eta}\,\partial_r \dot{\sigma}_E = 0 \quad .$$

And, as for $\dot{\varepsilon}(t_L) = \dot{\varepsilon}(t_u) = 0$:

$$\boxed{\int_{t_L}^{t_u} e^{\beta t}\,\partial_r \dot{\sigma}_E(y,t)\,dt = 0} \qquad (8)$$

The R - V P_{unload} boundary is implicitely defined by *(8)*. It is interesting to notice that the plastic case can be obtained by setting $\eta = 0$ or $\beta = \infty$ in *(8)* :

$$\partial_r \dot{\sigma}_E(y,t) = 0 \quad .$$

This purely formal deduction happens to be correct (see [1]).

4.4- Reloading viscoplastic boundary :

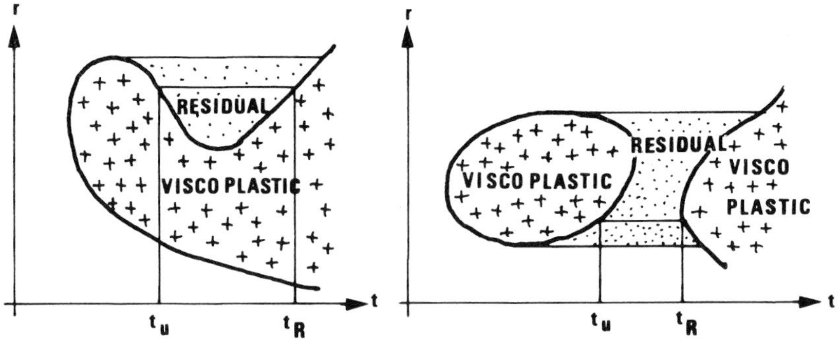

FIGURE 4 :
Reloading viscoplastic boundary.

A boundary separating a residual zone and a viscoplastic *progressing* zone is referred to as a R-V P_{reload} boundary, $z = z(t_R)$ (see figure 4). The sign of ω may change between $t = t_u$ and $t = t_R$ (see case 2).
t_u and t_R are defined as follows :

$$t_u \leqslant t \leqslant t_R \qquad \dot{\varepsilon}^{vp}(z,t) = 0 \quad ,$$

$$t < t_u \text{ and } t > t_R \qquad \dot{\varepsilon}^{vp}(z,t) \neq 0 \quad .$$

For $t_u \leqslant t \leqslant t_R$ (7) holds :

$\dot{\varepsilon}(t_u) = \dot{\varepsilon}(t_R) = 0$; $\varepsilon(t_u) = \varepsilon(t_R)$; $\partial_r\ \sigma_p(z,t_u) = 4\ \omega_u\ C/z$ and $\partial_r\ \sigma_p(z,t_R) = 4\ \omega_R C/z$.
t_R is then implicitely defined by :

$$\boxed{\partial_r\ \sigma_E(z,t_u) - 4\ \omega_u\ C/z = \partial_r\ \sigma_E(z,t_R) - 4\ \omega_R\ C/z} \qquad (9)$$

(9) holds in the special case of a plastic behavior, $\eta = 0$. The reason is that, in the plastic case, though $\dot{\varepsilon}^{vp}(z,t_u)$ does not vanish to zero, $\partial_r\ \sigma_r(z,t_u) = 4\ \omega(z,t_u)\ C/z$ (see [1]).

5. STRESS COMPUTATION

The calculation of stresses -and especially of residual stresses- is of major importance for engineering purposes :

Let :

$$I(r,t) = \sigma_r(r,t) - \sigma_E(r,t) = \int_r^{\infty} \frac{E\ \varepsilon^{vp}\ (u,t)}{1 - \nu}\ \frac{du}{u}$$

and a, b, be the boundaries of a viscoplastic zone. For any r, $a \leqslant r \leqslant b$, (4') and (5) can be combined in order to eliminate ε^{vp} :

$$\beta\ \partial_r\ I + \beta\ \partial_r\ (\sigma_E - \sigma_p) + \partial_t\ \partial_r\ I = 0 \quad .$$

It is easy to check that for any function $\Phi(r,t)$, $\partial_r\ \Phi = 0$ implies $[\partial_t\ \Phi]_a^b = 0$. Then :

$$\Big[\ \beta\ \partial_t\ \sigma_E + \beta\ \partial_t\ I + \partial_t\ \partial_t\ I\ \Big]_{a(t)}^{b(t)} = 0 \quad .$$

Several viscoplastic zones, respectively bounded by a_i and b_i, $i \in [1,n]$ can simultaneously exist, $a_1 \leqslant b_1 \leqslant a_2$, etc... For $r = b_n$, $\partial_t\ I = \partial_t\ \partial_t\ I = 0$. Furthermore, let $I_0 = I(a_1,t) = I(0,t) = [\sigma_r + \theta]\ (0,t)$, as for $I(r,t)$ is constant in an elastic zone

$$\frac{\partial^2\ I_0}{\partial t^2} + \beta\ \frac{\partial I_0}{\partial t} = \beta \sum_1^n \Big[\ \partial_t\ \partial_E\ \Big]_{a_i(t)}^{b_i(t)} \qquad (10).$$

As for $a_i(t)$ and $b_i(t)$ can be computed, I_0 and the residual stress at the origin of the coordinates are known.

APPENDIX

RESIDUAL STRESSES IN A NUCLEAR UNDERGROUND WASTE DISPOSAL

In France, and in many other countries, underground disposal of high level radio-active wastes is generally considered as the safest option. Those wastes must be kept confined during a very long period of time ; then, special care must be kept to the integrity of the so called "geological barrier", i.e. the rockmass surrounding the canisters.

Large amounts of heat are generated by the wastes during several centuries. Then the temperature of the rockmass will increase. Thermal stresses will appear; when viscoplastic material are considered, as rocksalt or clay for instance, large creep will be favoured by high temperatures. Then the concept of compact disposal (many canisters in a small volume) should be favorable to the creation of an adequate sealing of the hottest zone in which the canisters are placed.

But one must keep in mind that cooling will take place after heating. As for rheological behavior includes major irreversible effects, severe residual stresses can remain after cooling, including tensile stresses in the surrounding of the previously hottest zones. Rocks generally exhibit poor tensile strength ; then existence of fissures and loss of imperviousness due to thermal contraction must be discussed.

As for rocks have a rather poor thermal conductivity, the cooling will be slow (several centuries). Then if the rheological behavior is viscoplastic, rather than purely plastic, the residual stresses may be reduced to an acceptable level.

In the following we will show that a very simple model allows for a discussion of the major features of this problem.

DECREASING HEAT SOURCES IN THE SPHERE $r < a$:

Heat is produced at the constant rate A_0 per unit time per unit volume in the sphere $0 \leqslant r < a$. Let $t' = \frac{4\,k\,t}{a^2}$, $r' = \frac{r}{a}$, $\beta' = \frac{a^2\,\beta}{4\,k}$

The functions θ_0 and σ_{E_0} can be expressed as follows (see [2]) :

$$\theta_0(t') = \frac{2\,E\,\alpha}{3(1-\nu)}\,\frac{A_0\,a^2}{4\,K}\left\{ t' + 2\,\frac{t'}{r'}\left[i^2\mathrm{erfcx}(\varepsilon) + t^{\frac{1}{2}}\, i^3\mathrm{erfcx}(\varepsilon)\right]_{\varepsilon=-1}^{\varepsilon=+1} + \right.$$

$$\left. + \frac{2}{3}\,Y(r'-1)\,\frac{(r'-1)^2\,(2+r')}{r'} \right\}$$

$$\sigma_{E_0}(t') =$$

$$-\frac{2\,E\,\alpha}{3(1-\nu)}\,\frac{A_0\,a^2}{4\,K}\left\{ t' - 6\,\frac{t'^{\frac{5}{2}}}{r'^3}\left[\varepsilon\,\frac{r'}{t'}\,i^3\mathrm{erfcx}(\varepsilon) + x(\varepsilon)\;i^4\mathrm{erfcx}(\varepsilon) + i^5\mathrm{erfcx}(\varepsilon)\right]_{\varepsilon=-1}^{\varepsilon=+1}\right.$$

$$\left. + \frac{2}{5}\,Y(r'-1)\,\frac{(r'-1)^3\,(r'^2 + 3\,r'+1)}{r'^3} \right\}$$

$x(\varepsilon) = (1 + \varepsilon\, r')\,/\,t'^{\frac{1}{2}}$, $\varepsilon = \pm\,1$. Y is the Heaviside function.

Consider now the case of decreasing heat production, $A = A_0\,\exp\,(-\gamma\,t)$. In this case, the functions $\theta(\gamma,t)$ and $\sigma_E(\gamma,t)$ are obtained as follows :

$$\gamma' = \frac{a^2\,\gamma}{4\,k}$$

$$\theta(\gamma',t') + \int_0^{t'} \theta(\gamma',\tau)\,d\tau = \theta_0(t')\;\;;\;\theta(\gamma',0) = 0$$

$$\sigma_E(\gamma',t') + \int_0^{t'} \sigma_E(\gamma',\tau)\,d\tau = \sigma_{E_0}(t')\;;\;\sigma_{E_0}(\gamma',0) = 0\quad .$$

Those two functions are easy to compute ; then *(6)*, *(8)*, *(9)* allow for computation of the development of viscoplastic zones.

A- First, let us examine the case $\gamma' = \beta'$ (which means that the rates of thermal and viscous phenomena are of the same order of magnitude).

The influence of cohesion on both development of viscoplastic zones and intensity of residual stresses is of major importance.

For high values of cohesion (case 1) a viscoplastic zone appears and develops in the vicinity of the surface $r = a$, then smoothly regresses (fig. 5.a) and in the final state exists an unic annular ER zone (fig. 5.b).

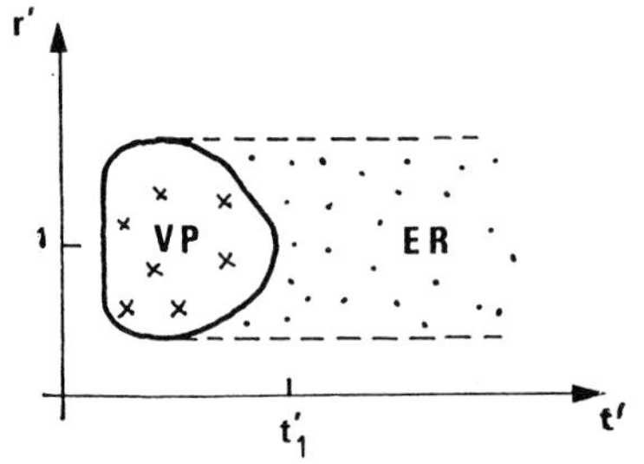

Fig. 5.a

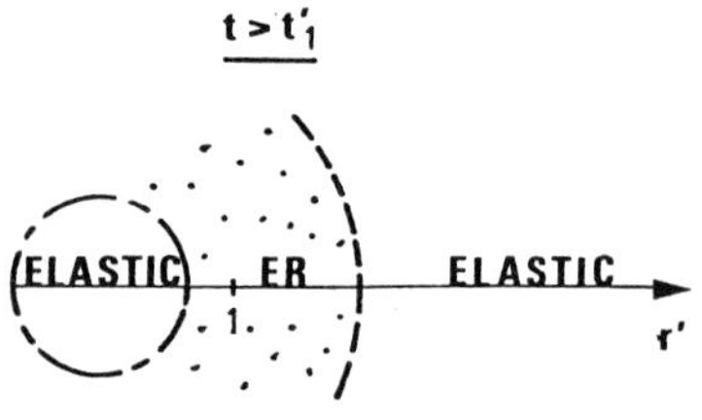

Fig. 5.b

FIGURE 5 :
Case 1 - High values of cohesion.

For smaller values of cohesion (case 2) a second viscoplastic zone appear during thermal unloading, in which the major principal stresses are switched ($\sigma_r > \sigma_\varphi$ or $\omega = -1$). This phenomenon is similar to Bauschinger effect (see figure 6). For $t = t'_2$ seven different zones must be distinguished (figure 6.b).

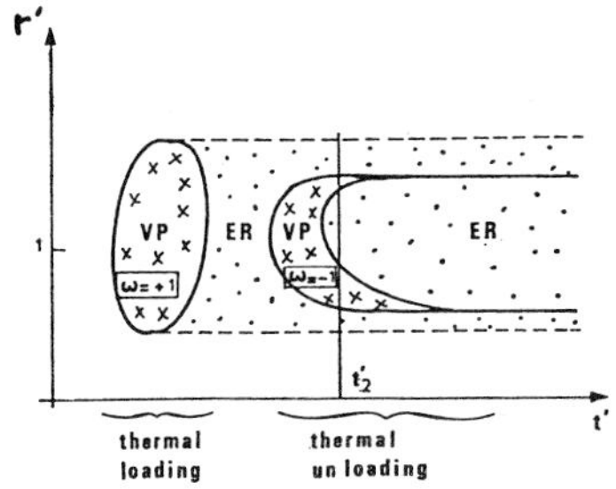

Fig. 6.a

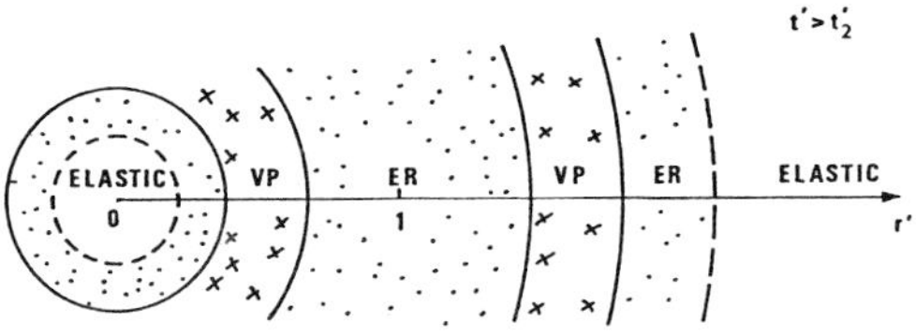

Fig. 6.b

FIGURE 6 :
Case 2 - Small values of cohesion.

The residual stress at the origin of coordinates, $\sigma_0 = \sigma_r$ (r=0 , t=∞) is a tensile stress. Values of the residual stress σ_0, versus cohesion c, are shown on figure 7. The existence of a maximum must be noticed.

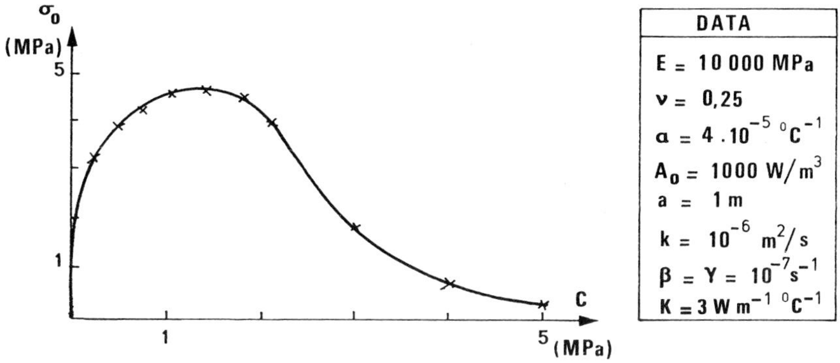

FIGURE 7 :
Residual stress, versus cohesion c.

B- Results are quite similar when $\gamma' \neq \beta'$; first a viscoplastic zone $\omega=+1$ develops then vanishes and in some cases a second viscoplastic zone $\omega=-1$ appear during the thermal loading, after the cooling phase has begun.

The residual stress σ_0 is strongly dependent of the ratio β'/γ'. When $\beta'/\gamma' \ll 1$ (which means that the viscosity is very high) the medium behaves as if it was purely elastic : the viscous response is too slow, as compared to the rate of thermal loading and its effects are practically negligible ; the residual stress is very small. Conversely, when $\beta'/\gamma' \gg 1$, the rate of thermal loading is very small and the medium behaves as if it was purely plastic.

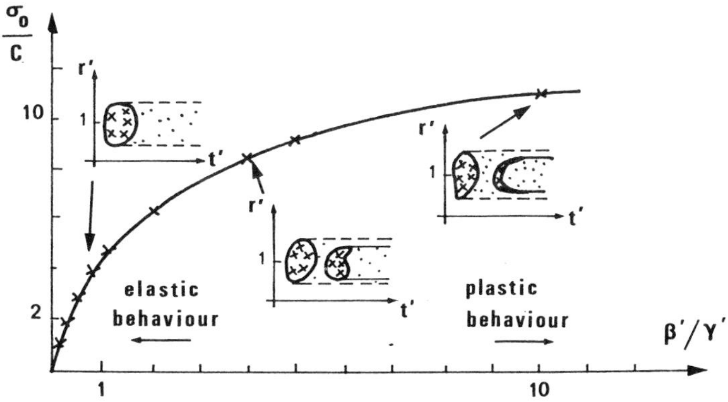

FIGURE 8 :
Residual stress, versus β'/γ'.

CONCLUSION

Equation *(6)* and Eqs. *(8)*-*(9)* which follow from it make possible the calculation of the development of the various zones created by a thermal load ; equation *(10)* allow for calculation of the residual stresses. Though the viscoplastic problem is evidently non-linear, it is noticeable that the thermal load affects Eq. *(6)* only through a function which is a linear form for the temperature distribution considered.

ACKNOWLEDGEMENTS

We should like to thank A. Giraud, Ph. D. Student, who performed the numerical computations and the parametric study discussed in the appendix.

The work described in this paper forms a part of the overall programme in Nuclear Waste Disposal currently in progress at Laboratoire de Mécanique des Solides, Ecole Polytechnique. It is supported by Agence Nationale pour la Gestion des Déchets Radioactifs (A.N.D.R.A.).

REFERENCES

[1] Bérest, P., Point Source of heat in an elastoplastic medium, Journal of Thermal Stresses, 1986.

[2] Carslaw, H.S., and Jaeger, J.C., Conduction of heat in solids (Oxford at the Clarendon Press, 1947).

[3] Ishikawa, H., Transient thermoelastoplastic stress analysis for a hollow sphere using the incremental theory of plasticity, Int. J. of Solids and Structures, vol. 13, pp. 645-655, 1977.

[4] Ishikawa, H., Thermoelastoplastic creep stress analysis for a thick walled tube, Int. J. of Solids and Structures, vol. 16, pp. 291-299, 1980.

[5] Wickens, L.M., The pore water and effective-stress response of clay sediments to a heat source, Report n° TP 966, Theoretical Physics Division, AERE Harwell, Oxon, 1982.

[6] Gamer, U., Elastoplastic deformation of a centrally heated disk, Journal of Thermal Stresses, vol. 8, n° 1, pp. 41-51, 1985.

Thermomechanical Couplings in Solids
H.D. Bui and Q.S. Nguyen (Editors)
Elsevier Science Publishers B.V. (North-Holland)

MODELING OF DIMENSIONAL CHANGE IN METAL MATRIX COMPOSITE SUBJECTED TO THERMAL CYCLING

Minoru Taya* and Tsutomu Mori**

* Department of Engineering Science, University of Oxford, Parks Road, Oxford, OX1 3PJ, U.K.,
On leave from Department of Mechanical Engineering
University of Washington FU-10, Seattle, WA 98195
U.S.A.
** Department of Materials Science and Engineering
Tokyo Institute of Technology
4259 Nagatsuda, Midori-ku, Yokohama 227
Japan

1. INTRODUCTION

Metal matrix composites(MMCs) are rapidly becoming the strongest candidates for many high temperature applications. The main objective using an MMC system is to replace the existing superalloys and increase the service temperature above that of the superalloys. Thus, the service temperature range of MMCs overlaps that of superalloys at the low temperature end and ceramics and ceramic matrix composites at the high temperature end. In order to better design the MMC components in high temperature application, one must assess a complete picture of its performance in severe environments. The severe environments that MMCs will presumably experience are high temperature excursions(creep and thermal cycling), an oxidation environment at high temperature, and high strain-rate impact. Among the above severe environments, thermal cycling is well known to cause significant degradation for most of MMC systems[1-16].

The degradation of as-thermally cycled MMCs has been attributed to the mismatch of the coefficients of thermal expansion(CTEs) between the matrix and fiber, and also to the reaction products at the matrix-fiber interfaces. These contributions result in the microscopic degradation in MMCs; the formation of microvoids in W/Cu composite[3,4], the roughning and cracking of the interfaces in B/Al composite[5-7], in FP/Al composite[8], in FP/Mg composite[9], in SiC/Ti composite[10], and C/Al composite[13]. The above microscopic level degradation which are always associated with the matrix-fiber interfaces leads to the reduction in the mechanical properties, stiffness, strength and toughness, and also to the dimensional change of the MMC specimens.

The dimensional changes have been observed in several continuous fiber MMC [3,7,9,11,13,16] and short fiber MMC systems[12,15]. Though the majority of the above works reported that the major dimensional change occurred along the fiber axis, our recent results on continuous carbon fiber /aluminum composite have not confirmed this[13]. Namely, we have observed the finite swelling along the transverse direction(perpendicular to the fiber axis), but not along the fiber axis(longitudinal direction). In the case of short fiber MMCs, only limited data on the dimensional changes have been reported to date, one under pure thermal cycling[12] and the other under a combination of constant applied stress and thermal cycling[15]. Unlike the case of continuous fiber MMC system, no clear evidence of the debonding of the matrix-fiber interface was observed for short fiber MMCs[12,15].

Some attempts have been made to model the dimensional changes in the as-thermally cycled MMC[2,4,14,16]. Most of the above models are based on one-dimensional stress flow in the matrix, thus applicable to continuous fiber MMC. In these models the matrix-fiber interface is assumed to be perfectly bonded, thus preventing it from sliding(or accelerated deformation), while the model proposed by Wakashima and his co-workers can account for the viscous flow of the interface and it can predict the longitudinal swelling of W/Cu composite reasonably well[4], but its applicability to other MMC systems remains to be investigated. Unlike continuous fiber MMC, modeling for short fiber MMC is not straightforward due to the three-dimensional geometry of fiber and its misorientation. Derby has recently proposed a model to predict the longitudinal strain-rate of a short fiber MMC subjected to both thermal cycling and constant applied stress[14]. The Derby's model can explain the experimental results[15] well. However, it fails to predict the longitudinal swelling of the as-thermally cycled MMC. For the strain-rate in the Derby's formula is proportional to the applied stress, and it becomes zero for thermal cycling only(i.e, no applied stress), which is not consistent with most of the existing experimental results. On the other hand in the model by Wakashima et al the matrix-fiber interface is assumed to behave a Newtonian viscous fluid and it can simulate well the matrix-fiber interfaces of the compatible type such as W/Cu and eutectic composites when they are subjected to high temperatures. The Wakashima et al model, however, may not be applicable to the matrix-fiber interfaces of imcompatible type where the interfaces consist of brittle compounds and thus its behavior at high temperatures is different from Newtonian viscous flow. In fact, we have recently observed such a case in continuous carbon fiber/aluminum(C/Al) composite[13] where the matrix-fiber interfaces were broken during the early stage of thermal cycling and thereafter the transverse swelling occurred predominantly. The thermal cycling test was conducted between room temperature and 623 K. Fig.1(a) and (b) reveal the interfaces of C/Al composite before and after thermal cycling(1000 times),respectively. Whether the interface is compatible type or not can be considered to be dependent on the temperature.Namely, at very high temperature the incompatible interface can be changed to a compatible type.

The above brief review on the previous works on the dimensional changes in the as-thermally cycled MMCs has led us to believe that no analytical model has been proposed for a short fiber MMC. In this paper we propose such a model, which will be described in Section 2. In Section 3 a parametric study based on the present model is conducted to examine the effect of several key material parameters on the dimensional change. Then in Section 4, the analytical results are compared with our previous experimental results on SiC whisker/2124 aluminum composite[12],followed by discussion on the validity of the present model. Finally the concluding remarks are given in Section 5.

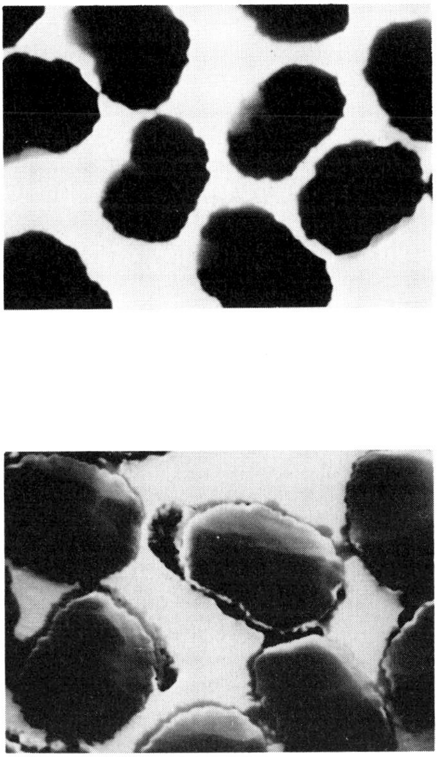

(a)

(b)

FIGURE 1

SEM photos revealing the matrix-fiber interfaces of C/Al composite, before (a) and after the thermal cycling (N=1000).

2. ANALYTICAL MODEL

2.1 Assumptions

To facilitate the computation, we assume the following:

(1) the temperture-time relation is of the step function type as shown in Fig.2 where initially a MMC specimen is subjected to high temperature(T_h) and it is stress-free.

(2) at low temperature(T_l), the stress relaxation occurs by plastic deformation in the matrix. This implies that the temperature change $\Delta T = T_h - T_l$, is large enough to cause the yielding of the matrix.

(3) at high temperature the stress relaxation occurs by first matrix creep and followed by the diffusion around the matrix-fiber interfaces.

(4) all short fibers are modeled by prolate spheroid of the same size and are aligned along the x_3-axis ,otherwise distributed randomly in the matrix(Fig.3). Hence the composite system of Fig.3 becomes transversely isotropic.

(5) the matrix metal is isotropic(E_m, ν_m, α_m), so is the fiber material(E_f, ν_f, α_f) where E, ν and α are Young's modulus, Poisson's ratio and coefficient of thermal expansion(CTE), and subscripts m and f denote the matrix and fiber phase, respectively. It should be noted here that the present model can account for anisotropy of both matrix and fiber in principle, but the resulting computation will become rigorous.

2.2 Formulation

Referred to Fig.2, the present formulation can be divided into two cases, one for the cooling process(a→b→c) and the other for the heating process(c→d→e), which will be discussed seperately below.

2.2.1 Formulation for the cooling process:a→b→c

First we consider purely the cooling process(a→b) in order to determine the critical temperature T_y at which the matrix metal yields first. In order to make use of the Eshelby's model[17], we assume that the plastic strain in the matrix is uniform. By using the Eshelby's model modified for large volume fraction of fiber[18,19], we can establish the equations to compute the thermal stress field in a short fiber MMC $\underset{\sim}{\sigma}$ which are given by[18]:

$$\begin{aligned} \underset{\sim}{\sigma} &= \underset{\sim}{C}_f \cdot (\bar{\underset{\sim}{e}} + \underset{\sim}{e} - \underset{\sim}{\alpha}^*) \\ &= \underset{\sim}{C}_m \cdot (\bar{\underset{\sim}{e}} + \underset{\sim}{e} - \underset{\sim}{e}^*) \end{aligned} \qquad (1)$$

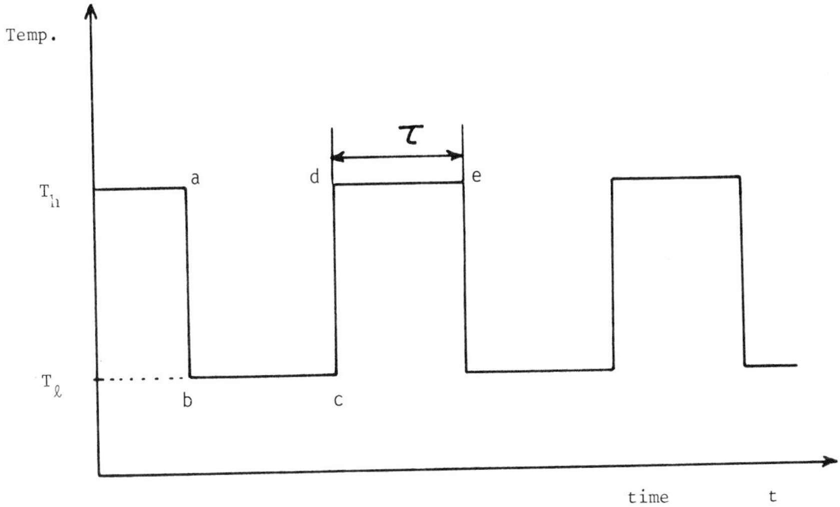

FIGURE 2
Temperature - time curve for thermal cycling.

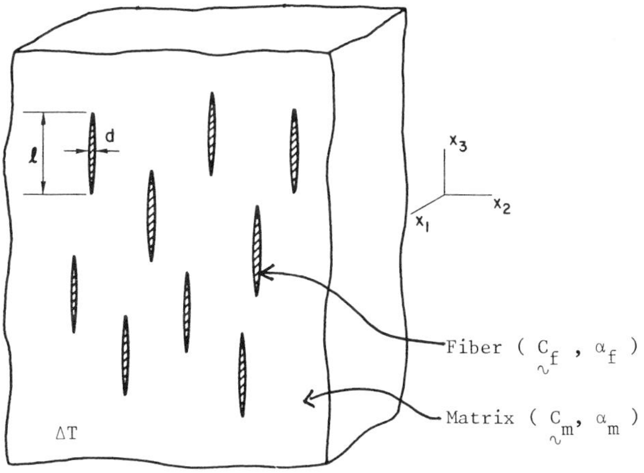

FIGURE 3
Analytical model for a short fiber MMC

where $\underset{\sim}{C}_f$ and $\underset{\sim}{C}_m$ are the stiffness tensors of fiber and matrix, respectively, $\underset{\sim}{\alpha}^*$ is the mismatch strain due to the difference of CTEs between fiber and matrix and given by

$$\underset{\sim}{\alpha}^* = (\ \alpha_f - \alpha_m,\ \alpha_f - \alpha_m,\ \alpha_f - \alpha_m,\ 0,\ 0,\ 0\ )\Delta T \tag{2}$$

In eq.(1), $\bar{\underset{\sim}{e}}$ is the strain disturbance due to all fibers and averaged over the matrix domain, and it is related to the average stress disturbance in the matrix $<\underset{\sim}{\sigma}>_m$ as

$$<\underset{\sim}{\sigma}>_m = \underset{\sim}{C}_m \cdot \bar{\underset{\sim}{e}} \tag{3}$$

where $<\ >_m$ denotes the quantity averaged over the matrix domain. In eq.(1) $\underset{\sim}{e}$ is the strain disturbance due to a fiber and $\underset{\sim}{e}^*$ is eigenstrain of the equivalent inclusion which has the same material properties as the matrix[19], and it is related to $\underset{\sim}{e}$ as[17]

$$\underset{\sim}{e} = \underset{\sim}{S} \cdot \underset{\sim}{e}^* \tag{4}$$

where $\underset{\sim}{S}$ is the Eshelby's tensor and a function of Poisson's ratio of the matrix and fiber aspect ratio (l/d). In the above equations underneath tildas and dots denote tensors and inner product, respectively. Since $\int_D \underset{\sim}{\sigma}\, dv = 0$ where D denotes the entire composite domain[19],

$$\bar{\underset{\sim}{e}} + f\ (\ \underset{\sim}{e} - \underset{\sim}{e}^*) = 0. \tag{5}$$

and where f is the volume fraction of fiber. It should be noted here that $\bar{\underset{\sim}{e}} = 0$ for $f = 0$, hence it is called "interaction term" which increases with f. From eqs.(1),(4) and (5), we obtain

$$(\ \underset{\sim}{C}_f - \underset{\sim}{C}_m) \cdot \{ (1 - f)\ \underset{\sim}{S} \cdot \underset{\sim}{e}^* + f\ \underset{\sim}{e}^* \} + \underset{\sim}{C}_m \cdot \underset{\sim}{e}^* = \underset{\sim}{C}_m \cdot \underset{\sim}{\alpha}^* \tag{6}$$

Eq.(6) will be used to solve for unknown eigenstrain $\underset{\sim}{e}^*$ which has two independent non-vanishing components, $e^*_{11} = e^*_{22}$ and e^*_{33} due to the transeverse isotropy of the composite(Fig. 3). Once $\underset{\sim}{e}^*$ is solved, the averaged stress in the matrix $\langle \tilde{\sigma} \rangle_m$ and in the fiber $\langle \tilde{\sigma} \rangle_f$ are calculated from eqs.(1),(4) and (5) and they are given by

$$\langle \tilde{\sigma}^T \rangle_m = - f \, \tilde{C}_m \cdot (\tilde{S} \cdot \tilde{e}^* - \tilde{e}^*)$$

$$\langle \tilde{\sigma}^T \rangle_f = (1 - f) \, \tilde{C}_m \cdot (\tilde{S} \cdot \tilde{e}^* - \tilde{e}^*) \tag{7}$$

where superscript T denotes that the stresses are induced by temperature change. In order for the matrix to yield due to the temperature change, the following yield criterion(Von Mises type) must be satisfied:

$$3/2 \langle \tilde{\sigma}^{T'} \rangle_m \cdot \langle \tilde{\sigma}^{T'} \rangle_m = Y^2 \tag{8}$$

where prime denotes deviatoric tensor and Y is the matrix yield stress which is assumed to be constant. Since $\langle \tilde{\sigma}^{T'} \rangle_m$ is a linear function of $\Delta T_y (= T_y - T_h)$, we can solve for T_y from eq.(8). T_y is expected to be $T_l \leq T_y < T_h$.

The temperature difference between T_y and T_l will be used to furhter yield the matrix. Then, the uniform plastic strain in the matrix at T_l, $\tilde{e}^P$:($-e_p/2$, $-e_p/2$, e_p ,0, 0, 0) will be computed in the following way; first we calculate the corresponding eigenstrain $\tilde{e}^*$ due to this plastic strain by replacing $\tilde{\alpha}^*$ by $-\tilde{e}^P$ in eq.(1)[20]

$$\tilde{C}_f \cdot (\bar{\tilde{e}} + \tilde{e} + \tilde{e}^P) = \tilde{C}_m \cdot (\bar{\tilde{e}} + \tilde{e} - \tilde{e}^*) \tag{9}$$

and then the average stress $\langle \tilde{\sigma} \rangle$ due to the plastic strain are obtained as

$$\langle \tilde{\sigma}^P \rangle_m = -f \, \tilde{C}_m \cdot (\tilde{S} \cdot \tilde{e}^* - \tilde{e}^*)$$

$$\langle \tilde{\sigma}^P \rangle_f = (1 - f) \cdot (\tilde{S} \cdot \tilde{e}^* - \tilde{e}^*) \tag{10}$$

The plastic strain $\tilde{e}^P$ in the matrix can be determined by the yield condition:

$$\langle \sigma_{33} \rangle_m - \langle \sigma_{11} \rangle_m = Y \tag{11}$$

where $\langle \sigma_{ij} \rangle_m$ is a component of the total average stress in the matrix, $\langle \tilde{\sigma}^T \rangle_m + \langle \tilde{\sigma}^P \rangle_m$. Once the plastic strain $\tilde{e}^P$ is solved, the average stress field due to this plastic strain ,$\langle \tilde{\sigma}^P \rangle_m$ and $\langle \tilde{\sigma}^P \rangle_f$ can be solved explicitly.

The macroscopic plastic strain at stage d, along the x_3-axis($\langle\varepsilon_L^P\rangle$)and along the x_1-axis($\langle\varepsilon_T^P\rangle$) are obtained as

$$\langle\varepsilon_L^P\rangle = (1 - f)\, e_p$$
$$\langle\varepsilon_T^P\rangle = -f\, e_p/2 \tag{12}$$

where the subscripts L and T denote the longitudinal(x_3-axis) and transverse(x_1-axis) component,respectively.

2.2.2 Formulation for the heating process:c→d→e

The process for pure heating c→d will be considered first. The thermal stress induced during the process c→d is just negative of that during the process a→b(eq.(7)). Thus, at stage d, the only stress in the composite is that due to the plastic strain, $\langle\tilde{\sigma}^P\rangle_m$ and $\langle\tilde{\sigma}^P\rangle_f$(eq.(10)) which give rise to the flow(equivalent) stress in the matrix σ_m ,given by

$$\sigma_m = \langle\sigma_{33}^P\rangle_m - \langle\sigma_{11}^P\rangle_m \tag{13}$$

The sign of σ_m is expected to be negative since $\alpha_f<\alpha_m$.

Next, we consider the process d→e, i.e,$T=T_h$ for duration of τ for which the matrix creep and diffusional relaxation around the matrix-fiber interfaces are assumed to take place. As for the matrix creep ,the power-law of Dorn type is used[21] and it is given by

$$\dot{e}_c = \pm A(\sigma_m/G)^n GbD_0/(\kappa T_h)\cdot e^{-Q_v/RT_h} \tag{14}$$

where

$\dot{e}_c$: creep rate

A : Dorn constant

σ_m: flow stress

b : Burgers vector

κ : Boltzman's constant

D_0: pre-exponential constant for self diffusion

R : gas constant

T_h: Temerature for the process d→e

The sign $\pm$ in front of A in eq.(14) depends on $\pm$ of the flow stress σ_m and in the present case, (-) sign will be used since $\sigma_m< 0$. Due to its dependance on flow stress, the creep strain-rate $\dot{e}_c$ will be calculated incrementally. To this end, the total time span for the process d→e is subdivided into M segments of the same time interval, $\Delta t=\tau/M$. For the i-th step, the creep strain-rate

$$\Delta e_c(t_i) = -A\{\sigma_m(t_{i-1})/G\}^n GbD_0/(\kappa T_h)\cdot e^{-Q_v/RT_h} \quad (15)$$

where

$$t_i = i\Delta T$$
$$\sigma_m(t_{i-1}) = \sigma_m + \Delta\sigma_m(t_1) + \cdots\Delta\sigma_m(t_{i-2}) \quad (16)$$

and where $\Delta\sigma_m(t_{i-1})$ is computed from eqs.(9) and (10) and the equation similar to eq.(13) where $\tilde{e}^P$ is replaced by $(-\Delta e_c(t_{i-1})/2, -\Delta e_c(t_{i-1})/2, \Delta e_c(t_{i-1}), 0, 0, 0)$. At the end of this process, $\sigma_m(t_M)$ is expected to be non-zero and the total creep strain gives rise to the following macroscopic strain:

$$\langle\varepsilon_L^C\rangle = (1 - f)e_c$$
$$\langle\varepsilon_T^C\rangle = -fe_c/2 \quad (17)$$

where e_c is the sum of all incremental creep strain computed by eq.(15).

The above non-zero flow stress of the matrix at stage e, however, can be reduced to zero if the diffusional relaxation at the interfaces occurs[22]. This diffusional relaxation can be modeled by Eshelby type model if eigenstrain $\tilde{e}^D:(-e_d/2, -e_d/2, e_d, 0, 0, 0)$ is introduced into the fiber domain. The stress field within the fiber due to $\tilde{e}^D$, $\langle\tilde{\sigma}^D\rangle_f$ can be computed from eqs.(9) and (10) where $\tilde{e}^P$ is replaced by $-\tilde{e}^D$. From the requirement of the diffusional relaxation, $e_{11}^D + e_{22}^D + e_{33}^D = 0$, the total stress must satisfy the following[22]:

$$\langle\sigma_{11}^P\rangle_f + \langle\sigma_{11}^C\rangle_f + \langle\sigma_{11}^D\rangle_f = \langle\sigma_{33}^P\rangle_f + \langle\sigma_{33}^C\rangle_f + \langle\sigma_{33}^D\rangle_f \quad (18)$$

Eq.(18) implies that the total stress in the fiber becomes hydrostatic, resulting in the minimum Gibbs free energy. From eq.(18), $\tilde{e}^D$ is solved as

$$\tilde{e}^D = \tilde{e}^P + \tilde{e}^C \quad (19)$$

The macroscopic strain due to $\tilde{e}^D$, $\langle\tilde{\varepsilon}^D\rangle$ is then given by

$$\langle\varepsilon_L^D\rangle = fe_d = f(e_p + e_c)$$
$$\langle\varepsilon_T^D\rangle = -fe_d/2 = -f(e_p + e_c)/2 \quad (20)$$

Finally, the total macroscopic strain at the end of one cycle (at stage e), $\langle \underset{\sim}{\varepsilon} \rangle$ is given as the sum of eqs.(12),(17) and (20), i.e,

$$\begin{aligned} \langle \varepsilon_L \rangle &= (1 - f)(e_c + e_p) + f(e_c + e_p) \\ &= e_p + e_c \\ \langle \varepsilon_T \rangle &= -(1 - f)(e_p + e_c)/2 - f(e_p + e_c)/2 \\ &= -(e_p + e_c)/2 \end{aligned} \tag{21}$$

In the above equations, the superscripts L and T again denote the longitudianl(along the x_3-axis) and transverse(along the x_1-axis) component, respectively.

At the conclusion of each cycle the macroscopic strain given by eq.(21) remains as permanent one in the composite and the state of stress in the fiber becomes hydrostatic which in turn induces hydrostatic state of stress in the composite. Thus, this hydrostatic stress will not have any effect on the matrix yielding for the next thermal cycle. Hence, the permanent strain accumulated at the end of N cycles can be estimated as N times $\langle \underset{\sim}{\varepsilon} \rangle$.

3. PARAMETRIC STUDY

In order to examine the effect of various parameters on the dimensional change, we have conducted a parametric study by using the model developed in the previous section. The target short fiber MMC is SiC whisker/2124 aluminum(SiCw/Al) composite since its basic material data can be obtained from a standard material book and we have recently conducted some experiment to collect the data of the dimensional change[12] and the exponent n in its power law creep behavior. The basic material data of SiCw/Al composite are given in Table 1 and the data related to the matrix creep are given in Table 2.

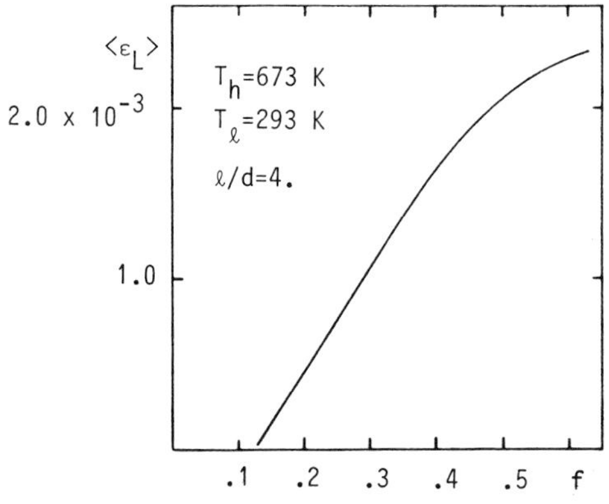

FIGURE 4
Dimensional change $\langle \varepsilon_L \rangle$ vs. fiber vol. frac. f.

parameters	symbols	units	matrix(2124Al)	fiber(SiCw)
shear modulus	G	GPa	25.4 at 300 K	182.5
Poisson's ratio	ν	1	0.33	0.17
CTE	α	/K	$24.7*10^{-6}$	$4.3*10^{-6}$
yield stress	Y	MPa	50.8(.2% str.)	-
fiber aspect ratio	l/d	1	-	2 ~ 100
fiber vol frac.	f	1	-	0.1~0.55

Table 1 Material data of SiCw/2124 Al composite

parameters	symbols	units	data	ref.
Dorn parametr	A	1	$3.4x10^{6}$	24
shear mod.at Th	G	GPa	20.2	25
pre-exponent	D_0	m^2/sec	$1.7x10^{-4}$	24
activation energy	Q_v	KJ/mole	142.	24
Burgers vector	b	m	$2.86x10^{-10}$	24
gas constant	R	J/mole/K	8.314	24
Boltzman's const	κ	J/K	1.381	24
exponent	n	1	$n=5+.025\sigma_m$	23

Table 2 Creep data of 2124 Al

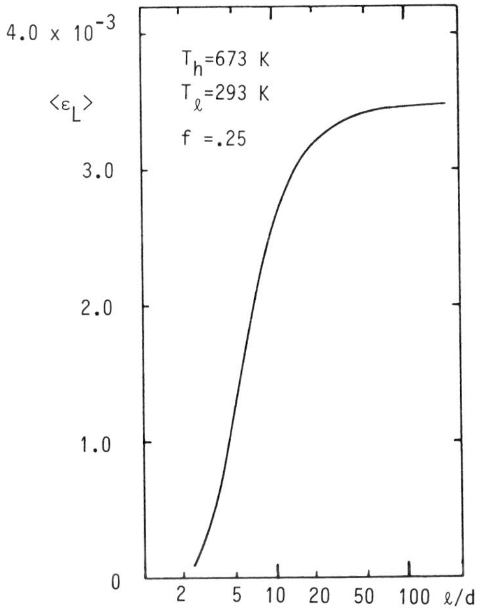

FIGURE 5
Dimensional change $\langle\varepsilon_L\rangle$ vs. fiber aspect ratio ℓ/d.

The exponent n in the power law creep of 2124 Al was found to be a function of stress[23], which is approximated by $n=5 + .025\sigma_m$ for $T_h=673$ K. Though our creep test is based on T=573 K, similar dependance of n on σ_m was found for other aluminum alloys for the temperature range 573~673 K[24], but its accuracy remains to be investigated.

First the effect of fiber volume fraction f on the dimensional change along the fiber axis, $\langle\varepsilon_L\rangle$ was studied for l/d=4, $T_h=673$ (T_l is always fixed as room temperaure, 293 K in this computation) with the other data given by Tables. The results of $\langle\varepsilon_L\rangle$(per cycle) are plotted against f in Fig.4. It follows from Fig.4 that higher the fiber volume fraction is , the larger the dimensional change becomes. Dependance of $\langle\varepsilon_L\rangle$ on l/d is also examined for f=.25 and the results of $\langle\varepsilon_L\rangle$ are given as a function of l/d in Fig.5. It is clear from Fig.5 that the dimensional change per cycle increases rapidly with l/d, but its rate of increase becomes smaller at higher l/d. Next the effect of high temperature T_h on the dimensional change was examined for f=.15 and the other data fixed. The results of $\langle\varepsilon_L\rangle$ for $T_h=673$ and 773 K are 1.576×10^{-4} and 5.24×10^{-4}, respectively. This indicates that the higher temperature induces larger permanent strain. Garmong has also predicted similar results for the effect of f and T_h by his one-dimensional model[1,2].

4. COMPARISON WITH EXPERIMENT AND DISCUSSION

An attempt is made here to compare the analytical and experimental results. In our experiment[12] SiCw/Al plate of length(L)=87 mm, width(W)=40 mm and thickness(TH)=6.35 mm were shuttled between hot($T_h=673$ K) and cold fluidized bath($T_l=293$ K) for N=1000 cycles. Then, the dimensional changes of the as-thermally cycled specimen were measured along three directions, along the length(ε_L), width(ε_W) and thickness(ε_{TH}). The measured dimensional changes are summarized in Table 3 where the corresponding analytical prediction for two different exponents n is also given for comparison.

ε at 1000 cycles	experiment[12]	analytical results n=5.08	n=4.9
ε_L	0.055	0.158	0.129
ε_W	0.02	-0.079	-0.0645
ε_{TH}	-.079	-0.079	-0.0645

Table 3 Comparison between the analytical and experimental results of 15% SiCw/2124 Al composite with T=673 K

It follows from Table 3 that the analytical results overestimate the experimental ones. This is mainly due to the fact that short fibers(SiC whiskers) in the present model are assumed to be all aligned along the x_3-axis(L-direction), requiring the transverse isotropy of the composite(W=TH), while in actual composites though the majority of short fibers are aligned along the L-direction, some fraction are still misoriented in-plane, and all short fibers are perpendicular to TH-direction. Thus, it is expected that the analytical results for TH-direction agree with the experimental ones and the measured ε_L is smaller than the analytical one, which is indeed the case in Table 3. It is noted from Table 3 that exponent n=5.08 was generated from the approximation formula(see Table 2) and it may not be accurate. Hence, if one assume a different value of n, for example, n=4.9, then, the predicted dimensional changes become closer to the experimental results. This illustrates how sensitive n is.

In our previuos experiment[12], we have also measured the dimensional changes of 15% SiCw/2124 Al composite which was subjcted to higher temperature,T_h=773 K. The experimental results are puzzling; the dimensional changes along the three directions(L-,W- and TH-direction) are much smaller than those of T_h=673 K. This can be attributed to two reasons: (1) at T_h=773 K,recrystalization of 2124 Al is expected to occur, thus resulting in the elimination of most of the damages(including dimensional changes) caused by the thermal cycling, and (2) power-law break down due to higher stress. The first reason is considered to be the major one and the state of stress and temperature seem to support the fact that the recrystalization has actually occurred according to the deformation mechanism map[24]. The second reason becomes important when the matrix flow stress reaches at least $0.1G_m$ where G_m is the shear modulus of the matrix[24]. Under such a circumstance, the creep behavior may be described by either exponential-law or sinh-law type. In the present model, however, variable exponent n was used to account for the higher stress in the matrix creep of power-law type. Thus, the accurate measurement of n for the range of higher stress is needed.

We have also conducted thermal cycling tests on another MMC system, continuous carbon fiber/5056 aluminum(C/Al) composite with f=.7[13]. The dimensional changes observed in C/Al composite are completely different from those in other continuous fiber MMCs[3,4] and short fiber MMC[12] which have been discussed above. Namely, the matrix-fiber interfaces seem to have been broken at early stage of the thermal cycling(say N_c) and thereafter the dimensional change has occurred along the transverse direction. In this case, the present model can still be used to predict such a threshold number of cycles N_c at which the interface breaks down due to the permanent strain accumulated that can be computed from eq.(21). This task, however, requires additional information on the fracture strain of the interface which may be difficult to obtain. It was also found in the above experiment that the interfacial bonding if the fiber is treated properly can be improved, resulting in the delay in the break down of interfaces.

In the present model we have assumed the plastic strain induced by $\Delta\alpha\Delta T$ in the matrix is uniform whenever the equivalent stress reaches the matrix yield stress. In reality, the plastic strain is likely to occur first in localized regions, at fiber-ends in a short fiber MMC[18,26], or in the matrix adjacent to the interfaces in a continuous fiber MMC[27]. However, as ΔT increases, the further plastic deformation can take place in a low dislocation density area[28]. This implies that the plastic strain will become more uniformly distributed as temperature differential continue to increase.

It is also assumed in the present model that the plastic deformation does not take place at high temperature. If we allow the plastic deforamtion to take place at high temperature(stage d in Fig.2), then this would induce additional plastic strain $\underset{\sim}{e}^{P*}:(-e_p^*/2,\ -e_p^*/2,\ e_p^*,\ 0,\ 0,\ 0)$ in the matrix with $e_p^* < 0$ due to the lower yield stress at high temperature. Thus, the plastic strain at stage d would be smaller than the present results, and this smaller plastic strain would then induce the smaller flow stress σ_m which in turn give rise to smaller creep strain. Therefore, the net permanent strain per cycle would become smaller than that based on the present model, resulting in the better prediction.

5. CONCLUDING REMARKS

An attempt is made to construct an analytical model which can predict the dimensional changes in the as-thermally cycled short fiber MMC. Then, the numerical results based on the present model are compared with the existing experimental results, resulting in a qualitative agreement. A parametric study was also conducted to examine the effect of several parameters on the dimensional changes. This has led to the following concluding remarks:

(1) The larger, the fiber volume fraction, fiber aspect ratio and the higher T_h are, the larger the dimensional changes become.

(2) The predicted dimensional change is quite sensitive to exponent n in the matrix creep of power-law type. Thus, the more accurate measurement of n will be needed to better predict the dimensional change by the present model.

(3) A comparison between the analytical and experimental results indicates that the present model can predict the dimensional changes reasonably well as fas as T_h and σ_m remain in the region for which the matrix creep of power-law type is valid.

(4) The present model needs further modifications to better predict the dimensional changes, for examples, to account for the misorientation of short fibers and the dependance of the yield stress on temperature.

ACKNOWLEDGEMENTS

One of the present authors(M.T.) is greatly thankful to the Royal Society for its partial support for this work, and we are to Mr. T. Kyono of Toray Inc. for his help in providing SEM photos.

REFERENCES

[1] Garmong, G.,Metall. Trans.,5 (1974) 2183.
[2] Garmong. G.,ibid, 5 (1974) 2199.
[3] Wakashima, K.,Kawakubo, T. and Umekawa S., ibid, 6A (1975) 1755.
[4] Yoda, S.,Takahashi, R.,Wakashima, K. and Umekawa, S.,ibid, 10A (1979) 1796.
[5] Grimes, R.A. and Maisel, J.E.,ibid, 8A (1977) 1999.
[6] Olsen, G.C. and Tompkins, Continuous and Cyclic Exposure Induced Degradation in Boron Reinforced 6061 Aluminum Composites, in: Failure Modes in Composites IV,(TMS-AIME, Warrendale,PA,1977) 1.
[7] Wright, M.A.,Metall. Trans., 6A (1975) 129.
[8] Kim, W.H.,Koczak,M.J. and Lawley, A., Effect of Isothermal and Cyclic Exposures on Interface Structure and Mechanical Properties of FP-Al_2O_3/Aluminum Composites, in: Kuhlman-Wilsdorf, D. and Harrigan, W.C.,Jr.(eds) New Development and Application in Composites,(TMS-AIME, Warrendale, PA,1979) 40.
[9] Bhatt, R.T. and Grimes, H.H.,Thermal Degradation of the Tesile Properties of Unidirectional Reinforced FP-Al_2O_3/EZ Magnesium Composite, in: Hack, J.E. and Amateaux, M.F.(eds), Mechanical Properties of Metal Matrix Composites (TMS-AIME, Warrendale, PA,1983) 51.
[10] Park, Y.H. and Marcus, H.L., Influence of Interface Degradation and Environment on the Thermal and Fracture Fatigue Properties of Titanium-Matrix/Continuous SiC Fiber Composites, ibid,65.
[11] Wolff, E.G., Min, D.K. and Kural, M.H.,J. Mater. Sci.,20 (1985) 1141.
[12] Patterson, W.G. and Taya, M., Thermal Cycling Damage of SiC Whisker/2124 Aluminum, in: Harrigan, W.C.,Jr.,Strife, J. and Dhingra, A.K. (eds), Proc. ICCM-V (TMS-AIME, Warrendale, PA,1985) 53.
[13] Kyono, T., Hall I.W. and Taya,M., Thermal Cycling Damage in Carbon Fiber/ Aluminum Composites, in: Kawata, K. et al (eds), Proc. 3rd Japan-U.S Conf. on Comp. ,Mater.,(Springer Verlage, 1986) in press.
[14] Derby, B. Scripta Metall., 19 (1985) 703.
[15] Wu, M.Y. and Sherby, O.D.,ibid, 18 (1984) 773.
[16] Wakashima, K., Choi, B. and Lee, S.H.,Temperature Cycling-Induced Superplasticity in Metal Matrix Fiber Composite, in: the same as Ref.13,in press.
[17] Eshelby, J.D., Proc. Roy. Soc. A241 (1957) 376.
[18] Takao, Y. and Taya, M.,J. Appl. Mech., 107 (1985) 806.
[19] Mura, T.,Micromechanics of Defects of Solids, (Martinus Nijhoff Publication, The Hague, 1982).

[20] Tanaka, K. and Mori T.,Acta Metall., 18 (1970) 931.
[21] Mukherjee, A.K.,Bird, J.E. and Dorn, J.E.,Trans. ASM, 62 (1969) 155.
[22] Mori, T., Okabe, M. and Mura, T.,,Acta Metall., 28 (1980) 319.
[23] Taya, M. and Lilholt, H., in preparation.
[24] Frost, H.J. and Ashby, M.F.,Deformation-Mechanism Maps,(Pergamon Press, Oxford, 1982).
[25] Gittus, J.,Creep, Viscoelasticity and Creep Fracture in Solids,(Applied Science Publishers, London, 1975).
[26] Taya, M. and Mori, T.,Acta Metall. in press.
[27] Dvorak, G.J.,Rao, M. and Tarn, J.Q., J. Comp. Mater., 7 (1973) 194.
[28] Arsenault, R.J. and Taya, M.,Acta Metall.,in press.

Thermomechanical Couplings in Solids
H.D. Bui and Q.S. Nguyen (Editors)
Elsevier Science Publishers B.V. (North-Holland)

THERMO-HYGRO-MECHANICAL COUPLINGS IN WOOD TECHNOLOGY AND RHEOLOGICAL BEHAVIOURS

Christian HUET *

Ecole Nationale des Ponts et Chaussées et Groupement Scientifique Rhéologie du Bois (GS CNRS n° 81)
ENPC - CERAM - La Courtine - B.P. 105 - F 93194 NOISY-LE-GRAND Cedex

ABSTRACT

Main thermo-hygro-mechanical problems and phenomena involved in the processing and utilization of wood are reviewed. A review of recent research made in this field or now in progress is presented, dealing both with experimental and theoretical approaches. A few theoretical representations of the observed behaviours used to predict some of their consequences through structural and process calculations are given. A few physical mechanisms relating to the constitution of wood as a polymeric natural composite with anatomic specific features, and considered as responsible for some of the phenomena involved are described.

1. INTRODUCTION

Wood is one of the oldest and one of the most widely used materials. But it appears that the problem of its mechanical behaviour and constitutive equations has, in France and in the last forty years, received much less attention from the mechanicists than other materials like steel, polymers, ceramics, rocks, soils, or concrete.

In fact, although very common as a material, wood is from several point of view a very complicated one, exhibiting at several levels anisotropy, heterogeneity, and sensibility to environmental conditions. Further, as a natural material, the large scatter observed on its properties may discouraged the scientific approach or at least involve very large and expansive experimental programs that can be managed only through sophisticated statistical methods.

Although a lot of work has already been done by wood scientists and technologists disseminated almost everywhere in the world, no complete set of constitutive equations allowing the treatment of thermo-mechanical problems in concrete situations is yet available.

In this lecture, which is given the character of a review, we shall first present the practical situations where the thermo-hygro-mechanical couplings in wood can be involved. Then we shall give some information about the various aspects and levels of the internal constitution of wood. In the following sections,

---

* Past Assistant-Director for Research of The French Technical Centre for Wood and Furnitures (CTBA).

we shall describe the main features of the various rheological properties exhibited by wood and the interactions of its mechanical response with physical parameters like temperature and moisture-content. In the last sections, we shall consider briefly the problems of the physical interpretations given and the modelization efforts made in the present state of the art.

The opportunity to enter this interesting field of research was given to us by the Centre Technique du Bois et de l'Ameublement (French Technical Institute for Wood and Furnitures). This Institute has begun since a few years to promote in France a large movement of research in this area. It did it by creating, with the help of the French Ministry of Research and of the French Agency for the Mastering of Energy, its own team and facilities on this subject and by taking, with the French Centre National de la Recherche Scientifique, the initiative of creating a Scientific Grouping (GS CNRS n° 81) specially devoted to the Rheology of wood. This french Grouping involves at present time nine research teams, and has relationships with others, in France and out of France.

Let us mention also that the 19th Annual Colloquium of the French Group of Rheology, held in Paris in November 1984 [1] and devoted to the Rheology of anisotropic materials, provided us with the state of the art for the french research, since it gathered, in its second part specially devoted to wood, one general review lecture and thirteen communications, eleven of which were of french origin.

For completeness of the information about the french situation, let us mention also that two scientific societies have been created recently in France to promote teaching and research in the fields of science and technology of wood. One, named ARBOLOR is located in Nancy and gathers teams of the northeastern part of France [2]. The other, named ARBORA, is located in Bordeaux, and does the same for the southwestern part.

2. THERMO-HYGRO-MECHANICAL COUPLINGS IN WOOD GROWTH, PROCESSING AND UTILIZATION

In wood, couplings between thermal, hygroscopic and mechanical factors and effects are involved at the various stages of the life of the tree, then during its harvesting, its processing into wood products or structural elements and, finally, during most of its utilizations.

During the life of the tree, the successive addition of new cells near its external surface, their progressive transformation into dead cells inside the tree, and the exchanges of heat and mass with the surroundings results into phenomena that are known as growth-stresses (cf. for instance [3]). These can be considered, at least for a part, as responsible for the damages that may occur during the life of the tree, as well as during the harvesting process. In some cases, these damages are spectacular, and may appear as large and deep helicoidal cracking on the trunck of the living tree, or as sudden bursting during felling, under the sudden redistribution of the stresses induced by the transverse cutting of the trunck.

These growth stresses have still an influence in the subsequent operations of sawing and of drying, where they can result in non acceptable curvatures or twisting of the pieces of lumber. Furtherly, in the case of drying, additional factors for such damages are the non uniform distribution of moisture involved by the very process of drying, and the non uniform shrinkage that results of it. Thus, an additional field of stresses, the drying stresses, occurs in order to grant the compatibility of the field of the total deformation. This can result in various drying damages like cracking, checking, honeycourb, collapse, and various types of warp, like bow, crook, twist, oval, diamond and cup (cf. [4] [5][6] for instance).

When, as it is often the case in practice for instance in outdoor utilizations, the pieces of wood are used in conditions where the temperature and humidity of the surroundings is varying with time, additional coupling effects are observed. They may have drastic consequences upon the serviceability and the lifetime of the structure in which they are inserted. These must be accounted for in wood building codes or regulation (cf. for instance [7][8]). The most spectacular of such couplings is the so-called mechano-sorptive effect (GROSSMAN [9]), appearing in the form of drastic changes in creep when moisture, temperature or both are varied. It is striking to observe that this effect, which will be described in more details in the sequel, turns out to present a lot of similarities with the so-called drying-creep well known for concrete (cf. for instance BAŽANT et alii [10]).

Other industrial processes may involve thermo-mechanical couplings, or thermo-hygro-mechanical couplings depending on the value of the temperature and the initial state of the wood pieces. It is the case for instance for the phenomena involved in high temperature drying under load (cf. for instance [11][12][13]), allowed for by the thermo-viscoelasticity exhibited by wood. It is also the case for the phenomenon of stress reversal in timber drying : because of the coupling between thermal and moisture gradients appearing in the drying process on one hand, thermo-hygro-rheological properties of the material and self equilibration of the stress distribution on another hand, it turns out that the stresses at the surface of the wood piece, that are initialy tensile, become compressive after a time and conversely for the stresses at the center. One may see for instance [14] for a numerical modeling of this effect through a finite element method and some constitutive assumptions, and also [15] for other topics related to the drying stresses.

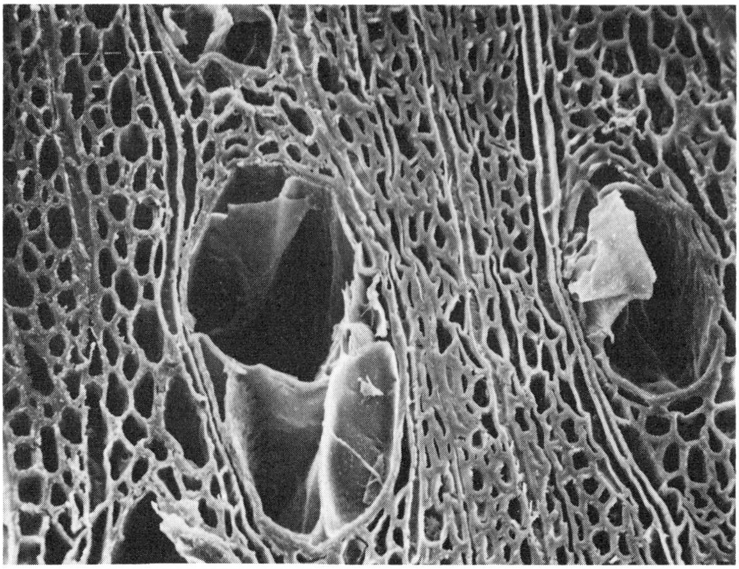

FIG. 1. Collapse in hardwood under drying. The hollow cells, about 10 m in diameter, experience transverse buckling under the action of capillary forces. The larger tubes are sap vessels (S.E.M. observation by Y. TRENARD, C.T.B.A., private communication, 1986).

Among these industrial processes, let us mention also wood bending and wood forming and densification by thermocompression, that both involve large deformations (cf. [16]).

In wood bending (see for instance [17]), green wood planks are applied upon a curved mold with some initial precompression and then dried while their shape is mechanically maintained. After drying, irreversible curvature is obtained that allows the piece to be taken off from the mold and used.

In thermocompression [18], the softening of wood at temperatures between 100 and 200°C is used to apply large transverse deformations to wood. With cooling under mechanical restraint, these deformations become also irreversible, and this restraint can be released without further change of shape.

Lastly, among industrial processes involving thermo-mechanical couplings, we shall quote the processes of grinding, pulping and refining performed in order to obtain chips or particles for boards, as well as cellulose fibers for paper and fiber-boards. In these operations, that involve the behaviour of the material in shear at high deformations and high frequencies (up to a few thousand Hertzs), the influence of temperature on properties and the dissipative effects are of importance (cf. for instance [19]).

3. THE VARIOUS LEVELS OF THE INTERNAL CONSTITUTION OF WOOD

As seen above, the behaviour of pieces of wood of practical or commercial size is largely affected by the influence of temperature and humidity. Since temperature is our macroscopic feeling of the movements of the kinetic units at the molecular or atomic level, and since humidity is involved through the interactions of the solid part of wood with the water molecules, it turns out that the constitution of wood at the molecular level will be of primary importance in the occurence of the couplings the macroscopic consequences of which have been described above.

Basically, the solid substance constituting wood is the skeleton, in form of a hollow tube, of a living cell. This skeleton remains after the death of the cell which occurs when the tree is still alive. The wall of this tube is made of several anisotropic layers, with symmetry axes oriented in several directions. Each of these layer is made of a macromolecular material of biological origin with three main components : cellulose in the form of semi-cristalline macromolecules, hemicelluloses and lignin.

The cellulose is constituted by very long, up to about 5 micrometers, macromolecular chains packed into bundles. From place to place, segments of such chains order themselves in cristallites involving segments of one and the same chain, as well as segments of neighbouring cells. Other parts of the chain exhibit desorder, like in a linear synthetic polymer, and are thus amorphous. The degree of polymerisation of the long-chain cellulose polymer extents in the range 5,000 to 10,000. The width of such a chain is only 0.8 nanometers (8 Å). These chains are formed by the linear polymerisation of glucose anhydride cyclic molecules, bonding together by oxygen linkages. The glucose anhydride is formed, in the living sheath of the tree, the cambium, from the unit of glucose through the re.oval of a water molecule. The glucose itself is provided in the leaves, through the photosynthesis process, by a combination of water and carbon dioxyde in the presence of sunlight, with free oxygen as a by-product.

Other sugars are produced by the photosynthesis process. Along with glucose they provide the hemicelluloses, that are polymers of relatively low molecular weight. Most of these are branch-chain polymers, generally made of 150 or less basic sugar anhydrides.

In the cell wall, parallel cellulose chains are packed into long bundles, surrounded by hemicelluloses, that form the microfibrils. Within the microfibril, ten or more cristalline regions can be crossed by a single cellulose chain, and about 60 to 70 % of the cellulose in the cell wall is in cristalline form. The hemicelluloses take place in the space between the microfibrils and, to some extent, within the amorphous regions of the latter.

Lignin, a phenolic polymer of complex constitution and high molecular weight, occurs in wood in a variety of forms. It has the properties of a thermoplastic and is most often considered as playing the role of a matrix connecting and rigidifying the assembly of microfibrils, between which it is deposited, and, within these, in the amorphous regions. In fact recent electronic microscopy observations [20] seem to show that lignin may be present in the cell wall under the form of a network of separate strings, of about 7 to 10 nanometers in diameter, inserted between the cellulose microfibrils.

Within the thickness of the cell wall, the microfibrils are arranged with a high degree of order, as for most artificial composites. Four layers are observed denoted by P , the outer one or primary wall, and S_1 , S_2 , S_3 forming together the secondary wall. In P , formed at the beginning of the cell life, the microfibrils form a rather loose random network. The three layers S_1 , S_2 , S_3 are then progressively built up as deposites of microfibrils, spiraled around the cell interior. The outer one S_1 and the inner one S_3 exhibit a spiral angle almost perpendicular to the long axis (50 - 70° from it). The intermediate and thicker one S_2 form with this axis a much smaller angle (10 to 30°). Each layer S_1 or S_2 involves 4 to 6 layers of clustered microfibrils of uniform thickness called lamellae. Within the S_2 layer, the number of lamellae extends from 30 to 40 in thin-walled cells (early wood, grown in spring), to 150 or more in thick-walled cells (late wood, grown in summer). The total ticknesses are 0.1 micrometer for S_1 and S_3 , and 0.6 micrometer for S_2 .

The cells themselves can be of a few millimeter long and are sticked together by the middle lamella, made of lignin. Their geometry exhibits other details, like pits providing passageways for flow between the cell lumen. These are directly important for the circulation of fluids within the living tree, and also in wood, during the stages of drying or rehumidification for instance, and are thus involved in the couplings. Hardwood, which loose their leaves in winter season, exhibit special cells with large lumens disposed in series, forming thus vessels that conduct sap along the height of the tree.

Under temperate climate, the growth of the tree by its outer sheath gives to wood a further morphologic structure, in form of annual rings visible at the macroscopic level. Their thickness can extend from about 1 millimeter to one centimeter or more. As mentionned above, each ring involves two main layers called earlywood (thin walls, large lumen, low gross density), and latewood (thick walls, large gross-density).

In the trunck, the arrangement of the cells is modified around regions forming clamps for the branches. The cross-section of these regions appear as knots in the pieces of lumber.

Although the qualitative behaviour of wood is mainly governed by its constitution at its molecular or polymolecular level, the magnitudes of its mechanical or rheological parameters can be highly affected by the other levels of its internal structure. As we shall see in section 7, this implies that these various levels of heterogeneity must be taken into account in the modelization of wood properties. This has also direct consequences, like a broad scattering of the magnitudes of the mechanical characteristics, and also very pronounced scale effects.

4. QUASI-INSTANTANEOUS BEHAVIOUR AND THERMO-HYGRO-ELASTIC PROPERTIES FOR WOOD

Two of the most striking features of the quasi-instantaneous behaviour of wood under moderate loading are its anisotropy and its sensibility to environmental conditions, and specially to temperature and humidity. By quasi-instantaneous response, we mean the one that can be observed in current testing practice, in a few minutes or seconds.

For moderate loadings, the quasi-instantaneous response is elastic, with a linearity domain in the stress space depending on the orientation. We have described in [16] the main features of the quasi-instantaneous response under high loading that are known up to now. It turns out that, in this range, wood can exhibit various forms of instability that reacts on the shape of the load-deflection curve. For instance, under compression in the transverse direction, the progressive buckling of the shells constituted by the hollow cells results, in a first stage, into a decreasing tangent modulus and, in a second stage, into an increasing one. The S shape of the curve thus obtained resembles the one obtained for instance for rubber in tension.

In the linear elastic range, it is customary to say that wood is orthotropic with longitudinal Young modulus of elasticity ranging from 10 000 to 20 000 MPa, and transverse modulus being 10 to 20 times lower. In fact, as we have shown with our coworkers in [21], this is not exactly true since wood exhibits effects like piezoelectricity [22] that are forbidden for materials with a center of symmetry. The examination of the microstructure of wood, and specially of the coiling of the microfibrils always in the same sense, leads to the conclusion that wood belongs to another class of symmetries of the orthorhombic system, possessing three orthogonal axes of symmetry of order 2. Thus, strictly speaking, wood is not orthotropic. Nevertheless this has no influence upon the number, equal to 9 as well known, of independent coefficients in the tensor of elastic moduli or compliances, since they are the same for the three classes of symmetry of the orthorhombic system.

But one must be aware that this may be of influence for the tensors expressing the couplings with thermal or hygroscopic phenomena when the corresponding couplings tensors are of odd order. For instance, if one should consider couplings between the stress tensor and the vectors expressing the gradients of the temperature or humidity, or, conversely, between the moisture flux and the stress, one would have to consider the true class of symmetry involved in wood, since the corresponding coupling tensors are of order three. To our knowledge it seems that this point has not been considered up to now in the litterature.

Other thermo-hygro-elastic parameters are the thermal dilatation coefficient α , the moisture swelling (or shrinkage) coefficient β , the specific heat capacity c and the sorption heat capacity r . The two first are tensors of order two. Because of the orthorhombic symmetry, those tensors are diagonals, but not isotropic : they involve three independent components that, for instance, are β_L , β_R and β_T for the shrinkage coefficients in the longitudinal, radial and tangential directions, respectively. Typical values of β_T and β_R range, according to the species, from 6.10^{-2} to 6.10^{-1} and from 3.10^{-2} to 4.10^{-1} respectively, by percent of moisture content (cf. for instance [23]). The latter is counted in weight, with the oven-dry state as reference state. Thus, because of the low-density of wood, relative moisture contents higher than 100 % are frequently observed in the green state. The shrinkage coefficients are almost constants from the oven-dry state to moisture-content up to the so-called "fiber saturation point". According to the species, the magnitude of the latter turns around 25 % . Above this point, the shrinkage coefficients become almost zero. The longitudinal coefficient is 10 to 20 times lower than the tangential one. It has thus much less practical consequences. It is also the case for the thermal dilatation coefficient, with values turning around $10.10^{-6}\ K^{-1}$ depending on the species and the direction. This can be seen easily when noting that wood equilibrium moisture content in practical conditions of use ranges from about 8 % for indoor conditions to 18 % for outdoor conditions. Although the anisotropy of the materials influences both the elastic moduli tensor and the shrinkage tensor, it has been pointed out by GUITARD [24], that the stress obtained when shrinkage is fully restrained, as calculated through appropriate multiplication of the first tensor by the second one, is about to be isotropic.

A typical value for wood heat capacity is about $1.7\ kJ\ kg^{-1}K^{-1}$. A typical value for the sorption heat at low moisture content is about $10^4\ kJm^{-3}$. It falls at $10^3\ kJm^{-3}$ around the fiber saturation point. These values are highly dependent on moisture content.

A direct consequence of the values taken by the coupling parameters is that temperature-change under loading may be observed for wood [25], even for ramp loadings applied during a few minutes, and leading to axial compressive strains around 2 to 8.10^{-3}. For strains up to 5.10^{-3}, the temperature increase is a linear function of the strain. For a loading of 60 MPa applied in 6 minutes, a temperature rise of about 10^{-1} K may be observed. When the slope of the loading ramp is increased, the slope of the temperature rise in the proportional range increases too, and tends to the slope that can be calculated in adiabatic conditions from the above thermo-elastic properties. Thus, the behaviour observed in the proportionality range can be considered as thermoelastic. Beyond the proportionality limit, the temperature increases much faster. This can be attributed to dissipative effects coming from viscoelastic or viscoplastic deformations, and from damages occuring in the material beyond some yield-point. Of course, the results are also influenced by the thermal conductivity of wood, which is also a diagonal second order tensor λ, involving three independent components λ_L, λ_R, λ_T, with values depending also on moisture content.

As shown by BOEHLER et al. [26], anisotropy involves non homogeneous strains in tests performed in simple tension out of the grain. This has been shown to hold also for wood and requires specific testing disposals [27].

5. DELAYED RESPONSE IN CONSTANT ENVIRONMENTAL CONDITIONS

Another very important character of wood behaviour is its ability to exhibit delayed response to sustained loadings, that result for instance in phenomena like creep and relaxation. These properties of wood are known since a very long time, as it is attested for instance by the work of VICAT [28] at the beginning of the XIXth century. Although writing about half a century or more before MAXWELL, KELVIN, BOLTZMANN and VOLTERA laid the first theoretical bases for viscoelasticity, VICAT, at the beginning of his paper devoted to the first experiments made about the creep of iron, quoted creep and relaxation of wood as phenomena well known at that time.

When considering the delayed response of wood to sustained load, it is generally accepted that, under moderate loadings -up to about 50 % of the quasistatic instantaneous breaking strength-, the behaviour of wood in equilibrium with a constant environment is linearly viscoelastic. For higher loadings, nonlinearities and even delayed failure may be involved. Since the breaking strength of wood is a material parameter exhibiting rather large scatter, delayed failure during creep experiments may occur sooner than expected. This involves appropriate disposals when planning and performing the experiment.

A great variety of forms have been proposed for the creep function (see for instances [29] to [39]). Use is most often made of classical mechanical models, such as the generalized MAXWELL or KELVIN models, built from springs and dashpots, and of modified expressions obtained from the previous ones, as well as of expressions built on the power-law model. Since the latter has been successfully applied to a lot of other materials like bitumen [40], polymers, metals and cement concrete [41], it is an interesting fact that it seems also well fitted to the representation of wood creep. This is related to the very high initial rate of creep, followed by a rapid decrease of this rate, which can be observed on very long times. For this reason, use of logarithmic time scales in order to graphically represent the results must be preferred.

It is known that creep and relaxation, linear or not, can be observed on materials the behaviour of which is not pure viscoelasticity, but involves additional features, like plasticity or physico-chemical changes. Thus it seems valuable to use the criterion proposed by MANDEL [42] and which is based on shape restoration (restitution de forme in French). In this experiment, the strain is suddenly set back to zero. The response of a material to such an experiment is, first a reversal of the stress, second a progressive decrease of its magnitude, the so-called stress self-rubbing out phenomenon (effacement in French). If this stress rubbing out is total (the stress magnitude goes progressively to zero), one can conclude that the behaviour of the material is, linearly or not, truly viscoelastic. If it goes to a finite value different from zero when time goes to infinity, one must conclude that it is not viscoelastic, but viscoplastic, or that some chemical or physical change has occured during the experiment (as it is the case for instance for concrete which exhibits aging, or for polymers in the stage of polymerization).

We have performed recently [43] one first such test on a specimen of spruce, using, at room temperature, a testing machine that we have just designed for CTBA in cooperation with our coworkers B. FELIX and A. HADJ HAMOU. Up to the precision of the measurements, we observed that, at room temperature, the rubbing out is total after a few hours for a relaxation period of about one hour followed by shape restoration at zero strain. Thus, we can conclude that, at least in a certain domain of loading and environmental conditions, the behaviour of spruce is truly viscoelastic in MANDEL's sense. But this first observation requires to be confirmed and extended through a systematic experimental program.

In constant environmental conditions, creep of wood, and other correlative viscoelastic responses, are largely affected by the temperature and the moisture content. Higher are the magnitudes or these parameters, higher (or faster) is the creep.

Further, according to experiments performed in bending on pine timber, plywood, chipboards, waferboard, and fibre boards by DINWOODIE et alii [37], higher are the temperature and the relative humidity of the surrounding, more pronounced are their influence on the creep results. For instance, the following ordering sequence can be observed for the relative positions of the creep curves : (20°C, 90 % RH) $\gg$ (30°C, 68 % RH) $\gg$ (20°C, 65 % RH) > (20°C, 30 % RH) > (10°C, 62 % RH).

Quantitative representations of the influences of temperature and moisture content of wood, the latter being, in equilibrium conditions, related to relative humidity through the well known sorption isotherms, see for instance [44], have been proposed by RANTA-MAUNUS [45]. They are based on the particularization to constant conditions of FRÉCHET developments by multiple integrals of the behaviour functional (these developments are analogous to the ones we have studied in [46][47][48], and for which we have given a generalization of the CARSON-LAPLACE transform method, and relationships between generalized creep and relaxation kernels).

The expressions proposed by RANTA-MAUNUS for the creep factor of birch and spruce plywoods at constant temperature and various moisture contents are shown on Table 1, concurrently with expressions derived from a strain-hardening behaviour. In the latter, the creep rate $\dot{\varepsilon}$, at a given instant, is supposed to be a function of the stress, the temperature, the humidity and the total creep already performed. In all these expressions there is evidence of a strong coupling between creep, moisture content and stress level for birch, while no dependency upon the stress level is apparent for spruce. Thus the non-linearities and coupling phenomena may depend strongly on the wood species. RANTA-MAUNUS observed also that better fitting is obtained with the FRÉCHET development than with the strain-hardening model.

Table 1

SPECIES DEPENDENT COUPLING BETWEEN STRESS LEVEL AND MOISTURE CONTENT FOR THE CREEP FACTOR OF PLYWOODS IN CONSTANT CLIMATE (from RANTA-MAUNUS [45])

Fitting from Fréchet development

(2.1) $$\frac{E(t)}{E(o)} = 1 + 0{,}094\, t^{0{,}2} + 31\, u^{3}\, s\, t^{0{,}5}$$ (birch)

(2.2) $$\frac{E(t)}{E(o)} = 1 + 0{,}66\, u\, t^{0{,}254 + 1{,}42\, u^{2}}$$ (spruce)

Fitting from a strain-hardening law

(2.3) $$\frac{E(t)}{E(o)} = 1 + 3{,}04\, u^{1{,}35}\, s^{0{,}31}\, t^{0{,}26}$$ (birch)

(2.4) $$\frac{E(t)}{E(o)} = 1 + 1{,}215\, u^{1{,}33}\, t^{0{,}29}$$ (spruce)

with

s stress level = $\dfrac{\sigma}{\sigma_{breaking}}$

u moisture content = mass of water/mass of oven dry solid.

A thorough study of the influence of temperature on creep at constant moisture content has been performed recently by SALMÉN [49] between 20 to 140°C. Spruce specimen were submitted to sinusoidal loadings in water saturated conditions, and the two components E' and E'' of the longitudinal and transversal YOUNG complex moduli were measured for frequencies ranging from 5.10^{-2} to 2.10^{1} Hz. TOBOLSKY curves in terms of the temperature at a given frequency were plotted for the storage modulus E' and the loss factor $\tan\delta$.

Evidence of a pronounced glass transition appearing between 70°C and 110°C has been obtained. Above 110°C, a less pronounced decrease of the stiffness is observed, but no evidence of a rubbing plateau can be seen. Under 70°C, the storage modulus is also a decreasing function of temperature with a slope less than between 70°C and 110°C. SALMÉN found that master-curves analogous to the ones currently observed on artificial polymers could be obtained, and that the well-known W.L.F. equation [50] is valid in the above interval, with a glass transition temperature T_g equal to 72°C at low frequency, but with a range of validity of 55°C above T_g only, in place of about 100°C for industrial polymers. The corresponding apparent activation energy ΔH in the vicinity of T_g has been found to be of 450 kJ mol^{-1}. Below T_g, the experimental results appear to follow an ARRHENIUS law, with an activation energy of 395 kJ mol^{-1}. Above 130°C, the temperature dependency seems again to follow an ARRHENIUS law, with an activation energy of 570 kJ mol^{-1} (136 kcal/mole). Between 25°C and 75°C, an activation energy of about 67 kJ $mole^{-1}$ (16 kcal/mole) has been obtained recently in CTBA [51] for spruce at 12 % moisture content. These results have also confirmed that in general, wood creep increases more for moisture contents ranging from 8 to 18 %, than for temperatures ranging from 25 to 75°C.

6. DELAYED RESPONSE IN VARYING ENVIRONMENTAL CONDITIONS

In addition to the influence of constant temperature and moisture content on the delayed response of wood, it was observed about 25 years ago by ARMSTRONG and KINGSTON in Australia [52][53] that wood response may be largely modified when the values of these parameters are changed while the wood specimen is kept under load. One striking observation is that creep drastically increases when the moisture content is increased for the first time. Another one is that a further creep increase can result from a decrease of moisture content following the first increase. In fact, in cyclic moisture changes, subsequent drying periods appear to have more pronounced effects on the creep increase than the humidification periods. The latter may even, after the first one, result into some recovery of the strain, although, in constant conditions, creep is, as we have seen, a monotonic increasing function of moisture content.

To some extent, similar behaviour results from temperature changes. Evidence of many other striking features exhibited by wood behaviour with temperature and/or moisture content changes under load had been obtained in subsequent studies by various authors [54] to [64]. They have been reviewed extensively by GROSSMAN [58], who coined the term "mechano-sorptive effects" to name this class of wood behaviour. More recent studies have been made or are now in progress, namely by HUNT [65] in the United Kingdom, and by GRIL in France. A modelization of these phenomena elaborated recently by GRIL is communicated to this Symposium [66].

These effects may have serious consequences since the deflection exhibited by a loaded beam under extensive cyclic loading may reach 20 times the instantaneous deflection produced by the load at the beginning of the test.

A few characteristics of this behaviour are that :

- after correction of the corresponding -determined at zero load- swelling or shrinkage, a moisture content change under load may result in a net increase of the creep whatever the sign of the change ;
- after the first one, a decrease in moisture has generally a lower effect that a decrease ;
- the final deformation increase during a given step in moisture content depends on the step magnitude, but not on the duration of the change ;
- water movement induced by a moisture gradient involves no effects as far as the moisture content spatial distribution remains stationary in time ;
- on unloading, the supplementary deformation obtained through moisture content changes appear to be non recoverable when no further increase of moisture content are involved during or after unloading ;
- the permanent set thus obtained can be largely removed by a subsequent increase in moisture content ;
- this moisture induced recovery is not influenced by a delay of several weeks in the moisture increase after unloading, nor by a temporary increase in moisture content before unloading ;
- a rise in temperature up to 80°C during drying in a deformed state may involve a loss in recovery capacity ;
- the effects may vary from one kind of loading (bending, tension, compression, torsion) to another ;
- relaxation of stress in bending and in torsion during absorption may be greater than at constant moisture content ;
- in bending, the stress during absorption may first relax and then increase slightly after a time ;
- after a large moisture increase up to or near the fibre saturation point, subsequent cycling of the moisture content at lower levels do not involve further pronounced effect.

As it can be seen, these effects correspond to a complicated behaviour. In fact discrepancy between observations may still be found, depending on the author.

Only a few proposals have been made to represent these effects by constitutive equations, based on phenomelogical or physical consideration. Two main models exist. They have been provided by RANTA-MAUNUS in 1975 [57] and BAZANT in 1985 [63]. RANTA-MAUNUS used again a FRÉCHET development of the functional, that was then particularized in order to fit with experimental evidence. This resulted mainly in the use of a hydroviscoelastic coefficient a with three sets of values : a^{++} for first increase at any moisture content level, a^{-} and a^{+} for further decreases, or increases without trespassing the maximum moisture content reached previously. These parameters have been extensively studied by HUNT [60][62][65] in a recent past. He showed namely that their magnitudes and even signs may depend of both the moisture content and the previous moisture history, in the sense of the order of succession at least, since no chronology effects seem to be involved.

Adapting to wood an approach he previously developed for behaviour of cement concrete, BAŽANT used two kinds of generalized MAXWELL chain models built through thermodynamic arguments. The first kind was with varying viscosities and uncoupled thermal dilatation and hygroscopic swelling. The second was with non varying viscosities, but coupled thermal dilatation and hygroscopic swelling coefficients. He quoted the first as representing moisture induced creep, the second as stress induced shrinkage, and demonstrated the equivalence of these two concepts as two ways of factorizing the coupling product $\sigma \dot{h}$, where σ is the stress, h the humidity, and dot denotes the time derivative.

Because of length limitations, we cannot give explicit account of the two important models thus developed by RANTA-MAUNUS and BAŽANT, and that are described in full details in the original papers quoted above. Because of the complexity of these phenomena, the amount of available experimental evidence is far from being sufficient.

7. MAIN IDEAS PROPOSED FOR INTERPRETATION AND MODELLING

In the past twenty years, a lot of attempts have been made to give, from several points of view and at various levels of the wood constitution, explanations and prediction tools for the various phenomena exhibited by wood behaviour. We shall distinguish mainly the molecular or physico-chemical level, the anatomic or micromechanical level and the macroscopic level. We shall mention also various attempts made to identify by experiments the various internal variables that are thought to be responsible for the phenomena under consideration.

It is now widely accepted that the qualitative features of the various couplings involved in the mechanical response of wood take their origin in the existence of hydrogen bonding between the various elements of the macromolecules and/or the water molecules constituting wood.

Based on this basic ideas, several -more or less refined- models have been elaborated. For instance, as soon as 1966, KAUMAN [67] formulated the hypothesis that the same molecular linkages may be involved by swelling with water absorption and by mechanical deformations. Using simple arguments of statistical mechanics and the absolute chemical reaction rate theory, he arrived at the conclusion that the creep of wood, the swelling on sorption, the temperature and moisture dependencies may well be explained by a rate process with experimental activation energy of 15 to 35 kJ $mole^{-1}$ (4 to 8 kcal/mole) and that several linkages with that energy must be broken simultaneously in a cooperative process. More or less developed theories based on similar ideas have been developed since that time by NISSAN [68], CAULFIELD [69] and BAZANT [63] in the United States, by VAN DER PUT [70] in the Netherlands, and by GRIL [66] in France. NISSAN introduced the concept of the class of "H - bond dominated solids", i.e. the class of solids in which the density and characteristics of the H-bond primarly govern mechanical properties, while covalent, ionic or Van der Waals bonds or entropic mechanisms play a minor role. Paper and ice, belong to this class. It turns out that, at least to some extent, it is also the case for wood as well as for cement concrete. For such materials, an important parameter is the total number of intermolecular H-bonds per unit volume of solids. Other important parameters are the energy U required to break the bond, and the force constant of stretching the bond given by $k_R = \partial^2U/\partial R^2$ $(R = R_o)$, where R is the total distance between the two atoms linked by the H atom, and R_o its equilibrium value, corresponding to the minimum of U. The values of k_R may be evaluated theoretically from quantum mechanics and give for instance 13.5 to 18 N/m for ice and 15.3 to 19.9 N/m for cellulose inter -or intra- molecular bonds, with energies ranging from 4.5 to 5.0 kcal/mole. These figures for k_R are of the same order of magnitude as ones obtained for ice from spectroscopic measurements.

The concept of a population of H-bonds breaking and reforming randomly in a fashion influenced by water molecules arrival or departure from a sorption site, as well as by temperature and external mechanical action, has also been used by IRVINE [71], BAZANT [63], HUNT [65], GRIL [66] and others.

Both CAULFIELD [69] and VAN DER PUT [70] make use of the absolute reaction rate equation for activation over an energy barrier of some conceptual "flow unit" with no further specification. The application of an external force introduces a bias in the curve describing the space variation of the potential energy, breaking the symmetry between the forward and backward transitions. The facts that one direction becomes thus preferred results, after a time, in a macroscopic effect. Higher is the frequency of the transitions, i.e. the temperature, sooner the effect is visible and faster is the creep. These considerations have been extended to delayed fracture under constant load, and other loading programs, through the introduction of the surface energy involved at a crack tip [70]. From experimental data obtained for Douglas fir, CAULFIELD derived the size of the hypothetical moving element and the height of the potential energy barrier. He arrived for delayed fracture of Douglas fir at about 5.6×10^{-27} m^3 (5 600 $Å^3$) and 117 kJ/mole (28 kcal/mole), quoting that these magnitudes have also been reported for paper in tensile rupture. He derives from this value of the energy that, as far as the breaking process must be attributed to the breaking of hydrogen bonds, about 6 such bonds must break cooperatively as proposed by NISSAN and KAUMAN.

At the level of the cell wall, it has been recognized [72][65] that the angle of the microfibrils in the S_2 layer has a marked influence on shrinkage, creep and delayed fracture and on the mechano-sorptive effect as well. The composite structure of the cell-wall has also been introduced into models [73].

At the level of the cellular micro-structure of wood, several analytical or numerical models, more or less sophisticated have already been elaborated in the past twenty years [74][75].

At the macroscopic level, analytical or numerical analyses of the strains and stresses distribution inside wood pieces of finite size under the influence of cutting, sawing or drying have been performed. We may quote for instance [76][77][78] for application of analytical methods to loss of the redistribution of growth stresses in log ends due to crosscutting, and [79] for application of finite element methods to the stress distribution around a knot. Most of these modelizations are made with a linear orthotropic elastic constitutive law. Recently, classical linear viscoelastic models have been used for the calculation of drying stresses [79].

Experimental analysis has also been used, through various methods, to determine the distributions of the temperature, the moisture content, the strains and the stresses inside specimens of finite size, see for instance [81][82]. This generally involves serious experimental difficulties, namely because, for precise measurements, destructive methods are needed. For instance, a method has been developed in Nancy [83] to measure the growth-stresses through small "carots". Another method developed in CTFT and CTBA by CHARDIN et alii makes use of a full destruction of a trunck through progressive cutting off long slices of wood on which various measurements are performed (cf. SALES [84]).

Other various experimental techniques are used in order to determine the internal parameters of wood constitution. These involve X-Ray absorption for the determination of the local bulk density [85], or X-Ray diffraction to determine the microfibrillar angles, as well as the amount of crystalline cellulose in the various cell wall layers [86]. They may involve also spectroscopic or other physico-chemical methods like IR or RMN analysis.

These methods are important to provide information for the modelization of the wood behaviour at the micro-mechanical level. We are now beginning a program on this topic in ENPC-CERAM in cooperation with CTBA and other laboratories.

8. CONCLUSION

As seen from the indications presented in this review lecture, the couplings between mechanical, thermal and hygroscopic phenomena that can occur in wood are numerous. We have sketched a few ones in some more details than the others, although still briefly.

Although a lot of work has been done, as may be seen from the above review, the present situation is that one is still far from having available tools capable of reliable predictions on the behaviour of wood pieces or structures in variable loadings and variable environmental conditions. For instance, prediction formulas proposed in the new Eurocode or various national building codes lack still of sufficient -scientifically established- grounds, and also, as a consequence, of general acceptance.

The tools presently available for modelization of industrial processes, for instance at the level of the design of industrial plants or facilities and of process control, are still insufficient, mainly because not enough quantitative knowledge on the constitutive equations of the material in terms of the numerous parameters involved are up to now available for use in the powerful numerical methods of analysis that exist nowadays.

This results in a gap that will be of more and more consequences for the wood industries, that must compete with other materials and improve their capability of answering to the evolution of the requirements in quality and/or costs.

As it can be seen from the elements given in this lectures, it turns out that the problems involved in wood thermo-hygro-rheology and thermo-hygro-mechanics are largely interdisciplinary. Physico-chemistry is needed as well as pure mechanics, numerical analysis or statistics. Thus good specialists of these various disciplines, each one being ready to grasp something from the others, must be gathered in teams associated to one project. This gives the opportunity of other couplings that are expected to be fruitful and to provide the needed progress.

ACKNOWLEDGEMENTS

This work has been done partly in Ecole Nationale des Ponts et Chaussées, Centre de Recherches en Analyse des Matériaux, in the framework of the CNRS Scientific Group n° 81 "Rheology of Wood", and with the support of the French Technical Center for Wood and Furnitures (CTBA), the French Ministry for Research, the French Agency for the Mastering of Energy (AFME), and the Economic European Community, and partly under a consultant contract with CTBA.

REFERENCES

[1] HUET, C., BOURGOIN, D., RICHEMOND, S., (Eds). Rheology of anisotropic materials. Proceedings of the 19th Colloquium of the French Group of Rheology (Editions CEPADUES, F Toulouse, 1986).

[2] ARBOLOR, Rapport d'activité, Juin 1985 (CNRF Champenoux, F 54280 SEICHAMPS).
[3] SALES, C., Growth stresses : main results obtained on the study of some Guyana species and consequences for end use (communication to I.U.F.R.O. Congress, Ljubljana, September 1986).
[4] C.T.B.A., Fiche d'information sur le séchage du bois (Fiche technique n° 41. Centre Technique du Bois et de l'Ameublement, Paris, 1984).
[5] SIMPSON, W.T., Drying Technology Int. Jal, 2(2) (1983-84), 235-264.
[6] SIMPSON, W.T., Drying Technology Int. Jal, 2(3) (1983-84), 353-368.
[7] French D.T.U., Règles de calcul et de conception des Charpentes en Bois C.B. 71 (Eyrolles, Paris, 1972).
[8] CRUBILÉ, P., EHLBECK, J., BRÜNIGHOFF, H., LARSEN, J. and SUNLEY, J., Eurocode 5. Common unified rules for timber structures. Report for the European Communities (1985).
[9] GROSSMAN, P.U.A., Wood Sc. and Tech. 10 (1976) 163-168.
[10] BAŽANT, Z.P. (Ed.). Draft State of the Art Report on creep and shrinkage of Concrete : mathematical modelling. In : BAZANT, Z.P. (Ed.) Preprints of the Fourth RILEM Int. Symp. on creep and shrinkage of concrete (Northwestern University, Evanston, ILL., August 1986) pp. 134-141.
[11] STMITH, W.B. and SIAU, J.F., J. Inst. Wood Sc. 8-45 (1979) 129-133.
[12] BASILICO, C. and MARTIN, M., Int. J. Heat and Mass transfer 27-45 (1984) 657.
[13] BASILICO, C., MOYNE, C., High Temperature convective Drying of softwood and hardwood : drying kinetics and product quality interactions. In Proc. 4th Int. Drying Symposium (Kyoto, July 1984).
[14] MORGAN, K., THOMAS, H.R. and LEWIS, W., Wood Sc. 15-2 (1982) 139-149.
[15] BAJOLET, D., SOBUE, N. and PLUVINAGE, G., Contribution to the study of the effect of drying stress on fracture toughness of wood. In : HUET, C., BOURGOIN, D., RICHEMOND, S. (Ed.) Rheology of anisotropic materials (Editions CEPADUES, Toulouse) pp.577-586.
[16] HUET, C., Physical bases for the thermo-hygro-rheological behaviours of wood in finite deformations. In : GITTUS, J., NEMAT-NASSER, S. and ZARKA, J., (Eds) Large deformations of solids, Physical Bases and Mathematical Modelling (Elsevier, London, 1986) pp. 409-438.
[17] SEIGNEUR, P. and BARADUC, A., Cintrage du Bois massif. Internal report H 809 (CTBA, Paris, 1985).
[18] TRIBOULOT, P., GUITARD, D. and DARYANTO, H., Thermocompression dans le Bois massif. Ecole d'été "Séchage". Le Kleebach 30-31 Mai 1985 (ARBOLOR, Nancy, 1985).
[19] HÖGLUND, H., SOHLIN, U. and TISTAD, G., Tappi, 59-6 (1976) 144-147.
[20] RUEL, K. and BARNOUD, F., Aspects ultrastructuraux et macromoléculaires sur la lignine d'épicéa. In : W.G - KAUMAN (Ed.). Colloque Sciences et Industries du Bois du Ministère de la Recherche et de la Technologie, Grenoble 20-22 Sept. 1982 (C.T.B.A., Paris) Vol. 3, pp. 139-149.
[21] LE GOVIC, C., TRENARD, Y. and HUET, C., Material symmetries in wood : morphological data and physical consequences. In : HUET, C., BOURGOIN, D., RICHEMOND, S., Rheology of anisotropic materials (Editions CEPADUES, Toulouse, 1986) pp. 413-426.
[22] FUKADA, E., J. Phys. Soc. Japan, 10-2 (1955) 149-154.
[23] JOLY, P., MORE-CHEVALIER, F., Théorie, pratique et économie du séchage des bois (Vial, F Dourdan, 1980).
[24] GUITARD, D., SALES, C., PRESIOZA, M., Contraintes internes de séchage dans le matériau bois : le tenseur des coefficients de contrainte de séchage comme caractéristique isotrope de la matière ligneuse. Bois et Forêt des Tropiques (to appear).
[25] IMAYAMA, N., Mokuzai Gakkaishi, J. Japan Wood Res. Soc., 27-7 (1981) 529-534.
[26] BOEHLER, J.P., EL AOUFI, E., Heterogeneity of stress and strain fields in anisotropic materials. In : HUET, C., BOURGOIN, D., RICHEMOND, S., Rheology of anisotropic materials (Editions CEPADUES, Toulouse, 1986) pp. 131-150.
[27] FELIX, B., HADJ HAMOU, A., LE GOVIC, C., ROUGER, F., HUET, C., Incidence de l'anisotropie du bois sur son comportement mécanique (Communication to ARBORA Colloquium, Bordeaux, Oct. 1986, to appear).

[28] VICAT, L.J., Annales de Chimie et de Physique, 54 (1833) 35-40.
[29] KITAZAWA, G., Relaxation of wood under constant strain. Technical Publ. n° 67 (The New-York state College of Forestry, Syracuse, N.Y., 1947).
[30] KITAHARA, K. and OKABE, N., J. Jap. Wood Res. Soc. 5-1 (1959) 12-18.
[31] SCHNIEWIND, A.P. and BARRETT, J.D., Wood Sc. and Tech. 6 (1972) 43-57.
[32] RANTA-MAUNUS, A., Viscoelasticity of plywood under constant climatic conditions (Publication 3, Technical Research Center of Finland, Building Technology and Community Development, Helsinki, 1972).
[33] HOYLE, R.J., GRIFFITH, M.C., ITANI R.Y., Wood and Fiber Sc. 17-3 (1985) 300-314.
[34] GRESSEL, P., Holz Roh-Werkstoff 44 (1986) 133-138.
[35] KOLLMANN, F., Holz Roh-Werkstoff 3 (1961) 73-80.
[36] MUKADAI, J., Wood Sc. Technol. 17 (1983) 203-216.
[37] DINWOODIE, J.M.W., PIERLE, C.B. and PAXTON, B.H., Wood Sc. Technol. 18 (1984) 205-224.
[38] REINHART, H.W., Holz Roh-Werkstoff 31 (1973) 352-355.
[39] PIERLE, C.B., DINWOODIE, J.M. and PAXTON, B.H., Wood Sc. Technol. (1985) 83-91.
[40] HUET, C., Annales Ponts et Chaussées, 6 (1965) 373-429.
[41] BAŽANT, Z.P. and OSMAN, E., Matériaux et Constructions 9-49 (RILEM, Paris, 1976) 3-11.
[42] MANDEL, J., Rheol. Acta 12 (1973) 393-397.
[43] HUET, C., unpublished result (Centre Technique du Bois et de l'Ameublement, 1986).
[44] CHRISTENSEN, G.N. and KELSEY, K.E., Aust. J. Appl. Sci. 9 (1958) 265.
[45] RANTA-MAUNUS, A., cf. [32].
[46] HUET, C., Rheol. Acta 12 (1973) 279-88.
[47] HUET, C., SERVAS, J.M. and MANDEL, J., Comptes-Rendus Acad. Sc. Paris 277 A (1973) 1003-5.
[48] HUET, C., J. Rheology 29-3 (1985) 245-257.
[49] SALMÉN, L., J. Materials Sc. 19 (1984) 3090-6.
[50] FERRY, J.D., Viscoelastic properties of polymers (John Wiley and Sons, N.Y., 1980).
[51] LE GOVIC, C., HADJ HAMOU, A., FELIX, B. and HUET, C., Fluage de l'épicéa en flexion circulaire : étude expérimentale des facteurs humidité et température (communication to ARBORA Colloquium, Bordeaux, Oct. 1986).
[52] ARMSTRONG, L.D. and KINGSTON, R.S.T., Nature 185-4716 (1960) 862-863.
[53] ARMSTRONG, L.D. and KINGSTON, R.S.T., Aust. J. Appl. Sc. 13-4 (1962) 257-76.
[54] RACZKOWSKI, J., Holz Roh-Werkstoff 27-6 (1969) 232-237.
[55] URAKAMI, H. and FUKUYAMA, M., Mokuzai Gakkaishi, J. Japan Wood Res. Soc. 15-2 (1969) 71-75.
[56] ARMSTRONG, L.D., Wood Science 5-2 (1972) 81-86.
[57] RANTA-MAUNUS, A., Wood Sc. Technol. 9 (1975) 189-205.
[58] GROSSMAN, P.U.A., cf. [9].
[59] ARIMA, T. and GROSSMAN, P.U.A., J. Inst. Wood Sc. 8-2 (1978) 47-52.
[60] HUNT, D.G., J. Int. Wood Sc. 9 (1982) 136-138.
[61] POŽGAJ, A., Holztechnologie 6 (1984) 318-320.
[62] HUNT, D.G., J. Materials Sc. 19 (1984) 1456-7.
[63] BAŽANT, Z.P., Wood Sc. Technol. 19 (1985) 159-177.
[64] BAŽANT, Z.P. and MEIRI, S., Wood Sc. Technol. (1985) 179-182.
[65] HUNT, G.D., J. Materials Science 21 (1986) 2088-96.
[66] GRIL, J., A uniaxial model for hygro-thermo-mechanical loading of wood, this volume.
[67] KAUMAN, W.G.K., Holz Roh-Werstoff 24-11 (1966) 551-556.
[68] NISSAN, A.H., Wood Sc. Technol. 11 (1977) 147-151.
[69] CAULFIELD, D.F., A chemical kinetics approach to the duration of load problem in wood (Private communication, Forest Product Laboratory, Madison, WISC., 1983).
[70] VAN DER PUT, T.A.C.M., A model of deformation and damage processes based on deformation kinetics (Private communication, Stevin Laboratory, NL Delft, 1986).

[71] IRVINE, G.M., Wood Sc. Technol. 19 (1985) 139-149.
[72] CAVÉ, I.D., Wood Sc. Technol. 6 (1972) 284-292.
[73] CAVÉ, I.D., J. Microscopy 104-1 (1975) 47-52.
[74] PERKINS, R.W., Forest Products J. 17-5 (1967) 59-70.
[75] STUPNICKI, J., Analysis of the behaviour of wood under external load based on a study of the cell structure (Acta Polytechnica Scandinavia, Civil Eng. and Bdng. Const. Series n° 53, Trondheim 1968).
[76] BYRNES, F.E. and ARCHER, R.R., Wood Sc. 10-2 (1977) 81-84.
[77] SALES, C., Contribution à l'analyse des contraintes de séchage dans le bois (Thèse D.D.I., Nancy I, Juin 1984).
[78] SALES, C., BORDONÉ, P., Residual stresses induced by drying : mathematical model, measurements, and consequences on the drying process. In : HUET, C., BOURGOIN, D., RICHEMOND, S. (Ed.) Rheology of anisotropic materials (Editions CEPADUES, F Toulouse, 1986), pp. 561-576.
[79] CRAMER, S.M. and GOUDMANS, J.R., Wood and Fiber Sc. 15-4 (1983) 338-349.
[80] RAZAFINDRAKOTO, J.C. and VALENTIN, G., Rheological problems in models of softwood drying. In : HUET, C., BOURGOIN, D., RICHEMOND, S. Rheology of anisotropic materials (Editions CEPADUES, F Toulouse, 1986), pp. 543-560.
[81] BAI, G.L., GARRAHAN, P., Wood Sc. Technol. 18 (1984) 121-135.
[82] FUTÓ, L.P., Holz Roh-Werkstoff 40 (1982) 45-50.
[83] POLGE, H., KELLER, R. and NEPVEU, G., (Private communication, C.N.R.F., Champenoux, F. Seichamps, 1984).
[84] SALES, C., cf. [3].
[85] POLGE, H., (Private communication, CNRF, Champenoux, F Seichamps, 1984).
[86] PAAKARI, T., SERIMAA, R., Wood Sc. Technol. 18 (1984) 79-85.

Nota Bene : Although already rather long, this list contains only the references quoted in the text as significant examples, with privilege given to the oldest, the newest or the french ones. It must not be considered as exhaustive. Many other relevant papers on these subjects that are available are not quoted here because of length limitations. Nevertheless, we feel that the present references list can be helpful for people that want to enter this field.

Thermomechanical Couplings in Solids
H.D. Bui and Q.S. Nguyen (Editors)
Elsevier Science Publishers B.V. (North-Holland)

UNEXPECTED PHENOMENA IN CYCLIC HIGH-TEMPERATURE PLASTICITY

F.D. FISCHER
Prof. of Mechanics, Univ. for Mining and Metallurgy,
A-8700 Leoben, Austria

F.G. RAMMERSTORFER
Prof. of Light Weight Structures, Techn. Univ. of Vienna,
A-1040 Vienna, Austria

F.B. BAUER
Chief Research Engineer, Vereinigte Edelstahlwerke,
A-2634 Ternitz, Austria

H.J. BOEHM
Asst. Prof., Inst. of Light Weight Structures,
Techn. Univ. of Vienna, A-1040 Vienna, Austria

1. INTRODUCTION

The deformation behaviour of highly alloyed steel specimens under cyclically alternating temperatures is studied both experimentally and by computer simulation. The motivation for this project arose from the fact that certain components of heat treatment facilities tend to be deformed very much in the course of their operating life. Such components, e.g. hangers, are periodically heated to high temperatures (typically about 900°C) and then quenched by dipping into water or oil. This cyclic temperature loading of the hangers leads to accumulated damage. In such cases failure typically occurs due to large cracks initiated internally, which totally differ in character from the well known chill cracks in casting dies etc.

2. EXPERIMENTAL PROGRAM

Cylindrical specimens of different geometries and of three different types of steel were used in the experimental program. The specimen studied in detail in this paper is a cylinder with a radius R=20 mm and a length L=90 mm made of an austenitic steel X15 CrNiSi 25 20. The experimental setup is shown in Fig. 1 and consists of two furnaces and two quenching pools. The specimens were heated in a furnace to about 900°C within 20 min. Then they were pulled out and transferred into the quenching pools within 7 sec, during which period they were subjected to cooling by the surrounding air. The subsequent quenching in water lasted 20 min, following which the specimens were pulled out of the pool and remained in the air for a further 20 min, before the cycle was started again with heating in the furnaces. Each specimen was subjected to more than 2000 such cycles unless failure occurred earlier.

FIGURE 1
Test Setup

It turned out that some specimens - typically those manufactured from ferritic steels - broke into pieces after about 200 cycles, while austenitic specimens tended to endure about 2000 cycles. Some of the results of this experimental program are reported by Kohl et al. [1] and are reproduced in Figs. 2 to 5.

Fig. 5 shows the deformed shape of an austenitic specimen after 1660 cycles. It can be seen that under the same heating/cooling regime the austenitic specimen did not alter its shape excessively and became "barrel shaped". Austenitic-ferritic specimens tended to assume a more spherical shape, their cratered surface giving

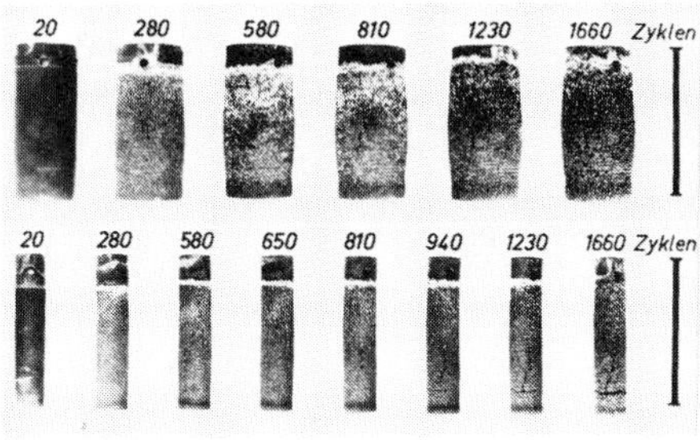

FIGURE 2
Shape deformation of an austenitic steel specimen as function of the number of cycles;
First row: R=20 mm, L=90 mm
Second row: R=10 mm, L=90 mm

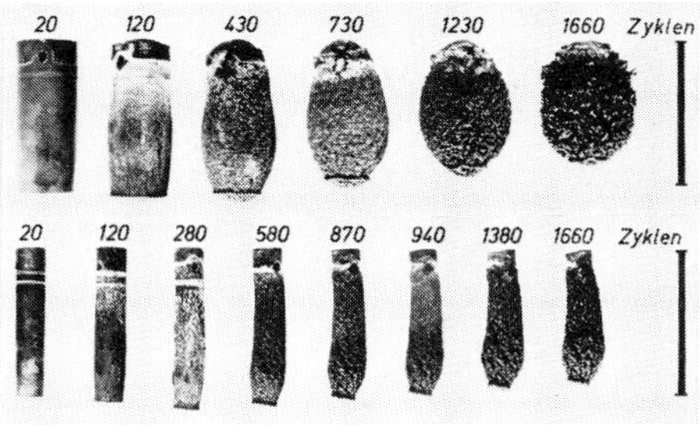

FIGURE 3
Shape change of an austenitic/ferritic steel specimen as function of the number of cycles;
First row: R=20 mm, L=90 mm
Second Row: R=10 mm, L=90 mm

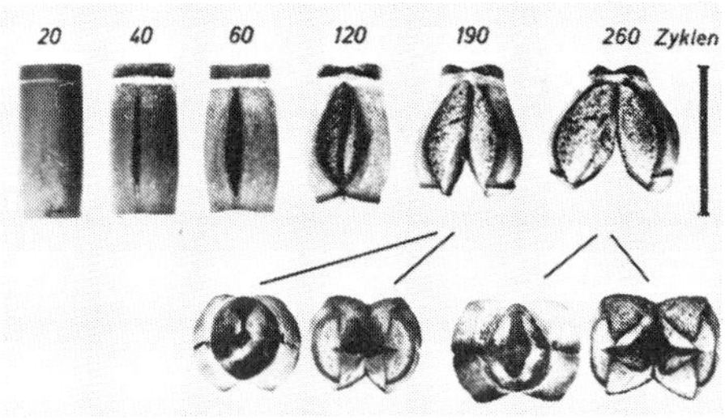

FIGURE 4
Shape deformation of a ferritic steel specimen as function of the number of cycles;
R=20 mm, L=90 mm

FIGURE 5
Shape of an austenitic specimen after 1660 cycles

them the air of a bizarre sculpture. Ferritic specimens typically failed after the development of large, axially oriented cracks.

3. COMPUTATIONAL ANALYSIS

3.1 Temperature Distribution

The temperature fields were computed using the finite element code ADINAT [2], the main emphasis being on the cooling phase, during which high temperature gradients were expected to appear. For this analysis a constant temperature of 900 C was assumed for the initial temperature field, corresponding to the situation after 20 min of heating. The specimen loses heat through convection and through radiation, corresponding to heat fluxes of the type

$$q_r = -f.\varepsilon.h_r.(T_s^4 - T_e^4)$$

and

$$q_c = -h_c.\beta.(T_s - T_e)$$

respectively.

Here q_r and q_c are the heat fluxes due to radiation and convection, respectively, f is a shape factor, ε the emissivity and h_r the Stefan-Boltzmann constant. T_s and T_e are the surface and environmental temperatures measured in Kelvin, respectively, and h_c is the convection coefficient, which, due to the formation of steam bubbles on the specimen's surface, depends on the surface temperature in a rather complicated manner.

The h_c data actually employed for the computation are based on the h_c-distribution described by Mitsutsuka et al. [3] suitably modified by using a position-dependent correction factor β. The original curve given by Mitsutsuka was derived experimentally for plates submerged in cold water and is shown in Fig. 6.

The correction factor β for the convection coefficient h_c was found by dividing the submerged surface of the sample into four zones each having its own correction factor and setting the appropriate values of β in such a way that the calculated temperatures correspond closely to the temperatures measured experimentally in four points.

By writing the heat flux due to radiation in the form

$$q_r = -f.\varepsilon.h_r.(T_s - T_e)(T_s + T_e)(T_s^2 + T_e^2)$$

one can formally define an "effective radiative heat transfer coefficient"

$$h_c^r = f.\varepsilon.h_r.(T_s + T_e).(T_s^2 + T_e^2)$$

with

$$q_r = -h_c^r.(T_s - T_e)$$

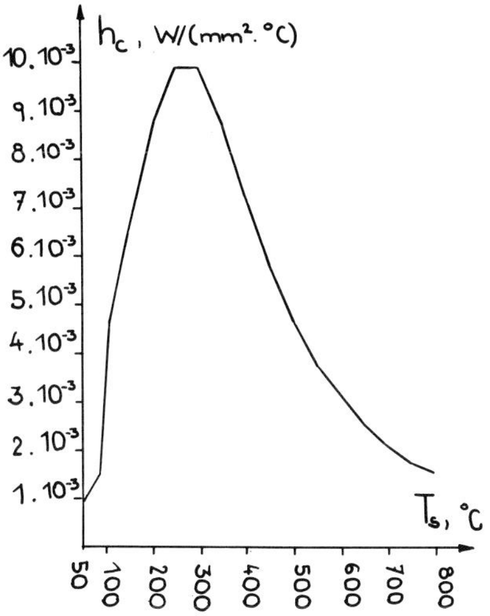

FIGURE 6
Convection coefficient for plates submerged in cold water (from Mitsutsuka et. al. [3])

which allows a comparison of the relative importance of radiative and convective heat transfer. For T_e=293 °K values for h_c^r typically range between 10 and 100 W/(m^{2}K), which are at least one order of magnitude smaller than the convection coefficients as shown in Fig. 6.

The above heat transport model incorporates material nonlinearities as well as nonlinear boundary conditions and the temperature fields, of course, had to be found iteratively. The calculated and measured temperature-time-histories for selected points can be compared in Figs. 7a and 7b.

3.2 Calculation of Deformations and Stresses

Deformation and stress calculations were carried out employing a thermo-elastic-plastic material model incorporating creep and kinematic hardening [4], the well known finite element code ADINA developed by Bathe et al. [5] being used. The mechanical material properties used in the computations are shown in Figs. 8a and 8b.

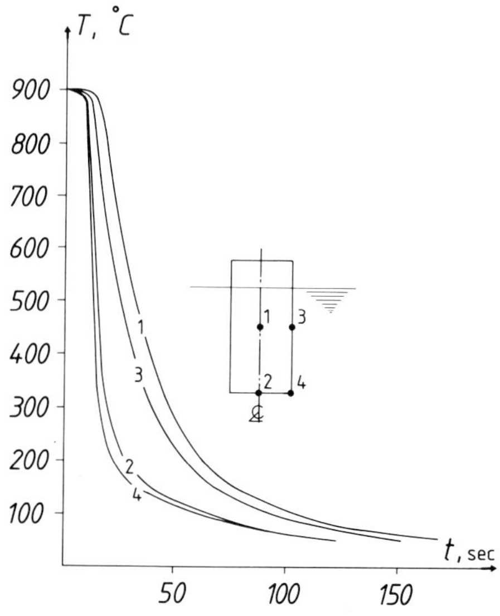

FIGURE 7a
Temperature-time history, measured

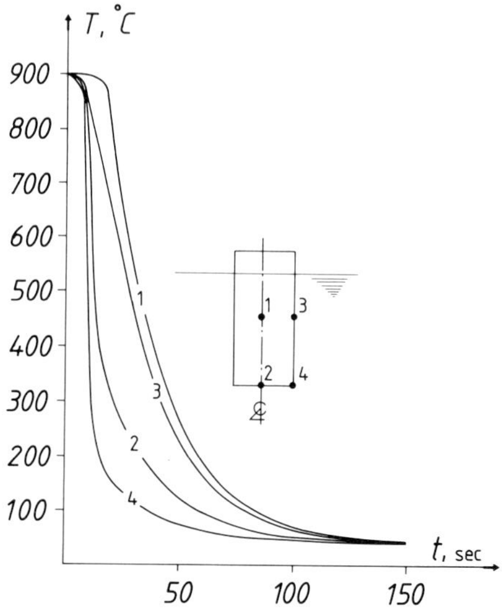

FIGURE 7b
Temperature-time history, computed

3.3 Discussion of the Stress State and the Deformed Shapes

After about 150 sec of cooling the specimen has an average temperature of about $40^\circ C$ and can be treated as cooled down. Figs. 9a and 9b show the three residual stress components, σ_r, σ_ϑ and σ_z as well as the accumulated equivalent plastic strain ε_v^P, which is defined via the plastic work W^P by the relations

$$\Delta\varepsilon_v^P = \tfrac{2}{3}\left(\Delta\underline{\varepsilon}^P\right)^T.\Delta\underline{\varepsilon}^P, \quad \Delta W^P = \underline{\sigma}^T.\Delta\underline{\varepsilon}^P,$$

where $\Delta\underline{\varepsilon}^P$ is the vector of the incremental plastic strain components. σ_r, σ_ϑ and σ_z are depicted along the r-direction at z=0. The accumulated equivalent plastic strain surpasses values of 2% in the vicinity of the specimen's axis, thus indicating a zone of high material straining within each cooling cycle.

Following a further cooling phase of 20 min the heating of the specimen to $900^\circ C$ within 20 min was modelled and the redistribution of the residual stresses by creep was studied. It was found that above a temperature level of about $600^\circ C$ a few minutes were sufficient for the stresses to vanish. Figs. 10a and 10b show the relative deformations of the surface of the specimen at room temperature and above $600^\circ C$, respectively.

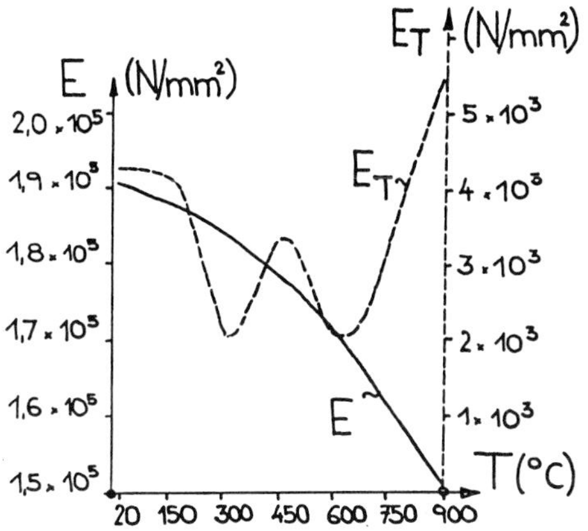

FIGURE 8a
Temperature dependence of Young's modulus E and strain hardening modulus E_T

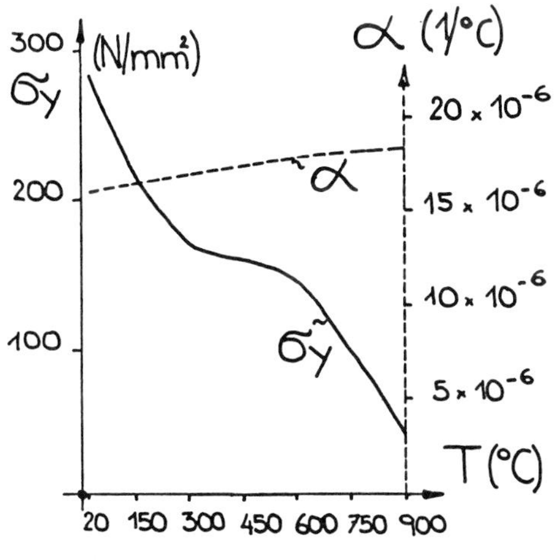

FIGURE 8b
Temperature dependence of the initial yield stress σ_y and the mean value of the coefficient of thermal expansion α

The relative deformations are defined as the irreversible contributions to the total displacements, which remain after subtracting the displacements caused directly by unrestrained thermal expansion. The relative deformation vector

$$\underline{u}_r^T = (u_r, w_r)$$

has the components

$$U_r = U - \alpha(T_t) \cdot (T_{ref} - T_t) \cdot r,$$
$$W_r = W - \alpha(T_t) \cdot (T_{ref} - T_t) \cdot (z - z_0)$$

where T_t is the actual temperature of the sample, $\alpha(T_t)$ is the mean value of the thermal expansion coefficient at the current temperature T_t, T_{ref} is the stress free reference temperature and z is the axial coordinate of an immovable reference point.

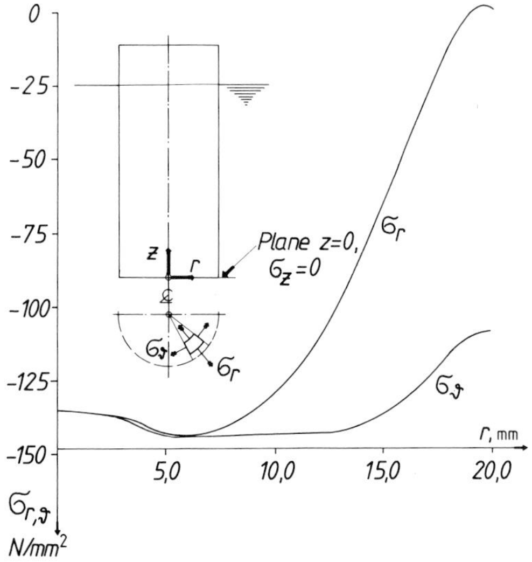

FIGURE 9a
Stress components at the base after
150 sec of cooling

The two radial deformation shapes in Figs. 10a and 10b do not differ very much. Due to the elastic part of the stress state the volume of the sample is increased slightly at room temperature. In the nearly stress free state after some minutes of creep the original volume and the "deformed" volume are nearly exactly the same. The length of the specimen is reduced by about 0.0055 mm per temperature cycle. The radial shape changes tend to make the specimen more barrel shaped, the maximum increase of the radius being about 0.0045 mm per cycle. If this value is compared to the increase of the radius of about 17% found experimentally after 870 cycles, which corresponds to a mean increase of the radius of 0.0045 mm per cycle, the agreement between experiment and calculation can be seen to be surprisingly good.

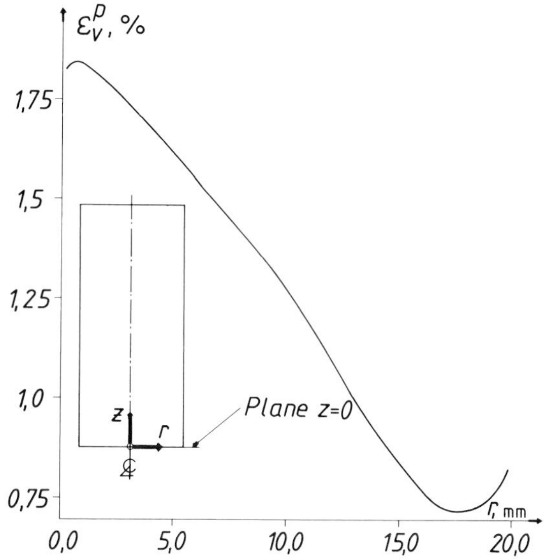

FIGURE 9b
Accumulated equivalent plastic strain
at the base after 150 sec of cooling

Thus, in the course of each cooling-down/heating-up cycle residual stresses arise due to position dependent cooling in the quenching phase and are practically totally relaxed by creep during the respective heating phase. These stresses, however, cause the sample to undergo shape changes, which are qualitatively similar for each cycle and which accumulate. Of course, crack systems arising during the temperature cycling influence the resulting shape, too. Such crack systems tend to start at the centre of the sample's bottom. Here the radial stress σ_r and the hoop stress σ_ϑ are equal and thus the stress path is angled at 45 in the stress plane, compare Fig. 11a.

Following the shifting of the yield curve, the size of which changes in dependence of the actual temperature and therefore of time, at first the stress state is "tension" until about 12 sec after the onset of quenching, see point "1" in Fig. 11a. This is followed by 5 sec of unloading, until point "2" is reached and then plastification under "compression" takes place as far as point "3", i.e. until 150 sec into the cooling cycle, at which

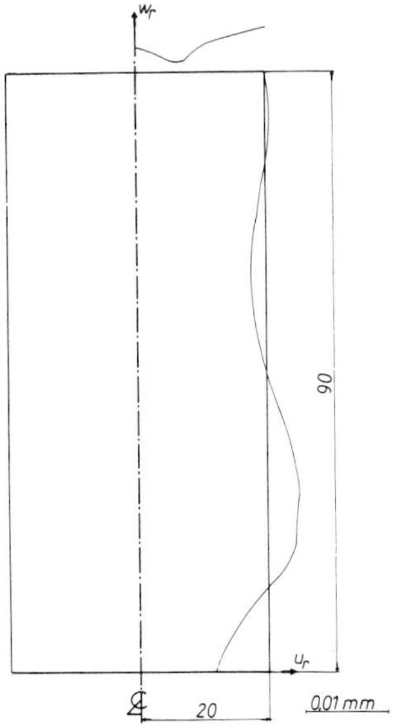

FIGURE 10a
Relative deformations after cooling to room temperature

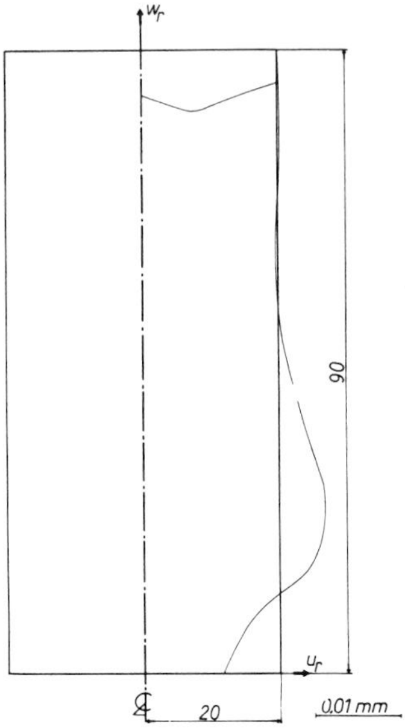

FIGURE 10b
Relative deformations after reheating at temperatures in excess of 600°C

time the sample has reached room temperature. Finally, during the heating phase, creep/relaxation effects cause the stress state to change to unloading again. If creep effects are neglected, the stress point reaches position "4" in Fig. 11a at the end of the cycle. Consideration of creep causes the stress point to return to the origin of the stress plane at the end of the cycle, but the origin of the actual yield surface may now be shifted and does not necessarily coincide with the origin of the stress plane.

Qualitatively the plastic work can be calculated from Fig. 11b by studying the area under the curve for σ_V, the equivalent stress, in relation to the accumulated equivalent plastic strain. The dashed lines in Fig. 11b give a qualitative estimate of the influence of creep. The specific work dissipated by plastification lies between 2.5 and 3.8 Nmm/mm^3. If the same amount of plastic work is invested into a sample under uniaxial tension, 1.5% to 2% plastic strain will arise for the same material at a temperature of 15°C.

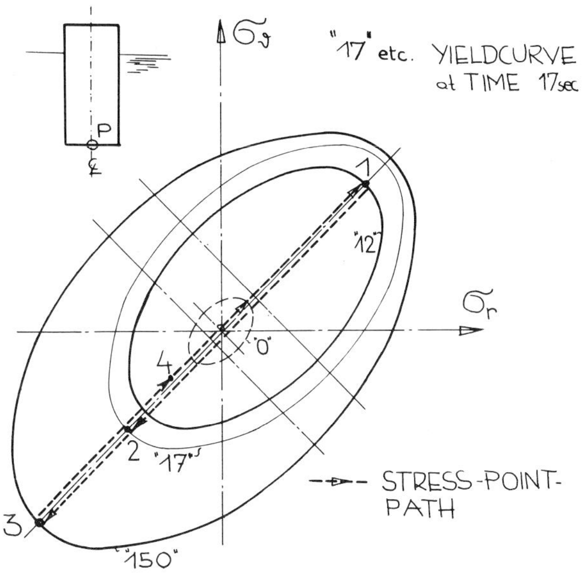

FIGURE 11a
Path of the stress point in the stress plane at r=0, z=0

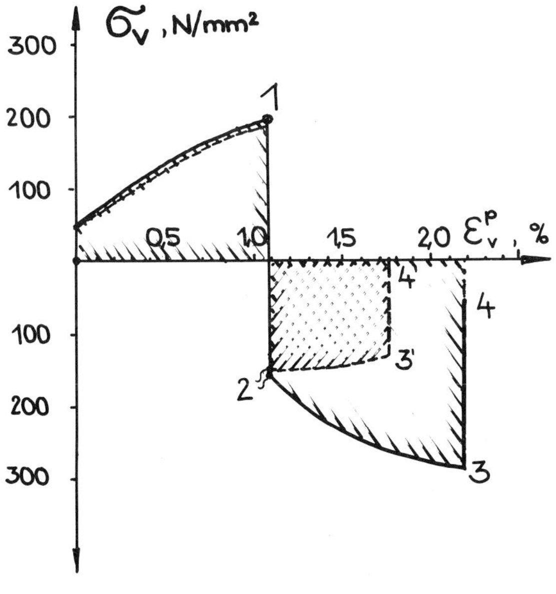

FIGURE 11b
Equivalent stress versus accumulated equivalent strain at r=0, z=0

Observing the Manson-Coffin-Rule [6]

$$N \cdot (\Delta\varepsilon^{p})^{n} = const$$

where N is the number of load cycles leading to failure and n is an exponent ~ 3, it appears to be qualitatively justified to conclude that the high amount of plastic work invested cyclically is responsible for the development of cracks in a low number of cycles, which was observed for each specimen of the type under consideration.

REFERENCES

[1] Kohl H., Bauer F.: Schaedigung hitzebestaendiger Staehle und Legierungen durch staendigen Temperaturwechsel, Proc. VDI-Werkstofftechnik, Tagung "Werkstoff und Schaden", Munich 1983.

[2] ADINAT: A Finite Element Program for Automatic Dynamic Incremental Nonlinear Analysis of Temperatures, Rept. AE 81-2, ADINA Engng. Inc., Watertown, MA, 1981.

[3] Mitsutsuka M., Fukuda K.: Trans. Iron Steel Inst. Japan, 21 (1981) 689.

[4] Snyder M.D., Bathe K.J.: Formulation and Numerical Solution of Thermo-Elastic-Plastic and Creep Problems, MIT-Rept. 82 448-3, Dept. of Mech. Engng., MIT, Cambridge, MA, 1977.

[5] ADINA: A Finite Element Program for Automatic Dynamic Incremental Nonlinear Analysis, Report AE 81-2, ADINA Engng. Inc., Watertown, MA, 1981.

[6] Kocanda S.: Fatigue Failure of Metals, Sijthoff and Noordhoff Intl. Pub., Alphen van den Rijn, Netherlands, 1978.

SUMMARY

First an extended experimental program is described reporting the results of cyclic heating/cooling of highly alloyed steel cylinders. Specimens having the same initial shape and undergoing the same heat treatment reached totally different shapes depending on the steel type. In a second step a computational analysis is presented for a specific specimen. The conclusion is reached that the change of the samples' shape is due to the plastic strains accumulated during the cooling phase. The high level of residual stresses at the lowest temperatures vanishes again during the following heating phase due to creep/relaxation. The results of a series of heat treatments are approximated by adding a number of the above cycles. Finally it should be emphasized that this is a project under research and that a final resumee for different materials and geometries cannot be given at present.

Thermomechanical Couplings in Solids
H.D. Bui and Q.S. Nguyen (Editors)
Elsevier Science Publishers B.V. (North-Holland)

METALLIC STRUCTURE UNDER MECHANICAL AND CYCLIC THERMAL LOADING

J.F. JULLIEN - M. COUSIN - S. IGNACCOLO
Laboratoire BETONS et STRUCTURES
INSTITUT NATIONAL DES SCIENCES APPLIQUEES
I.N.S.A.
20 Avenue A. Einstein
69621 VILLEURBANNE Cedex - France

Abstract :

The study concerns the behaviour of metallic structures submitted to the combination of a primary loading and a secondary cyclic thermal sollicitation.

More and more complex models allow the production of the cyclic plastic behaviour of materials. Nevertheless, many difficulties still remain in analysing the behaviour of structures. The very conservative nature of the rules of construction shows the uncertainties within this field. A better comprehension of the phenomena is likely to be achieved by experimental studies. Most of the experimentations that have been developed over these last few years concern essentially the structure of "type BREE". As a matter of fact, concerning the question of the three bars as well as for the test of tensile-twisting or tensile-bending, the sollicitation applied are such that they produce a state of stresses which is the same on the whole structure. Therefore, such studies bring out behaviour of material more than behaviour of structures. Tests on structures are still very rare and especially tend to observe the total behaviour. It has become necessary to increase experimentations where the combination of primary and secondary sollicitations touches a restricted volume of the structure and where an adapted device of measurements allows the study of the local and global behaviours of structure. It is for this reason that we have developed two experimental studies whose results are compared with those stemming from numerical calculation.

The first study concerns the behaviour of cylindrical shell under tensile constant load and cyclic axial thermal step. The intensity of the thermal gradient can reach 350°C./cm. This value corresponds to an intensity of thermal stress three times the yield stress of the material. The axial step in temperature is obtained by forced internal convection. The escape gases of a burner are aired into the cylinder where a deflector forces them to hit the internal side very locally at high speed. A watering by means of jets focused on the same level allows cooling and supply of the desired temperature. The radial thermal step is insignificant, considering the small thickness of the cylinder and the mode of heating used here. The interest in the chosen method for creating thermal sollicitation is to keep the external side of the cylinder quite unoccupied for the measurements. The instrumentation must enable the measure of residual displacements around important displacements caused by the primary and secondary sollicitations. This necessity has obliged us to conceive and set up transducers of axial and radial displacements adapted to the measure under rigorous thermal conditions.

Characterisation'tests under monotonous tensile and cyclic strains imposed at different temperatures allow us to point out the material behaviour law according to different types of modelization.

We have selected the CASTEM System whose INCA part especially gives the possibility of taking a cyclic hardening of the material into account with a development of the parameters with temperature. The comparison of the results brings an imperfect determination of displacement values for the ultimate state by means of numerical modelization. However, the correct description of the behaviour in the course of the first cycles is very promising for the capacities of the calculation codes in order to find the solution for such problems.

The second study concerns the interaction of mechanical and cyclic thermal stresses on the instability of cylindrical shells.

Thin cylindrical shells are submitted to a lateral external pressure and an axial cyclic thermal gradient. The cylindrical shells are obtained by electrodeposition of nickel on an aluminium mandrel (diameter : D = 150mm, Length : L = 150mm, thickness e = 0.120mm to 0.250mm). This process leads to near-perfect cylinders.

The thermal secondary loading is obtained by an internal circular inductor. In order to improve the axial thermal gradient the lower part of the cylinder is cooled by a maintained level of water. Due to the severe thermal conditions, the behaviour of the structure is studied by using an automatic rotative scanning system with a non-contact transducer. Several combinations of loadings are studied. If a small number of thermal cycles is applied to the cylinder before applying a lateral pressure loading, the buckling pressure load is slightly reduced. If cyclic thermal loading is applied simultaneously with a maintained pressure load, the pre-critic radial displacements increase with the number of thermal cycles until buckling occurs.

The main purpose of this work is to bring out points of reference not only useful for the validation of the codes, but also for the simplified methods which remain a finality within this field.

Thermomechanical Couplings in Solids
H.D. Bui and Q.S. Nguyen (Editors)
Elsevier Science Publishers B.V. (North-Holland)

DYNAMIC BEHAVIOUR OF A CARBON-RESIN COMPOSITE

M.NOUILLANT, F.JOUBERT, J.M. DELAS
Laboratoire de Génie Mécanique de l'I.U.T. "A"
351, cours de la Libération
Université de Bordeaux I - 33405 Talence Cedex
FRANCE

I. Introduction

Many static studies have been achieved concerning the behaviour of composite materials made of organic matrix reinforced with high modulus fibers : values of stiffness matrix parameters, fracture analysis. But, concerning the dynamic behaviour, few experiments have been performed to show the dissipative phenomena, mostly due to the internal damping.

As for as usual metallic alloys are concerned, both the internal damping value stays low (about 10^{-3}) still the yield point, and temperature's rising of some degrees may be considered as admissible. But concerning composite materials, this is no more true.

Actually, the application of cyclic loading, even at low level, in particular material directions leads to a significant value of the complex part of the elastic modulus, then creates a source distribution inside the material. Depending upon the geometrical parameters, the thermal characteristics of the material, the global exchange conditions and this kind of loading, this source distribution leads to a temperature field. If the loading and the thermal exchange conditions lead to a temperature's field, where, in every point of the structure, the thermomechanical coupling is low, then the working conditions are stable.

But if, locally, exist sites with high thermomechanical coupling, then a local resin melting may occur. The melting zone may be or not extensive according to the kind of applied loading.

The aim of this paper is on one hand to propose methods to establish, from experimental results in tension - compression, relations between the internal damping and the following parameters : stress (σ), frequency (F), temperature (T), and fibers directions (θ) ; on the other hand, to forecast, using an iterative method applied to the heat equation, the working limits of a structure under a cyclic loading.

II . Experimental devices for mechanical energy measurement in tension - compression.

To access to the mechanical energy, two main kinds of devices are usually used.

II .1 Classical test machines :

These machines may be considered as an excitation system by free extremity. In this case, the phase-shift between the displacement and the force is equal to the loss angle (δ). A measurement of this phase shift allows to accede directly to the internal damping : $\frac{\Delta W}{W} = 2\pi \; tg \; \delta$

That kind of device is appropriate as for as the internal damping of composite materials is significant, and allows the study of their behaviour under strong stress and low frequency (5 Hz to 20 Hz)

II .2 Resonant machines :

These machines are considered as a system composed by two identical masses connected by the specimen. It is excited by a sinusoîdal force (generated by an electrodynamic vibrator) applied to the first mass. Then, in oscillation resonance, the phase shift between the exciting force and displacement speed is equal to 90°.

In this conditions, the internal damping value is given by :

$$\frac{\Delta W}{W} = \frac{\pi}{M} \cdot \frac{F}{\gamma}$$

M : the first mass
F : exciting force
γ : acceleration of the first mass

This device works in a diferent range of frequency (60 to 300 Hz)

II .3 Specimens :

The tests have been performed on hollow parallelipipedic bars made of carbone - epoxy.

The coordinate system is described on the following figure.

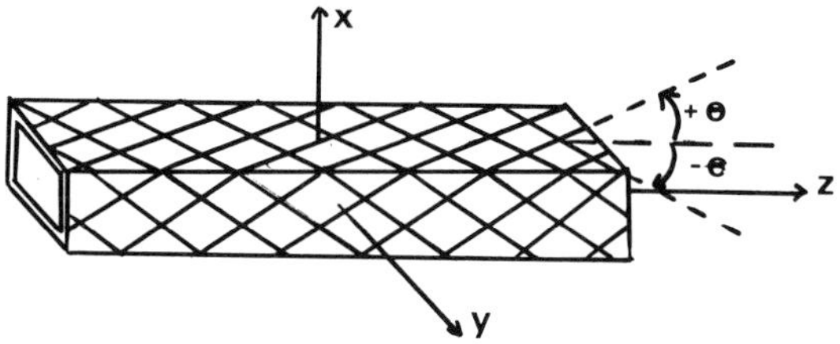

Oz : longitudinal direction
Ox, Oy : transverse directions
2L : specimen total length
V : specimen total volume
θ : fiber orientation (in longitudinal direction Oz)

III Determination of the law : $\Delta w = f(T, \sigma_{cte}, F_{cte}, \theta_{cte})$

Various methods may be imagined to establish such a relation.

III.1 : 1st method.

The specimen is maintained at a constant temperature (T_r) using a cooling device (forced convective rule).

The measurement of the value $(\frac{\Delta w}{w})$ for different temperatures allows to establish a general function :

$\frac{\Delta w}{w} = f(T)$ for σ = cte ; F = cte ; θ = cte

This method allows measurements at high stresses and low temperatures.

III.2 : 2nd method :

This method tends to determine the relation $\frac{\Delta w}{w} = f(T)$ from the heat equation upon a conduction rule (radially insulated specimen).

Assuming that there is no transverse heat flux ($\phi_T = 0$) and that the source distribution is not uniform, then the heat equation is written :

$$\lambda(\theta) \,.\, \Delta T(z) + q(z) = 0$$

or since : $q(z) = - \lambda(\theta) \,.\, \Delta T(z)$

The experimental knowledge of the temperature distribution ($T(z) = f_2(z)$) allows then, performing a double derivation, to accede to the source distribution : $q(z) = f_1(z)$

Substituing the variable z between these two relations yields to :

$$q(T) = f_1 \circ f_2^{-1}(T) \quad (1)$$

Also, assuming the whole injected mechanical energy is heat transformed, then :

$$\Delta w_{meca.} = \Delta w_{ther.} \quad (2)$$

From relations (1) and (2), we can write a law such as :

$$\frac{\Delta w}{w} = f(T, \sigma_{cte}, F_{cte}, \theta_{cte})$$

This method is quick and easy but somehow acute : the source distribution obtained by a double derivation of the experimental temperature profile may be erroneous if the interpolation function doesn't take the errors relative to the data acquisition into account.

III . 3 3rd method

$\frac{\Delta W}{W} = \sum_{i=0}^{n} a_i T^i$ is the supposed form of the $\frac{\Delta W}{W}$ (T) relation.

The specimen is subjected to tension - compression cyclic loading and the whole injected mechanical energy is assumed to be heat transformed :

$\Delta w_{meca.} = \Delta w_{therm.} = \Delta w$

Let us call W the maximal elastic energy, and share the whole volume V in p elementary volumes V_i.

$$V = \sum_{i=1}^{p} V_i$$

As, Δw and w are extensive variables, and as the stress field is considered as uniform in tension - compression.

$$\frac{\Delta W}{W} = \frac{1}{p} \sum_{i=1}^{p} \left(\frac{\Delta W_i}{W_i}\right)$$

then : $$\frac{\Delta W}{W} = \frac{1}{p} \sum_{i=1}^{p} \sum_{j=0}^{n} a_j T_i^j$$

So using different temperature fields and the associated $(\frac{\Delta W}{W})$ measurement, we get a system such as :

$$\left[\frac{\Delta W}{W}\right] = [\,M\,] \cdot [a]$$

The resolution of this system allows to calculate every a_i and set up a relation such as :

$$\frac{\Delta W}{W} = \sum_{i=0}^{n} a_i T^i$$

IV An iterative approach of the working limit

Assuming that the stress distribution is known and constant, the $\frac{\Delta W}{W}(T)$ relation has been determined, we proposed an iterative method applied to the heat equation for acceding to the temperature distribution of a working structure.

The working system stability is depending on the convergence of iterative method. For more clearness of this paper we assume the specimen is subjected to a pure conduction rule.

Hypothesis : - pure conduction : $\Phi_T = 0$

- known geometrical characteristics and boundary conditions
- $\Delta W_{meca.} = \Delta W_{therm.}$
- constant working stress : $\sigma_T = cte$
- constant frequency anf fiber directions
- initial specimen temperature T

* 1st step :

An uniform stress distribution σ_T is applied to the specimen.

As the initial temperature is constant then the source distribution too : $q(z) = q_0 \quad -L \leq z \leq +L$

$$\text{Since} \begin{cases} \dfrac{\Delta W_1}{W} = f(T_0) \\ q_0 = \Delta W_1 \end{cases} \Rightarrow \quad q_0 = W \,.\, f(T_0)$$

Solving the heat equation yields to the temperature distribution at this stage :

$$\Delta(T_1(z)) = \frac{q_0 L^2}{2\Lambda}\left(1 - \frac{z^2}{L^2}\right) \quad ; \; -L \leq z \leq +L$$

* 2nd step

The sample still under σ_T is set under a $T_1(z)$ temperature distribution. Then the source density may become a z function.

As locally $\begin{cases} \left[\dfrac{\Delta W_2}{W}\right]_{local} = f[T_1(z)] \\ q_1(z) = (\Delta W_2)_{local.} \end{cases} \Rightarrow q_1(z) = W_{local} \cdot f[T_1(z)]$

($W_{local.}$: elastic energy density)

Solving the heat equation yields to a new temperature distribution at this stage (solution gived in these conditions, by CARSLAW and JAEGER)

$$\Delta T_2(z) = \frac{4.L}{\pi^2.\lambda} \sum_n \frac{1}{n^2} \cos\left(\frac{n\pi}{2L} z\right) . \int_{-L}^{+L} q_1(x) . \cos.\left(\frac{n\pi}{2L} x\right) dix$$

* i^{th} step

The sample still under σ_T is set under a T_{i-1} (z) temperature distribution. Then the source distribution may become a z function.

As locally :

$$\begin{cases} \dfrac{(\Delta W_i)}{W}_{local.} = f\,[T_{i-1}(z)] \\ q_{i-1}(z) = (\Delta W_i)_{local.} \end{cases} \Rightarrow q_{i-1}(z) = W_{local} \cdot f[T_{i-1}(z)]$$

Solving the heat equation allows the knowledge of the new temperature distribution at this stage :

$$\Delta T_i(z) = \frac{4.L}{\pi^2.\lambda} \sum_n \frac{1}{n^2} \cos\left(\frac{n\pi}{2L} z\right) . \int_{-L}^{+L} q_{i-1}(x) . \cos.\left(\frac{n\pi}{2L} x\right) dx$$

Thus, there is stability of the system if :

$$\forall z \in [-L ; +L] \qquad \lim_{i \to \infty} T_i(z) = T(z)$$

On a strict point of view, the convergency is then defined by :

$\forall\ \varepsilon > 0,\ \exists\ I(\varepsilon,z)\ /\ i > I(\varepsilon,z) \Rightarrow |T_i(z) - T(z)| < \varepsilon\ ,\ \forall z \in [-L ; +L]$

V Conclusion

The present paper gives experimental methods available for particular conditions of the $\frac{\Delta W}{W}(T, \sigma, F, \theta)$ relation.

The iterative method which is proposed to study the stability problems is generally available. However, except for simple cases (as the one we studied) that's to say for complex geometries under mixed thermal conditions (conduction, convection, radiation) the use of numerical methods is necessary to solve the heat equation.

Thermomechanical Couplings in Solids
H.D. Bui and Q.S. Nguyen (Editors)
Elsevier Science Publishers B.V. (North-Holland)

THE THERMODYNAMICS OF CREEP DAMAGE§

A.C.F. Cocks* and F.A. Leckie+

*Department of Engineering, Leicester University, Leicester LE1 7RH, England
+Department of Theoretical and Applied Mechanics, University of Illinois at Urbana-Champaign, 216 Talbot Laboratories, 104 S.Wright St., Urbana, IL 61801 U.S.A.

An internal state variable theory is presented for the tertiary stage of creep based on an understanding of the internal mechanisms responsible for the degradation of the material. It is shown that a scalar potential exists from which the inelastic strain-rate and rate of increase of damage can be derived. It is recognized, however, that for certain classes of materials and certain types of loading this may not be the most appropriate form for the constitutive laws.

1. INTRODUCTION

When a component is stressed at temperatures in excess of 0.3 of the absolute melting point it may creep and eventually fail. A typical creep curve obtained from a test conducted at a constant stress, σ, is shown in Fig.1. The quantities that are important to the Engineer are the steady-state creep-rate, $\dot{\varepsilon}_{ss}$, the time to failure, t_f, and the strain to failure, ε_f. A measure of the extent of the tertiary stage of creep is given by the quantity

$$\lambda = \frac{\varepsilon_f}{\dot{\varepsilon}_{ss} t_f} \tag{1.1}$$

which has been termed the creep damage tolerance by Ashby and Dyson [1]. A high value of λ indicates that the material can readily redistribute stress away from points of initial stress concentration in structural situations, while the response of components composed of materials with low values of λ can be very sensitive to the presence of any stress concentration features.

The effect of stress state on the time to failure is given by the shape of the isochronous surface, Fig.2. These surfaces connect those stress states that lead to the same failure times. Leckie and Hayhurst [2] identified the two extremes of behaviour shown in Fig.2, where the time to failure is either determined by the maximum principal stress or the von Mises effective stress. It is important that any constitutive laws we develop are capable of accurately reflecting these uniaxial and multiaxial features of material behaviour.

§This work was supported by the Department of Energy through contract DOE 1198 with the Materials Research Laboratory at the University of Illinois.

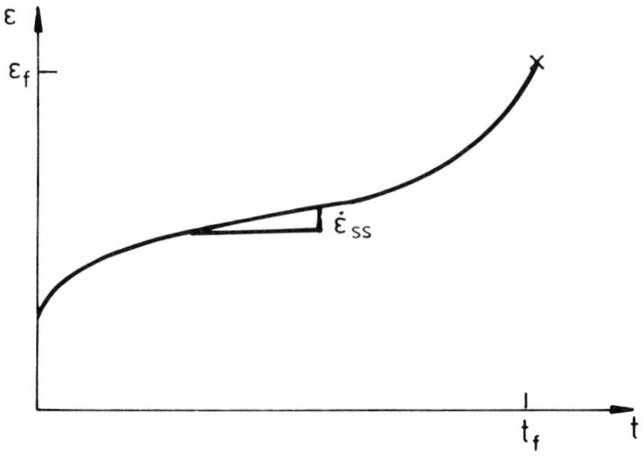

Fig.1 Uniaxial creep curve

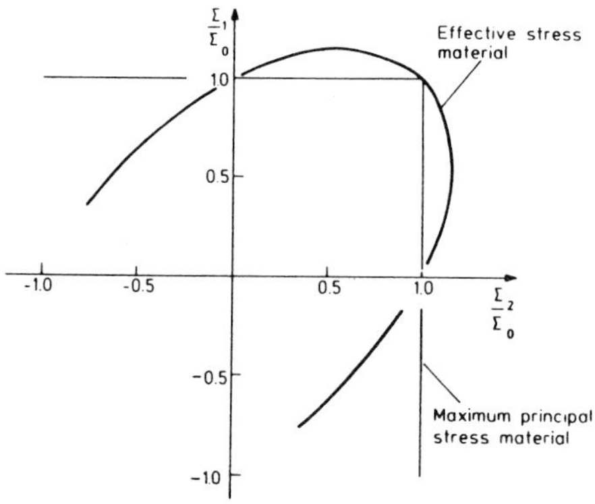

Fig.2 Isochronous surfaces in plane stress space for maximum principal stress and von Mises effective stress materials

The increasing creep-rate during the tertiary stage of creep is due to the accumulation of damage in the material. This damage can take several forms : voids can nucleate and grow within the material ; dislocation networks can form around precipitate particles and accelerate the creep rate by aiding the climb of dislocations over the particles; or the contribution made by precipitate particles to the yield strength and creep resistance of the material can decrease as the material deforms plastically. In each case we can describe the state of the material at a given instant in time by its Helmholtz free energy. The contribution from the voids and dislocations are included in terms of internal state variables. This thermodynamic description is presented in Section 2 where we identify the conditions under which it is possible to obtain a potential from which the inelastic strain-rate and damage-rate can be derived. In Section 3 we use these results to obtain the structure of the constitutive relationships when the damage is in the form of intergranular cavities which grow by grain-boundary diffusion. The resulting structure for other mechanisms of damage are briefly discussed and compared with the results of this void growth mechanism. At this stage the theory can deal with any number of internal state variables, for example, when the damage is in the form of intergranular cavities a state variable can be assigned to each grain-boundary in the material. We must, however, impose certain constraints on the final form of constitutive relationships, recognizing that any set of equations must be verifiable experimentally and must be suitable for use in component analysis. These constraints require the constitutive relationships to contain only a limited number of state variables. In Section 4 we reduce the equations for the void growth mechanism to a single state variable theory and examine how the predictions of the model compare with general observations of material behaviour. It is found that values of λ predicted by this model are much less than those obtained experimentally. The predictions can be improved by increasing the number of state variables or by remodelling the material so that it is possible to use a single state variable theory. This second possibility is examined, which yields a slightly different structure for the constitutive laws.

2. THERMODYNAMIC FORMALISM

In this section we outline the approach used by Cocks and Leckie [3] in obtaining constitutive relationships for damaging materials, which was based on the work of Rice [4]. For each mechanism of deformation and damage we can identify a set of internal state variables, α_k, which either represent the position of dislocations within the material or the distribution and size of any voids. For a given internal structure and applied elastic strain, E_{ij}^e, the Helmholtz free energy is given by

$$\psi = \left(\frac{1}{2} C_{ijk\ell} E_{k\ell}^e E_{ij}^e + f(\alpha_k)\right) V \qquad (2.1)$$

Where $C_{ijk\ell}$ is the elastic stiffness tensor, which may be a function of the α_k's, and $f(\alpha_k)$ is a function of the state variables, which either represents the additional elastic stored energy due to the presence of the dislocations or the internal surface energy due to the presence of the voids.

The thermodynamic forces associated with the state variables E_{ij}^e and α_k can be found by differentiating eqn.(2.1):

$$\dot{\psi} = (\Sigma_{ij} \dot{E}_{ij}^e + S_k \dot{\alpha}_k) V \qquad (2.2)$$

where

$$\Sigma_{ij} = \frac{1}{V} \frac{\partial \psi}{\partial E_{ij}^e}$$

is the remote stress, and

$$S_k = \frac{1}{V}\frac{\partial \psi}{\partial \alpha_k}$$

The repeating index of eqn.(2.2) implies summation over all the internal variables.

We now accept the Clausius-Duhem inequality as the proper statement of the second law of thermodynamics for irreversible processes:

$$V\frac{dS}{dt} + \frac{\partial}{\partial x_i}\left(\frac{qi}{T}\right) = \dot{\Theta} \geq 0 \qquad (2.3)$$

where S is the entropy per unit volume, qi is the rate of heat flux out of an element of material, T is the absolute temperature, x_i is the distance from a datum and Θ is the rate of entropy production. For isothermal processes eqn.(2.3) becomes, after making the usual manipulations [5],

$$\Sigma_{ij}\dot{E}^T_{ij} V - \dot{\psi} \geqslant 0$$

where $\dot{E}^T_{ij}$ is the total strain-rate. Substituting eqn.(2.2) into eqn.(2.3) gives

$$\Sigma_{ij}\dot{E}^P_{ij} - S_k\dot{\alpha}_k \geqslant 0$$

where $\dot{E}^P_{ij} = \dot{E}^T_{ij} - \dot{E}^e_{ij}$ is the inelastic strain-rate. This strain results from the motion of dislocations through the material or from the growth of voids, so that

$$\dot{E}^P_{ij} = g_{ijk}\,\dot{\alpha}_k \qquad (2.4)$$

Combining this with eqn.(2.4) gives

$$(\Sigma_{ij}g_{ijk} - S_k)\,\dot{\alpha}_k \geqslant 0$$

or

$$F_k\,\dot{\alpha}_k \geqslant 0 \qquad (2.5)$$

where $F_k = (\Sigma_{ij}g_{ijk} - S_k)$ is the thermodynamic driving force for the process.

To proceed further we need expressions for the rate of increase of the internal state variables. For each damaging mechanism described by Ashby and Dyson [1] and examined by Cocks and Leckie [3] these relationships are of the form

$$\dot{\alpha}_k = \dot{\alpha}_k(F_k, \alpha_k) \qquad (2.6)$$

i.e. the rate of increase of α_k is only a function of the affinity, F_k, associated with it and the present value of α_k. In these situations Rice [3] has shown that a scalar potential, Φ, exists from which the inelastic strain-rate can be derived:

$$\dot{E}^P_{ij} = \frac{\partial \Phi}{\partial \Sigma_{ij}} \tag{2.7}$$

where $\Phi = \Phi\ (\Sigma_{ij},\ S_k,\ \alpha_k)$

Cocks and Leckie [3] further demonstrate that the rate of increase of the internal state variables are derivable from the same potential.

$$\dot{\alpha}_k = -\frac{\partial \Phi}{\partial S_k} \tag{2.8}$$

If a composite material is formed from a number of elements of volume V_ℓ where the response of each element can be represented by a potential Φ_ℓ, then a potential

$$\Phi = \frac{1}{V}\ \Sigma \Phi_\ell V_\ell$$

exists, which is the volume average of the potentials Φ_ℓ, such that eqn.(2.7) gives the inelastic strain-rate for the composite and the rate of increase of each state variable is given by

$$\dot{\alpha}_k^\ell = -\frac{V}{V_\ell}\ \frac{\partial \Phi}{\partial S_k{}^\ell}$$

where the superscripts ℓ indicate that the internal state variables are associated with the ℓth element of the composite

3. STRUCTURE OF CONSTITUTIVE EQUATIONS FOR VOID GROWTH BY GRAIN-BOUNDARY DIFFUSION

The most common form of material degradation in engineering materials arises from the nucleation and growth of intergranular cavities. After nucleating, the cavities can grow by one of a number of mechanisms : power-law creep; grain-boundary diffusion; surface diffusion; or by a coupling of any two or all three of these mechanisms [6]. Here we consider the situation where the number and spacing of the voids remains fixed as they grow by grain-boundary diffusion. Nucleation and the other mechanisms of void growth are considered elsewhere [3].

A schematic drawing of a section of a creeping material is shown in Fig.3. As the voids grow on the grain-boundaries the resulting deformation is accommodated by deformation of the grains and by sliding between them. So that the final potential contains information about these deformation and sliding processes, as well as the void growth process.

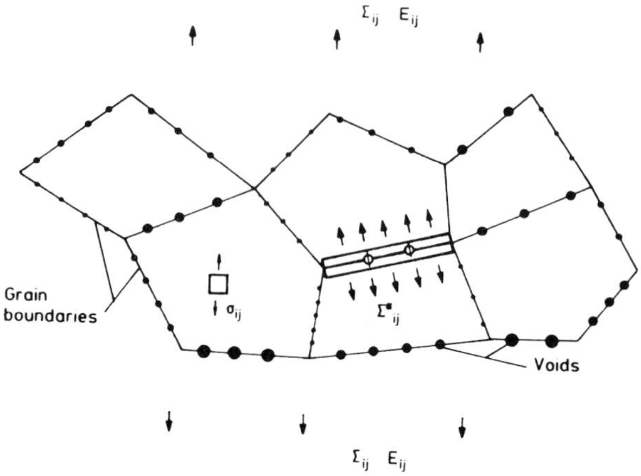

Fig.3 An element of material containing a number of cavitated grain boundaries subjected to a stress Σij

We facilitate the analysis of this situation by isolating a grain-boundary element of material, Fig.4, and assume that the major source of inelastic deformation within this element arises from the growth of the voids. The element experiences a mean local stress Σ_{ij}^{α} and its orientation within the body is indicated by the two outward normals to the grain-boundary $\underline{n}^{\alpha}$ and $\underline{n}_1^{\alpha}$.

The mechanism of void growth we are considering involves the diffusive transport of material away from a void by grain-boundary diffusion which is uniformly deposited on the grain-boundary. We assume that surface diffusion is sufficiently fast that the voids maintain a spherical shape as they grow. As the voids grow the material deforms plastically in the direction of the outward normals to the grain boundary.

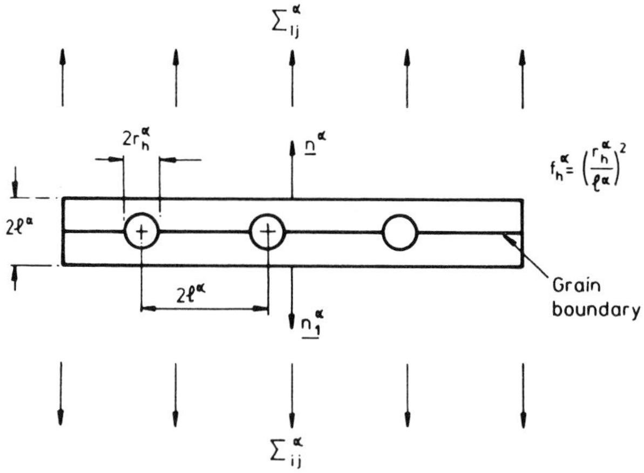

Fig.4 A typical grain-boundary element of Fig.3 subjected to a local stress Σ_{ij}^{α}.

The state of the element of material is described by the elastic strain $E_{ij}^{e\alpha}$ and the area fraction of voids on the grain boundary

$$f_h{}^{\alpha} = \left(\frac{r_h{}^{\alpha}}{\ell^{\alpha}} \right)^2$$

where $r_h{}^{\alpha}$ is the void radius and $2\ell^{\alpha}$ is the cavity spacing. The Helmholtz free energy is obtained by following the conceptual reversible path of Fig.5. A cut is made along the grain-boundary; material is scooped out to form the voids and spread evenly along the grain boundary. The surfaces are then rejoined and the resulting change of free energy is

$$\psi = \frac{2\gamma_s f_h{}^{\alpha}}{\ell^{\alpha}} V^{\alpha}$$

where γ_s is the surface energy per unit area and V^{α} is the volume of the element of material.

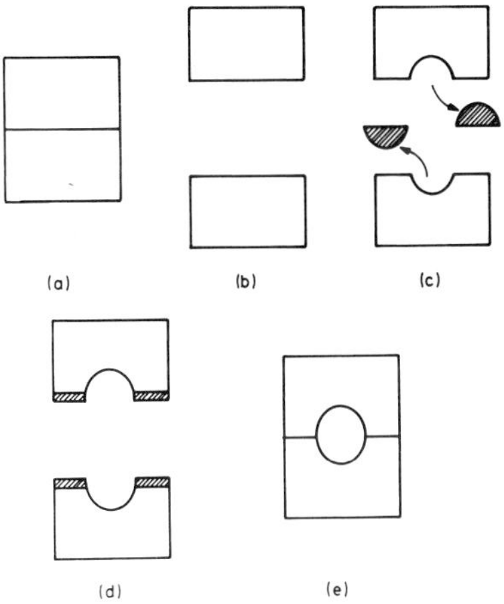

Fig.5 Conceptual reversible process of forming a void on the grain boundary: (a) initial void free material; (b) a cut is made along the grain boundary; (c) material is scooped out to form a void; (d) this material is spread evenly along the grain boundary; (e) the two pieces of material are rejoined.

Application of the stress Σ_{ij}^{α} gives the total free energy

$$\psi = \left(\frac{1}{2} C_{ijk\ell} E_{k\ell}^{e\alpha} E_{ij}^{e\alpha} + \frac{2\gamma_s}{\ell^{\alpha}} f_h^{\alpha} \right) V^{\alpha} \qquad (3.1)$$

This equation is the analogue of eqn.(2.1) with $f_h{}^{\alpha} \equiv \alpha_k$ and $\frac{2\gamma_s}{\ell^{\alpha}} f_h{}^{\alpha} \equiv f(\alpha_k)$.

Differentiating eqn.(3.1) w.r.t. the state variables $E_{ij}^{e\alpha}$ and $f_h{}^{\alpha}$ gives the thermodynamic forces -

$$\left.\begin{aligned} \Sigma_{ij}^{\alpha} V^{\alpha} &= \frac{\partial \psi}{\partial E_{ij}^{e\alpha}} \\ \Sigma_h^{\alpha} V^{\alpha} &= \frac{\partial \psi}{\partial f_h^{\alpha}} = \frac{2\bar{\gamma}s}{\ell^{\alpha}} V^{\alpha} \end{aligned}\right\} \tag{3.2}$$

The only additional piece of information we need is the relationship between the inelastic strain-rate $\dot{E}_{ij}^{\alpha p}$ and $\dot{f}_h^{\alpha}$. This relationship is

$$\dot{E}_{ij}^{\alpha p} = f_h^{\alpha^{\frac{1}{2}}} n_i^{\alpha} n_j^{\alpha} \dot{f}_h^{\alpha} \tag{3.3}$$

and eqns.(2.4) and (2.5) give the thermodynamic driving force

$$F_h^{\alpha} = \left(\Sigma_{ij} n_i^{\alpha} n_j^{\alpha} f_h^{\alpha\frac{1}{2}} - \Sigma_h^{\alpha} \right) \tag{3.4}$$

A detailed analysis of this mechanism of void growth gives [6]

$$\dot{f}_h^{\alpha} = \frac{\Omega \left(\Sigma_{ij}^{\alpha} n_i^{\alpha} n_j^{\alpha} f_h^{\alpha\frac{1}{2}} - \Sigma_h^{\alpha} \right)}{\ell^{\alpha 3} f_h^{\alpha} \ell n \left({}^{1}/f_h^{\alpha} \right)} \tag{3.5}$$

This equation is of the form, eqn.(2.6), which allows us to prove the existence of a scalar potential, Φ_h^{α}, such that

$$\dot{E}_{ij}^{\alpha P} = \frac{\partial \Phi_h^{\alpha}}{\partial \Sigma_{ij}^{\alpha}}$$

and (3.6)

$$\dot{f}_h^{\alpha} = \frac{\partial \Phi_h^{\alpha}}{\partial \Sigma_h^{\alpha}}$$

where

$$\Phi_h^{\alpha} = \frac{\Omega \left(\Sigma_{ij}^{\alpha} n_i^{\alpha} n_j^{\alpha} f_h^{\alpha\frac{1}{2}} - \Sigma_h^{\alpha} \right)^2}{2 \, \ell^{\alpha 3} f_h^{\alpha} \ell n \left({}^{1}/f_h^{\alpha} \right)}$$

The overall potential for the material depicted in Fig.3 can be obtained by combining these potentials for void growth with those for the deformation of the grains and those for sliding along grain boundaries. Cocks and Ponter [7] show that for conditions of slowly changing stress the inelastic strain of a microscopic element of a grain interior, Fig.3, is given by -

$$\dot{\varepsilon}_{ij}^{P} = \frac{\partial \phi}{\partial \sigma_{ij}} \tag{3.7}$$

where ϕ is generally assumed to be

$$\phi = \frac{1}{n+1}\dot{\varepsilon}_o\sigma_o\left(\frac{\sigma_e}{\sigma_o}\right)^{n+1} ,$$

σ_e is the von Mises effective stress and $\dot{\varepsilon}_o$, σ_o and n are material constants. The inelastic strain-rate of a grain-boundary element of material, Fig.4, due to grain-boundary sliding is given by [3].

$$E_{ij}^{\alpha s} = \frac{\partial\Phi_s^{\alpha}}{\partial\Sigma_{ij}^{\alpha}} \tag{3.8}$$

where
$$\Phi_s^{\alpha} = \frac{\tau_{max}^2}{4\ell^{\alpha}\eta} ,$$

τ_{max} is the maximum resolved shear stress in the plane of the grain-boundary and η is the viscosity of the boundary.

The potentials of eqns.(3.6)-(3.7) can be combined using the procedure of Section 2 to give a macroscopic potential for the response of the material. Examples of simple forms of these potentials are given at the beginning of the next section.

The above analysis was developed for a set of discrete state variables. Onat and Leckie [8] give a procedure for describing the damage in terms of a series of even order tensors. For the situation of Figs.3 and 4 the distribution of damage can be represented by a surface Γ which is constructed by placing one end of an outward normal to a grain boundary at the origin of a linear co-ordinate system so that the other end lies at a point on a sphere, A, of unit radius. Extending the normal by an amount which scales as f_h^{α} gives a point on the surface Γ, Fig.6. Onat and Leckie [8] show that this surface can be represented by a series of even order tensors

$$\left.\begin{aligned} D &= \frac{1}{4\pi}\int_A f_h^{\alpha}\,dA \\ D_{ij} &= \frac{15}{8\pi}\int_A f_h^{\alpha}(n_i^{\alpha} n_j^{\alpha} - \frac{1}{3}\delta_{ij})\,dA \end{aligned}\right\} \tag{3.9}$$

etc.

Cocks and Leckie [3] define a series of stress like quantities -

$$\Sigma^A = \frac{1}{4\pi} \int_A \Sigma_h^\alpha \, dA$$

$$\Sigma_{ij}^A = \frac{15}{8\pi} \int_A \Sigma_h^\alpha (n_i n_j - \frac{1}{3} \delta_{ij}) dA$$

etc.

and show that

$$\left.\begin{aligned} \dot{E}_{ij}^P &= \frac{\partial \Phi}{\partial \Sigma_{ij}} \\ \dot{D} &= -\frac{\partial \Phi}{\partial \Sigma^A} \\ \dot{D}_{ij} &= -\frac{\partial \Phi}{\partial \Sigma_{ij}^A} \end{aligned}\right\} \quad (3.10)$$

Giving the same basic structure of constitutive law as when the damage is described in terms of discrete quantities. A similar structure is found for the other mechanisms of void growth and the other types of damage. When the damage is in the form of voids the material becomes anisotropic as it deforms and a large number of state variables are required to model the material's response to an arbitrary history of loading. This number can be reduced if the range of loading experienced by a material is restricted. The structure of these laws for proportional loading are examined in the next section.

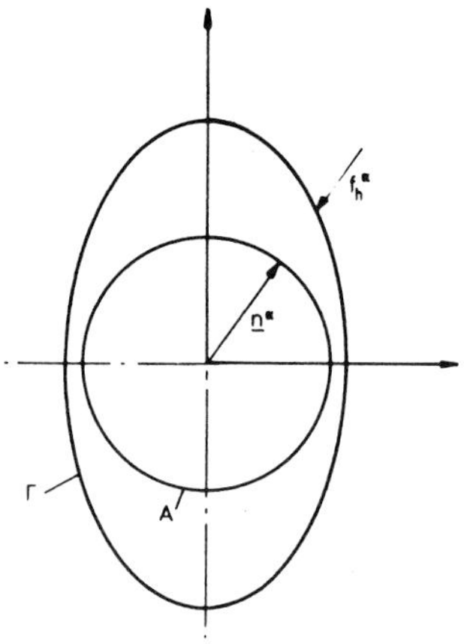

Fig.6 The surface Γ is the locus of the end points of a series of radial lines of length fh^α in the direction of $\underline{n}^\alpha$, which originate on a sphere of unit radius A.

A more general result is obtained when the material damages as a result of the formation of dislocation networks around precipitate particles or due to a reduction in strength of the precipitate particles. In these situations a single scalar state variable gives an adequate description of the material's response for all loading histories [3]. The aluminium alloys tested by Leckie and Hayhurst [2] and the nickel based super alloys tested by Dyson [9] damage according to this mechanism. Experiments performed by Trampczynski and Hayhurst [10] using non-proportional loading histories support this conclusion for this class of materials when they found a single scalar state variable adequately describes the response of these materials.

4. A SINGLE STATE VARIABLE THEORY FOR PROPORTIONAL LOADING

For proportional loading histories we might expect to be able to model the material's response using a single state variable theory, even when the damage is in the form of intergranular voids. When the damage is represented by a series of tensorial quantities, eqn.(3.9), the simplest single state variable theory would assume that the damage, D, is uniformly distributed on the grain boundaries, so that the material remains isotropic as it creeps. The anisotropic behaviour found in practice with this class of materials is retained when the discrete state variable description is reduced to a single state variable. We therefore concentrate on this description of the material in the remainder of this section.

We consider the situation where the voids all lie on those grain-boundaries which lie normal to the direction of maximum principal stress Σ_I. Then

$$\Sigma_{ij}\, n_i\, n_j = \Sigma_I$$

We further restrict our discussions to those situations where the voids grow by grain-boundary diffusion. There are two limiting cases for the overall response of this type of material. At high stresses (not too high so that the mechanism of void growth becomes controlled by power-law creep) the surrounding creeping material can readily accommodate the deformation that results from the plating of material on the grain boundaries. The macroscopic potential is then given by

$$\Phi = \frac{\dot{\varepsilon}_o \sigma_o (C_1 \Sigma_e)^{n+1}}{(n+1)} + \frac{\Omega \upsilon}{2\ell^3\, \ell n(1/f_h)} (\Sigma_g - \Sigma_h f_h^{-\frac{1}{2}})^2 \qquad (4.1)$$

where $$\Sigma_g = C_2\, S_{ij}\, n_i n_j + \Sigma_m \qquad (4.2)$$

is the local stress normal to the grain boundaries. This stress depends on the geometry of the grains and the extent of grain-boundary sliding. The constants C_1 and C_2 reflect the effect of grain-boundary sliding on the behaviour of the material. If the grain boundaries slide freely they are unable to support any shear stress and the material creeps faster than when there is no sliding. Ghahremani [11] gives values of C_1 for these instances.

It ranges from 1 for no sliding to about 1,15 for free sliding. Grain-boundary sliding also influences the stress normal to the grain boundary, with the constant C_2 ranging from 1 for no sliding to a maximum of 4 for free sliding. The stress Σ_e of eqn.(4.1) is the von Mises effective stress and S_{ij} represents the components of the deviatoric stress tensor. The quantity υ is the total volume fraction of grain-boundary elements which lie normal to $\underline{n}$.

The inelastic strain-rate and rate of increase of damage are obtained by differentiating eqn.(4.1):

$$\dot{E}_{ij} = \frac{\partial\Phi}{\partial\Sigma_{ij}} = \frac{3}{2}\,\dot{\varepsilon}_o\left(\frac{C_1\Sigma_e}{\sigma_o}\right)^n \frac{C_1 S_{ij}}{\Sigma_e} + \tag{4.3}$$
$$\frac{\Omega\upsilon}{\ell^3\,\ell n(1/f_h)}\,(\Sigma_g - \Sigma_h f_h^{-\frac{1}{2}})\left(C_2(n_i n_j - \frac{1}{3}\delta_{ij}) + \frac{1}{3}\delta_{ij}\right)$$

$$\dot{f}_h = -\frac{1}{\upsilon}\frac{\partial\Phi}{\partial\Sigma_h} = \frac{\Omega}{\ell^3 f_h^{\frac{1}{2}}\,\ell n(1/f_h)}\,(\Sigma_g - \Sigma_h f_h^{-\frac{1}{2}}) \tag{4.4}$$

Cocks and Leckie [3] have examined the predictions of this model over the range of temperature and stress where it is applicable. They show that λ lies between 1 and 2 (and is generally closer to 1) and the time to failure is inversely proportional to 1. The isochronous surface for this material depends on the value of C_2, but for $C_2 = 1$ it is the surface of maximum principal stress, Fig.2. The copper tested by Leckie and Hayhurst [2] fails according to a maximum principal stress criterion but gives values of $\lambda \approx 4$ with the time to failure varying as $\Sigma^{-\upsilon}$, where $\upsilon \approx 5.6$. There are therefore major features of the material behaviour that the model is unable to explain. The stress dependence can be improved by including the effects of the nucleation of cavities [12,13] but models of this type, where it is assumed that the entire behaviour can be described by the behaviour of those grain-boundaries which lie normal to the maximum principal stress, always lead to low values of λ

Dyson [14] recognized that at low stresses void growth can be constrained by the deformation of the surrounding material. Hutchinson [15] has analysed the material response in the limit of fully constrained growth in the absence of any grain-boundary sliding. In this limit the stress supported by the damaged grain boundaries is much less than the applied stress and the material creeps as if there were cracks on the damaged grain boundaries. Tvergaard [16] has modified the analysis to include the effects of grain-boundary sliding. The resulting potential for a single set of damaged grain-boundaries is

$$\Phi = \frac{\dot{\varepsilon}_o\sigma_o}{n+1}\left(\frac{C_1\Sigma_e}{\sigma_o}\right)^{n+1}\left(1 + C_3\upsilon N\left(\frac{\Sigma_g - \Sigma_h f_h^{-\frac{1}{2}}}{\Sigma_e}\right)^2\right) \tag{4.5}$$

where C_3 is a constant reflecting the constraint imposed by the deformation of the grains and by grain-boundary sliding and N is the number of cavities on a cavitated grain-boundary. The stress Σ_g is again given by eqn.(4.2), but, in this instance, it is not the stress supported by the grain boundaries. It is the stress that would have been supported by the grain boundaries if they were not damaged. The inelastic strain-rate and damage-rate are then -

$$\dot{E}^{P}_{ij} = C_1\dot{\varepsilon}_o\left(\frac{C_1\Sigma_e}{\sigma_o}\right)^n\left\{\frac{3}{2}\,\frac{S_{ij}}{\sigma_o} + C_3\upsilon N\left[\frac{n-1}{n+1}\,\frac{3S_{ij}}{2\Sigma_e}\left(\frac{\Sigma_g-\Sigma_h f_h^{-\frac{1}{2}}}{\Sigma_e}\right)^2 + \frac{2}{n+1}\,\frac{(\Sigma_g - \Sigma_h f_h^{-\frac{1}{2}})}{\Sigma_e}\left(C_2(n_i n_j - \tfrac{1}{3}\delta_{ij}) + \tfrac{1}{3}\delta_{ij}\right)\right]\right\} \qquad (4.6)$$

and
$$\dot{f}_h = \frac{C_3\upsilon N}{f_h^{\frac{1}{2}}}\left(\frac{\Sigma_e}{\sigma_o}\right)^{n-1}\left(\frac{\bar{\Sigma}_g - \Sigma_h f_h^{-\frac{1}{2}}}{\sigma_o}\right) \qquad (4.7)$$

The inelastic strain-rate only depends weakly on the damage f_h, so $\lambda\sim 1$. The rate of growth of the damage is now controlled by the rate of deformation of the surrounding material. Eqn.(4.7) predicts that the stress dependence for the time to failure is the same as the creep exponent and gives an isochronous surface which is represented by the equation

$$\left(\frac{\Sigma_e}{\Sigma_o}\right)^{n-1}\frac{\Sigma_I}{\Sigma_o} = 1$$

for $C_2 = 1$, where Σ_o is the uniaxial tensile stress that gives the same time to failure. For large values of n this surface is well approximated by the von Mises ellipse for plane stress loading situations, provided $\Sigma_m/\Sigma_e > -0.3$. These predictions of the stress dependence of the time to failure and the shape of the isochronous surface are in accord with many observations of the behaviour of Engineering materials, but the value of λ is much less than is found in practice. The reason for this low value of λ is again due to the fact that only one set of grain boundaries have been considered in the analysis.

Each of the models discussed so far in this section could be improved by including the effect of damage on other grain boundaries. Their effect can only be included by introducing other state variables if the potential form of Section 3 is retained, but we would like, if possible, to model the material in such a way that we need only use a single state variable.

In the above models, where the area fraction of voids on the grain-boundaries was used as the measure of damage in the material, it was found that the growth of this damage does not strongly affect the strain-rate of the material. It is only when the voids link to form a physical crack or the growth of the voids becomes constrained, and stress is shed from the grain boundary so that the material creeps as if a crack were present, that the creep rate is strongly affected. The number of crack-like features, which can be represented by the quantity υ of eqn.(4.6), therefore determines how fast the material creeps.

Cocks and Leckie [3] analysed the situation where the void damaged grain boundaries all lie nearly normal to the direction of maximum principal stress. Then the potential of eqn.(4.5) can be used to obtain the strain-rate of the material where υ is the measure of the damage in the material. As the material creeps the grain-boundaries which lie normal to the maximum principal stress become damaged first (i.e. behave as if they were cracked), then with the accumulation of creep strain more grain boundaries damage and υ increases. Cocks and Leckie [3] showed that the growth of this damage is not derivable from the strain-rate potential in the form of eqn.(4.5) but additional equations are required to determine how fast it grows.

5. CONCLUSIONS

It has been shown that constitutive laws for the creep deformation of engineering materials which include the effect of damage can be presented in potential form, where both the inelastic strain-rate and rate of increase of damage are derivable from the same potential. If, however, the main concern is to obtain a set of simple constitutive laws one is not constrained to express them in this form. In certain situations it might be more suitable to obtain a potential which gives the inelastic strain-rate for a given distribution of damage, and have another set of relationships which give the rate of growth of this damage.

REFERENCES

[1] Ashby, M.F. and Dyson, B.F., National Physical Laboratory Report DMA (A) 77 (1984)
[2] Leckie, F.A. and Hayhurst, D.R., Acta Met 25 (1977) 1059
[3] Cocks, A.C.F. and Leckie, F.A., Creep Constitutive Equations for Damaged Materials, in print
[4] Rice, J.R., Jnl.Mech.Phys. Solids 19 (1971) 433
[5] Ponter, A.R.S., Bataille, J. and Kestin, J., Jnl.de Mec. 18 (1979) 511
[6] Cocks, A.C.F., and Ashby, M.F., Prog.Mat.Sci. 27 (1982) 189
[7] Cocks, A.C.F., and Ponter, A.R.S., Leicester University Engineering Department Report 85-2 (1985)
[8] Onat, E.T., and Leckie, F.A., A Continuum Description of Creep Damage, this volume
[9] Dyson, B.F. and McLean, M., Acta Met. 31(1983)17
[10] Trampczynski, W.A., and Hayhurst, D.R., Creep Deformation and Rupture Under Non-Proportional Loading, in Ponter, A.R.S. and Hayhurst, D.R., (eds.), Creep in Structures (Springer-Verlag, Berlin, 1980) pp.388-405
[11] Ghahremani, F., Int.Jnl. Solids Structures 16 (1980) 847
[12] Cocks, A.C.F., Acta Met., 33 (1985) 129
[13] Argon, A.S., Inhomogeneities in Creep 1983 ASM Materials Science Seminar, in Flow and Fracture at Elevated Temperature (ASM, Metals Park, Ohio, 1985) pp.121-148

[14] Dyson, B.F., Canad.Met.Quart. 18 (1979) 31
[15] Hutchinson, J.W., Acta Met. 31 (1983) 1079
[16] Tvergaard, V., Jnl.Mech.Phys. Solids 33 (1985) 447

Thermomechanical Couplings in Solids
H.D. Bui and Q.S. Nguyen (Editors)
Elsevier Science Publishers B.V. (North-Holland)

ON A THERMOMECHANICAL CONSTITUTIVE THEORY AND ITS APPLICATION TO CDM, FATIGUE, FRACTURE AND COMPOSITES

Jinghong Fan

Institute of Constitutive Theory, Department of Engineering Mechanics, Chongqing University, Chongqing, PRC, 630044

While the irreversible thermodynamic process of damaged material is described in the ε_{TqD} space, the thermodynamic state variables including entropy are defined in its ε_T subspace. Based on the proposed physical mechanism and model for the coupling effects of damage on the inelastic internal variable $q^{(\alpha)}$, the evolution equations of $q^{(\alpha)}$ and $dD^{(\beta)}/d\bar{\eta}$ have been studied in the ε_{Tq} and ε_{TD} subspace respectively. By this approach the second order coupling effects are neglected and the Clausius-Duhem inequality could be easily satisfied. A new type of constitutive equation of damaged material has then been developed. Among its many salient features the constitutive equations are also to follow the form invariance law of dissipative material proposed by the author recently. Many applications of the present work, including the analysis of cyclic elastoplastic strain and damage distribution of a plate with two cracks made of AlCu and low cycle fatigue like prediction, have been reviewed and discussed. The latest work about the application to dilative behavior of material induced by hole growth of large deformation in front of crack are also simply mentioned.

1. INTRODUCTION

The thermomechanical coupling in degradated material take an important role in its stress response and failure. Inspite of the success of the conceptual development about the important role of damage on creep rupture (Kachanov, 1958) and the recent explosive development of CDM, the mechanisms of coupling effect between the microcrack growth and dislocation kinetics are still not quite clear. It turns out macroscopically that the more realistic constitutive equation and evolution equation to suitably account for this coupling effect should be further developed. For instance, the equivalent strain hypothesis to calculate the damage effect on constitutive equation needs to be verified and modified [1] . In some theoretical formalism of CDM [2], [3], [4], the Helmholtz free energy has been separated as elastic free energy Ψ_e, plastic free energy Ψ_p and damage free energy Ψ_D, which is the artificial decoupling and may result in conceptual confusing and remarkable descripency with realistic material behavior. Moreover some formalism of CDM followed the way of classical plasticity by introduction of a dissipative potential [5], which may not be easy to choose corectly in two or three dimensional problems and may make the

CDM analysis quite complicated.

The above short comment shows the existed need for developing a new CDM system. In the present work we first describe the irreversible thermodynamic process of damaged material in ε_{TqD} space and its subspace, and then proposed a physical model to analyze damage effect on the inelastic responses, which make it possible to describe the evolution equation of plastic internal variable in ε_T subspace. Then we develop a new type of constitutive equations for CDM and introduce new intrinsic scales $Z_D^{\star}$, $Z_H^{\star}$ for damage material, by which the changed property of material and the effect of damage history could be accounted for.

2. ε_{TqD} SPACE AND ITS SUBSPACE IN IRREVERSIBLE THERMODYNAMICS

We are dealing with a thermomechanical constitutive theory of solids undergoing irreversible deformation process, which, by nature, should be considered as an irreversible thermodynamic process. In the irreversible thermodynamics with internal variable formalism the thermodynamic state of material undergoing an irreversible process can be specified by a set of independent and measurable quantities such as strain and temperature as well as a number of inelastic internal variables $q^{(\alpha)}$ ($\alpha = 1,2,\ldots,n$) and damage internal variables $D_{ij}^{(\beta)}$ ($\beta = 1,2,\ldots,m$), which control the irreversibilities in the sense that the change of the material internal structure is described by its increments $dq_{ij}^{(\alpha)}$ and $dD_{ij}^{(\beta)}$, and if we denote Q_{ij}^{α}, Y^{β} as the generized internal frictional forces which oppose the corresponding internal generized rate $\hat{\dot{q}}_{ij}^{(\alpha)}$ and $\hat{\dot{D}}_{ij}^{(\beta)}$ then the dissipative work rate in the irreversible deformation and damage process is positive and equal to :

$$\delta \dot{W} = \sum_{\alpha = 1}^{m} Q_{ij}^{(\alpha)} \hat{\dot{q}}_{ij}^{(\alpha)} + \sum_{\beta = 1}^{n} Y^{(\beta)} \hat{\dot{D}}^{(\beta)} \geqslant 0 \qquad (1) \quad .$$

According to Kestin and Rice [6], "The central difficulty in extending thermostatics to the study of irreversible process turns on the existence of an entropy in terms of which the second part of the second law could acquire a physical meaning". One way to circumvent this difficulty was given by Valanis based on the integrability of the differential form of the first law using extended form of the Caratheodory conjecture [7], but the inevitable comment on which is still existed even "partial integrability as the basis for the existence of entropy in irreversible systems" has been given [8]. The point of these comments are how can we fix the internal variables in a irreversible thermodynamic process. In this paper we try to answer this problem by giving the definition of entropy for irreversible thermodynamics in a subspace ε_T, which belongs to the total space ε_{TqD} consisted of strain tensor ε_{ij}, temperature T, internal variables $q^{(\alpha)}$, ($\alpha = 1,2,\ldots,m$) and $D^{(\beta)}$, ($\beta = 1,2,\ldots,n$). Specifically speaking the thermodynamic state in irreversible thermodynamic process is defined by those state variables of a particular point P in subspace ε_T, where P is the intersecting point of the subspace ε_T with the irreversible thermodynamic path traced in the ε_{TqD} space (see figure 1). It is then obvious that the definition of state variable by using subspace ε_T is one thing (internal variables $q^{(\alpha)}$, $D^{(\beta)}$ are fixed in this case), and the irreversible thermodynamic process is another (internal variable is changed). Further the irreversible thermodynamic process is defined as a process in which certain energy is dissipated to overcome generalized frictional force for crossing the ε_T subspace and then change the internal variable. Based on this conceptual development, the existence of entropy as a state variable in irreversible thermodynamic process could be strictly proved, and then we could obtain the following thermodynamic relations [9] :

$$\eta = \eta\,(\varepsilon_{ij}, T, D^{(\beta)}, q^{(\alpha)}) \quad ; \quad \Psi = \Psi\,(\varepsilon_{ij}, T, D^{(\beta)}, q^{(\alpha)} \qquad (2),(3),$$

$$\sigma_{ij} = \partial\Psi/\partial\varepsilon_{ij} \;,\; \tau_{\alpha\beta} = \partial\Psi/\partial E_{\alpha\beta} \;,\; \eta = -\,\partial\Psi/\partial T \qquad (4.a\text{-}b\text{-}c),$$

$$\delta\,\dot{W} = -\left(\sum_{\alpha=1}^{m} \frac{\partial\Psi}{\partial q_{ij}^{(\alpha)}}\,\dot{q}_{ij}^{(\alpha)} + \sum_{\beta=1}^{n} \frac{\partial\Psi}{\partial D^{(\beta)}}\,D^{(\beta)}\right) \geqslant 0 \qquad (5)\;.$$

where eq. (4.a) is for small deformation, η denotes entropy, Ψ is Helmoholtz free energy per unit undeformed volume, E_{ij} — Green strain tensor, $\tau_{\alpha\beta}$ — second Piola-Kirchhoff stress. Eq. (5) is Clausius-Duhem inequality in terms of internal variables. From eqs. (1) and (2) we could easily obtain the following expressions for the generized frictional forces :

$$Q_{ij}^{(\alpha)} = -\,\partial\Psi/\partial q_{ij}^{(\alpha)} \quad ; \quad Y^{(\beta)} = -\,\partial\Psi/\partial D^{(\beta)} \qquad (6.a\text{-}b).$$

It should be to take notice that the generized rate $\hat{\dot{q}}^{(\alpha)}$ and $\hat{\dot{D}}^{(\beta)}$ are not with respect to the Newtonian time t, but some intrinsic time scales Z proposed by Valanis in 1971 [10], which are not only determined by the changed material property but also the accumulation of inelastic strain in the irreversible deformation process, in other words which have definite physical content and are material dependent positive scales. Specifically speaking the intrinsic time measures ζ_D and ζ_H are defined respectively as :

$$d\zeta_D = \|\, de_{ij}^{P} \,\| \;,\; d\zeta_H = |\; d\zeta_{KK}^{P} \;| \qquad (7.a\text{-}b),$$

ζ_D and ζ_H will be further related to the changed material property by hardening functions f_D and f_H to get the intrinsic time scales Z_D, Z_H as :

$$dZ_D = d\zeta_D / f_D(\zeta_D, \dot{\zeta}_D) \quad ; \quad dZ_H = d\zeta_H / f_H(\zeta_H, \dot{\zeta}_H) \qquad (7.c\text{-}d).$$

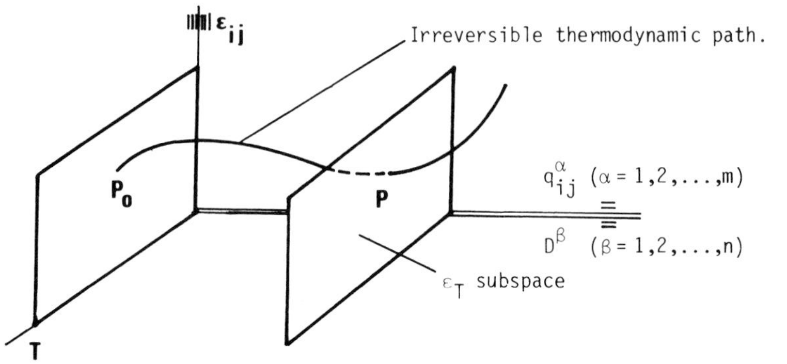

FIGURE 1 : ε_{TqD} space and ε_T subspace.

3. FORM INVARIANCE LAW OF CONSTITUTIVE EQUATION OF DISSIPATIVE MATERIAL

After some review of the concept of the generized time to describe the deformation history of materials with memory, we have proposed and proved mathematically a form invariance law of constitutive equation of dissipative materials [11]. In simple words is says, "if some kinds of intrinsic time Z could be defined suitably such that the relation between $Q^{(\alpha)D}_{ij}$ and $dq^{(\alpha)D}_{ij}/dz$ in dissipative materials has the same mathematical form for the relation between $Q^{(\alpha)V}_{ij}$ and $dq^{(\alpha)V}_{ij}/dt$ in viscoelastic material, then the form of the constitutive equation at hand is just the same as that of the viscoelastic material". This law is extremely important because it provides a theoretical foundation and a guide line to develop the new and more realistic models including finite deformation by directly and suitably utilizing the plentiful and elegant results in viscoelasticity. In that paper the introduction of intrinsic time, fundamental concept of endochronic plasticity proposed by Valanis in 1971, is closely related to the generized frictional forces, by which the suitable definition of intrinsic time may be searched. This concept opens possibilities of profound importance, at least in principle, to relate the dissipative micromechanisms to nonlinear constitutive behavior. The examples for application of the form invariance law have been shown [12]. It includes the developing of constitutive equations of inelastic anisotropic constitutive equation, applying the simplified form of which in the case of transversal isotropy to unidirectional composites, the nonlinear stress-strain relations of these materials under various off-axis tensile loads have been got. For the nonlinear numerical analysis the remarkable advantage of the form invariance law is to easily get the numerical algorithm and code for some different dissipative materials because we have already developed a numerical algorithm for this kind of constitutive equation [13], and it will be basically the same according to the law.

4. MECHANISM AND MODEL FOR EFFECT OF DAMAGE ON CONSTITUTIVE EQUATION

Even though the form invariance law of dissipative materials is important, the basic assumption of which in the original version [11] being that the internal variables are independent. From theoretical point of view if we have N internal variables which are not independent and relate each other by M equations, we could do some algebrac elimination and get (n-m) independent variables and satisfy the condition of form invariance law. However, this situation may not be suitable for damaged materials, for the internal variables $q^{(\alpha)}$ and $d^{(\beta)}$ may be related and the elimination may take the new independent variables lost the physical meaning. We, therefore, should prove that in what conditions the form invariance law may still be valid in this case. To build up the new analytical system for damaged material we should start from some physical model of damage effect on constitutive behavior.

For ductile metal, microcracks and voids occur and grow during loading and some environmental conditions. It is reported that there are some voidlike damage in crystals at low stress level, but the creep damage of the usual polycrystalline metals at high stress mainly due to the nucleation and the growth of micro-cracks and voids at grain boundaries [14]-[15], and the holes take on a variety of shapes depending on their spacing and on the temperature and strain rate [14]. From a view point of physics both plastic and creep deformations are consequences of the thermo-activated slip motions of dislocations, so we could put forward following physical model of effects of damage on constitutive relation of inelastic material in a unified fashion :

1- The first effect is to release the resistance in damaged grain boundary and to deduce the residual stress in crystals to make the generized internal frictional force for slip motions of dislocations decreasing. This softing effect corresponding to deviatoric response will be taken care by some monotonic decreasing softing function W_D which may be dependent on the stress state and the intrinsic time scale ζ_D.

2- The second effect of all kinds of damage hole could make some volume change (dilatation) even the matrix is plastically incompressible. This softing effect corresponding to hydrostatic response will be taken care by some other softing function W_H, which may be mainly dependent on the stress state and the shape and dimension of the hole as well as ζ_H.

3- The third effect is to damage the fundamental elastic frame consisted of polycrystal structure to make some elastic constants, for instance G and K, decreasing.

To explain the above consideration about the damage effect, we use the usual parallel mechanical model in figure 2 for illustration. In this figure the displacement $q^{(\alpha)}$ of the α-th dashpot representsthe inelastic strain associated with the α-th mechanism of plastic or creep flow, the displacement ε denotes the total strain, E^{α}— the elastic modulus of α-th spring. If the black box EFGH is in constant temperature, then the Helmholtz free energy Ψ equal to the elastic energy stored in all springs and we have :

$$\Psi = \sum_{\alpha=1}^{m} \frac{1}{2} E^{(\alpha)} \left(\varepsilon - q^{(\alpha)}\right)^2 \qquad (8.a)\ .$$

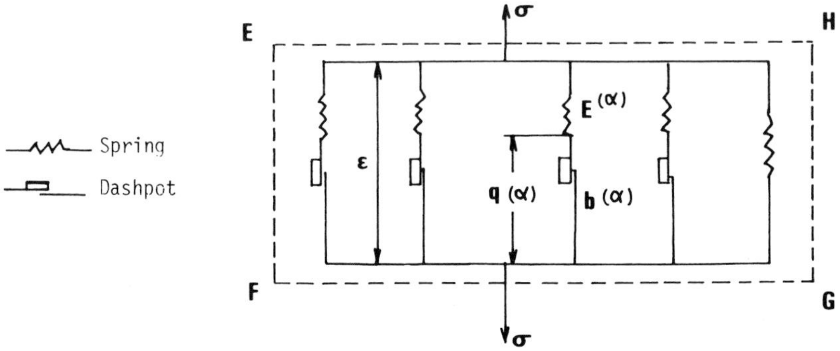

FIGURE 2 : Damage model in one dimension.

While the above third effect is to make the spring to damage and deduce the elastic modulus from $E^{(\alpha)}$ to $\overline{E}^{(\alpha)}$, the first and second effects are to make the dashpot resistance decreasing. From this model we could easily get the free energy $\overline{\Psi}$ of the damaged material and the total elastic modulus $\overline{E}$ for the damaged elastic frame as :

$$\overline{\Psi} = \sum_{\alpha=1}^{m} \frac{1}{2} \overline{E}^{(\alpha)} \left(\varepsilon - q^{(\alpha)}\right)^2 \qquad (8.b)$$

$$\overline{E} = \sum_{\alpha=1}^{n} \overline{E}^{(\alpha)} \qquad (9)\ .$$

Based on this analysis of mechanism of effect of damage on constitutive equation, we could extend the damaged mechanical model to three dimensional case. Following Valanis and remembring that at constant temperature the free energy, in essence, is the elastic energy stored in the crystals, we may assume that the free energy, $\psi^{(\alpha)}$, associated with the α-th mechanism of inelastic flow is the square of elastic strain of that mechanism 16], i.e.,

$$\psi^{(\alpha)} = \frac{1}{2}\left(\underset{\sim}{\varepsilon} - \underset{\sim}{q}^{(\alpha)}\right) \cdot \underset{\approx}{E}^{(\alpha)} \cdot \left(\underset{\sim}{\varepsilon} - \underset{\sim}{q}^{(\alpha)}\right) \qquad (10),$$

where $q^{(\alpha)}$ is the inelastic strain associated with the α-th mechanism of inelastic flow. Since usually the crystal structure between slip planes does not deform substantially even in some damage state, we may ignore interaction between glide systems. Then with good accuracy we could write Helmoltz free energy Ψ as :

$$\Psi = \sum_{\alpha=1}^{m} \Psi^{(\alpha)} = \sum_{\alpha=1}^{m} \frac{1}{2}\left(\underset{\sim}{\varepsilon} - \underset{\sim}{q}^{(\alpha)}\right) \cdot \underset{\approx}{E}^{(\alpha)} \cdot \left(\underset{\sim}{\varepsilon} - \underset{\sim}{q}^{(\alpha)}\right) \quad (11)$$

and for damaged material as :

$$\overline{\Psi} = \sum_{\alpha=1}^{m} \overline{\Psi}^{(\alpha)} = \sum_{\alpha=1}^{m} \frac{1}{2}\left(\underset{\sim}{\varepsilon} - \underset{\sim}{q}^{(\alpha)}\right) \cdot \underset{\approx}{\overline{E}}^{(\alpha)} \cdot \left(\underset{\sim}{\varepsilon} - \underset{\sim}{q}^{(\alpha)}\right) \quad (12),$$

where $\underset{\approx}{E}^{(\alpha)}$ and $\underset{\approx}{\overline{E}}^{(\alpha)}$ are the generized elastic modulus corresponding to α-th mechanism for undamaged and damaged material respectively. To know the physical meaning of $\underset{\approx}{E}^{(\alpha)}$ and $\underset{\approx}{\overline{E}}^{(\alpha)}$ we let $q^{(\alpha)} \to 0$ in eqs. *(11)* and *(12)* and comparing it with the following two equations :

$$\Psi = \frac{1}{2} \underset{\sim}{\varepsilon} \cdot \underset{\approx}{E} \cdot \underset{\sim}{\varepsilon} \qquad ; \qquad \overline{\Psi} = \frac{1}{2} \underset{\sim}{\varepsilon} \cdot \underset{\approx}{\overline{E}} \cdot \underset{\sim}{\varepsilon} \quad (13\text{-}a.b) \quad ,$$

we then obtained :

$$\underset{\approx}{E} = \sum_{\alpha=1}^{m} \underset{\approx}{E}^{(\alpha)} \qquad ; \qquad \underset{\approx}{\overline{E}} = \sum_{\alpha=1}^{m} \underset{\approx}{\overline{E}}^{(\alpha)} \quad (14\text{-}a.b).$$

These expressions show $\underset{\approx}{E}^{(\alpha)}$ ($\underset{\approx}{\overline{E}}^{(\alpha)}$) are components of elastic modulus $\underset{\approx}{E}$ ($\underset{\approx}{\overline{E}}$) and it express different aspects of material elasticity. For isotropic material we have :

$$E_{ijkl} = \lambda\, \delta_{ij}\, \delta_{kl} + \mu\, (\delta_{ik}\, \delta_{jl} + \delta_{il}\, \delta_{jk}) \quad (15\text{-}a)$$

$$\overline{E}_{ijkl} = \overline{\lambda}\, \delta_{ij}\, \delta_{kl} + \overline{\mu}\, (\delta_{ik}\, \delta_{jl} + \delta_{il}\, \delta_{jk}) \quad (15\text{-}b)$$

$$E^{(\alpha)}_{ijkl} = \lambda^{(\alpha)}\, \delta_{ij}\, \delta_{kl} + \mu^{(\alpha)}\, (\delta_{ik}\, \delta_{jl} + \delta_{il}\, \delta_{jk}) \quad (15.c)$$

$$\overline{E}^{(\alpha)}_{ijkl} = \overline{\lambda}^{(\alpha)}\, \delta_{ij}\, \delta_{kl} + \overline{\mu}^{(\alpha)}\, (\delta_{ik}\, \delta_{jl} + \delta_{il}\, \delta_{jk}) \quad (15\text{-}d) \quad ,$$

and from eq. *(15-a.b)*, we have :

$$\lambda = \sum_{\alpha=1}^{m} \lambda^{(\alpha)} \quad , \quad \overline{\lambda} = \sum_{\alpha=1}^{m} \overline{\lambda}^{(\alpha)} \quad , \quad \mu = \sum_{\alpha=1}^{m} \mu^{(\alpha)} \quad , \quad \overline{\mu} = \sum_{\alpha=1}^{m} \overline{\mu}^{(\alpha)}$$

$$K = \sum_{\alpha=1}^{m} K^{(\alpha)} \quad , \quad \overline{K} = \sum_{\alpha=1}^{m} \overline{K}^{(\alpha)} \quad (16).$$

Since different $\mu^{(\alpha)}$; $K^{(\alpha)}$ come from the same elastic frame of the polycristal structure, we have good reason to give the assumption that for isotropic damage of isotropic material the decreasing of $\mu^{(\alpha)}$, $K^{(\alpha)}$, $\lambda^{(\alpha)}$ for different number should be proportional, i.e.,

$$\frac{\overline{\mu}^{(1)}}{\mu^{(1)}} = \frac{\overline{\mu}^{(2)}}{\mu^{(2)}} = \frac{\overline{\mu}^{(3)}}{\mu^{(3)}} = \ldots = \frac{\overline{\mu}^{(m)}}{\mu^{(m)}} = \frac{\overline{\mu}}{\mu} = \chi_1 \quad (17),$$

$$\frac{\overline{K}^{(1)}}{K^{(1)}} = \frac{\overline{K}^{(2)}}{K^{(2)}} = \frac{\overline{K}^{(3)}}{K^{(3)}} = \ldots = \frac{\overline{K}^{(m)}}{K^{(m)}} = \frac{\overline{K}}{K} = \chi_0 \quad (18),$$

where χ_0, χ_1 are some parameters to express the elastic frame damage.

5. CONSTITUTIVE EQUATION OF DAMAGED MATERIAL

The key point to derive the constitutive equation of damaged material is how we could account for the coupling effect between $q^{(\alpha)}$ and $D^{(\alpha)}$. The Coleman equation *(4-a)* is valid for damaged material if Ψ is replaced by $\overline{\Psi}$, i.e.,

$$\sigma_{ij} = \frac{\partial \overline{\Psi}}{\partial \varepsilon_{ij}} \qquad (19).$$

From the analysis of first and second effect of damage on constitutive equation we may assume these effects could account for by bring the monotonic decreasing softing functions W_D and W_H into intrinsic time scales proposed by Valanis in 1978 [17], i.e.,

$$dZ_D = d\zeta_D / \overline{f}_D \quad , \quad dZ_H = d\zeta_H / \overline{f}_H \qquad (20\text{-}a.b),$$

$$f_D = W_D \, f_D \left(\zeta_D , \hat{\dot{\zeta}}_D\right) \qquad (21),$$

$$f_H = W_H \, f_H \left(\zeta_H , \hat{\dot{\zeta}}_H\right) \qquad (22).$$

Since the infinitesimal increase of damage internal variable $dD^{(\beta)}$ only have infinitesimal influence to the existed free energy $\overline{\Psi}$ and generized frictional force $Q_{ij}^{(\alpha)}$, we have good reason to neglect the second order effects of $dD^{(\beta)}$ on the evolution of internal variable $q^{(\alpha)}$, it means we could frozen $D^{(\beta)}$ temporarily to determine the evolution equation of $q^{(\alpha)}$ in the ε_{Tq} subspace. Based on this idea and keeping in mind that $q^{(\alpha)}$ are independent each other we could satisfy the Clausius-Duhem inequality (5) by :

$$Q_{ij}^{(\alpha)} \, \hat{\dot{q}}_{ij}^{(\alpha)} \geqslant 0 \qquad (23).$$

It is reasonable to assume that $\underset{\sim}{Q}^{(\alpha)}$ has a linear relation with the generized rate $\hat{\dot{q}}^{(\alpha)}$, i.e.,

$$Q_{ij}^{(\alpha)} = b_{ijkl}^{(\alpha)} \, \hat{\dot{q}}_{kl} \qquad (24).$$

Instead Ψ by $\overline{\Psi}$ in eq. *(6-a)* and then combining with eq. *(24)*, we obtain

$$\frac{\partial \overline{\Psi}}{\partial q_{ij}^{(\alpha)}} + b_{ijkl}^{(\alpha)} \, \hat{\dot{q}}_{kl}^{\alpha} = 0 \qquad (25).$$

For isotropic material, we have :

$$b_{ijkl}^{(\alpha)} = b_2^{(\alpha)} \, \delta_{ij} \, \delta_{kl} + b_1^{(\alpha)} \left(\delta_{ik} \, \delta_{jl} + \delta_{il} \, \delta_{jk} \right) \qquad (26).$$

$$q_{kl}^{(\alpha)} = p_{kl}^{(\alpha)} + q_H^{(\alpha)} \, \delta_{kl} \, ; \, \varepsilon_{kl} = e_{kl} + \varepsilon_H \, \delta_{kl} \, ; \, \sigma_{kl} = S_{kl} + \sigma_H \, \delta_{kl} \qquad (27\text{-}a.b.c).$$

Where :

$$q_H^{(\alpha)} = q_{KK}^{(\alpha)} / 3 \quad ; \quad \varepsilon_H = \varepsilon_{KK} / 3 \quad ; \quad \sigma_H = \sigma_{KK} / 3 \quad .$$

By using eqs. *(15-d)*, *(12)*, *(16)*, *(27)* and *(25)*, we obtain the evolution equations of internal variable as follows :

$$b_1^{(\alpha)} \frac{dp_{ij}^{(\alpha)}}{dZ_D} + \bar{\mu}^{(\alpha)} p_{ij} - \bar{\mu}^{(\alpha)} e_{ij} = 0 \qquad (28\text{-}a),$$

$$b_0^{(\alpha)} \frac{dq_H^{(\alpha)}}{dZ_H} + \bar{K}^{(\alpha)} q_H^{(\alpha)} - \bar{K}^{(\alpha)} \varepsilon_H = 0 \qquad (b_0^{(\alpha)} = b_2^{(\alpha)} + \frac{2}{3} b^{(\alpha)}) \qquad (28\text{-}b).$$

From eqs. *(17)* and *(18)*, we have $\bar{\mu} = \chi_1 \mu$, $\bar{K} = \chi_0 K$ then we rewrite eqs. *(27)*, *(28)* as :

$$b_1^{(\alpha)} \frac{dp_{ij}^{(\alpha)}}{dZ_D^*} + \mu^{(\alpha)} p_{ij}^{(\alpha)} - \mu^{(\alpha)} e_{ij} = 0 \qquad (29)$$

$$b_0^{(\alpha)} \frac{dq_H}{dZ_D^*} + K^{(\alpha)} q_H - K^{(\alpha)} \varepsilon_H = 0 \qquad (30).$$

Integrating the above equations and then combining with eq. *(19)*, we could easily get the constitutive equations as :

$$\bar{S}_{ij} = 2 \int_0^{Z_D^*} \mu \, (Z_D^* - Z_D^*) \frac{\partial e_{ij}}{\partial Z_D'^*} dZ_D'^* \qquad (31),$$

$$\bar{\sigma}_H = 3 \int_0^{Z_H^*} K \, (Z_H^* - Z_H'^*) \frac{\partial \varepsilon_H}{\partial Z_H'^*} dZ_H'^* \qquad (32).$$

Where :

$$\mu \, (Z_D^*) = \sum_{\alpha=1}^{m} \mu^{(\alpha)} e^{-\rho^{(\alpha)} Z_H^*} \quad , \quad \mu(0) = \mu \qquad (33\text{-}a.b),$$

$$K \, (Z_H^*) = \sum_{\alpha=1}^{m} K^{(\alpha)} e^{-\lambda^{(\alpha)} Z_H^*} \quad , \quad K(0) = K \qquad (34\text{-}a.b),$$

$$\bar{S}_{ij} = S_{ij} / \chi_1 \quad , \quad \bar{\sigma}_H = \sigma_H / \chi_0 \qquad (35\text{-}a.b),$$

and :

$$Z_D^* = \int_0^{Z_D} \chi_1 \, dZ_D' \quad , \quad Z_H^* = \int_0^{Z_H} \chi_0 \, dZ_H' \qquad (36\text{-}a.b).$$

It is interesting to notice that the constitutive equations *(31)*, *(32)* of damaged material have the following salient features :

(i) Its form is just the same as that of constitutive equation of linear isotropic viscoelastic material, i.e. the form invariance law is valid if S_{ij}, $\sigma_{\alpha\alpha}$ and Newtonian time t are replace by $\bar{S}_{ij}$, $\bar{\sigma}_{\alpha\alpha}$, $\bar{Z}_D^*$ and $\bar{Z}_H^*$ respectively.

(ii) The dependent of stress responses on damage history is taken account by Z_D^* and Z_H^*, which itself are functionals of damage parameters χ_0, χ_1 and damage softing functions W_D and W_H.

(iii) If there are no frame damage, i.e. $\chi_0 = \chi_1 = 1$, then $Z_D^* = Z_D$ and we get the same kind constitutive equations given by Valanis [17].

(iv) The effective deviatoric stress $\overline{S}_{ij}$ and effective hydrostatic stress $\overline{\sigma}_H$ are introduced analytically. In general $\overline{S}_{ij}$ and $\overline{\sigma}_H$ are different with the deviatoric and hydrostatic components of the effective stress proposed by Lemaitre, but under the condition that $\chi_0 = \chi_1$ (or Poisson's ratio being constant) they are the same. From this constitutive equations we also know that the equivalent strain hypothesis is not physically sound, because the damage induced softing effect described by W_D and W_H and the damage historic effect by Z_D^* and Z_H^* could not be included in that model.

(v) If we define :

$$e_{ij}^p = e_{ij} - S_{ij} / 2\overline{\mu} \qquad ; \qquad \varepsilon_H^p = \varepsilon_H - \sigma_H / 3\,\overline{K} \qquad (37\text{-}a.b),$$

and substitute e_{ij} and ε_H into eqs. (31)(32), we could get following constitutive equations by Laplace transformation :

$$\overline{S}_{ij} = S_y^0 \frac{de_{ij}^p}{dZ_D^*} + 2\,\mu_0 \int_0^{Z_D^*} \rho_1 \left(Z_D^* - Z_D^{*\prime}\right) \frac{\partial e_{ij}^p}{\partial Z_D^*} \, dZ_D^{*\prime} \qquad (38\text{-}a)$$

$$\overline{\sigma}_H = \sigma_H^0 \frac{d\varepsilon_H^p}{dZ_H^*} + 3\,K \int_0^{Z_H^*} \Phi_1 \left(Z_H^* - Z_H^{*\prime}\right) \frac{\partial \varepsilon_H^p}{\partial Z_H^{*\prime}} \, dZ_H^{*\prime} \qquad (38\text{-}b).$$

Where S_y^0 and σ_H^0 are deviatoric and hydrostatic yield stress.

The above formulae represent the constitutive equations of damaged material with yield surface. Following the same discussion in [17] we could show that eq. (3-a) represent a sphere in deviatoric stress space. The radius R and the center coordinate α_{ij} of which are :

$$R = S_y^0 \, W_D \, f\left(\zeta_D, \dot{\zeta}_D\right) \qquad , \qquad (39\text{-}a),$$

$$\alpha_{ij} = 2\,\mu_0 \int_0^{Z_D^*} \rho_1 \left(Z_D^* - Z_D^{*\prime}\right) \frac{\partial e_{ij}^p}{\partial Z_D^{*\prime}} \, dZ_D^{*\prime} \qquad (39\text{-}b).$$

From the above equations we can see that the effects of damage on yield surface are to make the radius deduced and the position of center changed. If the plastic strain occur immediately after loading (i.e., $S_y^0 = \sigma_H^0 = 0$), we get the constitutive equations without yield surface.

It is important to remember that even though the above constitutive equations include the general effects of damage, we may only consider some effects of them in particular cases. This view of point will be emphasized in following several sections.

6. DAMAGE EVOLUTION EQUATION AND DAMAGE FIELD ANALYSIS FOR INELASTICALLY INCOMPRESSIBLE MATERIAL

In this section, we only discuss the material whose hole volume growth is negligible and can be considered as inelastically incompressible. In this case the hydrostatic response is elastic and eq. (32) reduce to

$$\sigma_H = 3\ K\ \chi_0\ \varepsilon_H = 3\ \overline{K}\ \varepsilon_H \tag{40}$$

To propose the damage evolution equation, we assume approximatively that all the influence of increment of inelastic strain de^p_{ij} (or $d\underset{\sim}{q}^{(\alpha)}$) on damage evolution $dD^{(\beta)}$ could be accounted for by a damaged intrinsic time scale $\overline{\eta}$ defined as

$$d\ \overline{\eta} = g(\zeta)\ d\zeta \tag{41}$$

In other words we assume that $dD^{(\beta)} / d\overline{\eta}$ could be determined in ε_{TD} sub-space. We further assume $dD^{(\beta)} / d\overline{\eta}$ is a function of generized damage frictional force $Y^{(\beta)}$; i.e.,

$$\frac{dD^{(\beta)}}{d\overline{\eta}} = F_\beta \left(Y^{(\beta)}\right) \tag{42}$$

Keeping in mind that the following differential is conducted in ε_{TD} sub-space and using eq. (6-b) for one damage internal variable ($\beta = 1$) and reducing eq. (12) for isotropic material, we get :

$$dD = F(Y)\ d\overline{\eta} \quad , \quad Y = -\left(\frac{1}{\overline{\mu}}\ \overline{\Psi}_D\ \frac{\partial\overline{\mu}(D)}{\partial D} + \frac{1}{\overline{K}}\ \overline{\Psi}_H\ \frac{\partial\overline{K}(D)}{\partial D}\right) \tag{43-a.b}$$

Where $\overline{\Psi}_D$ and $\overline{\Psi}_H$ are deviatoric and hydrostatic free energy for the damaged isotropic material and $\overline{\Psi} = \overline{\Psi}_D + \overline{\Psi}_H$

$$\overline{\Psi}_D = \frac{1}{4\overline{\mu}(D)}\ S\ S \quad ; \quad \overline{\Psi}_H = \frac{1}{2\overline{K}(D)}\ \sigma_H^2 \tag{43-c.d}$$

If we take Lemaitre's definition of damage variable D, i.e.,

$$D = 1 - \overline{E}/E \tag{44}$$

and suppose $\nu = \text{const.}$ then

$$\overline{\mu} = \mu\ (1 - D) \quad ; \quad \overline{K} = K\ (1 - D) \tag{45-a.b}$$

$$\chi_1 = \chi_0 = \overline{E}/E \tag{46}$$

Substituting eqs. (45-a.b) into eq. (43-b) and taking F as linear function, we get :

$$dD = m\ Y\ d\overline{\eta} = m\ \frac{\overline{\Psi}}{(1 - D)}\ d\overline{\eta} \tag{47}$$

Where m is a material constant.

For solving a boundary value problem in engineering practice, it is necessary to determine material functions realistically and take a nonlinear numerical algorithm, which we used here are basically the same as [13], [18], [19].

By this approach the analysis of elastoplastic strain and damage distributions in a plate made of 2024 CuAl with two edge cracks under monotonic and cyclic loading have been conducted. All the experimental data are obtained from Ref. [20].

From figure 3 we could see the intrinsic time scale Z^* and damage parameter D are concentrated at the tip of crack, which may result in the crack propagation continuously starting at the tip. But from figure 4 we could see the interesting thing that the peak stress point is moving towards the front of the crack when loading increasing. Under the action of the high stress, inclusions will be fractured to cause the formation of voids or the voids could be initiated by decohesion along the particle-matrix interface. When the voids are growing the material is dilatable and we should apply simultaneously eq. (32) with eq. (31) to get the solution. It is important to notice that the softing function D_H should be dependent on the effective radius R of the hole and the stress state parameter α. By using Li's results [21], we could determine the function D_H in terms of R and α based on some semi-micro void model. Moreover this has been formulated in the corresponding constitutive equation for large deformation. Since the space is limited we should stop the discussion here to give some brief introduction to the other applications. The reader who interested in the ductile fracture by this constitutive theory is refered to Ref. [22], [23].

7. APPLICATION TO LOW CYCLE FATIGUE LIFE PREDICTION

In the steady state of cyclic process the hysteresis loop is the same for each cycle and the material has been saturated, we, therefore, can assume that the material function $f_{(\zeta)}$, $W_{D(\zeta)}$, $g_{(\zeta)}$ are constants. If we take the following kernel function

$$\rho(Z) = \rho_0 / \; Z^{\alpha} \qquad (48)$$

and do the same analysis as [24], we could easily get the explicit expressions for cyclic response by using the constitutive equations, it reads as pure shear :

$$\tau = \frac{2\,\rho_0\, f_0^{\alpha}}{\beta}\; \varepsilon^{\beta}_{p/2} \left(\frac{3}{2}\right)^{\beta - 1/2} \qquad (\beta = 1 - \alpha) \qquad (49)$$

tension and compression :

$$\sigma = \frac{2\,\rho_0\, f_0^{\alpha}}{\beta}\; \varepsilon^{\beta}_{p} \left(\frac{3}{2}\right)^{\beta + 1/2} \qquad (50).$$

Keep in mind that the damage in fatigue is a stochastic process and fatigue failure occurs abruptually, we may follow Valanis to write down the fatigue failure criterion as [25]

$$\int_0^{Z_c} p\,dZ = 100\% \qquad (51).$$

Where Z_c is a critical intrinsic time for fatigue failure. It may be reasonable to choose the probability density p of damage in the interval dZ as the function of $Y^{(\beta)}$ which is equal to the damage driving force in magnitude but opposite in direction we then have the following criterion :

$$\int_0^{Z_c} p\left(Y^{(\beta)}\right) dZ = 2 \tag{52}$$

If we take F as power function and $\beta = 1$, by using eq. (43-b) we get :

$$\int_0^{Z_c} \left[- \frac{1}{\bar{\mu}} \Psi_D \frac{\partial \bar{\mu}(D)}{\partial D} - \frac{1}{\bar{K}} \Psi_H \frac{\partial \bar{K}}{\partial D} \right]^s dZ = 1 \tag{53}$$

By the definition of D in eq. (44) , eq. (53) can be expressed as :

$$\int_0^{Z_c} \left(\frac{\Psi}{1 - D}\right)^s dZ = 1 \tag{54}.$$

If we neglect the contribution of transient process in the first several cycles and take the steady state as reference state we may assume $D = 0$ in eq. (52). Remembering each cycle consistes of two reversals in the symmetric strain control cyclic process and their contribution is the same in the steady state, we could write down the failure criterion for symmetrically fatigue :

$$2\, N_c\, g \int_0^{\Delta\zeta_c} \Psi^s\, d\zeta = 1 \tag{55}$$

where g is material constant, N_c the maximum cyclic number (fatigue life), $\Delta\zeta_c$ increment of intrinsic time scale in half cycle.

For simple tension-compression, the Helmholtz free energy in constant temperature is :

$$\Psi = \sigma^2/2E \tag{56}.$$

By using eqs. (50), (55) and (56), we get :

$$N_c \left(\frac{1}{1 + 2\, s\, \beta} \right) \varepsilon_p^{max} = C \tag{57}.$$

Which is just the Coffin-Manson's empirical formula. By the same approach we studied fatigue-creep interaction and fatigue path dependent behavior. The reader is refered to [26].

8. APPLICATION TO FIBER REINFORCED COMPOSITES

It is well known that the damage in matrix is very important to the constitutive behavior of fiber reinforced composites. To include this effect we could apply the constitutive equation *(31)* for the damaged matrix of composites and then develop a overall constitutive equation by the VFD geometric model proposed by Dvorak [27]. It is easily to show that the increment form of the damaged constitutive equation *(31)* is

$$dS_{ij} = \rho(0)\ de^{p}_{ij} + h_{ij}\ (Z^*)\ dZ^* \qquad (58),$$

where :

$$h^*_{ij} = \int_0^{Z^*} \hat{\rho}(Z^* - Z^{*\prime})\ \frac{\partial e_{ij}}{\partial Z^{*\prime}}\ dZ^{*\prime} \quad , \qquad \hat{\rho} = d\rho/dZ \qquad (59\text{-}a.b).$$

After some algebrac operation we obtain the following incremental constitutive equation in matrix form :

$$\left\{ d\sigma \right\} = \left[D_p \right] \left(\left\{ d\varepsilon \right\} + \left\{ dR_p \right\} \right) \qquad (60).$$

Fiber is assumed as transversely isotropic elastic material such that :

$$\left\{ d\varepsilon^f \right\} = \left[C^f \right] \left\{ d\varepsilon^f \right\} \qquad (61),$$

where C^f is elastic compliance matrix of fiber. Matrix is regarded as isotropic inelastic damage material being described by eq. *(60)*. Combing eqs. *(60)*, *(61)* and the governing equations of VFD Model [26], we finally get the required constitutive equation :

$$\{ d\bar{\varepsilon} \} = [\ \bar{C}\ ]\{ d\bar{\sigma} \} + \left\{ d\bar{R}_p \right\} \qquad (62).$$

Where $\{ d\bar{\sigma} \}$, $\{ d\bar{\varepsilon} \}$ are the overall stress and strain increment matrix, $[\bar{C}]$ overall elastic compliance matrix, $\{ dR_p \}$ a history dependent matrix. All the explicit expressions of the matrix component are omitted here. For the detail the reader is refered to [28], in which monotonic and cyclic stress responses have also been shown.

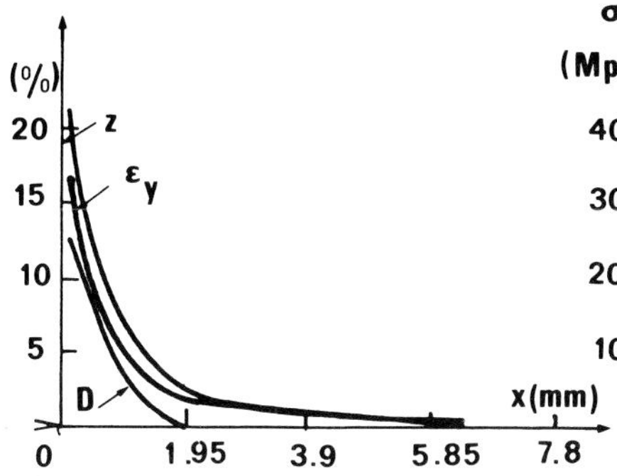

FIGURE 3 :

Distribution of ε_y,D,Z* along crack line (P = 9000 kg).

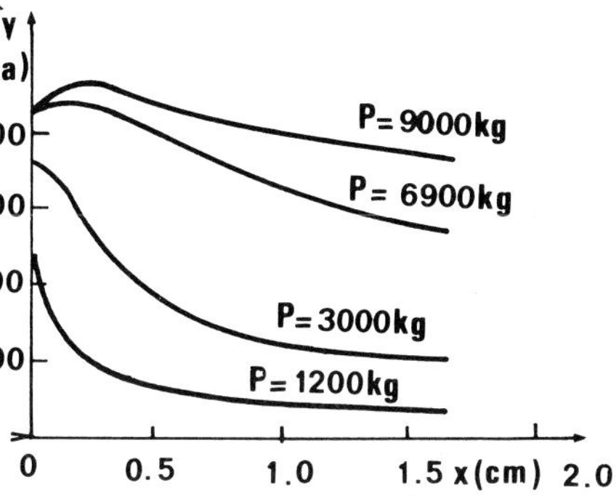

FIGURE 4 :

Distribution of σ_y along crack line in different loading.

REFERENCES

[1] Lemaitre, J., J. Engng. Mat. & Tech., (1985), pp. 83-89.
[2] Rousselier, G., IUTAM Symp. in 3-D Constitutive Relations and Ductile Fracture, 1980, pp. 331-335.
[3] Lee, H., et al., Proceedings of the Int. Conf. on Nonlinear Mechanics, 1985, pp. 566-573.
[4] Lee, H., et al., J. Huazhong University of Science and Technology, 1983, pp. 63-70.
[5] Krajcinovic, D., J. Appl. Mech., 1983, pp. 355-360.
[6] Kestin, J., & Rice, J.R., A Critical Review of Thermodynamics, 1970, pp. 275-297.
[7] Valanis, K.C., Int. J. Nonlinear Mech., 1971, pp. 337-360.
[8] Valanis, K.C., Z.A.M.M. 63, 1983, pp. 73-80.
[9] Fan Jinghong, Advances in Mechanics, 1985, pp. 273-290.
[10] Valanis, K.C., Archives of Mechanics, 1971, pp. 517-551.
[11] Fan Jinghong, Proceedings of the Int. Conference on Nonlinear Mechanics, 1985, pp. 108-113.
[12] Fan Jinghong & Zhong Junqian, J. Chongqing University, 1985, pp. 49-57.
[13] Valanis, K.C. and Fan Jinghong, Computers & Structures, 1984, pp. 714-724.
[14] Ashby, M.F. & Dyson, B.F., ICF-6 Symp., 1984, pp. 3-30.
[15] Carofalo, F., Fundamentals of Creep and Creep Rupture in Metals, (McMillan, New York, 1966).
[16] Valanis, K.C., Nuclear Engineering and Design, 1982, pp. 327-344.
[17] Valanis, K.C., Arch. Mech., 1980, pp. 171-191.
[18] Valanis, K.C. & Fan Jinghong, Proceedings of Int. Conference of Plasticity, 1983, pp. 153-173.
[19] Fan Jinghong & Peng Xianghe, Proceedings of Int. Conference of Computational Mechanics (In print).
[20] Lemaitre, J. & Dufailly, J., 3è Congrès Français de Mécanique, Grenoble, 1977.
[21] Li quo Chen & Howard, I.C., J. Mech. Phys. Solids, 1983, pp. 85-102.
[22] Fan Jinghong & Peng Xianghe (in print).
[23] Fan Jinghong a Zheng Xiangguo (in print).
[24] Valanis, K.C., A.M.D., vol. 47.
[25] Valanis, K.C., Journal de Mécanique, 1979.
[26] Fan Jinghong, Journal of Chongqing University, 1985, pp. 58-67.
[27] Dvorak, G.J., & Bahai-El-Din, Y.A., J. of Appl. Mech., 1982, pp. 327-334.
[28] Fan Jinghong & Zhang Junqin, Prof. of Int. Symposium on Composite Materials and Structures, 1986, pp. 1019-1025.

Thermomechanical Couplings in Solids
H.D. Bui and Q.S. Nguyen (Editors)
Elsevier Science Publishers B.V. (North-Holland)

PLASTIC BEHAVIOUR OF STEELS DURING PHASE TRANSFORMATIONS

J.B. LEBLOND(*), J. DEVAUX, G. MOTTET, J.C. DEVAUX

Centre de Calcul de FRAMATOME,
71380 Saint-Marcel, France

The anomalous plastic behaviour of steels during phase transformations is studied both theoretically (using the Hill-Mandel theory) and numerically (simulating a representative volume by a finite element mesh). Models for this behaviour are inferred from the results obtained.

1. INTRODUCTION

Several authors, including Denis et al. [1], Devaux [2], Giusti [3], Rammerstorfer et al. [4], Sjöström [5], have emphasized the need for models describing the anomalous plastic behaviour of steels during phase transformations for numerical calculations of residual stresses in welding or quenching operations. This behaviour can be subdivided into :

(*i*) *Classical plasticity*, i.e. response of the material (mixture of two phases) to variations of the applied stress or the temperature ;

(*ii*) *Transformation plasticity*, i.e. response of the material to variations of the phase proportions.

Transformation plasticity is currently attributed to two distinct mechanisms : that of Greenwood and Johnson [6], according to which the microscopic plastic strains due to volume incompatibilities between the two phases are "oriented" by the applied stress, and that of Magee [7] (for martensitic transformations), according to which the applied stress influences the orientation of the appearing martensite plates. It is evidenced experimentally by submitting a specimen to a constant stress during a transformation. In the most current case where the stress is small and uniaxial (let us say in the x direction), this strain is found to be of the form :

$$E^{tp}_{xx} = K\,\Sigma_{xx}\,\varphi(z) \qquad (1) \quad ,$$

where K is a constant, Σ_{xx} is the applied stress and φ is an increasing "normalized" function of the proportion z of the newly formed phase (phase 2)

(*) Also with the Laboratoire de Mécanique des Solides, Ecole Polytechnique, 91128 Palaiseau, France.

($\varphi(0) = 0$, $\varphi(1) = 1$). Giusti [3] and Leblond [8] have proposed the following generalization of this expression for the general case (variable, great or small, and triaxial stresses) :

$$\dot{E}_{ij}^{tp} = \frac{3}{2} K S_{ij} \varphi'(z) \dot{z} \qquad (2) ,$$

where S is the stress deviator. Recent experiments performed by Gigou [9] for biaxial stresses show that this is indeed a reasonable formula for multiaxial stresses. However this model lacks a theoretical foundation and an experimental validation in the case of high stresses.

Previous models for *classical plasticity* during phase transformations (Devaux [2], Giusti [3], Sjöström [5]) made the assumption of an ordinary plastic behaviour (ideal plasticity, or isotropic or kinematic hardening) with a "global" yield stress given by :

$$\Sigma^y = (1 - z) \sigma_1^y + z \sigma_2^y \qquad (3) ,$$

with obvious notations. This is an oversimplification ; indeed, because of microscopic plastic strains induced by volume incompatibilities, (macroscopic) plasticity must be expected to occur *right from the beginning of the stress-strain curve.*

The purpose of this paper is to fill the gaps evidenced above. Section 2 provides theoretical foundations for mathematical modelling of plastic behaviour during transformations ; the main problem here is to justify theoretically the existence of transformation plasticity, i.e. to show *without any a priori postulate* that the occurrence of a transformation induces the appearance of a new term in the macroscopic plastic strain rate. Section 3 studies both theoretically and numerically the response of phase-transforming steels to variations of the applied stress, in the case of ideal-plastic individual phases. Section 4 studies the response to variations of the phase proportions and the temperature under the same hypotheses. Finally, Section 5 extends the models proposed to include strain-hardening effects, and particularly the "regeneration" of strain hardening during transformations (due to the disappearance of old dislocations).

2. GENERAL RELATIONS

2.1- Micro- and macro-stresses and strains :

Following the theory of Hill [10] and Mandel [11], the macroscopic stress Σ and total strain E^t are identified with the average value of the corresponding microscopic quantities over a representative volume v :

$$\Sigma = \langle \sigma \rangle_v \quad ; \quad E^t = \langle \varepsilon^t \rangle_v \qquad (4) \quad .$$

The total microscopic strain is as usual decomposed into elastic, plastic and thermal parts :

$$\varepsilon^t = \varepsilon^e + \varepsilon^{th} + \varepsilon^p \qquad (5) \quad .$$

In this equation ε^{th} depends on which phase the point considered belongs to (see fig. 1) ; the thermal strain of a given phase at a given temperature is defined with respect to the strain of an arbitrary phase at an arbitrary temperature. The plastic strain includes (by definition) not only the ordinary plastic strain but also the pseudo-plastic strain associated with the transformation, i.e. the deviatoric part of the transformation strain (the isotropic part being included in ε_1^{th} and ε_2^{th}).

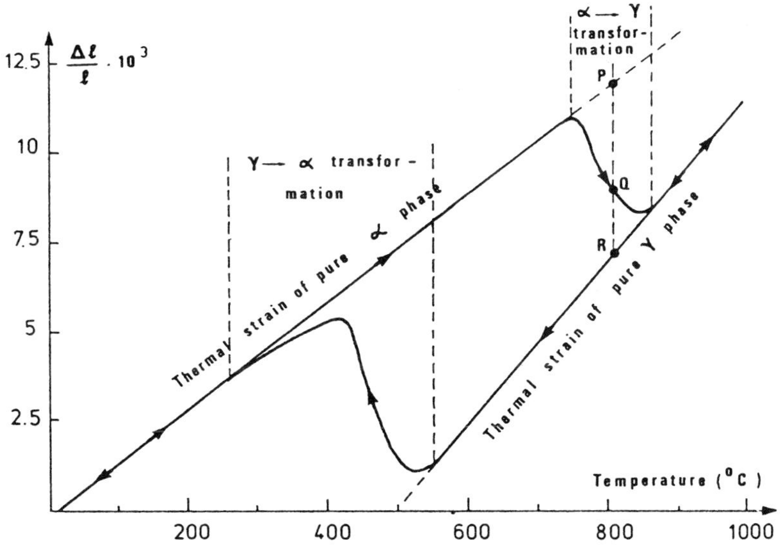

FIGURE 1 :

Dilatometric diagram of the A 508 Cl. 3 steel heated at 30°C.s^{-1} and cooled at -2°C.s^{-1}.

The following hypothesis is now introduced :

H.1 : The microscopic elastic compliance tensor m may be equated to the macroscopic overall elastic compliance tensor M (some of the following results still hold under a somewhat weaker assumption ; see Leblond et al. [12]).

The macroscopic elastic strain is then defined by :

$$E^e = M\,\Sigma = M < \sigma >_v = < M\,\sigma >_v = < m\,\sigma >_v = < \varepsilon^e >_v \qquad (6)\ .$$

The rest of the macroscopic strain, $E^t - E^e$, is decomposed into an isotropic part plus a traceless part :

$$\left.\begin{aligned} &E^t - E^e = E^{thm} + E^p \quad , \\ &E^{thm} = < \varepsilon^{th} >_v = (1-z)\,\varepsilon_1^{th} + z\,\varepsilon_2^{th} \quad , \\ &E^p = < \varepsilon^p >_v \quad . \end{aligned}\right\} \qquad (7)\ .$$

It is interesting to note that the "thermometallurgical" strain E^{thm} is precisely the total strain observed if *(i)* no macroscopic stress is applied and *(ii)* the material is *macroscopically isotropic*, i.e. all directions are physically equivalent on a macroscopic scale. Indeed E^e is then trivially zero, and E^p is also zero since it is isotropic (because of the condition of macroscopic isotropy) and traceless (because ε^p is traceless). As conditions *(i)* and *(ii)* are very generally fulfilled in dilatometry experiments (fig. 1), the strain observed in these experiments is precisely $(1-z)\,\varepsilon_1^{th} + z\,\varepsilon_2^{th}$, which means that the proportion of γ phase can be read from the ratio PQ/PR on figure 1. This rule is currently used to determine transformation kinetics from dilatometry diagrams but had not been justified theoretically up to now.

2.2- *Decomposition of macroscopic plasticity into classical plus transformation plasticity :*

Since E^e and E^{thm} are known by equations (6) and $(7)_2$, attention is now restricted to E^p. Differentiating equ. $(7)_3$ with respect to time and noting that ε^p is discontinuous through the transformation front F (because of the pseudo-plastic strain induced by the transformation), we obtain :

$$\dot{E}^p = (1-z) < \dot{\varepsilon}_1^p >_{v_1} + z < \dot{\varepsilon}_2^p >_{v_2} + \dot{z} < \Delta\varepsilon^p_{1\to 2} >_{F(U_n)} \qquad (8)$$

where v_1 and v_2 are the subvolumes of volume v occupied by phases 1 and 2, and where $< \Delta\varepsilon^p_{1\to 2} >_{F(U_n)}$ denotes the average value of the discontinuity of plastic strain $\Delta\varepsilon^p_{1\to 2}$ through F weighted by its normal velocity U_n (considered as positive when directed towards v_1) :

$$< \Delta\varepsilon^p_{1\to 2} >_{F(U_n)} = \frac{\int_F \Delta\varepsilon^p_{1\to 2}\, U_n\, dS}{\int_F U_n\, dS} \qquad (9)\ .$$

The expression of the microscopic plastic strain rate is of the type $\dot{\varepsilon}^p = (\ )\ \dot{\sigma} + (\ )\ \dot{T}$, *without any* $\dot{z}$*-term* because classical plasticity rules are supposed to be obeyed at the microscopic scale. However, when expressed in terms of macroscopic quantities, $\dot{\varepsilon}^p$ includes a term proportional to $\dot{z}$ in addition to terms proportional to $\dot{\Sigma}$ and $\dot{T}$, since a change of z implies a modification of the geometry and therefore some plastic strains, even if Σ and T are constant. Therefore equ. *(8)* can be put under the form (with obvious notations)

$$\dot{E}^p = \dot{E}^{cp} + \dot{E}^{tp} \qquad (10)$$

where :

$$\left.\begin{aligned}\dot{E}^{cp} = &\left\{ (1-z) < \frac{\delta\varepsilon_1^p}{\delta\Sigma} >_{V_1} + z < \frac{\delta\varepsilon_2^p}{\delta\Sigma} >_{V_2} \right\} \dot{\Sigma}\ + \\ &+ \left\{ (1-z) < \frac{\delta\varepsilon_1^p}{\delta T} >_{V_1} + z < \frac{\delta\varepsilon_2^p}{\delta T} >_{V_2} \right\} \dot{T}\end{aligned}\right\} \qquad (11)\ ;$$

$$\dot{E}^{tp} = \left\{ (1-z) < \frac{\delta\varepsilon_1^p}{\delta z} >_{V_1} + z < \frac{\delta\varepsilon_2^p}{\delta z} >_{V_2} + < \Delta\varepsilon^p_{1\to2} >_{F(U_n)} \right\} \dot{z} \qquad (12)\ .$$

These equations justify the decomposition of macroscopic plasticity into classical plus transformation plasticity, and also that of transformation plasticity into the two mechanisms proposed by Greenwood and Johnson [6] and Magee [7] (see the introduction) : indeed the term $\left\{ (1-z) < \delta\varepsilon_1^p/\delta z >_{V_1} + z < \delta\varepsilon_2^p/\delta z >_{V_2} \right\} \dot{z}$ in equ. *(12)* can be interpreted as corresponding to the mechanism of Greenwood and Johnson, and the term $< \Delta\varepsilon^p_{1\to2} >_{F(U_n)}\ \dot{z}$ to that of Magee (see [12]).

3. CLASSICAL PLASTICITY FOR IDEAL-PLASTIC PHASES

The subject of this section is E^{cp}, and more specifically :

$$\dot{E}^{cp}_{\Sigma} = \left\{ (1-z) < \frac{\delta\varepsilon_1}{\delta\Sigma} >_{V_1} + z < \frac{\delta\varepsilon_2}{\delta\Sigma} >_{V_2} \right\} \dot{\Sigma} \qquad (13)\ .$$

Attention is restricted from now on to $\gamma \to \alpha$ transformations (phase 1 = γ phase ; phase 2 = α phase), because the plastic behaviour during the $\alpha \to \gamma$ transformation is of no practical interest since the subsequent vanishing of stresses at high temperatures erases all memory of the previous mechanical history.

3.1- Theoretical approach :

In the case of *low stresses*, the treatment, which is detailed in Leblond et al. [13], is based on the following assumptions :

H.2 : Since the yield stress of the γ phase is small whereas that of the α phase is large, volume incompatibilities induce a complete plastification of the former but no plastification of the latter.

H.3 : Both phases are ideal-plastic and obey the von Misès criterion and the Prandtl-Reuss flow rule.

H.4 : $\varepsilon^{eq} = \int_0^t \left(\frac{2}{3} \dot{\varepsilon}^p_{ij} \dot{\varepsilon}^p_{ij} \right)^{1/2} dt$ and s denoting the microscopic equivalent plastic strain and stress deviator, correlations between $\delta\varepsilon^{eq}/\delta\Sigma$ and s can be neglected.

H.5 : The macroscopic stress deviator in phase 1, $S_1 = < s_1 >_{V_1}$, is almost equal to the overall stress deviator $S = < s >_V$.

H.6 : $\dot{E}^{cp}_\Sigma$ is non zero only if $\Sigma^{eq} = \left(\frac{3}{2} S_{ij} S_{ij} \right)^{1/2}$ varies, i.e. :

$$< \frac{\delta\varepsilon^{eq}}{\delta\Sigma_{ij}} >_V \dot{\Sigma}_{ij} \equiv < \frac{\delta\varepsilon^{eq}}{\delta\Sigma^{eq}} >_V \dot{\Sigma}^{eq} .$$

Equation *(13)* becomes, under these hypotheses (see [13]) :

$$\dot{E}^{cp}_\Sigma = \frac{3(1-z)}{2\,\sigma^y_1} < \frac{\delta\varepsilon^{eq}_1}{\delta\Sigma^{eq}} >_{V_1} S\, \dot{\Sigma}^{eq} \qquad (14) .$$

An estimate of $< \delta\varepsilon^{eq}_1/\delta\Sigma^{eq} >_{V_1}$, valid for medium or high values of z, can be derived by considering a special geometry (two bars in parallel) : the result is $< \delta\varepsilon^{eq}_1/\delta\Sigma^{eq} >_{V_1} \sim \frac{1}{zE}$, where E is Young's modulus. Inserting this value into equ. *(14)*, we get :

$$\dot{E}^{cp}_\Sigma = \frac{3(1-z)}{2\,\sigma^y_1} \frac{S}{zE} \dot{\Sigma}^{eq} \qquad (15) ,$$

i.e. in the case of an uniaxial stress (in the x direction) and a constant z :

$$\left(E^{cp}_\Sigma \right)_{xx} = \frac{1-z}{2\,\sigma^y_1} \frac{\Sigma^2_{xx}}{zE} \qquad (16) .$$

This equation describes the beginning of the stress-strain curve for medium or high values of z. It predicts a *parabolic* shape, which means that despite the fact that the stress-strain curve contains no elastic portion, it is nevertheless initially tangent to the elastic line with slope E. This phenomenon can be attributed to the progressive orientation of the microscopic stress deviators towards the direction of traction when the applied stress increases (see [13]).

In the case of *very high stresses*, it can be proved very simply that the ultimate stress Σ_u (maximum possible value of Σ^{eq}) is not given by a "mixture rule" analogous to *(3)*, but satisfies instead the inequality :

$$\Sigma_u \leqslant (1-z)\,\sigma_1^y + z\,\sigma_2^y \qquad (17)\ .$$

This inequality can be shown to be strict for $0 < z < 1$ (see [13]) but is difficult to refine theoretically.

3.2 - *Numerical approach* :

The numerical approach consists in simulating a representative volume by a cube divided into 125 (5 x 5 x 5) cubic elements, the physical characteristics of which correspond to the α or the γ phase in a random manner. These characteristics are representative of the austenitic and martensitic phases of the A 508 cl.3 steel at 350°C and are as follows : E = 182.000 MPa, ν = 0.3 for both phases (in accordance with hypothesis H.1) ; σ^y = 145 MPa for austenite and 950 MPa for martensite ; no strain hardening (in accordance with hypothesis H.3) ; thermal strain (in all directions) = - 0.42% for austenite and + 0.42% for martensite. Several proportions of martensite are studied : 12.5%, 25%, 50% and 75%. The cube is submitted to a traction force in the x direction and the resulting macroscopic stress-strain curve is recorded.

As an example, figure 2 shows the curve obtained for z = 50%, together with theoretical points (equ. (16)).

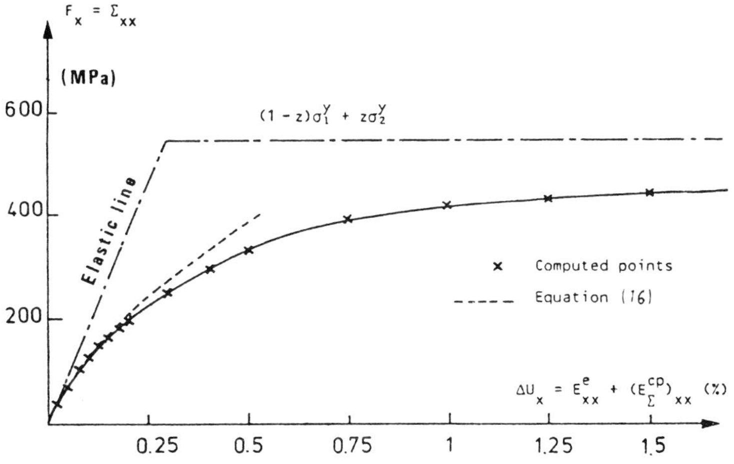

FIGURE 2 :

Stress-strain curve for a proportion of the harder phase equal to 50%.

It is observed that the agreement is quite good, even for stresses of the order of 0.5 times the ultimate stress, though equ. (16) is a priori supposed to be valid only for low stresses. The agreement is found to be still better for z = 75%, but poorer for lower values of z (12.5% and 25%), as expected theoretically (see § 3.1).

These calculations also provide an opportunity to test the validity of some of the hypotheses made in the treatment. As an example, figure 3 shows the relation between S_{1xx} and S_{xx} in the case where z = 50%. Hypothesis H.5 is well verified for small stresses (as expected theoretically). For large stresses S_{1xx} is observed to "saturate" at a value which is less than $\frac{2}{3}\sigma_1^y$; this can also be explained theoretically (see [13]).

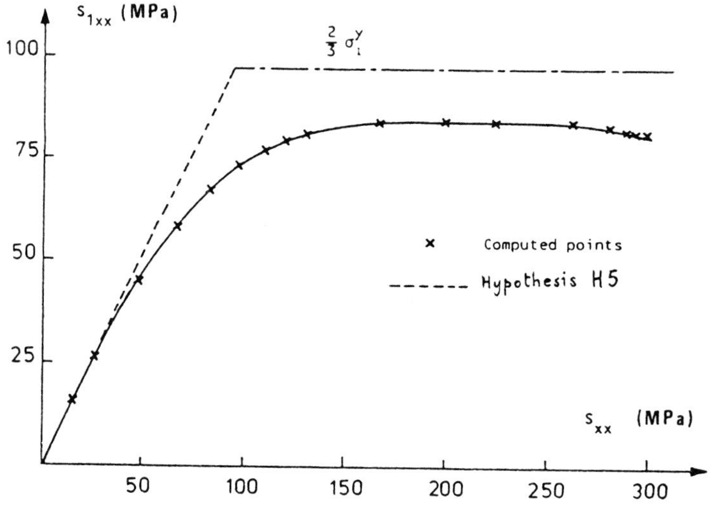

FIGURE 3 :
Relation between S_{1xx} and S_{xx} for z = 50%.

Finally figure 4 shows the ultimate stress Σ_u, as deduced from the stress-strain curve, versus z.

The difference between Σ_u and the upper bound $(1-z)\,\sigma_1^y + z\,\sigma_2^y$ is maximal for low values of z ; this is because the α phase is then dispersed into small isolated islands so that the ultimate stress is determined almost exclusively by the yield stress of the γ phase-matrix.

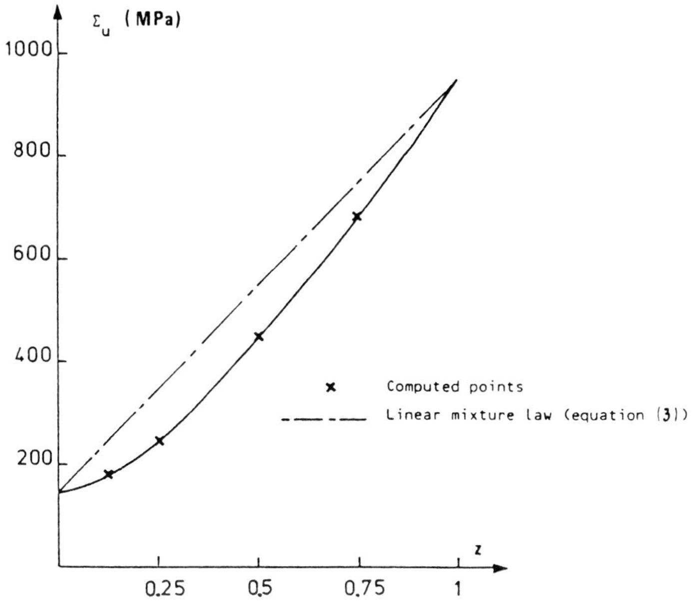

FIGURE 4 :
Ultimate stress as a function of the proportion of the harder phase.

3.3 - *A model for* $\dot{E}^{cp}_{\Sigma}$ *for ideal-plastic phases* :

Based on the previous results, the following model is proposed :

- Expression of Σ_u : $\Sigma_u = [\,1 - f(z)\,]\,\sigma_1^y + f(z)\,\sigma_2^y$ (18) ;

- Expression of $\dot{E}^{cp}_{\Sigma}$: $$\dot{E}^{cp}_{\Sigma} = \begin{cases} \dfrac{3(1-z)}{2\,\sigma_1^y}\,\dfrac{g(z)}{E}\,S\,\dot{\Sigma}^{eq} & \text{if } \Sigma^{eq} < \Sigma_u \\ \dot{\Lambda}\,S & \text{if } \Sigma^{eq} = \Sigma_u \end{cases} \qquad (19)\ ,$$

where the f and g functions are given in Table 1 and $\dot{\Lambda}$ is indeterminate.

z	0.	0.125	0.25	0.50	0.75	1.
f(z)	0.	0.044	0.124	0.391	0.668	1.
g(z)	0.	2.53	4.	2.76	1.33	1.

TABLE 1 : Values of the f and g functions.

4. TRANSFORMATION PLASTICITY FOR IDEAL-PLASTIC PHASES

The subject of this section is E^{tp}, the expression of which is given by equ. *(12)*, and also incidentally :

$$\dot{E}_T^{cp} = \left\{ (1-z) < \frac{\delta\varepsilon_1^p}{\delta T} >_{V_1} + z < \frac{\delta\varepsilon_2^p}{\delta T} >_{V_2} \right\} \dot{T} \qquad (20) \quad .$$

4.1 - Theoretical approach :

The theoretical approach is restricted to the case of low applied stresses, and similar to that leading to equ. *(14)*, replacing $\delta/\delta\Sigma$ by $\delta/\delta z$ or $\delta/\delta T$; the hypotheses made are those detailed above, plus the following extra assumption for transformation plasticity :

H.7 : The Magee mechanism is supposed to be negligible in the steels under consideration.

The equations obtained for $\dot{E}^{tp}$ and $\dot{E}_T^{cp}$ are similar to equ. *(14)* :

$$\dot{E}^{tp} = \frac{3(1-z)}{2\,\sigma_1^y} < \frac{\delta\varepsilon_1^{eq}}{\delta z} >_{V_1} S\,\dot{z} \qquad (21) \quad ;$$

$$\dot{E}_T^{cp} = \frac{3(1-z)}{2\,\sigma_1^y} < \frac{\delta\varepsilon_1^{eq}}{\delta T} >_{V_1} S\,\dot{T} \qquad (22) \quad .$$

An estimate of $< \delta\varepsilon_1^{eq}/\delta z >_{V_1}$ can be derived (see Leblond et al. [14]) by considering a typical geometry, that of an austenitic sphere in which a spherical core of α phase is growing ; the result is $< \delta\varepsilon_1^{eq}/\delta z >_{V_1} \simeq -2\,\Delta\varepsilon_{1\to2}^{th}(\ln z)/(1-z)$, where $\Delta\varepsilon_{1\to2}^{th}$ is the (scalar) difference of thermal strain between the two phases. Equ. *(21)* becomes then :

$$\dot{E}^{tp} = -\frac{3\,\Delta\varepsilon_{1\to2}^{th}}{\sigma_1^y} S\,(\ln z)\,\dot{z} \qquad (23) \quad .$$

This formula is of type *(2)*, with :

$$K = \frac{2\,\Delta\varepsilon_{1\to2}^{th}}{\sigma_1^y} \quad ; \quad \varphi(z) = z\,(1-\ln z) \qquad (24) \quad .$$

Equations *(23)* and *(24)* provide not only a theoretical justification of the phenomenological Giusti-Leblond model (equ. *(2)*), but also explicit expressions for K and φ. Table 2 and figure 5 show some comparisons between theoretical and experimental values of K and φ for several steels.

Steel	A533	35 NCD 16	Fe-Ni alloy (31% Ni)
$K_{th.}$ (MPa^{-1})	$9.3\ 10^{-5}$	$7.3\ 10^{-5}$	$9.4\ 10^{-5}$
$K_{exp.}$ (MPa^{-1})	1.10^{-4}	5.10^{-5}	1.10^{-4}

TABLE 2 : Theoretical and experimental values of K for some steels.

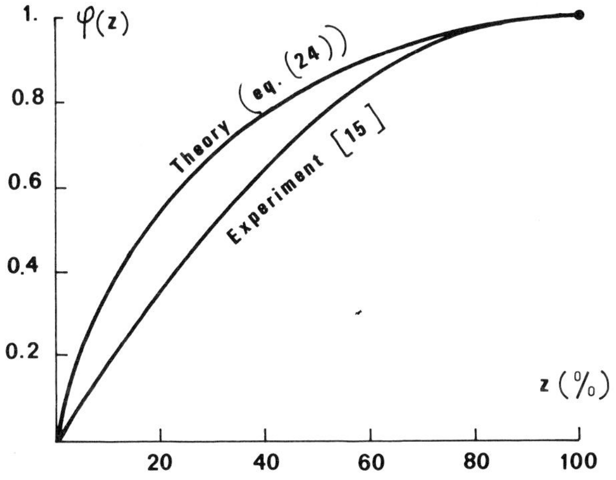

FIGURE 5 :

Theoretical and experimental (after [15]) values of the φ function.

Other comparisons can be made and yield generally a good agreement between theory and experiments. This justifies a posteriori the assumption that the Magee mechanism is often negligible.

Other authors [6], [16] have calculated E^{tp}, in the particular case of uniaxial and constant stresses. Greenwood and Johnson [6] found $K = 5\ \Delta\varepsilon^{th}_{1\to 2}/2\ \sigma^y_1$, which is roughly in accordance with the value given by equ. $(24)_1$; however their calculation relies on very doubtful assumptions, as analyzed by Abrassart [16]. Abrassart found $K = 3\ \Delta\varepsilon^{th}_{1\to 2}/4\sigma^y_1$, which disagrees with equ. $(24)_1$ and experiments ; the underestimation of K arises probably from the fact that the calculation is not incremental.

$< \delta\varepsilon^{eq}_1/\delta T >_{V_1}$ can also be estimated by considering a spherical geometry (see [14]) ; the result is $< \delta\varepsilon^{eq}_1/\delta T >_{V_1} = 2(\alpha_1 - \alpha_2)\ z \ln z/(1-z)$ where α_i is the coefficient of thermal expansion of phase i. Equ. (22) becomes then :

$$\dot{E}^{cp}_T = \frac{3(\alpha_1 - \alpha_2)}{\sigma^y_1}\ S\ z\ (\ln z)\ \dot{T} \qquad (25)$$

Inserting typical values of the parameters in this equation, one finds that the effect of variations of T is roughly 20 times smaller than that of variations of z.

4.2 - Numerical approach :

The principle of the numerical approach is the same as for classical plasticity ; the progress of the transformation is simulated by changing the physical characteristics (thermal strain, yield stress) of the elements one after another. For this simulation the 5 x 5 x 5 cube is found to yield unreliable results (see [14]), and a larger cube (10 x 10 x 10) is therefore used. Several loadings, physical characteristics and orders of transformation are studied. Figure 6 summarizes the results obtained for low and medium stresses.

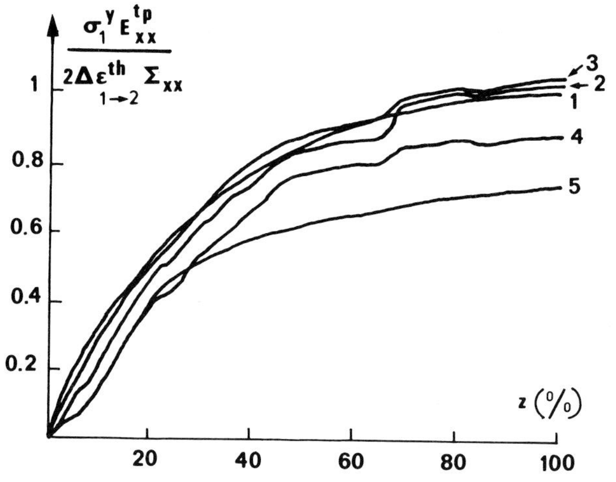

FIGURE 6 :

"Normalized" transformation plastic strain versus z for low or medium stresses.

(1) theory ; (2) ordered transformation, Σ_{xx} = 50 MPa ;
(3) ordered transformation, Σ_{xx} = 100 MPa ;
(4) ordered transformation, Σ_{xx} = 50 MPa, $\Delta\varepsilon_{1\to2}^{th}$ = 0.0042 instead of 0.0084 ;
(5) random transformation, Σ_{xx} = 100 MPa.

All curves are in rough agreement with the theoretical one. The order of transformation is seen to have some influence on the results ; the real curve is certainly comprised between those corresponding to an ordered transformation and a random one : indeed the real transformation is neither completely ordered nor completely random. It is also observed that E^{pt} is roughly proportional to

$\Delta\varepsilon^{th}_{1\to2}$ and Σ_{xx}, as predicted theoretically ; however E^{pt} seems to increase slightly more quickly than Σ_{xx}. This non-linear effect is confirmed experimentally (see e.g. Gigou [9]), and becomes quite notable when Σ_{xx} becomes close to σ^y_1.

Figures 7 and 8 allow for an investigation of the validity of hypotheses H.2 and H.5. It is observed that hypothesis H.2 is reasonably satisfied, except (of course) at the very beginning of the transformation. On the other hand hypothesis H.5 gives an acceptable approximation of S_1 only for $0 \leqslant z \leqslant 0.45$; that the theoretical results based on this hypothesis are nevertheless correct is readily explained by the fact that 80% of the effect of transformation plasticity occurs during the first half of the transformation.

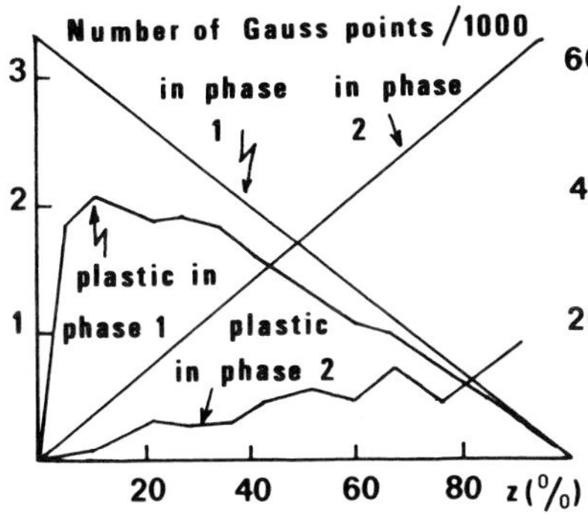

FIGURE 7 :

Verification of hypothesis H.2 (ordered transformation, Σ_{xx} = 50 MPa).

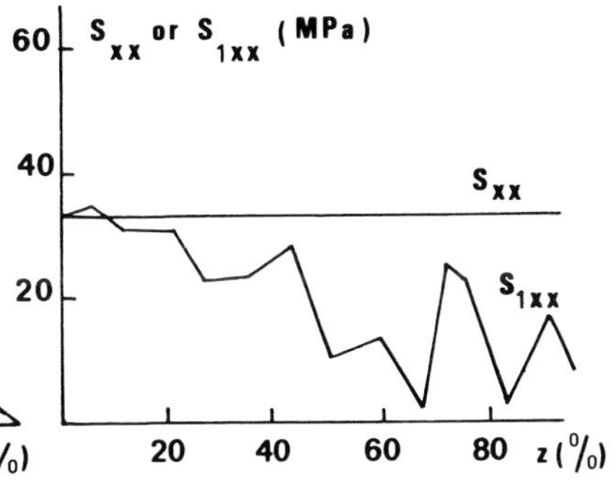

FIGURE 8 :

Verification of hypothesis H.5 (ordered transformation, Σ_{xx} = 50 MPa).

Figure 9 summarizes the results obtained for high stresses (greater than σ^y_1). Since these stresses are greater than Σ_u at the beginning of the transformation, they are applied only during the second half of the transformation, and the quantity plotted in fig. 9 is the increase of plastic deformation with respect to the "initial" state where z = 0.50, divided by the applied stress. The curve obtained for Σ_{xx} = 100 MPa is also plotted as a reference. This figure confirms the non-linear effect mentioned previously. However, this effect seems to depend rather on the ratio Σ_{xx}/Σ_u (Σ^{eq}/Σ_u in a triaxial case) than on the absolute value of Σ_{xx} : indeed, since the effect is weak between 100 and 260 MPa, if it depended on Σ_{xx} itself, it would be weak for stresses of the order of σ^y_1, which is not the case if these stresses are applied at the beginning of the transformation (see above).

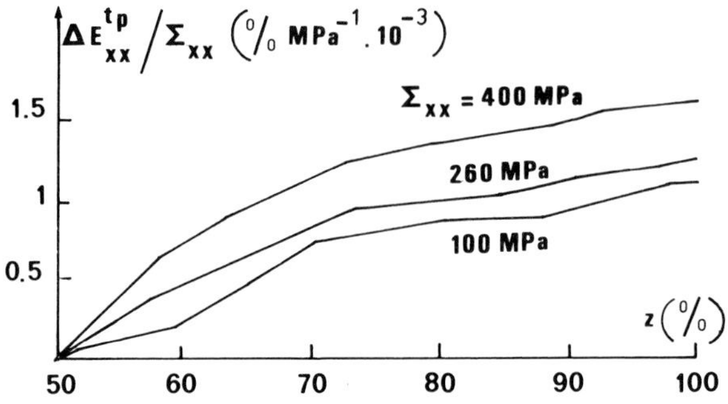

FIGURE 9 :

Transformation plastic strain divided by the applied stress versus z for high stresses.

4.3 - Models for $\dot{E}^{tp}$ *and* $\dot{E}^{cp}_T$ *for ideal-plastic phases :*

Based on the previous results, the following models are proposed :

$$\dot{E}^{tp} = - \frac{3\ \Delta\varepsilon^{th}_{1\to2}}{\sigma^y_1} S\ h\left(\frac{\Sigma^{eq}}{\Sigma_u}\right)(\ln z)\ \dot{z} \qquad (26),$$

where :

$$h\left(\frac{\Sigma^{eq}}{\Sigma_u}\right) = \begin{cases} 1 \text{ if } 0 \leqslant \Sigma^{eq}/\Sigma_u \leqslant 0.5 \\ 1 + 1.73\ (\Sigma^{eq}/\Sigma_u - 0.5) \text{ if } 0.5 \leqslant \Sigma^{eq}/\Sigma_u \leqslant 1 \end{cases} \qquad (27)\ ;$$

$$\dot{E}^{cp}_T = \frac{3(\alpha_1 - \alpha_2)}{\sigma^y_1} S\ z\ (\ln z)\ \dot{T} \qquad \text{(Equ. (25)).}$$

5. INTRODUCTION OF STRAIN HARDENING EFFECTS

Introduction of strain hardening effects requires :

(i) a generalization of the models proposed above for $\dot{E}^{cp}_\Sigma$, $\dot{E}^{tp}$ and $\dot{E}^{cp}_T$;

(ii) a formulation of evolution laws for the strain hardening parameters taking into account the "regeneration" during transformations.

5.1 - Models for $\dot{E}^{cp}_\Sigma$, $\dot{E}^{tp}$ *and* $\dot{E}^{cp}_T$ *with strain hardening effects :*

For isotropic hardening, microscopic scalar strain-hardening parameters ε^{eff}_1 and ε^{eff}_2 are introduced ; they are different from the usual parameters ε^{eq}_1 or ε^{eq}_2 because of the regeneration during transformations (see § 5.2 below). The theoretical treatment is exactly the same as for ideal plasticity, with the extra assumption that $\sigma_i^y(\varepsilon_i^{eff})$, which is non-uniform since the hardening parameter ε_i^{eff} is non-uniform, may be approximated by $\sigma_i^y(E_i^{eff})$ where :

$$E_i^{eff} = \langle \varepsilon_i^{eff} \rangle_{v_i} \qquad (28) \quad .$$

The equations for $\dot{E}_\Sigma^{cp}$, $\dot{E}^{tp}$ and $\dot{E}_T^{cp}$ are then the same as those for ideal-plasticity (equations *(18)*, *(19)*, *(25)*, *(26)*, *(27)*), replacing everywhere σ_i^y by $\sigma_i^y(E_i^{eff})$, except for the fact that $\dot{\Lambda}$ in equ. $(19)_2$ is no longer indeterminate but will be determined by the evolution laws of the hardening parameters.

For kinematic hardening, microscopic tensorial hardening parameters a_1 and a_2 are introduced. The treatment and the equations obtained are the same as for ideal plasticity, replacing everywhere s_i by $s_i - a_i$, S_i by $S_i - A_i$, S by S - A where :

$$A_i = \langle a_i \rangle_{v_i} \quad ; \quad A = \langle a \rangle_v = (1-z)\, A_1 + z\, A_2 \qquad (29) \quad ,$$

and $\dot{\Sigma}^{eq}$ in equ. $(19)_1$ by the derivative $\dot{\Sigma}_S^{eq}$ of Σ^{eq} for *a constant* A :

$$\dot{\Sigma}_S^{eq} = \frac{3}{2\,\Sigma^{eq}} (S_{ij} - A_{ij})\, \dot{S}_{ij} \qquad (30) \quad ;$$

furthermore $\dot{\Lambda}$ in equ. $(19)_2$ is here again no longer indeterminate.

5.2 - *Evolution laws for the hardening parameters* :

The principle of the following treatment is basically the same as that in the work of Sjöström [5] ; however the method of derivation and even the results are somewhat different.

For isotropic hardening, time differentiation of equ. *(28)* yields, taking into account the fact that volumes v_1 and v_2 are moving (see [17]) :

$$\left.\begin{aligned} \dot{E}_1^{eff} &= \langle \dot{\varepsilon}_1^{eff} \rangle_{v_1} + \frac{\dot{z}}{1-z} E_1^{eff} - \frac{\dot{z}}{1-z} \langle \varepsilon_1^{eff} \rangle_{F(U_n)} \quad ; \\ \dot{E}_2^{eff} &= \langle \dot{\varepsilon}_2^{eff} \rangle_{v_2} - \frac{\dot{z}}{z} E_2^{eff} + \frac{\dot{z}}{z} \langle \varepsilon_2^{eff} \rangle_{F(U_n)} \quad . \end{aligned}\right\} \qquad (31)$$

Three natural hypotheses are now introduced :

(i) if a point does not undergo a transformation, the evolution law for its hardening parameter is the same as usual ($\dot{\varepsilon}_i^{eff} = \dot{\varepsilon}_i^{eq}$) ;

(ii) there is no correlation between the location of the transformation front and the hardening parameter of the old phase 1, so that $\langle \varepsilon_1^{eff} \rangle_{F(U_n)} = \langle \varepsilon_1^{eff} \rangle_{v_1} = E_1^{ff}$;

(iii) the new phase 2 appears with a zero initial hardening parameter, so that $<\varepsilon_2^{eff}>_{F(U_n)} = 0$.

Equations (*31*) become thus :

$$\left.\begin{aligned} \dot{E}_1^{eff} &= <\dot{\varepsilon}_1^{eq}>_{V_1} \\ \dot{E}_2^{eff} &= <\dot{\varepsilon}_2^{eq}> - \frac{\dot{z}}{z} E_2^{eff} \end{aligned}\right\} \quad (32) \quad .$$

Two cases must now be distinguished. If $\Sigma^{eq} < \Sigma_u$, $\dot{\varepsilon}_2^{eq} = 0$ by hypothesis H.2, and

$$<\dot{\varepsilon}_1^{eq}>_{V_1} = <\frac{\delta\varepsilon_1^{eq}}{\delta\Sigma}>_{V_1}\dot{\Sigma} + <\frac{\delta\varepsilon_1^{eq}}{\delta z}>_{V_1}\dot{z} + <\frac{\delta\varepsilon_1^{eq}}{\delta T}>_{V_1}\dot{T}$$

can be evaluated as indicated in § *3.1* and *4.1* ; one gets thus

$$\left.\begin{aligned} \dot{E}^{eff} &= \frac{g(z)}{E}\dot{\Sigma}^{eq} - \frac{2\,\Delta\varepsilon_{1\to2}^{th}\ \ln z}{1-z}\, h\left(\frac{\Sigma^{eq}}{\Sigma_u}\right)\dot{z} + \frac{2\,(\alpha_1-\alpha_2)\ z\ln z}{1-z}\dot{T} \\ \dot{E}_2^{eff} &= -\frac{\dot{z}}{z} E_2^{eff} \end{aligned}\right\} \quad (33) \quad .$$

If $\Sigma^{eq} = \Sigma_u$, we assume that both phases are plastic and have uniform and identical plastic strain rates : $\dot{\varepsilon}_1^p = \dot{\varepsilon}_2^p = <\dot{\varepsilon}_1^p>_{V_1} = <\dot{\varepsilon}_2^p>_{V_2}$; then, neglecting Magee's mechanism, $\dot{\varepsilon}_1^p = \dot{\varepsilon}_2^p = \dot{E}^p$ by equ. (*8*), so that $<\dot{\varepsilon}_1^{eq}>_{V_1} = <\dot{\varepsilon}_2^{eq}>_{V_2} = \dot{E}^{eq}$ where $\dot{E}^{eq} = \left(\frac{2}{3}\dot{E}_{ij}^p\ \dot{E}_{ij}^p\right)^{1/2}$. We get thus :

$$\dot{E}_1^{eff} = \dot{E}^{eq} \quad ; \quad \dot{E}_2^{eff} = \dot{E}^{eq} - \frac{\dot{z}}{z} E_2^{eff} \quad (34) \quad .$$

For kinematic hardening, in the absence of transformation, a_1 and a_2 are of the form $Z_1(T)\ \varepsilon_1^p$ and $Z_2(T)\ \varepsilon_2^p$, so that the evolution laws are $a_i = Z_i(T)\ \dot{\varepsilon}_i^p + \frac{d Z_i}{Z_i\, dT} a_i\ \dot{T}$; in the presence of a transformation, a treatment analogous to that for isotropic hardening leads to the following evolution equations for A_1 and A_2 :

$$\left.\begin{aligned} &.\ \text{if } \Sigma^{eq} < \Sigma_u :\ \dot{A}_1 = Z_1\frac{\dot{E}^p}{1-z} + \frac{d Z_1}{Z_1\, dT}A_1\dot{T} \ ;\ \dot{A}_2 = \frac{d Z_2}{Z_2\, dT}A_2\dot{T} - \frac{\dot{z}}{z}A_2 \ ; \\ &.\ \text{if } \Sigma^{eq} = \Sigma_u :\ \dot{A}_1 = Z_1\dot{E}^p + \frac{d Z_1}{Z_1 dT}A_1\dot{T} \ ;\ \dot{A}_2 = Z_2\dot{E}^p + \frac{d Z_2}{Z_2 dT}A_2\dot{T} - \frac{\dot{z}}{z}A_2 \ . \end{aligned}\right\} \quad (35) \quad .$$

REFERENCES

[1] Denis, S., Gauthier, E., Simon, A., and Beck, G., Proc. Int. Symp. on the Calculation of Internal Stresses in Heat Treatment of Metallic Materials, Linköping (Sweden), (1984).
[2] Devaux, J.C., DGRST contract n° 7971095, final report (1982).
[3] Giusti, J., Thèse d'Etat, Paris (France), (1981).
[4] Rammerstorfer, F.G., Fischer, D.F., Mitter, W., Bathe, K.J. and Snyder, M.D., Comput. Struc., 13, 771, (1981).
[5] Sjöström, S., Thesis, Linköping (Sweden), (1982).
[6] Greenwood, G.W., and Johnson, R.H., Proc. Roy. Soc., A283, 403, (1965).
[7] Magee, C.L., Thesis, Pittsburgh (U.S.A.), (1966).
[8] Leblond, J.B., Note Interne Framatome n° TM/CDC/80.066, (1980).
[9] Gigou, D., Mémoire de DEA, Evry (France), (1985).
[10] Hill, R., J. Mech. Phys. Solids, 15, 79, (1967).
[11] Mandel, J., Proc. 11th Int. Cong. on Applied Mechanics, Munich (FRG)(1964).
[12] Leblond, J.B., Mottet, G., Devaux, J.C., J. Mech. Phys. Solids, 34, 395, (Part I), (1986).
[13] Leblond, J.B., Mottet, G., Devaux, J.C., J. Mech. Phys. Solids, 34, 411, (Part II), (1986).
[14] Leblond, J.B., Devaux, J., Devaux, J.C., submitted to J. Mech. Phys. Solids (Part III).
[15] Desalos, Y., IRSID report n° 95349401 MET 44 (1981).
[16] Abrassart, F., Thèse d'Etat, Nancy (France), (1972).
[17] Leblond, J.B., submitted to J. Mech. Phys. Solids (Part IV).

Thermomechanical Couplings in Solids
H.D. Bui and Q.S. Nguyen (Editors)
Elsevier Science Publishers B.V. (North-Holland)

SIMULATION OF QUENCHING PROCESS OF CARBURIZED STEEL GEAR UNDER METALLO-THERMO-MECHANICAL COUPLING

Tatsuo INOUE, Zhi-Gang WANG and Kohsuke MIYAO

Department of Mechanical Engineering, Kyoto University,
Sakyo-ku 606, Kyoto, Japan

A method to simulate carburized quenching process is presented in this paper when taking into account the effect of metallo-thermo-mechanical coupling. Here, emphasis is placed on the role of transformation plasticity. The processes of carburization and quenching for a gear of Ni-Cr-Mo steel (SNCM420HK) are simulated by use of finite element scheme, and the calculated results of carbon content, and temperature, metallic structure and stress are compared with experimental results. The distribution of the residual stress revealed that the transformation plasticity is to be introduced into proper method of such a simulation.

1. INTRODUCTION

Complicated coupling behaviour often plays an important role during the course of quenching of carburized steels as illustrated in Fig.1 [1]: When the temperature distribution (and heating or cooling rate in some cases) in a material varies, thermal stress (1) is caused in the body, and the phase transformation (2) depending on temperature affects the structure distributions such as austenite, pearlite and martensite. Phase changes brings out the transformation stress as well as the strain caused by transformation plasticity (3) and interrupt the stress and strain field of the body.

On the contrary to these phenomena, arrows with opposite direction indicate the coupling in the following manners. A part of mechanical work done by existing stress in the material converts into heat (4) to disturb the

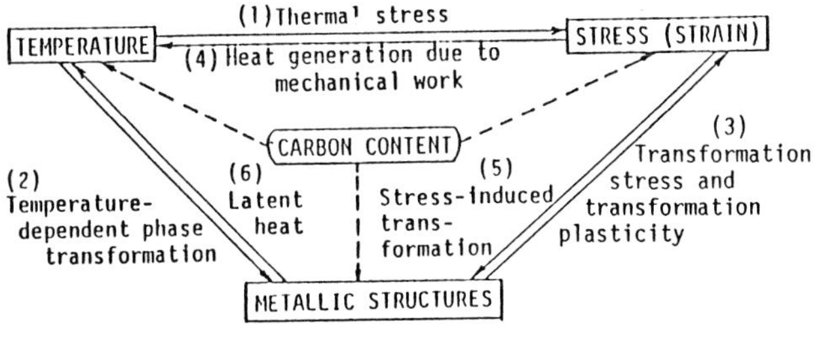

Fig. 1 Triangular diagram indicating metallo-thermo-mechanical coupling and the effect of carbon content.

temperature field. Accerelation of phase transformations by stress (or strain), called stress (or strain) induced transformation (5), has been discussed by metallurgists, and the effect is sometimes applied to improve the mechanical properties of metallic materials. The arrow with number (6) corresponds to the latent heat generation due to phase transformation which affects the temperature distribution as is discussed by the authors [2].

In addition to such effect of *metallo-thermo-mechanical coupling*, content of carbon in the carburized steel is considered to cause some influences on the fields as illustrated by broken lines in Fig.1.

As the first step of the analysis, in this paper, the diffused carbon content is determined by solving diffusion equation by finite element technique. In the second part of the paper, experimental results for Ni-Cr-Mo steel (SNCM20HK) are presented on the transformation plasticity during austenite-martensite and austenite-pearlite transformations to determine the evolution equation of transformation plastic strain rate, and the calculated results of stresses and metallic structures as well as temperature in a quenched steel gear are compared with measured data.

2. DISTRIBUTION OF CARBON CONTENT

A involute gear of 1.8%Ni-0.6%Cr-0.25%Mo steel with 0.2%C (SNCM420HK) is treated throughout the investigation. The dimensions are as follows: 29 teeth with module 9, 90 mm in width, 291 mm in pitch circle diameter and 200 mm in hole diameter.

The gear was held at a temperature of 930°C, carburized by butane gas for

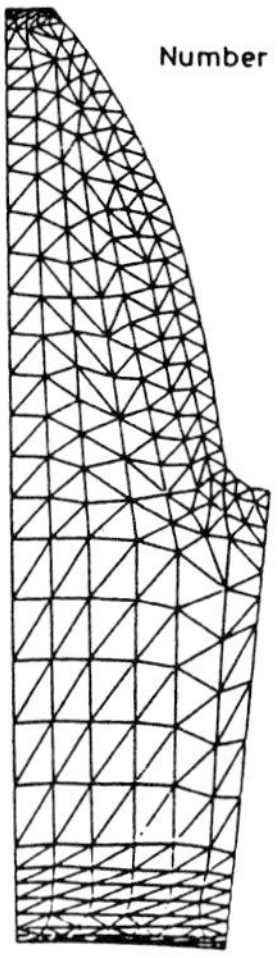

Fig. 2 Finite element division of a gear wheel.

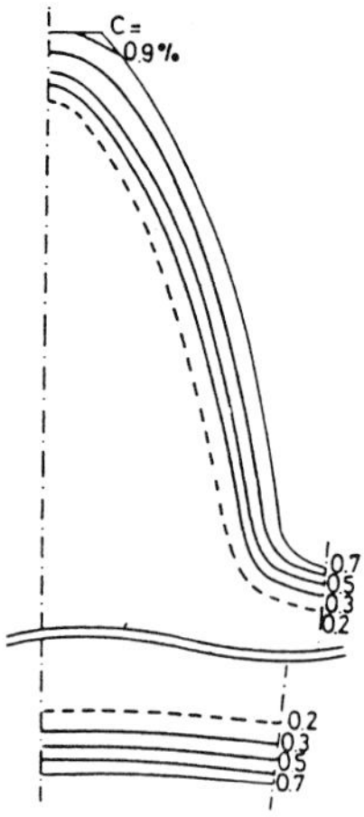

Fig. 3 Distribution of carbon content.

8 h and diffusion processes for 4 h. An ordinary diffusion equation for carbon content C

$$\dot{C} = \mathrm{div}\,(D\ \mathrm{grad}\ C) \tag{1}$$

was solved using finite element method based on Galerkin method [1] for diffusivity $D=\exp\{(C-12.52)\times 0.9215\}\ \mathrm{mm}^2/\mathrm{s}$. The finite element division of a part of the gear is illustrated in Fig.2, and the carbon content of the surface elements was assumed to be 1.17% during the carburization period, equivalent to the saturated value for the steel, whereas the gradient of C against surface normal was taken to be zero over the diffusion process. Figure 3 shows the calculated results of the distribution of carbon content after diffusion for 4 h.

3. GOVERNING EQUATIONS USED IN THE SIMULATION OF QUENCHING

3.1. Thermo-elastic-plastic Constitutive Equation considering Transformation Plasticity

Material under quenching is assumed to be composed of austenite, pearlite* and martensite with the volume fraction of ξ_A, ξ_P, and ξ_M respectively, (or, ξ_I (I=1, 2, 3)), and the physical and mechanical properties x is expressed in the form [2, 3]

$$x = \sum_{I=1}^{3} \xi_I x_I \quad \text{with} \quad \sum_{I=1}^{3} \xi_I = 1 \ . \tag{2}$$

Then the total strain rate is given

$$\dot{\boldsymbol{\varepsilon}} = \dot{\boldsymbol{\varepsilon}}^e + \dot{\boldsymbol{\varepsilon}}^p + \dot{\boldsymbol{\varepsilon}}^t \tag{3}$$

with elastic strain rate $\dot{\boldsymbol{\varepsilon}}^e$ involving thermal dilatation strain

$$\dot{\boldsymbol{\varepsilon}}^e = \frac{1+\nu}{E}\dot{\boldsymbol{\sigma}} - \frac{\nu}{E}(\mathrm{tr}\dot{\boldsymbol{\sigma}})\mathbf{1} + \{\frac{\partial}{\partial T}(\frac{1+\nu}{E})\boldsymbol{\sigma} - \frac{\partial}{\partial T}(\frac{\nu}{E})(\mathrm{tr}\boldsymbol{\sigma})\mathbf{1} + \alpha\mathbf{1}\}\dot{T}$$
$$+ \sum_{I=1}^{3}[\{\frac{\partial}{\partial \xi_I}(\frac{1+\nu}{E})\boldsymbol{\sigma} - \frac{\partial}{\partial \xi_I}(\frac{\nu}{E})(\mathrm{tr}\boldsymbol{\sigma})\mathbf{1} + \int_{T_0}^{T}\frac{\partial \alpha}{\partial \xi_I}dt\ \mathbf{1}\}\dot{\xi}_I] \ , \tag{4}$$

and plastic one

$$\dot{\boldsymbol{\varepsilon}}^p = \frac{9}{4H'\sigma_0^2}\{\mathrm{tr}(\boldsymbol{s}^*\dot{\boldsymbol{\sigma}}) - \frac{2}{3}\sigma_0\frac{\partial \sigma_0}{\partial T}\dot{T} - \frac{2}{3}\sum_{I=1}^{3}\frac{\partial \sigma_0}{\partial \xi_I}\dot{\xi}_I\}\boldsymbol{s}^*$$
$$\boldsymbol{s}^* = \{\boldsymbol{\sigma} - \frac{1}{3}(\mathrm{tr}\boldsymbol{\sigma})\mathbf{1}\} - \{\boldsymbol{\alpha} - \frac{1}{3}(\mathrm{tr}\boldsymbol{\alpha})\mathbf{1}\} \ . \tag{5}$$

Here, E, ν and α are Young's modulus, Poisson's ratio and thermal expansion coefficient, and σ_0 and H' stand for initial yield stress and hardening coefficient, respectively.

Strain rate incorporating with phase transformation $\dot{\boldsymbol{\varepsilon}}^t$ is assumed to be obtainable as the summation of isotropic strain rate $\dot{\boldsymbol{\varepsilon}}^m$ due to change in specific volume and anisotropic one $\dot{\boldsymbol{\varepsilon}}^{tp}$ owing to transformation plasticity as

---

* Other constituents like bainite and ferrite induced by the diffusion mechanism is included in *pearlite*.

$$\dot{\varepsilon}^{t} = \dot{\varepsilon}^{m} + \dot{\varepsilon}^{tp} \quad . \tag{6}$$

The phase change strain is reduced to

$$\dot{\varepsilon}^{m} = \sum_{I=2}^{3} \beta_I \dot{\xi}_I \mathbf{1} \quad , \tag{7}$$

with the dilatation coefficient β_I transformed from reference phase to the I-th phase.

Evolution of transformation plastic strain is generally expressed in the form [4]

$$\dot{\varepsilon}^{tp} = \sum_{I=2}^{3} \{\frac{2}{3} K_I h(\xi_I)\dot{\xi}_I \mathbf{s}^*\}, \quad h(\xi_I) = 2(1 - \xi_I) \ . \tag{8}$$

The parameters K_I appearing in the equation are determined by the experiments: Figures 4(a) and (b) show the measured dilatation-temperature diagrams during pearlitic and martensitic transformation under applied stresses, which give the approximately linear dependence of transformation plastic strain on applied stress and the results are $K_P=4.18\times10^{-5}$1/MPa and $K_M=5.08\times10^{-5}$1/MPa.

3.2. Heat Conduction Equation

A simplified equation

$$\rho c\dot{T} - \mathrm{tr}(\boldsymbol{\sigma}\, \dot{\varepsilon}^{p}) + \rho \sum_{I=1}^{3} \ell_I \dot{\xi}_I = \mathrm{div}(k \ \mathrm{grad}\ T) \tag{9}$$

is used to calculate the temperature by introducing heat generation by plastic work and latent heat ℓ_I. An expression in the form of Eq.(2) is used to

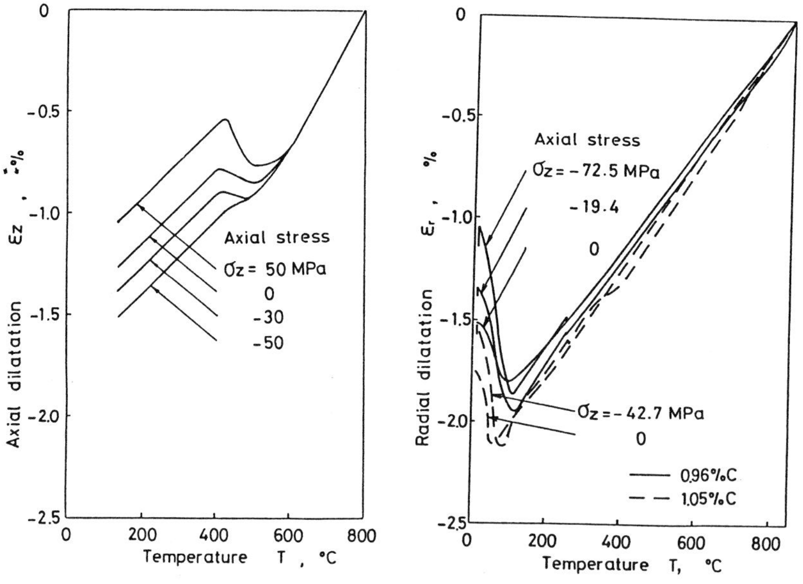

(a) Pearlitic transformation (b) Martensitic transformation

Fig. 4 Dilatation-temperature diagrams under applied stresses.

evaluate the density ρ, specific heat c, and heat conductivity k, depending on structure change as well as carbon content.

3.3. Equations of Transformation Kinetics

The diffusion type transformation of pearlite from austenite is assumed to follow the equation [1]

$$\xi_P = 1-\exp\{-\int_0^t f_1(T)f_2(C)f_3(\sigma)(t-\tau)^3 \, d\tau\}. \tag{10}$$

And, the kinetics of martensite formation is

$$\xi_M = 1-\exp\{\psi_1 T + \psi_2(C-C_0) + \psi_{31}\mathrm{tr}\sigma + \psi_{32}(\tfrac{1}{2}\mathrm{tr}\mathbf{s}^2)^{1/2} + \psi_4\}. \tag{11}$$

4. RESULTS OF SIMULATION OF QUENCHING

Figure 5 represents the calculated example of volume fraction of martensite in the 1-58 part of the gear, which shows the characteristic distribution due to the retained austenite in the region with high carbon content. The residual stresses along the tooth center line are plotted in Fig.6. Here, solid and broken lines indicate the results of calculation with and without transformation plasticity, respectively, whereas the open circles are the measured data by X-ray diffraction technique. It is concluded from the figures as well as other data of temperature and hardness that the proposed theory and the method of calculation considering the metallo-thermo-mechanical coupling givs the proper way of simulation of quenching process of carburized steel.

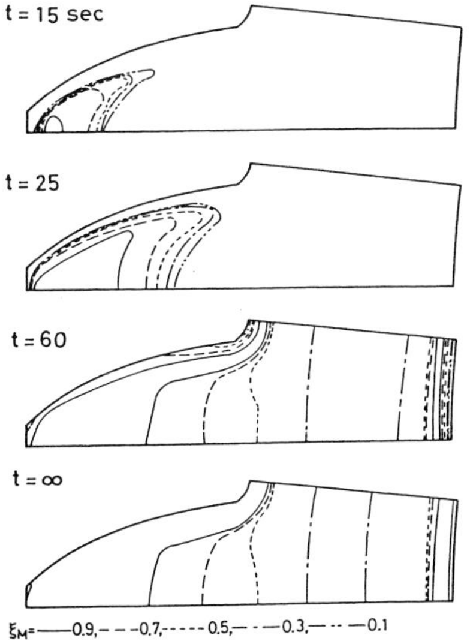

Fig. 5 Variation of volume fraction of martensite in the course of quenching.

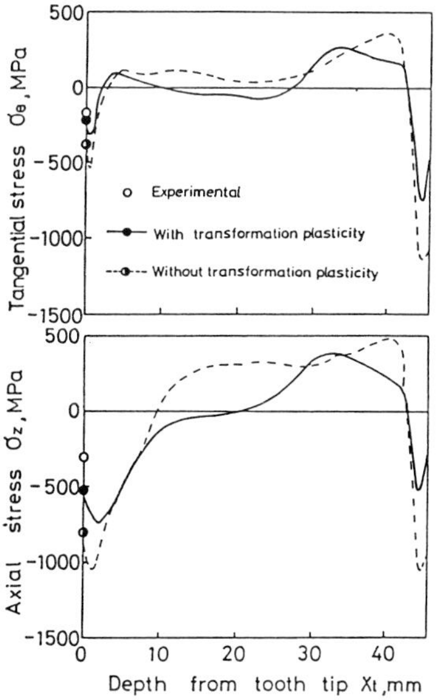

Fig. 6 Residual stress distributions.

5. CONCLUDING REMARKS

Metallo-thermo-mechanical coupling is carefully discussed in this paper to formulate fundamental equations governing the fields of metallic structure, temperature and stress relevant to simulate carburized quenching process of a steel gear wheel. Effects of transformation plasticity is also taken into account to inelastic constitutive equation. Comparison of the calculated results with experimental data indicates that the effect may affect some influences on the residual stress distribution after quenching.

ACKNOWLEDGMENTS

The authors should like to thank Manufacturing Engineering Research Center, Komatsu Ltd., for providing the testing material and cooperation with the experiments.

REFERENCES

[1] Inoue, T., Yamaguchi, T. and Wang, Z.G., Material Science and Technology, **1** (1985) 872.
[2] Inoue, T. and Wang, Z.G., Material Science and Technology, **1** (1985) 845.
[3] Bowen, R.W., Theory of Mixture (in Continuum Physics, Vol.3, ed. by A.C. Eringen) (1976) Academic Press.
[4] Desalos, Y., Giusti, J. and Gunsberg, F., Re-902 IRSID (1982).

Thermomechanical Couplings in Solids
H.D. Bui and Q.S. Nguyen (Editors)
Elsevier Science Publishers B.V. (North-Holland)

A UNIAXIAL MODEL OF WOOD FOR HYGRO-THERMO-MECHANICAL LOADINGS

J. GRIL*

Laboratoire de Mécanique des Solides, Ecole Polytechnique,
91128 Palaiseau-Cedex, France.

The objective of this paper is to formulate a uniaxial model of the mechanical behaviour of wood, in which the effects of both temperature and moisture changes are included. The model is based on a description of the ultrastructure and of its evolution, and thus physical interpretations can be given to the constants used. Numerical simulations show a good qualitative agreement with published data and help to clarify the apparent contradictions between them by introducing the concept of a limiting state of the deformation. Finally a new type of experimental procedure is proposed.

1. INTRODUCTION

Wood is a widely used material for both buildings and furnitures. It is highly sensitive to atmospheric conditions. Drying under a constant load induces an enhanced creep, although the opposite tendency would be expected from creep data at constant humidity. Furthermore, long term loading under changing temperature and humidity may lead to unexpectedly high deformations. Those phenomena suggest the occurrence of coupling effects between the mechanical stress and the sorption process. Some recent results [1], however, suggest that this "mechano-sorption" is of the same physical nature as mechanical creep. It can be viewed either as an acceleration of creep process or as a stress-induced hygrothermal expansion [2,3].

The following model is based on a qualitative interpretation of the mechano-sorptive effect by Grossman [4]. Simple analytical expressions will be proposed, and physical interpretations will be given to the constants used. Then typical experiments of creep under moisture change will be simulated using a numerical step-by-step analysis. In this manner, the gross features of the mechano-sorptive effect may be described.

2. THE UNIAXIAL MODEL

2.1. Creep and Sorption

Creep of wood is a thermally activated process. According to some authors

* Present adress: Wood Research Institute, Kyoto University, Uji 611 Japan

[5,6] its activation energy is about 20 to 25 kcal/mole. This value is intermediate between the binding energy of hydrogen bonds (5 to 8 kcal/mole) and the activation energy of segmental motion of amorphous high polymers in the glass transition region. Kauman [7] suggested that both creep and sorption involve breaking and remaking of the same hydrogen bonds in the amorphous zones of wood. The lower viscosity of wet wood is then easily explained because creep of wet wood involves breaking of polymolecular bonds with lower binding energies, while the creep of dry wood involves the breaking of stronger monomolecular bonds.

Wood is a water reactive material with high porosity at several levels. One should distinguish the mechanism of "macrodiffusion" between cell lumens from that of "microdiffusion" through the cell walls [3]. Characteristic half-time of diffusion through the cell wall should not exceed one minute at the most (extreme case of $T=300^{o}K$, $u=0\%$, thickness 0.01 mm). This is much lower than intervals allowed usually for moisture variation, so that the local microdiffusion time can be considered as instantaneous. Here "moisture content" will stand for the weight of bound water divided by the ovendry weight of wood.

2.2. The elementary mechanism

Our scale of representation is a small part of the cell wall. At this level (100nm) we have a juxtaposition of filaments with a cross section of approximately 4nm x 4nm called "microfibrils", made of crystalline cellulose. This cellulosic framework accounts for the memory effect during recovery [8]. It is embedded in an amorphous matrix containing hemicellulose, amorphous cellulose and lignin. It is not uniformly orientated, but most of the crystallites are orientated at a small angle to the cell wall axes. We shall see how far we can represent the macroscopic behaviour simply by modelling the behaviour of this small volume.

The very complex structure of the matrix is simplified according to a suggestion of Grossman [4]. Two long chains made of small elements of equal length are joined together at three points (Figure 1a). A-A' and C-C' are "strong" bonds and B-B' is a "weak" bond. The weak bond represents the hydrogen bonds involved in the process of both creep and sorption. It can be broken and reconstituted at the neighbouring position.

The cellulosic frame is represented here by an elastic bar parallel to the two chains (Figure 1c). This bar behaves like an elastic spring. It may count as the cellulose crystallites as well as the parts of the matrix not accessible to the water molecules. Thus it does not contribute to sorption or creep processes. As this is also the case for the strong bonds A-A' and B-B', they must be attached tightly to A" and B" respectively. So A-A'-A" or B-B'-B" represent the bonds ensuring the cohesion of the whole structure.

Figure 1b shows the natural state of the two components. If the bond between A-A' and A'' were broken and no load applied, the bar would have the extention d_0 and the chains the extension d_1. d_0 has a fixed value, but d_1 can change according to the location of the weak bond B-B'. To describe the configuration of the matrix, we define the following parameter:

(1) $X = d_1 - d_0.$

2.3. Length and Energy

The total energy of the single element described above is written:

$$(2)\quad G(d;F,X) = \underset{(i)}{\frac{1}{2}K_0(d-d_0)^2} + \underset{(ii)}{\frac{1}{2}K_1(d-d_1)^2} - \underset{(iii)}{F(d-d_0)} + \underset{(iv)}{\frac{1}{2}K_b(X-X_n)^2} + \underset{(v)}{W}$$

(v) is a dissipation term that is neglected here. (i) is the contribution of the elastic bar, and K_0 is the rigidity of the bar. (ii) is the elastic contribution of the chains. In contrast to the energy-induced elasticity of the crystalline regions (bar), the elasticity of the amorphous regions (chains) is entropy-induced, but it can be reduced to the assumed form (ii) for small extensions. Thus K_1 is an "effective" rigidity, that should be proportional to the absolute temperature and inversely proportional to the number of links in

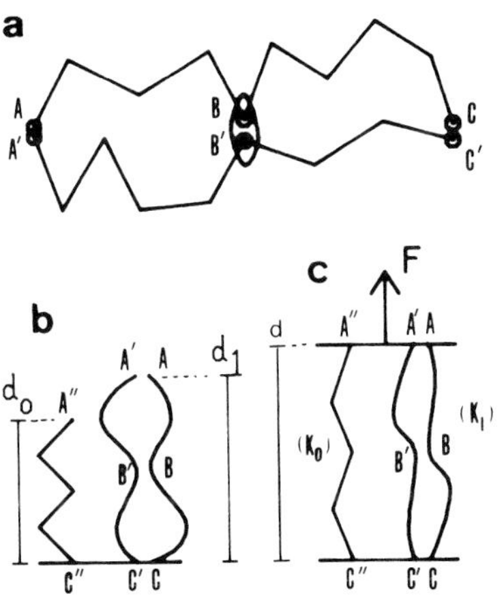

FIGURE 1
Modelisation of the microstructure.

(a) Matrix represented by two loose chains. (b) Natural state of the bar and the chains (indicated by 0 and 1 respectively). (c) Elementary mechanism with extension d under a load F.

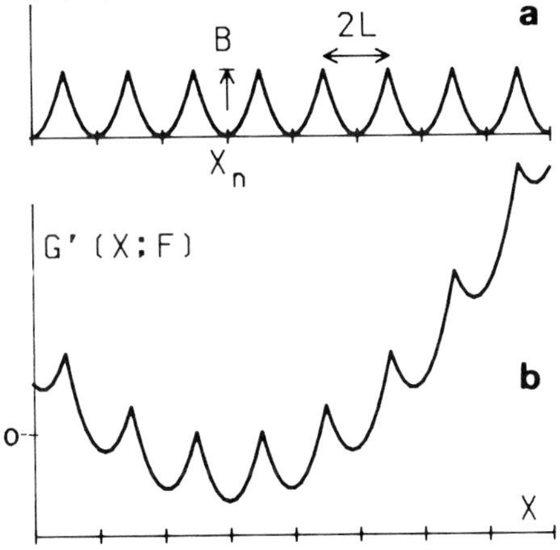

FIGURE 2
Potential energy as a function of the matrix configuration X.

(a) Energy of the weak bond (Note that here B stands for the breaking energy of the bond).
(b) Total energy (3) for a fixed load F.

the chains. (iii) is the potential energy of the load F applied on the mechanism. (iv) is an assumed expression for the energy of the weak bond B-B'. It is a periodical function of the matrix configuration X, with successive minima corresponding to the possible configuration (Figure 2a). Those minima are written X_n and we suppose that they are regularly spaced, 2L apart in length: $X_{n+1}=X_n+2L$, $n=1 \dots N-1$. Thus for a given X, the X_n appearing in (iv) is such that X is comprised between X_n-L and X_n+L. $B=K_bL^2/2$ is the binding energy of the weak bond. B is the energy required to break the bond, so that it may shift to the next position, and it should be a decreasing function of the moisture.

For a given applied load F and a given moisture content u we must look for the minimum of G(d;F,X), corresponding to the length and energy of constrained equilibria. We have:

$$(3) \qquad G'(F,X)= G(d_{min}(F,X),X;F) = \frac{1}{2}K_2\left(X-\frac{F}{K_0}\right)^2 - \frac{F^2}{2K_0} + \frac{1}{2}K_b(X-X_n)^2 + W$$

$$\text{where} \quad K_2 = \frac{K_0 \cdot K_1}{K_0+K_1} \quad \text{and} \quad d_{min}(F,X) = \frac{F+K_0d_0+K_1d_1}{K_0+K_1}$$

If the function G'(X;F) is drawn as a function of X for a fixed load we obtain a curve as shown on figure (2b). This curve has equilibria for many values Y_n of X, that may differ from the corresponding X_n:

$$(4) \qquad Y_n(F) = X_n + g\left(\frac{F}{K_0}-X_n\right) \qquad \text{with} \quad g = \frac{1}{1+K_b/K_2}$$

Here we suppose that all wells 1 ... N are always relative minima (this means that the load is limited in a certain range). In the later development the potential wells are characterised by the index n and we define barriers by:

$$(5) \qquad B_n^+(F) = G'(X_n+L,F)-G'(Y_n(F),F), \quad B_n^-(F) = G'(X_n-L,F)-G'(Y_n(F),F)$$

If the model lies in the potential well n, its length is given by:

$$(6) \qquad d_{min}(F,n) = d_0 + \frac{F + K_1Y_n(F)}{K_0+K_1}$$

2.4. Effect of moisture changes

Now as this weak bond is also involved in the sorption process it can be pictured as a small chain of water molecules placed between the two chains. The external stress induces a local stress, so that the next configuration would be more stable than the current one. But due to the rather high value of the breaking energy B the shift is very slow. If desorption occurs a water molecule leaves the site and the water chain is temporarily broken. This state is highly unstable because a hydroxyle group becomes unsaturated, and the structure tends to saturate it again by reforming a shorter water chain. This is a complex process involving the rearrangement of all the structure surrounding the link. It is represented in the present model by the reconfiguration of the two chains. Its typical duration is called "transition time" and written t_u.

So the matrix cohesion is lowered during t_u. This is interpreted by giving a lower value B_u to the binding energy B. Provided B_u is low enough and t_u long enough, the system will shift easily to the next configuration, so that creep is "activated" by desorption. This process occurs every time water molecules leave the site. Their total number is proportional to the decrease of moisture. Formally the effect of moisture change is described in the following way:

(7) $B=B_u$ during t_u after each moisture decrease of w_u

Here for simplification we shall admit that no activation of creep occurs during absorption. In Figure 3 the effect of a moisture cycle on the value of the breaking energy B is shown.

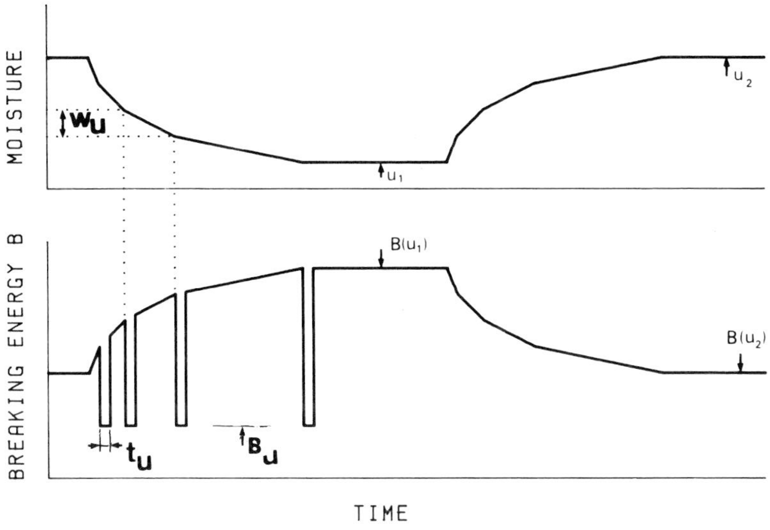

FIGURE 3
Effect of a moisture content cycle on the value of B

This example shows the two effects of moisture changes on creep: (i) direct effect of moisture content: B is supposed higher at low moisture, in order to predict a slower pure creep; (ii) activation of creep due to the state of non-equilibrium, resulting in a lower value of B during the desorption period, as described by (7).

2.5. Creep, yield and recovery as a result of thermal activation.

So far we intimated that the model lies in one of the potential wells. However, in reality the chains participate in the fluctuating thermal motions whose mean kinetic energy is proportional to the temperature T. Even though this mean kinetic energy may be smaller than the barrier closing the model in, its energy fluctuates and occasionally it may be big enough to overcome the barrier. The probability for such a transition per unit of time is given in statistical mechanics by a Boltzmann factor, so that we have:

$$(8) \qquad p_n^- = \frac{1}{t_b} \exp(-\frac{B_n^-}{kT}) , \qquad p_n^+ = \frac{1}{t_b} \exp(-\frac{B_n^+}{kT})$$

where $B_n^{\pm}$ is defined by (5) and t_b is a characteristic time that largely depends on the shape of the potential well and (weakly) on temperature.

To introduce this probabilistic feature into the model we have to speak about more than one element of the type described in section 3.1 and represent each pair of neighbouring crystalline and amorphous zones by such an element. Thus we have many such elements in parallel; each one shall carry the same fraction of the load while its length is subject to the thermal fluctuation. Even if initially all these elements had the same configuration, later the thermal fluctuation may have allowed some to leave their potential well and to hop to the next one. Thus in time there will be a distribution $Z_n(t)$, where Z_n is the proportion of particles which occupy the well n. Their length is given by d_{min} from (6) and (4), and we take their mean value as the length of the slab:

$$(9) \qquad D(t) = d_{min}(F,1).Z_1(t)+...+d_{min}(F,N).Z_N(t) \qquad Z_1(t)+...+Z_n(t) = 1 \;)$$

The distribution (Z_n) will be calculated by the assumption that it is determined by the equations:

$$(10) \qquad \frac{dZ_n}{dt} = p_{n-1}^+ Z_{n-1} - p_n^- Z_n - p_n^+ Z_n + p_{n+1}^- Z_{n+1} , \quad \text{with} \quad p_1^- = p_N^+ = 0$$

where the p's are the transition probabilities defined in (5). Thus Z_n grows because elements jump from the neighbouring wells into the well n and their numbers are proportional to $Z_{n\pm1}$. And Z_n decreases, because elements jump out of the well n into the neighbouring wells; their number is proportional to Z_n.

If the load and the ambient conditions remain constant, the system (10) tends towards a stationary solution Z_n^{eq} given by:

$$(11) \qquad p_n^+ Z_n^{eq} = p_{n+1}^- Z_{n+1}^{eq} \quad \text{for } n=1...N-1 \quad ; \quad Z_1^{eq}+...+Z_N^{eq} = 1$$

The system of differential equations must be solved for a given initial distribution which we may take as the equilibrium distribution corresponding to the zero load. At time $t_0=0$, a load is applied and this initial distribution is not the equilibrium any more. Therefore the system will tend towards the new equilibrium with a characteristic time t_{eq}.

Now we must discuss the values of the parameters, in order to simulate some typical cases of mechano-sorption.

3. NUMERICAL SIMULATIONS.

3.1 Geometry.

In order to fix the ideas we consider a slab of wood with a cross section 8nm x 8nm and a length of d_0 = 100nm. This scale suggests a portion of microfibril with its surrounding matrix. However, to fit with the basic picture of Figure 1, we imagine that the slab contains 25 elementary models, each one corresponding to two straight cellulose chains of length 100nm and two loose chains, three times as long.

Although the model gives the constitutive law of a small portion of the cell wall, it should represent, at least qualitatively, the macroscopic uniaxial behaviour in the direction of the fibers. Therefore instead of inducing the values of the spring constants K_0 and K_1 from hypothetical mechanical constants of wood constituents, we related them directly to expected instantaneous Young's modulus E_{el} and delayed Young's modulus E_{eq} in the fiber direction.

Instead of the length D and X we compute the corresponding (small) strains $(D-d_0)/d_0$ and $x=X/d_0$, and limit to the case of two configurations x_1 and x_2. Table I shows the set of "local" constants and table II the resulting "macroscopic" values, meant to apply to a softwood of density 0.5 at 300°K. The rigidities and the breaking energies vary linearly from the dry state (5%) to the wet state (25%). Their high values in table I are due to the shape of the potential wells and the shift of the equilibria in equation (4), and they result in table II in the desired "macroscopic" values.

TABLE I Constants for a single model

Geometry (nm)			
bar length	d_0	100	
chains length		300	
cross-section		1.6 x 1.6	
configuration	X_1, X_2	-6.6, +6.6	
Rigidities (N/m)		dry	wet
bar	K_0	0.33	0.27
chains	K_1	0.33	0.27
Energy of the Barriers (kcal/mole)		dry	wet
equilibium	B	125	100
transition	B_u	100	90
Time constants (s)			
rate process	t_b	10^{-8}	
transition time	t_u	60	
Water (nb. of molecules)			
Fiber saturation		400	
transition	(w_u)	13	
Load (N)			
linearity limit	F_{lin}	$1.0\ 10^{-10}$	

TABLE II "Macroscopic" Constants

Strain (%)			
	$(D-d_0)/d_0$		
	x_1, x_2	-6.6, +6.6	
Modulus (GPa)		dry	wet
elastic	E_{el}	13.9	11.1
delayed*	E_{eq}	4.0	3.8
Activation energies (kcal/mole)*		dry	wet
equilibium		25.0	20.0
transition		16.6	16.5
Time		dry	wet
pure creep*	t_{eq}	180 year	16 day
transition	t_u	1mn	
Moisture (%)			
range	u	5 to 25	
transition	w_u	1	
Stress (MPa)			
linarity		40	

* small stresses and temperature 300°K

3.2. Linearity.

This model predicts non-linearity when the load F approaches a typical value F_{lin}: Figure 4 shows the relationship between the stress and the "pure creep" response (i.e. at constant u and T) for two extreme values of the moisture u. The elastic response (t=0) is linear, while the delayed responses are not. In all the simulations the temperature was kept constant at 300°K. The model does not consider the moisture expansion.

The results are presented under two forms: (i) strain, stress and moisture versus time, and (ii) strain versus moisture. This second type was successfully used [1,10] to describe the basic features of the mechano-sorptive effect.

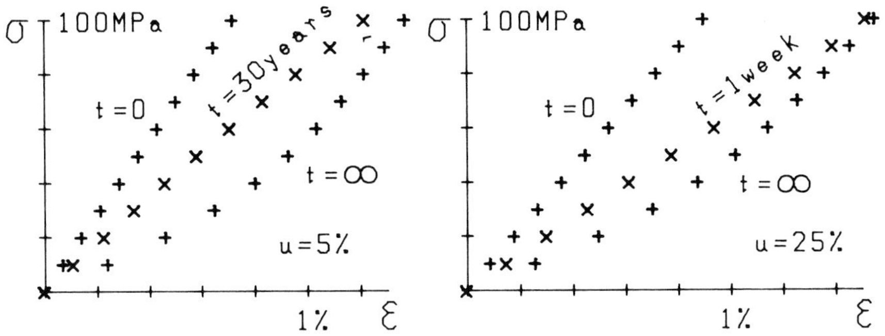

FIGURE 4
Non-linearity of the delayed response.

3.3. Drying under load in the dry range.

Figure 5 illustrates the simplest case of mechano-sorption. The simulated test is made of four parts. In (A) the specimen is dried slowly (2 months) from 12% to 6%: we observe an acceleration of creep although the elastic and delayed compliances decrease. For such a low range of moisture the model predicts almost no pure creep, so that the resulting increase of strain must be attributed to mechano-sorption only. In (B) the specimen is unloaded and kept at 6% for about one month, then u is cycled between 6% and 12% several times. This illustrates the so-called "drying set" of wood, that can be partially recovered through moisture cycling. In (C) this recovery process is completed by drastic moisture cyclings between 5% and 25%, so that the "specimen" is given its original state back and can be "used" again for a different experiment. In (D) it is dried under a load again from 12% to 6% but now in only one day. Like in (A) this results in an increase of creep, although more quickly, but with the same final strain. This simulation illustrates the early results of Schniewind [11].

On the strain-moisture graph only the parts corresponding to (A) and (D) are

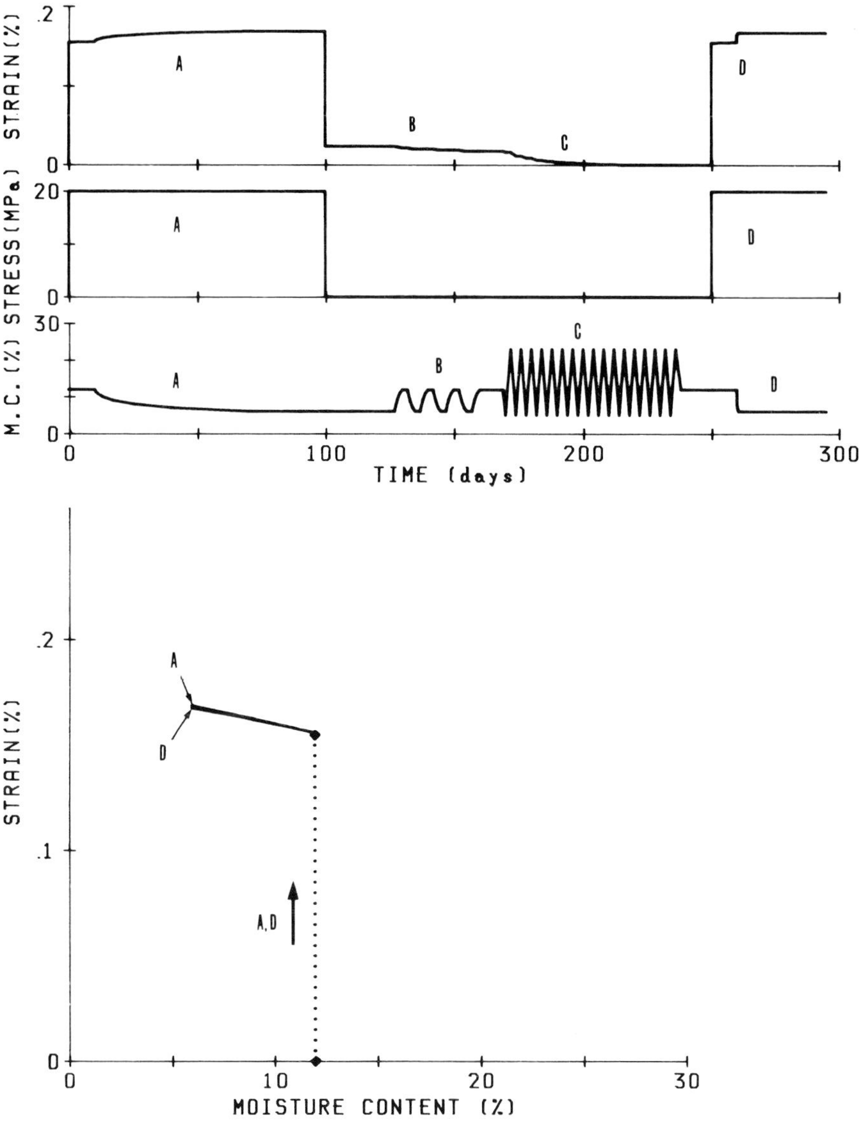

FIGURE 5. Simulation of drying under load in the dry range.
(Up) moisture content (m.c.), stress and strain versus time. (A) loading followed by a slow drying under load; (B) partial recovery through m.c. cycles after unloading; (C) complete recovery through energic m.c. cycles; (D) loading followed by a quick drying under load.
(Down) strain-moisture trajectories corresponding to (A) and (D); dotted lines for the loadings. Both rapid and slow drying under load result here in the same trajectories, so that for a small m.c. decrease, the additional mechano-sorptive component of strain is proportionnal to the m.c. decrement.

plotted. The two trajectories are almost superimposed. This confirms that these time-eliminating "trajectories" are good representations of the mechano-sorptive effect.

3.4. Cycling in the wet range and in the dry range.

Like in 3.3 Figure 6 shows two creep tests (A) and (D) separated by two recovery tests (B) and (C).

In (A) the specimen is loaded at 10 MPa and the moisture is cycled between 22% and 16%, while in (D) it is cycled between 12% and 6%. In both cases mechano-sorption is activated at each desorption ramp, while it is not during the absorption ramps, because the criterion (8) is never reached. In (D) this results in a much higher creep than what would have been expected at 12%, but in (A) pure creep at 22% would result in about the same strain as the creep activated by mechano-sorption.

This qualitative difference appears on the corresponding strain-moisture trajectories. Here the lower, middle and upper dotted lines show as function of u, respectively the elastic strain, the pure creep strain after the total duration of one test (50 days) and the delayed equilibrium strain. In (A) each desorption involves mechano-sorption, but the net result of one cycle is equivalent to a pure creep contribution at 22% during the same period of 5 days. This fact is illustrated by the vertical portions of trajectory corresponding to the two days at 22% at the end of each absorption ramps. Thus the case (A) fits with an observation by Hunt [1] that after a two-months test at varying moisture content the strain is not much more than the estimated pure creep strain at the maximum moisture of the test.

In (D), even after several cycles, the equilibrium lies far away. Therefore the desorption results each time in about the same additional strain. This accounts for the results by Ranta-Maunus [9], who considered that the slopes of the trajectories are constants of the material. However, according to our model, the slopes should decrease with the proximity of the equilibrium state as in reference [12]. This would have been obtained here by further cycling.

3.5 The Limiting State of Deformation.

The model was based on the hypothesis that creep and mechano-sorption are due to one single mechanism, driving the material to some limiting state depending on the actual stress only. However, with a usual creep test this process may take a very long time or many moisture cycles, and one can never be sure it is actually reached.

In Figure 7 we propose a different type of test, with a succession of stress levels, that would give quickly the limiting relationship between the stress and the delayed response, provided there is one. In the stress-strain repre-

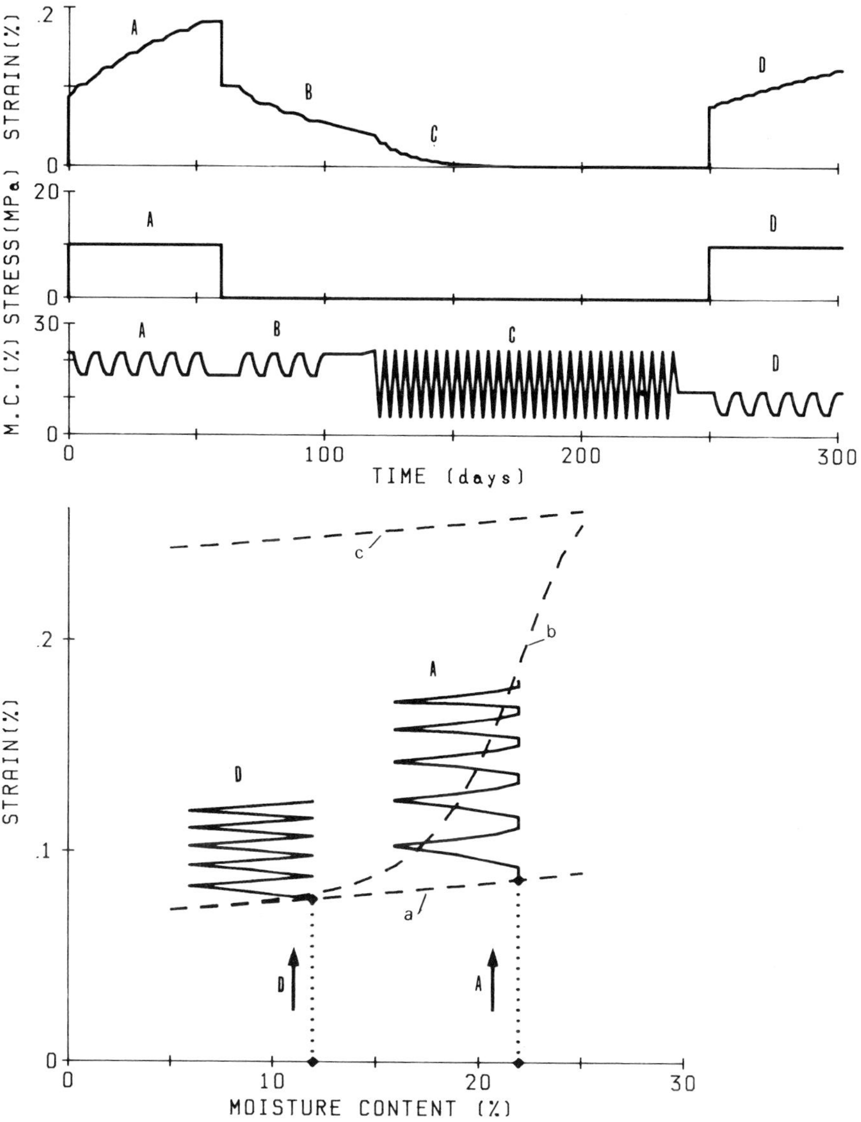

FIGURE 6. Simulation of a creep test under moisture cycles

(up) moisture content (m.c.), stress and strain versus time: (A) m.c. cycling in the wet range under load; (B) partial recovery through m.c. cycling after unloading; (C) complete recovery by more energic m.c. cycling; (D) m.c. cycling in the dry range under load.

(down) strain-moisture trajectories corresponding to (A) and (D). Dashed lines show the viscoelastic responses at constant m.c: in function of the moisture, for an applied load of 10 MPa: (a) elastic strain; (b) pure creep (i.e. at constant m.c.) after 50 days; (c) limiting creep (at infinitum time).

Pure creep effect is predominant in the wet case (A), while mechanosorptive activation is predominant in the dry case (D).

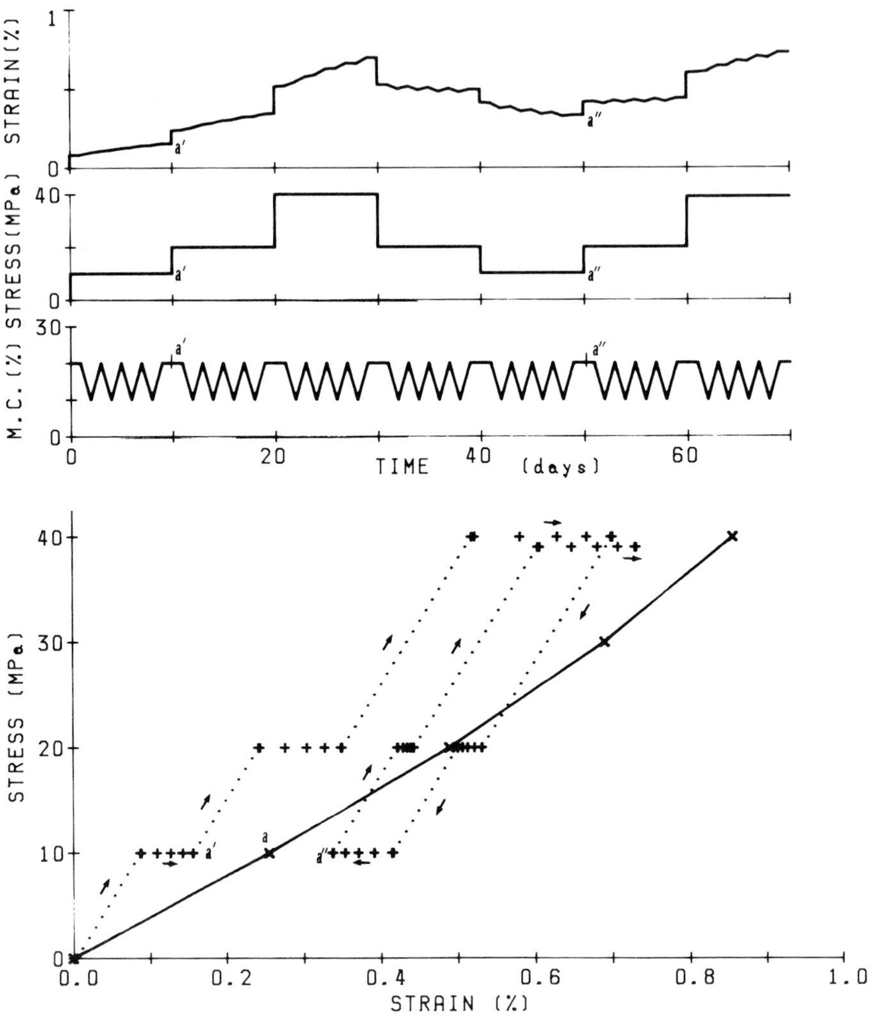

FIGURE 7. Simulation of a complex loading.

(up) Moisture content (m.c.), stress and strain versus time. A succession of stress levels are applied, and at each stage the m.c. is cycled between 20% and 10%, resulting each time in an enhanced creep or recovery.

(down) Corresponding stress-strain trajectory. The dotted lines show the elastic response at each loading or unloading. At each level of stress only strains attained at 20% m.c. are plotted.

This procedure aims to approximate the relationship between the stress and the limiting strain, indicated here by a continuous line. For example, the unknown limiting state (a), corresponding to a stress of 10 MPa, would be given, by the simulated experiment, both a lower bound (a') -highest strain reached in a positive creep stage- and an upper bond (a") -lowest strain reached in a recovery stage.

sentation, only the strains reached at u=20% are reported, and indicated by a (+). At each stress level the tendency to reach the corresponding equilibrium state of strain (limiting strain), represented by the continuous line, is enhanced by moisture cyclings. Due to the previous overloadings the limiting strain may be aproached by a recovery, so that if this complex loading were tried experimentally -and provided the model is true- they would give both lower and upper bounds of the limiting strain.

This stress-strain graph could be qualified as a "time and moisture eliminating representation". Even if this procedure does not give a rigorous proof of the limiting state hypothesis, at least it would test its consistency or its limits. This simulation shows that in some cases an apparently complex loading may reveal essential features that are hidden otherwise, and lead to a simple interpretation.

5. CONCLUSIONS

The above simulations have shown that the model proposed clarifies the apparent contradictions between different types of experimental results, by emphasizing the relationship between the time dependent creep and the moisture dependent mechano-sorption. The model is based on two separate set of parameters: the first one gives the shapes of the potential energy curves in the equilibrium conditions and should be identified using creep data at constant moistures and temperatures. The other set of data concerns the condition of non-equilibrium and can be identified through the slopes in the strain-moisture representation. However such hypothesis as the existence of the limiting state of strain cannot be checked without complex loadings of the type shown on Figure 7.

Indeed, the formulation of the condition of non-equilibrium proposed here could be generalized in the following way: "under some conditions, a moisture step w_u weakens the material during a transition time t_u". This weakening and the corresponding criterion can be imagined, as well as formally described, in many ways, but more or less one has to use this general type of description to ensure that the mechano-sorptive response is proportional to the moisture step (Figure 5).

ACKNOWLEDGEMENTS

This study has received a partial support from the "G.S. RHEOLOGIE DU BOIS du C.N.R.S."
The author wants also to thank Professor I. Müller, for his advices and his help in building this model, and Professor D. Hunt for useful dicussions and comments.

REFERENCES

[1] Hunt, D.G., J. of Materials Sci., 19 (1984) 1456.
[2] Schniewind, A.P., Wood Sci. and Technology 2 (1968) 188.
[3] Bažan, Z.P., Wood Sci. and Technology 19 (1985) 159.
[4] Grossman, P.U.A., Mechanosorptive Behaviour, in: General Constitutive relations for Wood and Wood-based Materials (USA, Syracuse, 1978), pp.313-25.
[5] Moriizumi, S., et al., Mokuzai Gakkaishi, 19 (1973) 109.
[6] Sawabe, O., Mokuzai Gakkaishi, 20 (1974) 517.
[7] Kauman, W.G., Holz als Roh- und Werkstoff, 24 (1966) 551.
[8] Boyd, J.D., in: Baas, P. (ed.), Nijhoff, M., and Junk, W., (publ.), New Perspectives in Wood Anatomy (La Hague, 1982) pp.171-222.
[9] Ranta-Maunus, A., Wood Sci. and Technology 9 (1975) 189.
[10] Hunt, D.G., J. of Materials Sci., 21 (1986) 2088.
[11] Schniewind, A.P., Holz als Roh- und Werkstoff, 24 (1966) 87.
[12] Hunt, D.G., J. Institute of Wood Sci., 9 (1982) 136.

Thermomechanical Couplings in Solids
H.D. Bui and Q.S. Nguyen (Editors)
Elsevier Science Publishers B.V. (North-Holland)

THERMO-MECHANICAL COUPLING IN LARGE DEFORMATIONS PARTICULARLY IN BIFURCATION PROBLEMS

TH. LEHMANN, Ruhr-University Bochum,
P.O. Box 102148, D-4630 Bochum

1. INTRODUCTION

We consider large non-isothermal inelastic deformations of solid bodies having in mind polycrystalline materials particularly. The phenomenological description of such processes shall be based on the following assumptions.

(A) The body is considered to be a classical continuum;

(B) the thermodynamical state of each material element is uniquely determined by a finite set of (external and internal) thermodynamical state variables defined in a so-called small state space; this may be valid even in such cases where the body as a whole is not in thermodynamical equilibrium.

These assumptions, of course, entail some restrictions concerning the kinds of processes which can be described in this way. Processes running very far from local thermodynamical equilibrium have to be treated in another way.

In the classical formulation of thermodynamics we only distinguish between reversible processes considered as a sequence of equilibrium states and irreversible (dissipative) processes characterized by non-equilibrium states. Within a small state space, however, we have to deal with four different kinds of processes:

1. Strictly reversible processes representing a sequence of equilibrium states governed by thermodynamical state equations like in classical thermodynamics.
2. Irreversible, dissipative processes characterized by non-equilibrium states and dynamical relaxation laws like in classical thermodynamics.
3. Dissipative processes appearing as a sequence of (constraint) equlibrium states. They occur if the relaxation times of the irreversible processes on the micro-level (large state space) are so small that on the macro-level (small state space) the changes of the state appear as a sequence of equilibrium states like, for instance, in plastic deformations [1 to 3].
4. Non-dissipative processes appearing as a sequence of (constraint) equilibrium states governed by state equations on the micro-level (large state space) but not on the macro-level (small state space). They occur when external work interacts with stochastic internal reversible processes on the micro-level leading to relevant changes of the internal state described in the small state space. This concerns, for instance, the energy stored in

micro-stress fields determined by the respective pattern of lattice defects.

The general mechanical and thermodynamical frame for the formulation of constitutive laws for large non-isothermal inelastic deformations following from the adopted basic assumptions is summarized in chapter 2 briefly. The particular scheme of the constitutive law depends on the kinds of thermo-mechanical processes which are involved in the considered problem. In chapter 3 some aspects connected with this subject are discussed. In the following the considerations are focussed mainly to stress induced (but temperature dependent) processes. This leads in chapter 4 to the definition of a rather general model of a constitutive law for elastic-plastic (and elastic-viscoplastic) materials which generalizes the classical concept of thermoplasticity. In chapter 5 some consequences are discussed following from such a generalized constitutive law with respect to certain bifurcation problems. In the concluding remarks some open questions are addressed.

2. GENERAL MECHANICAL AND THERMODYNAMICAL FRAME

Only the main aspects are summarized here. For more details see [4]. We introduce in the (rigid) space of observation an arbitrary space-fixed coordinate system x^α, with base vectors $\underset{\sim}{g}_\alpha(x^\rho)$, and metric $g_{\alpha\beta}(x^\rho)$. The initial position of a material point at time $\overset{\circ}{t}$ is

$$\overset{\circ}{x}{}^\alpha = x^\alpha(\overset{\circ}{t}) \ . \tag{1}$$

The motion of a particle is described by

$$x^\alpha = x^\alpha(\overset{\circ}{x}{}^\rho, t) \ . \tag{2}$$

The space-fixed coordinate system is needed for the definition of such quantities which are related to the space of observation originally like certain kinematic quantities and boundary conditions etc.

Additionally we introduce a body-fixed (co-moving) coordinate system ξ^i, with base vectors $\underset{\sim}{g}_i(\xi^r, t)$, and metric $g_{ik}(\xi^r, t)$. The body-fixed coordinate system simplifies the formulation of constitutive laws since all rigid body motions are eliminated from the very beginning. In the following all quantities are related to the base of the body-fixed coordinate system in the actual configuration.

The total deformation of the material elements can be measured by the quantity

$$q^i_k = \overset{\circ}{g}{}^{ir} g_{rk} \quad \text{or} \quad (q^{-1})^i_k = g^{ir} \overset{\circ}{g}_{rk} \tag{3}$$

where the superscribed $\circ$ denotes the respective quantity in the initial state [4 to 6]. From (3) we derive arbitrary strain tensors by means of isotropic tensor functions as, for instance, the Hencky strain tensor

$$\varepsilon^i_k = \frac{1}{2} (\ln q)^i_k \ . \tag{4}$$

The deformation rate is defined by

$$d^i_k = \frac{1}{2} (q^{-1})^i_r \, (\dot{q})^r_{\cdot k} \tag{5}$$

as can be realized easily, where $(\dot{\ })$ means the material time derivative (ξ^i kept constant).

The total deformation described by q^i_k can be decomposed into a reversible (elastic) part $\underset{(r)}{q}{}^i_k$ and into a remaining (inelastic) part $\underset{(i)}{g}{}^i_k$ by setting

$$q^i_k = \overset{\circ}{g}{}^{ir} g_{rk} = \overset{\circ}{g}{}^{im*} g_{mr} \overset{*}{g}{}^{rs} g_{sk} = \underset{(i)}{q}{}^i_r \, \underset{(r)}{q}{}^r_k \tag{6}$$

with
$$\underset{(r)}{g}{}^i_k = \overset{*}{g}{}^{is} g_{sk} \ . \tag{7}$$

The superscribed $*$ relates to a fictitious accompanying reference configuration defined by a fictitious, reversible process with frozen internal variables [4, 7, 8]. Analogously to (4) we define an elastic Hencky strain tensor

$$\underset{(r)}{\varepsilon}{}^i_k = \frac{1}{2} (\ln \underset{(r)}{q})^i_k \ . \tag{8}$$

From the decomposition (6) of the total deformation we obtain a corresponding additive decomposition of the deformation rate

$$d^i_k = \underset{(r)}{d}{}^i_k + \underset{(i)}{d}{}^i_k \tag{9}$$

with
$$\underset{(r)}{d}{}^i_k = \frac{1}{2} [(\underset{(r)}{q}{}^{-1})^i_r (\underset{(r)}{\dot{q}})^r_{\cdot k}]_s \tag{10}$$

where $[\ ]_s$ denotes the symmetrical part of the quantity in brackets. When the reversible deformations are not coupled with dissipative processes a corresponding decomposition of the specific work rate $\dot{w}$ results:

$$\dot{w} = \frac{1}{\overset{\circ}{\rho}} s^i_k d^k_i = \frac{1}{\overset{\circ}{\rho}} s^i_k \underset{(r)}{d}{}^k_i + \frac{1}{\overset{\circ}{\rho}} s^i_k \underset{(i)}{d}{}^k_i = \underset{(r)}{\dot{w}} + \underset{(i)}{\dot{w}} \ . \tag{11}$$

In these expressions s^i_k denotes the weighted Cauchy stress tensor which is related to the Cauchy stress tensor σ^i_k by

$$s^i_k = \frac{\overset{\circ}{\rho}}{\rho} \sigma^i_k \tag{12}$$

where ρ means the mass densitiy. When the reversible behaviour is isotropic as we assume in the following, s^i_k and $\underset{(r)}{\varepsilon}{}^i_k$ represent a conjugated pair of stress and strain. This means we can write with respect to the reversible (elastic) work rate

$$\underset{(r)}{\dot{w}} = \frac{1}{\overset{\circ}{\rho}} s_k^i \underset{(r)}{d_i^k} = \frac{1}{\overset{\circ}{\rho}} s_k^i \underset{(r)}{\overset{\nabla}{\varepsilon}{}_i^k} \tag{13}$$

where ∇ denotes the covariant time derivative which corresponds to the so-called Zaremba-Jaumann time derivative [9, 10].

The inelastic work rate $\underset{(i)}{\dot{w}}$ has to be decomposed once more into one part $\underset{(d)}{\dot{w}}$ which is dissipated immediately and into another part $\underset{(h)}{\dot{w}}$ which interacts with the energy stored in the micro-stress fields due to lattice defects. Therefore we have to put

$$\underset{(i)}{\dot{w}} = \frac{1}{\overset{\circ}{\rho}} s_k^i \underset{(i)}{d_i^k} = \underset{(d)}{\dot{w}} + \underset{(h)}{\dot{w}} \ . \tag{14}$$

The first law of thermodynamics states

$$\dot{u} = \dot{w} - \frac{1}{\rho} q^i\Big|_i + r = \underset{(r)}{\dot{w}} + \underset{(h)}{\dot{w}} + \underset{(d)}{\dot{w}} - \frac{1}{\rho} q^i\Big|_i + r \ . \tag{15}$$

In this equations denote: u specific internal energy, q^i heat flux, $|_i$ covariant derivative in the actual configuration, and r specific energy sources.

For simplicity we neglect energy fluxes apart from heat [4].

According to our basic assumption (B) u must be expressible as a unique function of a suitably chosen set of thermodynamical state variables.

We choose the specific entropy s, the reversible deformation $\underset{(r)}{\varepsilon_k^i}$ and a set a, α_k^i of internal variables as thermodynamical state variables, i.e. we assume

$$u = u\,(\underset{(r)}{\varepsilon_k^i}, s, a, \alpha_k^i) \ . \tag{16}$$

By a double Legendre-transformation (T: absolute temperature) we introduce the specific free enthalpy

$$\psi = u - \frac{1}{\overset{\circ}{\rho}} s_k^i \underset{(r)}{\varepsilon_i^k} - Ts = \psi\,(s_k^i, T, a, \alpha_k^i) \ . \tag{17}$$

From (17) we derive

$$\text{thermal state equation:} \quad \underset{(r)}{\varepsilon_k^i} = -\overset{\circ}{\rho} \frac{\partial\psi}{\partial s_i^k} = \underset{(r)}{\varepsilon_k^i}\,(s_k^i, T, a, \alpha_k^i) \ , \tag{18}$$

$$\text{caloric state equation:} \quad s = -\frac{\partial\psi}{\partial T} = s\,(s_k^i, T, a, \alpha_k^i) \ . \tag{19}$$

Comparing the two expressions for the changes of the specific free enthalpy following from (17)

$$\dot{\psi} = \dot{u} - \frac{1}{\overset{\circ}{\rho}} \overset{\nabla i}{s}_k \underset{(r)}{\varepsilon_i^k} - \frac{1}{\overset{\circ}{\rho}} s_k^i \underset{(r)}{\overset{\nabla k}{\varepsilon}_i} - \dot{T}s - T\dot{s} \tag{20a}$$

$$= \frac{\partial\psi}{\partial s_k^i} \overset{\nabla i}{s}_k + \frac{\partial\psi}{\partial T}\dot{T} + \frac{\partial\psi}{\partial a}\dot{a} + \frac{\partial\psi}{\partial \alpha_k^i} \overset{\nabla i}{\alpha}_k \tag{20b}$$

we obtain for the balance of remaining energy supply

$$\underset{(h)}{\dot{w}} + \underset{(d)}{\dot{w}} - \frac{1}{\rho} q^i\Big|_i + r = - T \left[\frac{\partial^2\psi}{\partial s_k^i \partial T} \overset{\nabla i}{s}_k + \frac{\partial^2\psi}{\partial T^2}\dot{T}\right]$$

$$+ \frac{\partial}{\partial a}\left[\psi - T\frac{\partial\psi}{\partial T}\right]\dot{a} + \frac{\partial}{\partial \alpha_k^i}\left[\psi - T\frac{\partial\psi}{\partial T}\right]\overset{\nabla i}{\alpha}_k \tag{21}$$

and for the balance of specific entropy (Gibbs equation)

$$T\dot{s} = \underset{(h)}{\dot{w}} + \underset{(d)}{\dot{w}} - \frac{1}{\rho} q^i\Big|_i + r - \frac{\partial\psi}{\partial a}\dot{a} - \frac{\partial\psi}{\partial \alpha_k^i}\overset{\nabla i}{\alpha}_k \, . \tag{22}$$

Decomposing the changes of the entropy into the reversible part $\underset{(r)}{\dot{s}}$ and the irreversible part $\underset{(d)}{\dot{s}}$ (entropy production), we obtain

$$T \underset{(r)}{\dot{s}} = \underset{(h)}{\dot{w}} - \frac{T}{\rho}\left(\frac{q^i}{T}\right)\Big|_i + r - \frac{\partial\psi}{\partial a}\dot{a} - \frac{\partial\psi}{\partial \alpha_k^i}\overset{\nabla i}{\alpha}_k - T\dot{\eta} \; , \tag{23a}$$

$$T \underset{(d)}{\dot{s}} = \underset{(d)}{\dot{w}} - \frac{1}{\rho T} q^i T\Big|_i + T\dot{\eta} \geqq 0 \; . \tag{23b}$$

$\dot{\eta}$ denotes the specific entropy production connected with dissipative internal processes and with energy supply by specific energy sources. We can also conclude from (21)

$$\underset{(h)}{\dot{w}} + r = \frac{\partial\psi}{\partial a}\dot{a} + \frac{\partial\psi}{\partial \alpha_k^i}\overset{\nabla i}{\alpha}_k + T\dot{\eta} \tag{24a}$$

$$- T\frac{\partial^2\psi}{\partial T^2}\dot{T} = c_p\dot{T} = \underset{(d)}{\dot{w}} - \frac{1}{\rho} q^i\Big|_i + T\frac{\partial^2\psi}{\partial s_k^i \partial T}\overset{\nabla i}{s}_k + T\frac{\partial^2\psi}{\partial a \partial T}\dot{a} + T\frac{\partial^2\psi}{\partial \alpha_k^i \partial T}\overset{\nabla i}{\alpha}_k + T\dot{\eta} \, . \tag{24b}$$

Within this thermodynamical frame given by the equations (17) to (24) the constitutive law for the material behaviour has to be formulated. It consists of

a) state function for the specific free enthalpy,
b) evolution law for inelastic deformations,
c) evolution laws for internal variables,
d) law of heat flux,
e) laws of entropy production.

It may be emphasized once more that we are disregarding energy fluxes apart from heat. Otherwise the ordinary evolution laws for internal variables have to be replaced by respective balance equations and, additionally, by laws for the other kinds of energy fluxes. Furthermore corresponding laws for entropy production have to be introduced.

3. A ROUGH CLASSIFICATION OF INELASTIC PROCESSES

Roughly we may distinguish between

(I) essentially thermally activated processes (but also depending on stress) like high-temperature (long-time) creep or relaxation, solid phase transformations, recrystallization, recovery etc., and

(II) mainly stress induced processes (but also depending on temperature) like plastic deformations which can be accompanied by damping (viscous phenomena) leading to short-time creep or relaxation phenomena like in visco-plastic behaviour.

In reality thermally activated and stress induced processes are always interacting. Therefore an exact separation of these two kinds of processes is impossible. However for practical reasons we may consider these processes separately. A certain survey which kind of processes is prevailing depending on the respective range of temperature and strain rate is given in [11, 12].

Thermally activated processes belong to the group of classical irreversible processes, whereas stress induced processes may belong to the 3.or 4. group of thermo-mechanical processes occuring in small state spaces as mentioned in the introduction. Idealized stress induced processes are characterized by the existence of a yield condition in the space of stress and temperature (depending on the internal state) which embraces the accompanying constraint equilibrium states, i.e. the elastic range.

In the following we restrict ourselves to such idealized stress induced processes, i.e. to elastic-plastic or elastic visco-plastic deformations in the narrow sense. However, even in such processes different yield mechanisms are interacting [13]. In a certain sense we may distinguish:

Local processes: They correspond to generation, dissolution, and redistribution of lattice defects within the single crystal grains but occuring at the same time in all concerned grains. These processes are influenced essentially by stress increments and mainly non-dissipative. This means the (small) mechanical work involved in these processes interacts with the energy stored in micro-stress fields connected with lattice defects.

Global processes: They are due to slip processes running through the whole body across the grain boundaries. They are mainly dissipative.

Local processes are prevailing in the beginning of inelastic deformations everytime when after (partial or total) unloading new inelastic deformations start. Particularly this can be observed in cyclic processes. Local processes also play an important rôle when the loading path changes more or less abruptly as it may occur in bifurcation problems at the bifurcation point.

Inelastic deformations due to local processes obey an evolution law different from that governing global processes. The following model of a constitutive law for an elastic-plastic or an elastic-viscoplastic material takes into account this difference.

4. A RATHER GENERAL MODEL OF A CONSTITUTIVE LAW FOR ELASTIC-PLASTIC AND ELASTIC-VISCOPLASTIC MATERIALS

4.1 Elastic behaviour

We assume that the elastic (reversible) behaviour is isotropic and independent of plastic deformations [14]. In this case the specific free enthalpy can be decomposed in the following manner [8, 13]

$$\psi\,(s^i_k,\, T,\, a,\, \alpha^i_k) = \psi^*(s^i_k,\, T) + \psi^{**}(T,\, a,\, \alpha^i_k)\,. \tag{25}$$

Furthermore we suppose that only the second invariant of α^i_k, i.e. $A = \alpha^i_k \alpha^k_i$ enters ψ^{**}, i.e.

$$\psi^{**} = \psi^{**}\,(T,\, a,\, A)\,. \tag{26}$$

The corresponding thermic state equation (18) reads

$$\underset{(r)}{\varepsilon^i_k} = -\,\overset{\circ}{\rho}\,\frac{\partial\psi^*}{\partial s^k_i} = \underset{(r)}{\varepsilon^i_k}\,(s^i_k,\, T)\,. \tag{27}$$

This leads to an incremental evolution law for the reversible part of deformations in the general form

$$\underset{(r)}{d^i_k} = \underset{(r)}{d^i_k}\,(s^i_k,\, T;\, \overset{\nabla}{s}{}^i_k,\, \dot{T})\,. \tag{28}$$

In most cases this can be approximated by a hypo-elastic tensor-linear law

$$\underset{(r)}{d^i_k} = \frac{1}{2G}\,\overset{\nabla}{t}{}^i_k + \left[\frac{1}{9K}\,\dot{s}^r_r + \alpha\dot{T}\right]\delta^i_k \tag{29}$$

with the following notations: t^i_k weighted stress deviator, G shear modulus, K bulk modulus, α coefficient of thermal expansion.

Concerning inelastic deformations for the present we restrict ourselves to elastic-plastic materials.

4.2 Inelastic local processes

We assume that local processes only occur if a yield condition of the form

$$F\,(s^i_k,\, T,\, a,\, \alpha^i_k) - (t^i_k - \alpha^i_k)(t^k_i - \alpha^k_i) - k^2\,(a,\, T) = 0 \tag{30a}$$

and the corresponding consistency condition

$$\dot{F} = \frac{\partial F}{\partial s^i_k}\,\overset{\nabla}{s}{}^i_k + \frac{\partial F}{\partial T}\,\dot{T} + \frac{\partial F}{\partial a}\,\dot{a} + \frac{\partial F}{\partial \alpha^i_k}\,\overset{\nabla}{\alpha}{}^i_k \tag{30b}$$

are fulfilled. This represents the mostly used simple approach for a yield condition taking into account isotropic and anisotropic hardening. It is not supposed that the yield condition embraces the point $s^i_k = 0$ in the space of stress and temperature since many experimental studies show that this is not the case in general (see, for instance, [15 to 20]). This causes some difficulties with respect to the second law of thermodynamics if we proceed in the usual manner. We come back to this point.

Concerning the evolution law for the inelastic deformations resulting from local processes we assume

$$\underset{(1)}{d}{}^i_k = \lambda \frac{\partial F}{\partial s^k_i} + \varkappa \overset{\nabla}{t}{}^i_k = 2\lambda\,(t^i_k - \alpha^i_k) + \varkappa \overset{\nabla}{t}{}^i_k \tag{31}$$

$$\text{with} \begin{cases} \varkappa = \varkappa(s^i_k, T, a, \alpha^i_k) \geqq 0, \quad \lambda \geqq 0 & \text{if } F = 0 \text{ and } \dot{F} = 0 \\ \varkappa = 0 \quad , \quad \lambda = 0 & \text{if } F < 0 \text{ or } \dot{F} < 0 \end{cases}$$

where λ has still to be determined. The second term renders the immediate influence of stress increments on local processes (see chapter 3) leading to deviations from the normality rule.

With respect to the evolution laws of internal variables we presume

$$a = \frac{\delta}{\overset{\circ}{\rho}}\,(t^i_k - \alpha^i_k)\,\underset{(1)}{d}{}^k_i - \beta\, a \sqrt{\underset{(1)}{d}{}^i_k\,\underset{(1)}{d}{}^k_i} \quad (32)\,; \qquad \overset{\nabla}{\alpha}{}^i_k = \zeta \underset{(1)}{d}{}^i_k - \gamma \alpha^i_k \sqrt{\underset{(1)}{d}{}^r_s\,\underset{(1)}{d}{}^s_r} \tag{33}$$

where the non-negative quantities δ, β, ζ, γ represent functions of the state variables subject to certain restrictions following from thermodynamical balance equations. The evolution laws (32) and (33) reflect certain interactions between hardening and softening processes. Similar approaches can be found in [21 to 24]. Particularly experimental results in cyclic processes suggest such approaches [24 to 26]. Both evolution laws can be interpreted on the physical background of dislocation theory. For simplification we may set

$$\sqrt{\underset{(1)}{d}{}^i_k\,\underset{(1)}{d}{}^k_i} \approx 2\lambda k + \varkappa \sqrt{\overset{\nabla}{t}{}^i_k \overset{\nabla}{t}{}^k_i}\;. \tag{34}$$

Then the evolution laws (32), (33) read

$$\dot{a} = \frac{\delta}{\overset{\circ}{\rho}}\,[2\lambda k^2 + \varkappa(t^i_k - \alpha^i_k)\overset{\nabla}{t}{}^k_i] - \beta a[2\lambda k + \varkappa \sqrt{\overset{\nabla}{t}{}^i_k \overset{\nabla}{t}{}^k_i}\;]\;, \tag{35}$$

$$\overset{\nabla}{\alpha}{}^i_k = \zeta[2\lambda(t^i_k - \alpha^i_k) + \varkappa \overset{\nabla}{t}{}^i_k] - \gamma \alpha^i_k\,[2\lambda k + \varkappa \sqrt{\overset{\nabla}{t}{}^r_s \overset{\nabla}{t}{}^s_r}\;]\;. \tag{36}$$

Putting the evolution laws (35, 36) into the consistency condition (30b) we obtain

$$\lambda = \frac{[1 - \varkappa\zeta - \frac{1}{2}\frac{\varkappa\delta}{\overset{\circ}{\rho}}\frac{\partial k^2}{\partial a}](t^i_k - \alpha^i_k)\overset{\nabla}{t}{}^k_i + \varkappa[\gamma(t^i_k - \alpha^i_k)\alpha^k_i + \frac{1}{2}\frac{\partial k^2}{\partial a}\beta a]\sqrt{\overset{\nabla}{t}{}^s_r \overset{\nabla}{t}{}^r_s} - \frac{1}{2}\frac{\partial k^2}{\partial T}\dot{T}}{k^2[2\zeta + \frac{\partial k^2}{\partial a}[\frac{\delta}{\overset{\circ}{\rho}} - \frac{\beta a}{k}] - \frac{2\gamma}{k}(t^m_n - \alpha^m_n)\,\alpha^n_m]} \tag{37}$$

With respect to the entropy production involved in local processes we may suppose

$$T\dot{\eta} = 0, \quad \underset{(1,d)}{\dot{w}} = \underset{(1)}{\xi}\,\frac{1}{\overset{\circ}{\rho}}\,t^i_k\,\underset{(1)}{d^k_i} \;\rightarrow\; \underset{(1,h)}{\dot{w}} = (1-\underset{(1)}{\xi})\,\frac{1}{\overset{\circ}{\rho}}\,t^i_k\,\underset{(1)}{d^k_i}\,. \tag{38}$$

In this case the balance equation (24a) reads

$$\underset{(1,h)}{\dot{w}} = (1-\underset{(1)}{\xi})\,\frac{1}{\overset{\circ}{\rho}}\,t^i_k\,\underset{(1)}{d^k_i} = \frac{\partial\psi^{**}}{\partial a}\,\dot{a} + \frac{\partial\psi^{**}}{\partial A}\,2\alpha^i_k\overset{\nabla}{\alpha}{}^k_i$$

$$= \frac{1}{\overset{\circ}{\rho}}\Big[\frac{\partial\psi^{**}}{\partial a}\,\delta(t^i_k-\alpha^i_k)+2\overset{\circ}{\rho}\zeta\frac{\partial\psi^{**}}{\partial A}\,\alpha^i_k\Big]\,\underset{(1)}{d^i_k} - \Big[\frac{\partial\psi^{**}}{\partial a}\beta a+\frac{\partial\psi^{**}}{\partial A}2\gamma A\Big]\Big[2\lambda k+\varkappa\sqrt{\overset{\nabla}{t}{}^r_s\overset{\nabla}{t}{}^s_r}\,\Big]. \tag{39}$$

This balance equation can be fulfilled, for instance, by setting

$$\delta\,\frac{\partial\psi^{**}}{\partial a} = 1 \rightarrow \frac{1}{\delta} = \frac{\partial\psi^{**}}{\partial a} \tag{40a}$$

$$2\overset{\circ}{\rho}\zeta\,\frac{\partial\psi^{**}}{\partial A} = 1 \rightarrow \frac{1}{2\overset{\circ}{\rho}\zeta} = \frac{\partial\psi^{**}}{\partial A} \tag{40b}$$

$$\underset{(1,d)}{\dot{w}} = \underset{(1)}{\xi}\,\frac{1}{\overset{\circ}{\rho}}\,t^i_k\,\underset{(1)}{d^k_i} = \Big[\frac{\beta a}{\delta} + \frac{\gamma A}{\overset{\circ}{\rho}\zeta}\Big]\,\Big[2\lambda k + \varkappa\sqrt{\overset{\nabla}{t}{}^r_s\overset{\nabla}{t}{}^s_r}\,\Big]\,. \tag{41}$$

From equations (24a) and (40a/b) it follows that δ and ζ have to satisfy the following integrability conditions among others

$$\frac{1}{2\overset{\circ}{\rho}}\,\frac{\partial}{\partial a}\Big(\frac{1}{\zeta}\Big) = \frac{\partial}{\partial A}\Big(\frac{1}{\delta}\Big), \tag{42a}$$

$$\frac{\partial}{\partial a}\Big(\frac{c_p}{T}\Big) = -\frac{\partial^2}{\partial T^2}\Big(\frac{1}{\delta}\Big), \tag{42b}$$

$$\frac{\partial}{\partial A}\Big(\frac{c_p}{T}\Big) = -\frac{1}{2\overset{\circ}{\rho}}\,\frac{\partial^2}{\partial T^2}\Big(\frac{1}{\zeta}\Big)\,. \tag{42c}$$

These conditions represent strong restrictions for the possible choice of δ and ζ in the evolution laws (35) and (36) of the internal variables. The quantities β, γ, and $\varkappa$, however, are only restricted by the requirement that λ according to equation (37) has always to be non-negative during inelastic deformations. This leads to very weak restrictions with respect to possible choices.

It has to be emphasized, that the entropy production in local processes described by equation (41) always is non-negative even if the external work rate $\underset{(1)}{\dot{w}}$ is negative. In this case the release of stored energy balances the demand for energy. Therefore the approach given above overcomes the difficulties which otherwise arise with respect to the second law of thermodynamics when the elastic range does not include the point $s^i_k = 0$.

4.3 Inelastic global processes

Concerning the description of global processes different possibilities exist. We may assume that these processes represent an independent mechanism governed by a separate yield condition (with internal variables different from those entering local processes) with corresponding evolution laws for global inelastic deformations and the respective internal variables. This is done in [13].

Another possibility exists assuming that the evolution laws for inelastic deformations (33) and internal variables (35),(36) remain unchanged even if global processes are involved. This means we put always

$$\underset{(i)}{d^i_k} = \underset{(l)}{d^i_k} \tag{43}$$

according to equation (31). However, in this case we have to suppose that the entropy production increases when global processes start. This can be done assuming

$$\underset{(d)}{\dot{w}} = \underset{(l,d)}{\dot{w}} + \underset{(g,d)}{\dot{w}} = (\underset{(l)}{\xi} + \underset{(g)}{\xi})\underset{(i)}{\dot{w}} = \bar{\xi}\underset{(i)}{\dot{w}} \tag{44a}$$

$$\text{with} \quad \underset{(g)}{\xi} \begin{cases} = 0 & \text{if } s^i_k \dfrac{\partial F}{\partial s^i_k} \leq 0, \text{ or } F < 0, \text{ or } \dot{F} < 0 \\ \geq 0 & \text{if } s^i_k \dfrac{\partial F}{\partial s^i_k} > 0, \; F = 0, \text{ and } \dot{F} = 0 \; . \end{cases} \tag{44b}$$

The condition (44b) ensures that the second law of thermodynamics is not violated.

This approach leads to a certain modification of the integrability conditions (42a/c) which shall not be discussed here.

It seems to be a better way to assume that global processes represent pure dissipative processes (i.e. not interacting with internal variables) leading to additional deformations. In this case we set

$$\underset{(i)}{d^i_k} = \underset{(l)}{d^i_k} + \underset{(g)}{d^i_k} \tag{45}$$

and assume (with F according to (30a))

$$\underset{(g)}{d^i_k} = \mu \frac{\partial F}{\partial s^k_i} , \tag{46a}$$

$$\text{with} \quad \mu(s^i_k, T, a, \alpha^i_k) \begin{cases} = 0 \text{ if } s^i_k \dfrac{\partial F}{\partial s^i_k} \leq 0, \text{ or } F < 0, \text{ or } \dot{F} < 0 \\ \geq 0 \text{ if } s^i_k \dfrac{\partial F}{\partial s^i_k} > 0, \; F = 0, \text{ and } \dot{F} = 0 \; . \end{cases} \tag{46b}$$

The quantity μ can be specified arbitrarily. The integrability conditions (42a/c) remain unchanged. Therefore this approach opens a great freedom in the formulation of the constitutive law.

4.4 Law of heat flux

The constitutive law is completed by the law of heat flux (Fourier's law)

$$q^i = -\bar{\lambda}\, T|_i \tag{47}$$

where $\bar{\lambda}$ represents the coefficient of heat conductivity.

4.5 Determination of the material functions entering the constitutive law

The laws determining the elastic behaviour and the heat flux are assumed to be known. Within the above given frame the inelastic behaviour is governed by the material functions $k^2(a,T)$, $\delta(a,A,T)$, $\zeta(a,A,T)$, $\gamma(s^i_k,T,a,\alpha^i_k)$, $\varkappa(s^i_k,T,a,\alpha^i_k)$, and $\mu(s^i_k,T,a,\alpha^i_k)$. They can be determined from a series of monotonic and cyclic (deformation or loading) processes with proportional and non-proportional paths at different temperature levels. In the case of a somewhat simplified constitutive law this is demonstrated in [27, 28].

4.6 Extension of the constitutive law to elastic-viscoplastic materials

The considerations can be extended to elastic-viscoplastic materials very easily when we adopt the so-called overstress approach for viscous damping. In this case we set

$$s^i_k = (s^i_k - \bar{s}^i_k) + \bar{s}^i_k \tag{48}$$

where $\bar{s}^i_k$ characterizes the accompanying constraint equilibrium state and $s^i_k - \bar{s}^i_k$ represents the viscous overstress.

Now, the yield condition reads

$$F(s^i_k,T,a,\alpha^i_k) = (t^i_k-\alpha^i_k)(t^k_i-\alpha^k_i) - k^2(a,T) = f^2 - k^2 > 0 \,. \tag{49}$$

The accompanying constraint equilibrium state is characterized by

$$\bar{F}(\bar{s}^i_k,T,a,\alpha^i_k) = (\bar{t}^i_k-\alpha^i_k)(\bar{t}^k_i-\alpha^k_i) - k^2(a,T) = \bar{f}^2 - k^2 = 0 \,. \tag{50}$$

Concerning the local processes we assume

$$\underset{(1)}{d^i_k} = \Lambda \left(\frac{F}{k^2}\right) \frac{\partial F}{\partial s^k_i} + \varkappa \overset{\nabla}{t}{}^i_k \tag{51a}$$

$$= \Lambda \frac{\partial \bar{F}}{\partial \bar{s}^i_k} + \varkappa \overset{\nabla}{t}{}^i_k \quad \text{with} \quad \lambda = \Lambda \sqrt{\frac{f^2}{k^2}} \,. \tag{51b}$$

The evolution laws for the internal variables read now

$$\dot{a} = \frac{\delta}{\overset{\circ}{\rho}} (\bar{t}^i_k - \alpha^i_k) \underset{(1)}{d^k_i} - \beta a [2\lambda k + \varkappa \sqrt{\overset{\nabla}{t}{}^r_s \overset{\nabla}{t}{}^s_r}\,] \tag{52}$$

$$\overset{\nabla}{\alpha}{}^i_k = \zeta \underset{(1)}{d^i_k} - \gamma \alpha^i_k \, [2\lambda k + \varkappa \sqrt{\overset{\nabla}{t}{}^r_s \overset{\nabla}{t}{}^s_r}\,]. \tag{53}$$

The further considerations follow the line sketched in 4.2.

Concerning the global processes we may assume

$$\underset{(g)}{d^i_k} = M \frac{\partial F}{\partial s^k_i} \tag{54a}$$

$$\text{with } M(s^i_k, T, a, \alpha^i_k) \begin{cases} = 0 \text{ if } (t^i_k - \alpha^i_k)t^k_i \leq 0 \text{ or } F \leq 0 \\ \geq 0 \text{ if } (t^i_k - \alpha^i_k)t^k_i > 0 \text{ and } F > 0 \;. \end{cases} \tag{54b}$$

In this way we obtain an analogous frame for the description of elastic-viscoplastic behaviour.

5. CONSEQUENCES OF THE THERMO-MECHANICAL COUPLING PARTICULARLY WITH RESPECT TO BIFURCATION PROBLEMS

The local changes of temperature are governed by equation (24b). The couple terms on the right hand side can be neglected in most cases. The dissipated work on the one hand and the divergence of heat flux on the other hand are usually the main influences on the temperature changes. Which of these phenomena is prevailing depends on the kind of the considered processes. At very low deformation rate the processes become isothermal approximately, at high deformation rate the processes become locally adiabatic approximately. In the first case dissipation of work and divergence of heat flux are counterbalancing. In the second case the heat flux can be neglected. In between the interaction of dissipated work and heat conduction has to be calculated carefully. In any case the processes become rate dependent even if the evolution law for inelastic deformations is rate independent. This is due to the time dependence of heat flux. Bifurcation problems (in the narrow sense) are characterized by the fact that at a certain stage of (one-parametrical) loading or deformation the process loses its uniqueness. This may happen before or after inelastic deformations occur. In the first case the thermo-mechanical coupling influences only the post-bifurcation behaviour and, maybe, the stability of the different branches. In the second case the bifurcation behaviour itself may be affected. This may be discussed taking the uniaxial tensile test of a cylindrical bar with shear stress free ends as an example.

In an assumed locally adiabatic process strain and temperature field remain homogeneous until onset of necking which represents a perfect bifurcation problem. The increase of temperature due to dissipation of mechanical work shifts the bifurcation point to an earlier stage of the loading or deformation process. This influence, however, is not very important [29, 30].

When the ends of the specimen are cooled and we take into account the heat flux to the cooled ends the temperature field does not remain homogeneous any more as soon as inelastic deformations start. We obtain a higher temperature in the center in comparison to the temperature at the ends of the specimen. Therefore also the deformation becomes inhomogeneous with the beginning of plastic deformations. The specimen loses its cylindrical shape already before the proper necking process starts [31 to 33]. It is an open question whether or not a perfect bifurcation problem still exists in this case. In any case the evolution of stresses and temperature in the region of necking is strongly influenced by the coupling of thermal and mechanical processes. The necking problem becomes quite rate sensitive [31 to 33].

At the onset of necking the stress distribution does not change abruptly. Therefore deviations from the normality rule play a minor role in this problem. The situation is different, for instance, in the buckling problem, of thin-walled tubes under combined loading. Already in isothermal plastic buckling processes this effect diminuishes the buckling load drastically [34,35]. It can be expected that the coupling of mechanical and thermal processes strengthens this tendency. The same is true for the formation of shear bands in uniaxial tension of a sheet.

6. CONCLUDING REMARKS

The coupling of thermal and mechanical processes in large inelastic deformations changes the mathematical character of the system of equations governing such deformation processes in comparison with isothermal processes. Variational formulations of such problems within the frame of convex funtional analysis are unknown. It is questionable whether such general formulations can be found at all. Therefore general methods for the proof of uniqueness etc. are still missing. This does not exclude that for particular constitutive laws a variational formulation can be given on which numerical methods can be based [36, 37].

REFERENCES

[1] J.H. Lambermont, Int. J. Eng. Sci. 12 (1974), 937.
[2] A.R.S. Ponter, J. Bataille, and J. Kesting, Journ. Méc. 18 (1979), 511.
[3] J. Kestin, Metal Plasticity as a Problem of Thermodynamics, this volume
[4] Th. Lehmann, in: The Constitutive Law in Thermoplasticity (ed.: Th. Lehmann). CISM Courses and Lectures No. 281, Springer-Verl. Wien/New York (1984) pp. 375-463.
[5] K. Thermann, in: The Constitutive Law in Thermoplasticity (ed.: Th. Lehmann). CISM Courses and Lectures No. 281, Springer-Verl. Wien/New York (1984) pp. 323-351.
[6] Th. Lehmann, Int. J. Engng. Sci. 20 (1982), 281-288.
[7] Th. Lehmann, ZAMM 63 (1983), T. 3.
[8] Th. Lehmann, in: Plasticity Today (ed.: A. Sawczuk and G. Bianchi). Elsevier Appl. Sci. Publ. London/New York (1985) pp. 115-134.
[9] Th. Lehmann, in: Proc. XI. Int. Congr. Appl. Mech. München (1964) (ed.: H. Görtler). Springer-Verl. Berlin (1966) pp. 376-382.
[10] Th. Lehmann, in: Mechanics of inelastic media and structures (ed.: O. Mahrenholtz and A. Sawczuk). Pol. Sci. Publ. Warszawa (1982) pp. 161-178.
[11] A. Rosenfield and G. Hahn, Trans. Am. Soc. Metals 59 (1966), 962.
[12] J. Campbell, Arch. Mech. 27 (1975), 407.

[13] Th. Lehmann, Acta Mech. 57 (1985), 1.
[14] E.H. Lee, Int. J. Sol. Struct. 17 (1981), 859.
[15] A. Phillips and R.J. Sierakowski, Acta Mech. 1 (1965), 29.
[16] M.A. Eisenberg and A. Phillips, Acta Mech. 11 (1971), 247.
[17] A. Phillips and H. Moon, Acta Mech. 27 (1977), 91.
[18] A. Phillips and Lee Chong-Won, Int. J. Sol. Struct. 15 (1979), 715.
[19] K. Ikegami, in: Proc. Euromech Coll. 115, Villard-de-Lans 1979 (ed.: J.-P. Boehler) Ed. CNRS Paris (1982), pp. 201-242.
[20] M.J. Michno and W.N. Findley, Int. J. Non-lin. Mech. 11 (1976), 59.
[21] P.J. Armstrong and C.D. Frederick, C.E.G.B. Report RD/B/N 731 (1966).
[22] J.L. Chaboche, Bull. Acad. Polonaise Sci., ser. Sc. Techn. 25 (1977),733.
[23] D. Besdo, ZAMM 60 (1980), T 101.
[24] Th. Lehmann, ZAMM 66 (1986), T 163.
[25] Th. Lehmann, B. Raniecki, and W. Trampczynski, Arch. Mech. 37 (1985),643.
[26] J.L. Chaboche, Int. J. Plasticity 2 (1986), 149.
[27] U. Zdebel, Diss. Ruhr-Univ. Bochum (1984), Mitt. Inst. Mech. RUB Nr. 43 (1984).
[28] U. Zdebel and Th. Lehmann, Some theoretical and experimental investigations on a constitutive law in thermoplasticity, in print.
[29] O. Bruhns and J. Mielniczuk, Ing. Arch. 46 (1977), 65.
[30] O. Bruhns and J. Mielniczuk, Bull. Acad. Pol. Sci. (Ser. Sci. Tech.) 26 (1978), 29.
[31] U. Blix, Diss. Ruhr-Univ. Bochum, Mitt. Inst. Mech. RUB Nr. 40 (1983).
[32] Th. Lehmann and U. Blix, Int. J. Plasticity 1 (1985), 175.
[33] Th. Lehmann and U. Blix, J. Therm. Stresses 8 (1985), 153.
[34] G. Zander, Diss. Ruhr-Univ. Bochum, Mitt. Inst. Mech. RUB Nr. 27 (1981).
[35] Th. Lehmann, Ing. Arch. 52 (1982), 391.
[36] J.F. Besseling, in: Trends in Solid Mechanics 1979 (ed. J.F. Besseling and A.M.A. van der Heijden); Delft Univ. Press (1979), pp. 53-78.
[37] J.F. Besseling, in: The Constitutive Law in Thermoplasticity (ed. Th. Lehmann) CISM Courses and Lectures No. 281, Springer-Verl. Wien/New York (1984), pp. 541-601.

Thermomechanical Couplings in Solids
H.D. Bui and Q.S. Nguyen (Editors)
Elsevier Science Publishers B.V. (North-Holland)

ASYMPTOTIC TRANSIENT THERMOELASTIC BEHAVIOUR

Gilles A. Francfort

Laboratoire Central des Ponts et Chaussées, 58 Bld. Lefebvre, 75732, PARIS CEDEX 15.

1. INTRODUCTION

The coupling between deformation and temperature fields in a linearly elastic solid is usually modelled with the help of a second order symmetric tensor α (the thermal dilation tensor). The resulting stresses, referred to as thermal stresses, are given by

$$\sigma^{th} = -\theta\alpha,$$

where θ is the increment of temperature with respect to the reference temperature field T_0 . The importance of the thermal stresses in stress analysis has long been acknowledged, although the temperature field is classically assumed to behave independently of its strain (stress) counterpart. From the standpoint of the mathematical analysis the effect of temperature changes reduces to that of a body force with density f^{th} defined as

$$\rho f^{th} = -\mathrm{div}(\alpha\theta),$$

where ρ is the volumic density of the material.

In time dependent problems however, "Biot's work has finally brought about the realization that [.....] the interaction of the thermal state and deformation of an elastic solid is such that the two effects cannot be treated separately" ([1], p.268).

The fully coupled system of linear thermoelasticity reads as

$$\rho\frac{\partial^2 u}{\partial t^2} = \mathrm{div}(a(e(u)-\alpha\theta)) + f,$$

(1)

$$\beta\frac{\partial\theta}{\partial t} = \frac{1}{T_0}\mathrm{div}(k\nabla\theta) - a\alpha\frac{\partial}{\partial t}e(u).$$

In (1), e(u) is the linearized strain field associated with the displacement field u, f represents the body loadings, T_0 the reference temperature and a, k, β are respectively the elastic tensor, the thermal conductivity tensor and the specific heat coefficient.

The functions $a(x)$, $\alpha(x)$, $k(x)$, $\beta(x)$, $\rho(x)$ are assumed to be bounded (measurable) functions and to satisfy, for almost any x,

$$a_{ijkh}(x) = a_{ijhk}(x) = a_{khij}(x),\ \alpha_{ij}(x) = \alpha_{ji}(x),\ k_{ij}(x) = k_{ji}(x),$$

(2)

$$\rho(x),\ \beta(x) > \alpha > 0,$$

$$a_{ijkh}(x)\xi_{ij}\xi_{kh} \geqslant \alpha\xi_{ij}\xi_{ij},\ \text{for all symmetric } \xi \text{ in } \mathbb{R}^9,$$

$$k_{ij}(x)\xi_i\xi_j \geqslant \alpha\xi_i\xi_i,\ \text{for all } \xi \text{ in } \mathbb{R}^3.$$

In this conference we focus on the behaviour of the system (1) in two separate settings, that of highly heterogeneous material [2] and that of a very thin sample [3]. In the first setting our goal is the description of the macroscopic evolution of such a medium; in the second setting an adequate two-dimensional (plate) model is sought.

When the thermoelastic material presents many heterogeneities (such as many inclusions in a matrix) the solution fields of any kind of initial boundary value problem tend to vary spatially at a scale ε comparable to the typical size of the heterogeneities. A laterally clamped plate maintained at ground temperature can be viewed as a thin three-dimensional thermoelastic body with thickness 2ε. We propose to investigate the limit fields u^0 and θ^0 of the displacement and temperature fields u^ε and θ^ε associated with any kind of initial boundary value problem as the scaling paramater ε tends to zero. Thus both problems are addressed in the spirit of *asymptotic analysis.*

The first problem was investigated in [2], [4] and the results obtained there are briefly recalled in Section 2. Attention is restricted to periodic distributions of heterogeneities although this restriction could be alleviated. The limit fields u^0 and θ^0 are found to be the solution fields associated with the response of a homogeneous linearly thermoelastic material submitted to the same boundary conditions, the same initial conditions on the displacement and velocity fields, but a different initial condition on the temperature field. This change is further shown to be a byproduct of fast time oscillations in the temperature field θ^ε. Specifically θ^ε varies in time as a function of $\frac{t}{\varepsilon}$. These oscillations are generally *aperiodic* and θ^0 represents their local average in time.

The second problem was investigated in a joint study with D. Blanchard [3] and the results obtained there are also briefly recalled in Section 3. Attention is restricted to an isotropic medium but this restriction is not essential to the analysis. The limit fields u^0 and θ^0 are found to be the solution fields associated with a flexional problem for the component of u^0 normal to the plate together with a coupled membrane-thermal problem for the other components of u^0 and for θ^0. Once again a change in initial condition generally occurs in the temperature field θ^0. The presence of fast time oscillations in the temperature field θ^ε is not however established. A stronger type of convergence of u^ε and θ^ε to u^0 and θ^0 is obtained whenever the initial conditions satisfy a compatibility condition. The compatibility equation is seen to be satisfied when the initial conditions on the displacement, velocity and temperature fields are such that *no* change of initial temperature occurs in the limit. It has yet to be decided whether compatibility always forces the initial temperature to remain unchanged.

In Section 4, a thermodynamical interpretation of the phenomenon of fast time oscillations of the temperature field based on the concept of accompanying equilibrium state [5] is proposed and the validity of the usually accepted linearization of the equations of thermoelasticity is discussed.

2. THE MACROSCOPIC BEHAVIOUR OF A HETEROGENEOUS THERMOELASTIC MATERIAL

Because of the periodicity assumption the microstructure is characterized by the periodic assembly of identical ε-scaled versions of an elementary volume $Y = (0,1)^3$. The mechanical and thermal characteristics of the microstructure are defined on Y, then scaled and extended by periodicity to all of $\mathbb{R}^3$. They are assumed to satisfy (2) for almost any x of Y. Their ε-periodic extension is then found to satisfy (2) for almost any x of $\mathbb{R}^3$. For example the ε-periodic extension $a^\varepsilon(x)$ of $a(x)$ is defined by

$$a^\varepsilon_{ijkh}(x) = a_{ijkh}\left(\frac{x}{\varepsilon}-[\![\frac{x}{\varepsilon}]\!]\right), \text{ for almost any x of } \mathbb{R}^3,$$

where ⟦t⟧ denotes the integer part of t and $a_{ijkh}(y)$ satifies (2) for almost any y of Y.

A bounded domain Ω with smooth boundary is considered. Its boundary $\partial\Omega$ is assumed to be clamped and maintained at ground temperature, although the analysis applies to other types of boundary conditions. At time t = 0, the displacement, velocity and temperature fields are given by

$$(3)\qquad \begin{aligned} u^{\varepsilon}(t=0) &= u_0, \\ \frac{\partial u^{\varepsilon}}{\partial t}(t=0) &= v_0, \\ \theta^{\varepsilon}(t=0) &= \theta_0. \end{aligned}$$

Note that oscillating initial conditions could be considered at the expense of lenghtening the presentation. The solution fields u^{ε} and θ^{ε} satisfy

$$\rho^{\varepsilon}\frac{\partial^2 u^{\varepsilon}}{\partial t^2} = \operatorname{div}(a^{\varepsilon}(e(u^{\varepsilon})-\alpha^{\varepsilon}\theta^{\varepsilon})) + f,$$

$$\beta^{\varepsilon}\frac{\partial \theta^{\varepsilon}}{\partial t} = \frac{1}{T_0}\operatorname{div}(k^{\varepsilon}\theta^{\varepsilon}) - a^{\varepsilon}\alpha^{\varepsilon}\frac{\partial}{\partial t}e(u^{\varepsilon}),$$

together with the appropriate initial conditions, *i.e.* (3) and boundary conditions, *i.e.*,

$$u^{\varepsilon} = 0,\ \theta^{\varepsilon} = 0 \text{ on } \partial\Omega.$$

As ε tends to zero, the fields u^{ε} and θ^{ε} are shown (*cf.* [2], Theorem or [4], Proposition 2) to converge (in an appropriate topology) to the solution fields u^0 and θ^0 of a system of the form (1) with constant coefficients: the macroscopic behaviour of the heterogeneous material that occupies the domain Ω is thermoelastic. The mechanical and thermal characteristics of the macroscopically equivalent (homogenized) material are defined in terms of the various concentration tensors. The concentration tensors compute the microscopic value of any relevant tensor at any point of the elementary volume Y in terms of its space average on Y. They are thus the solutions of static elastic or conduction problems on Y with periodic boundary conditions.

Specifically, if for any vector field ζ and any second order tensor field e defined on Y,

$$e^y(\zeta) = \frac{1}{2}(\nabla_y \zeta + \nabla_y \zeta^t),$$

$$\sigma^y(e) = ae,$$

we define χ^{ij} and ψ to be the periodic solutions on Y, unique up to a constant, of

$$\operatorname{div}(\sigma^y(e^y(\chi^{ij}))) = \operatorname{div}(\sigma^y(e^y(P^{ij}))),$$

$$\operatorname{div}(\sigma^y(e^y(\psi))) = \operatorname{div}(\sigma^y(ae)),$$

where

$$P_k^{ij} = y_i \delta_{jk}.$$

We also define θ^i to be the periodic solution on Y, unique up to a constant, of

$$\operatorname{div}(k\nabla_y(\theta^i - y_i)) = 0.$$

The various macroscopic coefficients are

$$a^0_{ijkh} = \int_Y (a_{ijkh}(y) - a_{pqkh}(y) e^y_{pq}(\chi^{ij}))\, dy,$$

$$\alpha^0_{ij} = (a^0_{ijpq})^{-1} \int_Y (a_{pqkh}(y)\alpha_{kh}(y) - a_{mnkh}(y) a_{kh}(y) e^y_{mn}(\chi^{pq}))\, dy,$$

$$k^0_{ij} = \int_Y (k_{ij}(y) - k_{mj}(y) \frac{\partial \theta^i}{\partial y_m}(y))\, dy,$$

$$\gamma_{ij} = \int_Y a_{ijkh}(y)\alpha_{kh}(y)\, dy - a^0_{ijkh}\alpha^0_{kh},$$

$$\kappa = \int_Y a_{pqkh}(y)\alpha_{kh}(y) e^y_{pq}(\psi)\, dy,$$

$$\underline{\beta} = \int_Y \beta(y)\, dy, \quad \underline{\rho} = \int_Y \rho(y)\, dy,$$

and the macroscopic system reads as .

$$\underline{\rho}\frac{\partial^2 u^0}{\partial t^2} = \operatorname{div}(a^0(e(u^0) - \alpha^0\theta^0)) + f,$$

$$(\underline{\beta}+\kappa)\frac{\partial\theta^0}{\partial t} = \frac{1}{T_0}\mathrm{div}(k^0\nabla\theta^0) - a^0\alpha^0\frac{\partial}{\partial t}e(u^0),$$

with boundary conditions

$$u^0 = 0,\ \theta^0 = 0 \text{ on } \partial\Omega,$$

and initial conditions

$$u^0(t=0) = u_0,$$

$$\frac{\partial u^0}{\partial t}(t=0) = v_0,$$

(4)

$$\theta^0(t=0) = \frac{\beta\theta_0+\gamma_{ij}e_{ij}(u_0)}{\underline{\beta}+\kappa}.$$

According to (3), (4), a change of initial temperature has occured during the asymptotic process. The understanding of this unusual phenomenon requires a more intimate analysis of the behaviour of the field θ^ε, at least for small times. A blow up of θ^ε around the initial time shows the absence of an initial layer. A multiple scales expansion of u^ε and θ^ε is performed in *both* space and time (*cf.* [2], Section 2). The leading term of the expansion of u^ε is precisely the homogenized displacement field u^0. The leading term τ^0 in the expansion of θ^ε is *not* the homogenized temperature field θ^0. It is seen to oscillate in time. Specifically, if

$$\theta^\varepsilon(x,t) = \Sigma\ \varepsilon^i\tau^i(x,\frac{x}{\varepsilon},t,\frac{t}{\varepsilon}),$$

τ^0 does not depend upon $\frac{x}{\varepsilon}$ but it depends upon $\frac{t}{\varepsilon}$. Its $\frac{t}{\varepsilon}$-dependence is intricate (*cf.* [4], Proposition 3); in particular it need not be periodic in $\frac{t}{\varepsilon}$, which contrasts with the rapid spatial oscillations of the fields due to the microstructure.

The fast time average of τ^0 is further shown to coincide with the homogenized temperature field θ^0, *i.e.*,

$$\lim_{T\to+\infty}\frac{1}{T}\int_0^T\tau^0(x,t,\delta)\,d\delta = \theta^0(x,t).$$

Thus the initial condition $\theta^0(t=0)$ is the initial average of the fast oscillations of θ^ε and it merely expresses at the initial time t=0 a phenomenon shared by all subsequent times.

A one-dimensional example, that of a "sandwich bar" is investigated in [4], Section 4. The change of initial temperature and the fast time oscillations of the temperature field are corroborated through the numerical simulation of a dynamic problem for that bar.

3. THE TWO-DIMENSIONAL BEHAVIOUR OF A THIN THERMOELASTIC SAMPLE

A thin three-dimensional flat plate of thickness 2ε is defined as

$$\Omega(\varepsilon) = \omega \times (-\varepsilon,\varepsilon),$$

where ω is a (smooth) bounded domain of $\mathbb{R}^2$. All coefficients or fields entering the problem are defined on $\Omega(\varepsilon)$ and appropriately rescaled to the reference domain

$$\Omega = \Omega(1) = \omega \times (-1,1).$$

To each point $x = (y,x_3)$ of Ω we associate the point $x^\varepsilon = (y,\varepsilon x_3)$ of $\Omega(\varepsilon)$. To each vector field $w^\varepsilon(x^\varepsilon)$ we associate the field $W^\varepsilon(x)$ defined as

$$W^\varepsilon_\alpha(x) = w^\varepsilon_\alpha(x^\varepsilon),\ W^\varepsilon_3(x) = \varepsilon w^\varepsilon_3(x^\varepsilon).$$

To each scalar field $z^\varepsilon(x^\varepsilon)$ we associate the field $Z^\varepsilon(x)$ defined as

$$Z^\varepsilon(x) = z^\varepsilon(x^\varepsilon).$$

To each tensor field $\tau^\varepsilon(x^\varepsilon)$ we associate the field $T^\varepsilon(x)$ defined as

$$T^\varepsilon_{\alpha\beta}(x) = \tau^\varepsilon_{\alpha\beta}(x^\varepsilon),\ T^\varepsilon_{33}(x) = \frac{1}{\varepsilon^2}\tau^\varepsilon_{33}(x^\varepsilon),$$

$$T^\varepsilon_{\alpha 3}(x) = \frac{1}{\varepsilon}\tau^\varepsilon_{\alpha 3}(x^\varepsilon),\ T^\varepsilon_{3\alpha}(x) = \frac{1}{\varepsilon}\tau^\varepsilon_{3\alpha}(x^\varepsilon).$$

The plate $\Omega(\varepsilon)$ is made of an inhomogeneous thermoelastic isotropic material. The Young's modulus $E^\varepsilon(x^\varepsilon)$, Poisson's ratio $\nu^\varepsilon(x^\varepsilon)$, thermal dilation coefficient $\alpha^\varepsilon(x^\varepsilon)$, heat conductivity coefficient $k^\varepsilon(x^\varepsilon)$, specific heat coefficient $\beta^\varepsilon(x^\varepsilon)$ and mass density $\rho^\varepsilon(x^\varepsilon)$ are given by

$$E^\varepsilon(x^\varepsilon) = E(x) > 0,$$

$$-1 < \nu^\varepsilon(x^\varepsilon) = \nu(x) < \frac{1}{2},$$

$$\alpha^\varepsilon(x^\varepsilon) = \alpha(x),$$

$$k^\varepsilon(x^\varepsilon) = k(x) > 0,$$

$$\beta^\varepsilon(x^\varepsilon) = \beta(x) > 0,$$

$$\rho^\varepsilon(x^\varepsilon) = \varepsilon^2\rho(x) > 0,$$

where E, ν, α, k, β, ρ are smooth on the closure of Ω and even in x_3. Note that the ε^2- dependence of ρ^ε upon ε is necessary for the resulting model to incorporate inertia effects (*cf.* [6]). The plate $\Omega(\varepsilon)$ is laterally clamped and maintained at ground temperature. It is submitted to the following set of body loadings f^ε, upper (lower) surface loading $g^{\pm\varepsilon}$:

$$f^\varepsilon_\alpha(x^\varepsilon,t) = f^0_\alpha(x,t) \;,\; g^{\pm\varepsilon}_\alpha(x^\varepsilon,t) = \varepsilon g^{\pm 0}_\alpha(x,t),\; \alpha = 1,2,$$

$$f^\varepsilon_3(x^\varepsilon,t) = \varepsilon f^{\,0}_3(x,t) \;,\; g^{\pm\varepsilon}_3(x^\varepsilon,t) = \varepsilon^2 g^{\pm 0}_3(x,t).$$

At time t=0, the displacement, velocity and temperature fields are taken to be

$$u^\varepsilon_\alpha(x^\varepsilon,t=0) = \underset{\sim}{u}^0_{0\alpha}(y) - x_3\frac{\partial u^0_{03}}{\partial y_\alpha},\; \alpha=1,2,$$

$$\varepsilon u^\varepsilon_3(x^\varepsilon,t=0) = u^0_{03}(y),$$

$$\varepsilon\frac{\partial u^\varepsilon}{\partial t}(x^\varepsilon,t=0) = v^0_0(x),$$

$$\theta^\varepsilon(x^\varepsilon,t=0) = \theta^{\,0}_0(y).$$

Once again, more general loadings and initial conditions could be considered at the expense of lenghtening the presentation (*cf.* [3]).

If the indices α, β and γ run from 1 to 2, the rescaled versions U^ε and Θ^ε of the solution fields u^ε and θ^ε satisfy

$$e_{\alpha\beta}(U^\varepsilon) - \alpha\Theta^\varepsilon\delta_{\alpha\beta} = \frac{1+\nu}{E}\Sigma^\varepsilon_{\alpha\beta} - \frac{\nu}{E}(\Sigma^\varepsilon_{\gamma\gamma}+\varepsilon^2\Sigma^\varepsilon_{33})\delta_{\alpha\beta},$$

$$e_{\alpha 3}(U^\varepsilon) = \varepsilon^2\frac{1+\nu}{E}\Sigma^\varepsilon_{\alpha 3},$$

$$e_{33}(U^\varepsilon) - \varepsilon^2\alpha\Theta^\varepsilon = \varepsilon^4\frac{1+\nu}{E}\Sigma^\varepsilon_{33} - \varepsilon^2\frac{\nu}{E}(\Sigma^\varepsilon_{\gamma\gamma}+\varepsilon^2\Sigma^\varepsilon_{33}),$$

$$\varepsilon^2\rho\frac{\partial^2 U^\varepsilon_\alpha}{\partial t^2} = \frac{\partial}{\partial x_j}\Sigma^\varepsilon_{\alpha j} + f^0_\alpha,$$

$$\rho\frac{\partial^2 U^\varepsilon_3}{\partial t^2} = \frac{\partial}{\partial x_j}\Sigma^\varepsilon_{3j} + f^0_3,$$

$$\beta\frac{\partial\Theta^\varepsilon}{\partial t} = \frac{1}{T_0}\left(\frac{\partial}{\partial y_\alpha}\left(k\frac{\partial\Theta^\varepsilon}{\partial y_\alpha}\right)+\frac{1}{\varepsilon^2}\frac{\partial}{\partial x_3}\left(k\frac{\partial\Theta^\varepsilon}{\partial x_3}\right)\right) - \frac{E\alpha}{1-2\nu}\left(\frac{\partial^2 U^\varepsilon_\alpha}{\partial t\partial y_\alpha}+\frac{1}{\varepsilon^2}\frac{\partial^2 U^\varepsilon_3}{\partial t\partial x_3}\right),$$

together with the appropriate rescaled boundary conditions, *i.e.*,

$$\Sigma^\varepsilon_{\alpha 3} = \pm g^{\pm 0}_\alpha \text{ on } \Gamma^\pm = \omega \times \{\pm 1\},$$

$$\Sigma^\varepsilon_{33} = \pm g^{\pm 0}_3 \text{ on } \Gamma^\pm,$$

$$\frac{\partial\Theta^\varepsilon}{\partial x_3} = 0 \text{ on } \Gamma^\pm,$$

$$U^\varepsilon = \Theta^\varepsilon = 0 \text{ on } \Gamma^1 = \partial\omega \times (-1,1),$$

and rescaled initial conditions, *i.e.*,

$$U^\varepsilon(t=0) = u^0_0,$$

$$\frac{\partial U^\varepsilon}{\partial t}(t=0) = v^0_0,$$

(5)

$$\Theta^\varepsilon(t=0) = \theta^0_0,$$

where

$$u^0_{0\alpha} = \underset{\sim}{u}^0_{0\alpha} - x_3 \frac{\partial u^0_{03}}{\partial y_\alpha},$$

(6)

$$\underset{\sim}{u}^0_0 \text{ and } u^0_{03} \text{ are independent of } x_3 .$$

In view of (5), (6) the initial displacement $U^\varepsilon(t=0)$ is a displacement field of the Kirchhoff-Love type. As ε tends to zero, the fields U^ε and Θ^ε are shown (*cf.* [3], Theorem 2) to converge (in an appropriate topology) to the solution fields u^0 and θ^0 associated with a flexional problem for the component of the displacement field normal to the plate together with a coupled membrane-thermal problem for the components of the displacement field in the plane of the plate and the temperature field. The mechanical and thermal characteristics of the plate problem are explicitly determined in terms of the original characteristics. Specifically the flexional problem reads as (*cf.* [3], Section 3)

$$\overline{\rho}\frac{\partial^2 u^0_3}{\partial t^2} + \frac{\partial^2}{\partial y_\alpha \partial y_\beta}\left(\left(\overline{\frac{E x_3^2}{1+\nu}}\right) \frac{\partial^2 u^0_3}{\partial y_\alpha \partial y_\beta} + \left(\overline{\frac{E\nu x_3^2}{(1-\nu)(1+\nu)}}\right) \Delta u^0_3 \delta_{\alpha\beta}\right) =$$

$$\overline{f^0_3} + (g^{+0}_3 + g^{-0}_3) + \frac{\partial}{\partial y_\alpha}(g^{+0}_\alpha - g^{-0}_\alpha) + \left(\overline{x_3 \frac{\partial f^0_\alpha}{\partial y_\alpha}}\right),$$

with

$$u^0_3 = \frac{\partial u^0_3}{\partial n} = 0 \text{ on } \partial\omega,$$

as boundary conditions and

$$u^0_3(t=0) = u^0_{03}, \quad \frac{\partial u^0_3}{\partial t}(t=0) = \frac{\left(\overline{\rho v^0_{03}}\right)}{\overline{\rho}},$$

as initial conditions, where the overbar $^{-}$ stands for $\int_{-1}^{1} dx_3$. If

$$\kappa = \overline{\left(\beta + \frac{E\alpha^2(1+\nu)}{(1-\nu)(1-2\nu)}\right)},$$

the coupled membrane-thermal problem reads as (*cf.*[3], Section 3)

$$\frac{\partial}{\partial y_\beta}\left(\overline{\left(\frac{E}{1+\nu}\right)} e_{\alpha\beta}(\underset{\sim}{u}^0) + \overline{\left(\frac{E\nu}{(1-\nu)(1+\nu)}\right)} \operatorname{tr} e(\underset{\sim}{u}^0)\delta_{\alpha\beta} - \overline{\left(\frac{E\alpha}{1-\nu}\right)}\theta^0\delta_{\alpha\beta}\right)$$

$$+ \overline{f^0_\alpha} + (g^{+0}_\alpha + g^{-0}_\alpha) = 0,$$

$$\kappa\frac{\partial\theta^0}{\partial t} = \frac{\partial}{\partial y_\alpha}\left(\overline{k}\ \frac{\partial\theta^0}{\partial y_\alpha}\right) - \overline{\left(\frac{E\alpha}{1-\nu}\right)} \operatorname{tr} e\left(\frac{\partial \underset{\sim}{u}^0}{\partial t}\right),$$

with

$$\underset{\sim}{u}^0 = \theta^0 = 0 \text{ on } \partial\omega,$$

and

(8) $$\theta^0(t=0) + \frac{1}{\kappa}\overline{\left(\frac{E\alpha}{1-\nu}\right)} \operatorname{tr} e(\underset{\sim}{u}^0(t=0)) = \frac{1}{\kappa}\left(\overline{\beta}\theta^0_0 + \overline{\left(\frac{E\alpha}{1-2\nu}\right)} \operatorname{tr} e(\underset{\sim}{u}^0_0)\right)$$

as initial condition.

The displacement field u^0 is of Kirchhoff-Love type, *i.e.* its two first components read as

$$u^0_\alpha = \underset{\sim}{u}^0_\alpha - x_3\,\frac{\partial u^0_3}{\partial y_\alpha}.$$

The initial value $\underset{\sim}{u}^0(t=0)$ can be shown to satisfy (*cf.* [3], Theorem 1)

$$\frac{\partial}{\partial y_\beta}\left(\overline{\left(\frac{E}{1+\nu}\right)} e_{\alpha\beta}(\underset{\sim}{u}^0(t=0)) + \left(\overline{\left(\frac{E\nu}{(1-\nu)(1+\nu)}\right)} + \frac{1}{\kappa}\overline{\left(\frac{E\alpha}{1-\nu}\right)}^2\right) \operatorname{tr} e(\underset{\sim}{u}^0(t=0))\delta_{\alpha\beta}\right)$$

(9)

$$= \frac{\partial}{\partial y_\alpha}\left(\frac{1}{\kappa}\overline{\left(\frac{E\alpha}{1-\nu}\right)}\left(\overline{\beta}\theta^0_0 + \overline{\left(\frac{E\alpha}{1-2\nu}\right)} \operatorname{tr} e(\underset{\sim}{u}^0_0)\right)\right) - \overline{f^0_\alpha}(t=0) - (g^{+0}_\alpha + g^{-0}_\alpha)(t=0).$$

It is easily deduced from (8) and (9) that a change of initial temperature generally occurs during the asymptotic process (take $u_0^0=0$, $f_\alpha^0=0$, $g_\alpha^{\pm 0}=0$, $\alpha(x)>0$). Unfortunately a further analysis of the temperature field θ^ε fails to produce the expected fast oscillations. Whether such a phenomenon presides over the change of initial temperature is an open question at this point.

The convergence of U^ε and Θ^ε to u^0 and θ^0 may be shown to be stronger (*i.e.* to take place in a smaller functional space) whenever a somewhat "twisted" compatibility condition is met by the initial conditions and loadings (*cf.* [3], Theorem 3). In the case where

$$v_0^0 = g_\alpha^{\pm 0}(t=0) = 0, \tag{10}$$

the compatibility equation becomes simpler. It is satisfied when $\underset{\sim}{u}{}_0^0$, $u_{0\,3}^0$ and θ_0^0 are such that

$$\alpha(1+\nu)\theta_0^0 = \nu \operatorname{tr} e(\underset{\sim}{u}{}_0^0), \qquad u_{0\,3}^0 = 0,$$

$$\frac{\partial}{\partial x_\beta}\left(\overline{\left(\frac{E}{1+\nu}\right)} e_{\alpha\beta}(\underset{\sim}{u}{}_0^0)\right) + \overline{f_\alpha^0(t=0)} = 0.$$

Under the above conditions, there is no change of initial temperature (or displacement), *i.e.*,

$$\theta^0(t=0) = \theta_0^0 \ , \ \underset{\sim}{u}{}^0(t=0) = \underset{\sim}{u}{}_0^0, \tag{11}$$

which leads us to state the following conjecture:

(11) is a necessary conditions for compatibility (i.e.*also for strong convergence) at least under the hypothesis (10).*

4. LINEARIZATION VERSUS ASYMPTOTIC BEHAVIOUR

Section 2 and 3 describe two apparently disconnected processes, which however share a shift in initial temperature. In Section 2 this change stems from fast time oscillations of the temperature field. The existence of such oscillations suggests that the thermomechanical response of a heterogeneous body to a given excitation sets this body out of thermodynamical equilibrium.

Whenever a system is far from thermodynamical equilibrium a

notion of "tangent" equilibrium state has to be defined if one is to apply the fundamental principles of thermodynamics to that system. In this spirit, Bataille and Kestin [5] introduced the notion of *accompanying equilibrium state*. It was subsequently revisited in [7] in the specific setting of a heterogeneous medium.

At any given time t, all state variables (except the temperature T(t)), the kinetic energy as well as the internal energy are frozen at their current values. The system is virtually driven to equilibrium through thermal dissipation. The final temperature $T_f(t)$ associated with the virtual equilibrium of the system is referred to as the thermodynamical temperature at time t.

In the context of a heterogeneous medium the statements of invariance concern the macroscopic averages of the relevant quantities. It is thus assumed that the average strain and average internal energy remain fixed during the virtual evolution. In a linearized setting the internal energy is proportional to the entropy since the free energy functional is quadratic in the state variables. It can be easily shown that the average strain and average entropy coincide with the limit strain and entropy during the limit process performed in Section 2. The homogenization process identifies with the virtual evolution towards the accompanying equilibrium state and the fast time oscillations of the temperature field describe the virtual evolution of that field from T(t) to $T_f(t)$ (the limit "homogenized" temperature).

This very simple analysis tends to demonstrate that the fast oscillations might be a mere byproduct of the linearization of the equations of thermoelasticity and it suggests that the resulting change in initial temperature (at least in the case of a heterogeneous material) is also a byproduct of the linearization.

These considerations prompt us to question the validity of the linearized equations of thermoelasticity in the context of an asymptotic analysis involving a small parameter. In a more general context the investigation of the conditions under which linearization and asymptotic behaviour commute may prove a worthy task (see [8] for example).

REFERENCES

[1] Chadwick, P., "Thermoelasticity, the Dynamical Theory", in: Progress in Solid Mechanics, Vol.1 (North-Holland, Amsterdam, 1960).

[2]Francfort, G.A., "Homogenization and Linear Thermoelasticity", SIAM J. Math. Anal., Vol.14, 4(1984) pp. 696-708.

[3]Blanchard, D. and Francfort, G.A., "Asymptotic Thermoelastic Behaviour of Flat Plates", to appear.

[4]Francfort, G.A., "Homogenization and Fast Oscillations in Linear Thermoelasticity", in: Lewis, R.W., Hinton, E., Bettess, P. and Schrefler, B.A., (eds.), Numerical Methods for Transient and Coupled Problems (Pineridge Press, Swansea, 1984) pp.382-392.

[5]Kestin, J. and Bataille, G., Thermodynamics of Solids, (Freudenstadt, 1979).

[6]Raoult, A., "Construction d'un modèle d'évolution de plaques avec terme d'inertie de rotation", Ann. Mat. Pura App., Vol.139 (1985) pp.361-400.

[7]Francfort, G.A., Nguyen, Q.S. and Suquet, P., "Thermodynamique et lois de comportement thermomécanique homogénéisées", CRAS, Série I, Vol.296 (1983) pp.1007-1010.

[8]Triantafyllidis, N. and Kwon, Y.J., "Thickness Effects on the Stability of thin Walled Structures, I", to appear.

Thermomechanical Couplings in Solids
H.D. Bui and Q.S. Nguyen (Editors)
Elsevier Science Publishers B.V. (North-Holland)

ANALYSIS OF THERMOMECHANICAL COUPLING BY BOUNDARY ELEMENT METHOD

M. PREDELEANU

Laboratoire de Mécanique et Technologie
E.N.S. de CACHAN/C.N.R.S./Université PARIS 6
61, Avenue du Président Wilson - 94230 CACHAN (France)

1. INTRODUCTION

During the last few years a considerable amount of interest has been shown in the Boundary Element Method (B.E.M.). Many theoretical investigations and computer programs have proved that this method has important advantages over other numerical methods in many fields of engineering applications.

The latest developments concerning new mathematical formulation, convergence and accuracy of solutions and also new numerical comparative studies have consolidated the B.E.M. as a very efficient tool for analysis.

Though in mathematical physics the first formulations by integral equations appeared a long time ago - in linear elasticity theory in the last century- today's boundary approach originated in mechanics of solids two decades ago with the work of Rizzo [1] in elastostatics. The method has been rapidly extended to linear time dependent problems [2]-[4] and finally to non-linear problems (plasticity and visco-plasticity) [5]-[8]. The formulation of the latter are based on the use of the fundamental solutions (kernel functions) of elastostatics or elastodynamics by considering the non-linear terms in governing equations as fictitions body forces. The draw-back of this "direct" approach is the need for an iterative procedure to obtain the solution, and consequently a domain discretization. Obviously the same procedure can be used for coupled problems (thermomechanical coupling problems, two phase media, etc...) with the same disadvantage.

An alternative mathematical formulation for linear thermomechanical coupled problems which avoids the iterative procedure was first given by the author [9]. The boundary integral equations are defined in terms of the coupled thermoelastic solutions, available in closed form in a Laplace transform space or frequency space for harmonic regime. The basic equations have been derived for a general class of linear viscoelastic bodies, defined by Riemann-Stieltjes integral convolution. New developments of the method have been obtained for the thermoelastic case. Thus, Fleurier and Predeleanu [10], [11] treat some numerical aspects concerning the implementation of the boundary element method given in [9] for the bidimensional and three-dimensional cases of the steady state problem in the frequency domain. To regularize strongly singular intergrals an indirect calculation method is proposed. For that conversion formulae of the domain integrals into boundary integrals are obtained.

By using the integral representations of the displacement and temperature fields, V. Sladek and J. Sladek [12] deduced new regularized boundary integral equations for the traction vector and heat flux which could serve also for the calculation of unknown quantities. M. Tanaka and K. Tanaka [13] proposed an integral equation formulation of the coupled thermoelastic problems in the time-space domain by using the classical fundamental solutions for elastostatics, elastodynamics and heat conduction theory but a formal solution of a Fredholm integral equation of the second kind had to be introduced.

2. BASIC RELATIONSHIPS

Let us consider a viscoelastic continuum, homogeneous and isotropic, initially undisturbed, occupying a finite region B with a regular boundary ∂B in the three-dimensional Euclidian space.

The constitutive relations for non-isothermal rheological behaviour can be deduced by means of two memory functionals defining Helmholtz free energy A and heat flux q as :

$$A = f_1 \, [\varepsilon(x,\tau) \ , \ T(x,\tau)]_{\tau=0}^{\tau=t} \tag{1}$$

$$q = f_2 \, [\text{grad}T(x,\tau)]_{\tau=0}^{\tau=t} \tag{2}$$

where f_1 is a scalar-valued bilinear functional and f_2 a vector-valued linear functional, T is the temperature difference from a constant temperature T_0 at a stress-free configuration, $\varepsilon = 1/2$ (grad u + grad^{t}u) is the linearized strain tensor and u is the displacement vector.

By using polynomial integral representation of the memory functionals f_1 and f_2, the linearized constitutive relations of coupled thermoviscoelasticity theory [14] can be expressed as :

$$s = 2\mu \bullet de \tag{3}$$

$$\text{tr}\sigma = 3(K \bullet d\,\text{tr}\varepsilon - \varphi \bullet dT) \tag{4}$$

$$\varrho S = \varphi \bullet d\, tr\varepsilon + m \bullet dT \tag{5}$$

$$q = -\, k \text{ grad } T \tag{6}$$

where s is the deviator of the CAUCHY stress tensor σ, e the deviator of the strain tensor ε, $tr\varepsilon$ the trace of ε, S the entropy difference from a constant initial entropy, ϱ the mass density. μ, K, φ and m are appropriate memory functions and k is a constant (thermal conductivity). In these relations $f \bullet dg$ means the Riemann-Stieltjes convolutions of two real valued-functions f and g defined by :

$$(f \bullet g)(t) = \int_{-\infty}^{t} f(t-\tau)\, dg(\tau) \tag{7}$$

provided the integral is meaningful (cf reference [15]). Consequently, the coupled heat conduction is :

$$k\Delta T - T_0 \frac{\partial}{\partial t} (m \bullet dT + \varphi \bullet d\, tr\, \varepsilon) + Q = 0 \tag{8}$$

where Q is the quantity of heat supplied per unit volum and unit time.

Special cases of the linear coupled thermoviscoelasticity theory can be obtained from the constitutive relations (3) - (6).

Case i) $\quad \varphi(t) = 3\alpha_T K(t)$ (9)

where K is the bulk memory function and α_T is the time-independant coefficient of linear thermal expansion. Then the equation (4) becomes :

$$tr\, \sigma = 3\, K \bullet d\, (tr\, \varepsilon - 3\alpha_T\, T) \tag{10}$$

Case ii) $\quad \varphi(t) = \gamma\, H\, (t)\, , \quad m(t) = m\, H(t)$ (11)

where H is the Heaviside function and m,γ are constants. Then the equation (4) becomes :

$$tr\, \sigma = -3\gamma\, T + 3K \bullet d\, tr\, \varepsilon \tag{12}$$

and the coupled heat conduction is written as :

$$k\, \Delta T - T_0 (m \frac{\partial T}{\partial t} + \gamma \frac{\partial}{\partial t} tr\varepsilon) + Q = 0 \tag{13}$$

As was remarked by Eringen [16] who developed a linear thermoviscoelasticity theory by assuming the relations (11), the difference between thermoelasticity and thermoviscoelasticity for isotropic solids occurs only in the stress-strain relations.

Case iii) The thermoelastic case can be obtained from the latter one by setting :

$$\mu(t) = \mu H(t) \quad , \quad K(t) = KH(t) \tag{14}$$

where μ is the elastic shear modulus and K the elastic bulk modulus. The equations (3) - (8) must be completed by the Cauchy's first law of motion :

$$\operatorname{div} \sigma + F = \varrho \frac{\partial^2 u}{\partial t^2} \tag{15}$$

where F is the body force vector.

If the displacement field u and the temperature field T are choosen as unknown fields, the governing coupled field equations can be expressed as :

$$u \bullet d(\Delta u) + (\lambda + \mu) \bullet d(\operatorname{grad} \operatorname{div} u) - \varphi \bullet d(\operatorname{grad} T) - \varrho \frac{\partial^2 u}{\partial t^2} + F = 0 \tag{16}$$

$$K \, \Delta T - T_0 \frac{\partial}{\partial t} [\, m \bullet dT + \varphi \bullet \operatorname{div} u \,] + Q = 0 \tag{17}$$

where $\lambda(t) = K(t) - \frac{2}{3}\mu(t)$.

The initial conditions are assumed be homogeneous and the boundary conditions can be mixed : on a part of the boundary $\partial B_\sigma \subset \partial B$ the traction vector $\sigma_n = \sigma.n$ is prescribed (n being the outward unit normal to ∂B) and on the complementary par $\partial B_u \subset \partial B$ the displacement vector u is prescribed. In addition the temperature field can be prescribed on $\partial B_T \subset \partial B$ and the heat flux $q_n = q.n$ on the complementary part $\partial B_q \subset \partial B$.

3. INTEGRAL EQUATION FORMULATION

3.1. Reciprocal theorem

A simple method to derive the integral formulations of the governing equations consists in using reciprocity relations such as those of Green type in potential theory and Betti type relations in elasticity theory. Such relations can be obtained directly by using the known results on Stieltjes convolutions or by introducing the Laplace transform space.

Let us consider the body subjected to two independent loading programs. Thus, the following Betti type reciprocity relation between the corresponding Laplace transform fields $\{\bar{u},\bar{T}\}$ and $\{\bar{u}',\bar{T}'\}$ can be shown

$$\int_{\partial B}(\bar{\sigma}_n\,\bar{u}' - \bar{\sigma}_n'\,\bar{u})\,ds - \frac{1}{pT_0}\int_{\partial B}(\bar{T}\,\bar{q}_n' - \bar{T}'\bar{q}_n)ds + \\ + \int_B(\bar{F}\bar{u}' - \bar{F}'\bar{u})dv - \frac{1}{pT_0}\int_B(\bar{Q}\bar{T}' - \bar{Q}'\bar{T})dv = 0 \tag{18}$$

where $\bar{f}(x,p)$ denotes the Laplace transform of a numerical function $f(x,t)$:

$$\bar{f}(x,p) = \int_0^\infty e^{-pt}\,f(x,t)\,dt \tag{19}$$

It is worth noting that the relation (18) do not contain neither the inertia terms and nor the rheological characteristics, his form being therefore the same in the quasi-static case and also for every class of viscoelastic constitutive laws, including obviously the elastic one [17]. This theorem generalize the results obtained for non-coupled thermal problem [18]-[20].

3.2. **Integral equations**

By using the the reciprocity relation (18) for two particular loading programs we obtain the integral representations of Somigliana type of the displacement field $\bar{u}$ and the temperature field $\bar{T}$ in every point inside domain B. The corresponding boundary integral equations are derived by a limiting procedure approaching the internal source point on the boundary ∂B of the body. As a consequence the resulting integral equations can be written as follows :

$$\begin{aligned} c(\xi)\bar{u}(\xi,p) = & \int_B\bar{F}(x,p)\bar{u}^*(x,\xi,p)dv(x) \\ & - \frac{1}{pT_0}\int_B\bar{Q}(x,p)\bar{T}^*(x,\xi,p)dv(x) \\ & + \int_{\partial B}[\bar{\sigma}_n(x,p)\bar{u}^*(x,\xi,p) - \bar{\sigma}_n^*(x,\xi,p)\bar{u}(x,p)]ds(x) \\ & - \frac{1}{pT_0}\int_{\partial B}[\bar{T}(x,p)\bar{q}_n^*(x,\xi,p) - \bar{T}^*(x,\xi,p)\bar{q}_n(x,p)]ds(x) \end{aligned} \tag{20}$$

$$\begin{aligned} c(\xi)\bar{T}(\xi,p) = & \int_B\bar{Q}(x,p)\bar{T}'(x,\xi,p)dv(x) \\ & - pT_0\int_B\bar{F}(x,s)\bar{u}'(x,\xi,p)dv(x) \\ & - pT_0\int_{\partial B}[\bar{\sigma}_n(x,p)\bar{u}'(x,\xi,p) - \bar{\sigma}_n'(x,\xi,p)\bar{u}(x)]ds(x) \\ & + \int_{\partial B}[\bar{T}(x,p)\bar{q}_n'(x,\xi,p) - \bar{T}'(x,\xi,p)\bar{q}_n(x,p)]ds(x) \end{aligned} \tag{21}$$

where the coefficient $c(\xi)$ has values

$$c(\xi) = \begin{cases} 1 & \text{, if } \xi \text{ is inside domain B} \\ \frac{1}{2} & \text{, if } \xi \text{ lies on boundary } \partial B \text{ (assumed smooth)} \\ 0 & \text{, if } \xi \text{ is outside domain B} \end{cases} \tag{22}$$

$\bar{u}^*(x,\xi,p)$ and $\bar{T}^*(x,\xi,p)$ are the Laplace transforms of the fundamental solutions of the coupled field equations (16)-(17) corresponding to an instantaneous point unit load at ξ and $Q = 0$. $\bar{\sigma}_n^*$ is the boundary traction vector determined by $\bar{u}^*$ and $\bar{q}_n^*$ is the normal heat flux determined by $\bar{T}^*$. $\bar{u}'(x,\xi,p)$ and $\bar{T}'(x,\xi,p)$ are the Laplace transforms of the fundamental solutions of the coupled field equations corresponding to an instantaneous point heat source and $F = 0$. $\bar{\sigma}_n'$ is the traction vector determined by $\bar{u}'$ and $\bar{q}_n'$ is the heat flux determined by $\bar{T}'$.

The singular integral are understood in the sense of Cauchy's principal value. By inversion of the Laplace transforms performed on the relations (20)-(21), integral equations for $u(\xi,t)$ and $T(\xi,t)$ can be obtained in time-space domain. It is important to note that the present formulation defined by equations (20)-(21) does not contain domain integrals in the absence of internal inputs (body forces and heat sources).

4. STEADY-STATE THERMOELASTIC PROBLEM

4.1. Governing equations

The field equations of the thermoelastic problem can be deduced from the equations (3)-(8) by putting $\mu(t) = \mu H(t)$, $K(t) = (\lambda+3\mu)\, H(t)$, $\varphi(t) = \gamma H(t)$, $m(t) = mH(t)$ where μ, λ are Lamé elastic moduli, $\gamma = (3\lambda + 2\mu)\alpha_T$, $m = c_\varepsilon/T_0$, c_ε being the specific heat at constant strain.

If steady state conditions are assumed i.e. the thermo-mechanical variables are harmonic functions of time, then the field equations in the frequency domain can be obtained from those in the Laplace transform domain by replacing the parameter p by $-i\omega$ and the Laplace transforms of all quantities by their amplitudes. So, let us consider the displacement and temperature field expressed as :

$$\begin{aligned} u(x,t) &= \mathrm{Re}\{\bar{u}(x,\omega)e^{-i\omega t}\} \\ T(x,t) &= \mathrm{Re}\{\bar{T}(x,\omega)e^{-i\omega t}\} \end{aligned} \tag{23}$$

where ω is the angular frequency. $\bar{u}(x,\omega)$ and $\bar{T}(x,\omega)$ are respectively the amplitudes of $u(x,t)$ and $T(x,t)$.

The field equations for $\bar{u}(x,\omega)$ and $\bar{T}(x,\omega)$ can be written in the scalar form as :

$$\mu\bar{u}_{j,kk} + (\lambda+\mu)\bar{u}_{k,kj} + \varrho\omega^2\bar{u}_j - \gamma\bar{T}_{,j} + \bar{F}_j = 0 \tag{24}$$

$$\bar{T}_{,kk} + \frac{i\omega}{\chi}\bar{T} + i\omega\eta\bar{u}_{k,k} + \frac{\bar{Q}}{k} = 0 \tag{25}$$

where $\bar{u}_j$ are the components of $\bar{u}$ with respect to a fixed rectangular cartesien coordinate system Ox_j, $j = 1,2,3$; x_j will denote the coordinate of the point $x \in B$, and a partial differentiation of a variable with respect to x_j is designated as $(\ )_{,j}$, $\chi = k/c_\varepsilon$, $\eta = \gamma T_0/k$.

The B.E.M. for the steady state thermoelastic problem is based on following coupled integral equations deduced from the relations (20) and (21) :

$$\begin{aligned}\frac{1}{2}\delta_{kj}\bar{u}_j(\xi,\omega) = &\int_B \bar{F}_j(x,\omega)\bar{u}^*_{jk}(x,\xi,\omega)dv(x) \\ &+ \frac{i}{\omega T_0}\int_B \bar{Q}(x,\omega)\bar{T}^*_k(x,\xi,\omega)ds(x) \\ &+ \int_{\partial B}[\bar{\sigma}_{nj}(x,\omega)\bar{u}^*_{jk}(x,\xi,\omega)ds(x) \\ &- \bar{\sigma}^*_{njk}(x,\xi,\omega)\bar{u}_j(x,\omega)]ds(x) \\ &+ \frac{i}{\omega T_0}\int_{\partial B}[\bar{T}(x,\omega)\bar{q}^*_{nk}(x,\xi,\omega) \\ &- \bar{T}^*_k(x,\xi,\omega)\bar{q}_n(x,\omega)]ds(x)\end{aligned} \tag{26}$$

$$\begin{aligned}\frac{1}{2}\bar{T}(\xi,\omega) = &\int_B \bar{Q}(x,\omega)\bar{T}'(x,\xi,\omega)dv(x) \\ &+ i\omega T_0\int_B \bar{F}_j(x,\omega)\bar{u}'_j(x,\xi,\omega)dv(x) \\ &+ i\omega T_0\int_{\partial B}[\bar{\sigma}_{nj}(x,\omega)\bar{u}'_j(x,\xi,\omega) \\ &- \bar{\sigma}'_{nj}(x,\xi,\omega)\bar{u}_j(x)]ds(x) \\ &+ \int_{\partial B}[\bar{T}(x,\omega)\bar{q}'_n(x,\xi,\omega) \\ &- \bar{T}'(x,\xi,\omega)\bar{q}_n(x,\omega)]ds(x)\end{aligned} \tag{27}$$

The kernel functions $\bar{u}^*_{jk}$, $\bar{T}^*_k$, with j,k = 1,2,3 are the fundamental solutions of the coupled equations (24), (25), regular at infinity, corresponding to the loading $\bar{F}_j(x,\xi) = \delta(x-\xi)\delta_{jk}$, $\bar{Q} = 0$, where δ is the Dirac function and δ_{jk} the Kroneker symbol. $\bar{u}^*_{jk}(x,\xi,\omega)$ denotes the displacement amplitude component in the j-direction at a point x due to a unit concentrated load acting in the k-direction at the point ξ. $\bar{T}_k$ is the corresponding induced temperature field. The second pair of kernel functions $\bar{u}'_j$, $\bar{T}'_i$, j = 1,2,3 is obtained as fundamental solutions at the same equations (24), (25) regular at infinity, and corresponding to the loading $\bar{F}_j = 0$, $Q(x,\xi) = \delta(x-\xi)$.

4.2. **Coupled fundamental solutions**

For the coupled thermoelastic solutions (24) and (25) there exist fundamental solutions in a closed form both for three and two dimensional cases (see Nowacki [21]).

4.2.1. **Three dimensional problem**

$$\bar{u}^*_{jk}(x,\xi,\omega) = -\frac{1}{4\pi\rho\omega^2}\left[\frac{\partial^2}{\partial x_j\partial x_j}(P(R,\omega)-P_0(R,\omega) - \frac{K_2^2}{R^2}\delta_{jk}\, e^{\pm i\omega R/c_2}\right] \tag{28}$$

$$\bar{T}^*_k(x,\xi,\omega) = \frac{\alpha\beta}{4\pi\tau\rho c_1^2(k_1^2-k_2^2)}\frac{\partial}{\partial x_k}\left[I_1(R,\omega) - I_2(R,\omega)\right] \tag{29}$$

$$\bar{u}'_j(x,\xi,\omega) = \frac{\tau}{4\pi k(k_1^2-k_2^2)}\frac{\partial}{\partial x_j}\left[I_2(R,\omega) - I_1(R,\omega)\right] \tag{30}$$

$$\bar{T}'(x,\xi,\omega) = \frac{1}{4\pi kR(k_2^2-k_1^2)}\left[(k_2^2-K_1^2)e^{ik_2R} - (k_1^2-K_1^2)e^{ik_1R}\right] \tag{31}$$

where :

$$I_s(R,\omega) = \frac{1}{R}e^{ik_sR} \quad , \qquad s = 1,2$$

$$P(R,\omega) = A_1 I_1(R,\omega) - A_2 I_2(R,\omega) - \frac{1}{R}$$

$$P_0(R,\omega) = R\,(e^{ik_2R} - 1)$$

$$R = |x-\xi|$$

$$A_1 = \frac{(k_1^2-\alpha)K_1^2}{k_1^2(k_1^2-k_2^2)} \quad , \qquad A_2 = \frac{(k_2^2-\alpha)K_1^2}{k_2^2(k_1^2-k_2^2)}$$

$$\alpha = \frac{i\omega}{\chi} \quad , \qquad \beta = \eta\tau\chi \quad , \qquad \tau = \frac{\gamma}{\lambda+2\mu}$$

$$c_1 = \left(\frac{\lambda+2\mu}{\varrho}\right)^{1/2} \quad , \quad c_2 = \left(\frac{\mu}{\varrho}\right)^{1/2}$$

$$K_1 = \frac{\omega}{c_1} \quad , \quad K_2 = \frac{\omega}{c_2}$$

k_1 and k_2 are the roots of the algebric equation

$$k^4 - k^2 \left[K_1^2 + \alpha(1+\beta)\right] + \alpha K_1 = 0 \tag{32}$$

4.2.2. **Two dimensional problem**

Consider the plane deformation problem of the steady state thermoelasticity theory for a domain D with a regular boundary ∂D in the two dimensional space. The governing equations and the fundamental solutions can be deduced from those presented in the previous section. For simplicity, we shall use here the same symbols for the functions introduced for the three-dimensional problem. Obviously, the boundary integral formulation is defined by the equations (24) and (25) which conserve the same form. On the other hand, the fundamental solutions for the two-dimensional case are written as :

$$\bar{u}^*_{jk} = - \frac{i}{4\pi\varrho\omega^2} \left\{ \frac{\partial^2}{\partial x_j \partial x_k} \left[A_1 H_0^1(k_1 r) - A_2 H_0^1(k_2 r) - H_0^1(K_2 r) \right] - K_2^2 \delta_{jk} H_0^1(K_2 r) \right\} \tag{33}$$

$$\bar{T}^*_k = \frac{i\alpha\eta X}{4c_1^2\varrho(k_1^2-k_2^2)} \frac{\partial}{\partial x_k} \left[H_0^1(k_1 r) - H_0^1(k_2 r) \right] \tag{34}$$

$$\bar{u}'_j = \frac{\tau i}{4k(k_2^2-k_1^2)} \frac{\partial}{\partial x_j} \left[H_0^1(k_1 r) - H_0^1(k_2 r) \right] \tag{35}$$

$$\bar{T}' = \frac{i}{4k(k_2^2-k_1^2)} \left[(k_2^2-K_1^2) H_0^1(k_2 r) - (k_1^2-K_1^2) H_0^1(k_1 r) \right] \tag{36}$$

where H_0^1 is the Hankel function of first kind and order zero and $r^2 = x_1^2+x_2^2$, $j,k = 1,2$.

4.3. **Conversion formulae**

As it was remarked, in the present B.E.M. formulation the unknown functions appear only in the boundary integrals. Without the internal inputs, only boundary element discretization is necessary for numerical treatment. Generally, if internal inputs are present domain integrals containing known functions must be calculated. Nevertherless, for some particular cases the domain integral can be converted into a boundary integral, more convenient numerically. Thus, the results of Stippess and Rizzo [22] for classical elastostatics have been generalized recently for coupled thermoelastic problem [10], [11].

Thus, if the body force field is derivable from a scalar potential whose Laplacian is at most a constant and the thermal source field is constant then the following conversion formula can be shown for 3D problem :

$$\int_B \bar{F}_i \bar{u}^*_{ik} \, dv = \int_{\partial B} \{\Omega \, (\frac{\partial^2 \phi}{\partial x_j \partial n} - \bar{u}^*_{ik} n_i) - \frac{\partial \phi}{\partial x_k} \frac{\partial \Omega}{\partial n} + k_0 \phi n_k\} \, ds \tag{37}$$

$$\int_B \bar{F}_i \bar{u}'_i = \int_{\partial B} \{\Omega \, (\frac{\partial G}{\partial n} - \bar{u}'_i n_i) - G \frac{\partial \Omega}{\partial n} + k_0 \frac{\partial G_1}{\partial n}\} \, ds \tag{38}$$

$$\int_B \bar{Q}\bar{T}^*_k dv = \int_{\partial B} \bar{Q} g n_k \, ds \tag{39}$$

$$\int \bar{Q}\bar{T}' dv = \int_{\partial B} \bar{Q} \frac{\partial f}{\partial n} \, ds \tag{40}$$

where : $\bar{F}_i = -\Omega_{,i}$, $\Delta\Omega = k_0$, $\bar{Q} = \text{const.}$

$$\phi(R) = -\frac{1}{4\pi\rho\omega^2 R} \left[A_1 e^{ik_1 R} - A_2 e^{ik_2 R} \right]$$

$$G(R) = -\frac{\tau}{4\pi k(k_2^2-k_1^2)} \left(\frac{e^{ik_1 R}}{R} - \frac{e^{ik_2 R}}{R} \right)$$

$$G_1(R) = -\frac{\tau}{4\pi k(k_2^2-k_1^2)R} \left(\frac{1}{ik_1} e^{ik_1 R} - \frac{1}{ik_2} e^{ik_2 R} \right)$$

$$g = \frac{i\eta\omega}{4\pi c_1^2 (k_1^2-k_2^2)R} (e^{ik_1 R} - e^{ik_2 R})$$

$$f = \frac{1}{4\pi k(k_2^2-k_1^2)R} \left[\frac{i(k_1^2-K_1^2)}{k_1} e^{ik_1 R} - \frac{i(k_2^2-K_1^2)}{k_2} e^{ik_2 R} \right]$$

An analogous result can be obtained for the two-dimensional case. For simplicity and for our purpose (to regularize some singular integral) we shall give here only the following conversion formulae suitable when the body force and thermal fields are both constants.

$$\int_D \bar{F}_i \bar{u}^*_{ik} ds = \int_{\partial D} [\psi(\frac{\partial^2 \phi}{\partial x_k \partial n} - \bar{u}^*_{ik} n_i) - \frac{\partial \phi}{\partial x_k} \bar{F}_{oi} n_i] dl \tag{41}$$

$$\int_D \bar{F}_i \bar{u}'_i ds = \int_{\partial D} [\psi(\frac{\partial G}{\partial n} - \bar{u}'_i n_i) + G\bar{F}_{oi} n_i] dl \tag{42}$$

$$\int_D \bar{Q}\bar{T}^*_k ds = \int_{\partial D} Q_0 \, g n_k dl \tag{43}$$

$$\int_D \bar{Q}\bar{T}' ds = \int_{\partial D} Q_0 \frac{\partial f}{\partial n} dl \tag{44}$$

where

$$\bar{F}_i(r) = \psi_{,i}(r) = \bar{F}_{oi} = \text{const.} \qquad , \qquad \bar{Q}(r) = \bar{Q}_0 = \text{const.}$$

$$\phi(r) = -\frac{i}{4\varrho\omega^2}\left[A_1 H^1_{0,1}(k_1 r) - A_2 H^1_{0,1}(k_1 r)\right]$$

$$G(r) = \frac{\tau i}{4\chi(k_1^2-k_2^2)}\left[H^1_{0,1}(k_1 r) - H^1_{0,1}(k_2 r)\right]$$

$$g(r) = \frac{-\eta\omega}{4c_1^2\varrho(k_1^2-k_2^2)}\left[H^1_{0,1}(k_1 r) - H^1_{0,1}(k_2 r)\right]$$

$$f(r) = \frac{-i}{4\chi(k_2^2-k_1^2)}\left[\frac{k_2^2-K_1^2}{k_2^2} H^1_{0,1}(k_2 r) - \frac{k_1^2-K_1^2}{k_1^2} H^1_{0,1}(k_1 r)\right]$$

By $H^1_{0,1}$ we have noted the derivative of the function H^1_0 with respect to the argument.

5. BOUNDARY ELEMENTS DISCRETIZATION

As usual (cf. reference [23]), the boundary ∂B of the domain is divided into p elements over which interpolation functions are adopted to approximate the unknown quantities as follows :

$$\{\bar{u}\} = [M^c]\{\bar{u}^c\} \quad , \quad \{\bar{\sigma}_n\} = [M^c]\{\sigma_n^c\}$$
$$\bar{T} = [M^c]\{\bar{T}^c\} \quad , \quad \bar{q}_n = [M^c]\{q_n^c\} \tag{45}$$

where $[M^c]$ are the interpolation (shape) functions and $\bar{u}^c$, $\bar{\sigma}_n^c$, $\bar{T}^c$ and $\bar{q}_n^c$ are the nodal values of displacement vector, traction vector, temperature and heat flux, respectively at nodal point ξc on ∂B. The summation convention is implied for $c = 1,2,\ldots, N_e$ where N_e is the number of nodal points belonging to the e-element. Omitting the internal inputs the discretized form of the equations (26) and (27) can be expressed as :

$$\frac{1}{2}\delta_{kj}\bar{u}_j(\xi,\omega) + \sum_{e=1}^{p}\sum_{c=1}^{N_e}\{u_j^c\int_{\partial B_e}\bar{\sigma}^*_{njk}(x(\zeta),\xi,\omega)M^c(\zeta)ds(\zeta)$$
$$-\sigma^c_{nj}\int_{\partial B_e}\bar{u}^*_{jk}(x(\zeta),\xi,\omega)M^c(\zeta)ds(\zeta) \tag{46}$$
$$+\frac{i}{\omega T_0}[q_n^c\int_{\partial B_e}\bar{T}^*_k(x(\zeta),\xi,\omega)M^c(\zeta)ds(\zeta)$$

$$- \bar{T}^c \int_{\partial B_e} \bar{q}^*_{nk}(x(\zeta),\xi,\omega) M^c(\zeta) ds(\zeta)]\} = 0$$

$$\frac{1}{2}\bar{T}(\xi,\omega) + \sum_{e=1}^{p} \sum_{c=1}^{N_e} \{i\omega T_0 [u^c_j \int_{\partial B_e} \bar{\sigma}'_{nj}(x(\zeta),\xi,\omega) M^c(\zeta) ds(\zeta)$$

$$- \sigma^c_{nj} \int_{\partial B_e} \bar{u}'_j(x(\zeta),\xi,\omega) M^c(\zeta) ds(\zeta)]$$

$$+ q^c_n \int_{\partial B_e} \bar{T}'(x(\zeta),\xi,\omega) M^c(\zeta) ds(\zeta) \tag{47}$$

$$- T^c \int_{\partial B_e} \bar{q}'_n(x(\zeta),\xi,\omega) M^c(\zeta) ds(\zeta)\} = 0$$

$\zeta = \{\zeta_i\}$, denotes the intrinsic coordinates on ∂B_e. Writing the equations (46) and (47) for all boundary (collocation) nodes and reorganizing all nodal variables, a matrix system of equation is obtained :

$$[A]\ \{X\} = \{B\} \tag{48}$$

where {X} groups all unknowns in $\bar{u}^c_i$, $\bar{\sigma}^c_{ni}$, $\bar{T}^c$, $\bar{q}^c_n$.

[A] and {B} are the boundary influence matrix, whose coefficients are deduced by calculating the integrals which appear in equations (46) and (47) and which are of the type :

$$I(\Phi) = \int_{\partial B_e} \Phi(x,\xi,\omega) M^c(\zeta) ds(\zeta) \tag{49}$$

where Φ is a weighting two point function.

If internal inputs are present, then the matrix {B} in addition contains terms defined by domain integrals or only boundary integrals if an alternative method of calculation described in previous section is adopted.

For the calculation of the integrals $I(\Phi)$ two cases have been considered [10]

a) If the source point does not belong to the boundary element ∂B_e, then classical Gauss procedure can be used and a computer subprogramme for the calculation of kernel functions.

b) If the source point belongs to the boundary element ∂B_e a similar procedure is employed for the integrals $I(\bar{u}^*_{jk})$, $I(\bar{T}^*_k)$, $I(\bar{q}^*_{nk})$, $I(\bar{u}'_j)$, $I(\bar{\sigma}'_{nj})$, $I(\bar{T}')$, by using the series expansions for the kernel functions.

On the contrary, the integrals :

$$I(\sigma^*_{njk}) = \int_{\partial B_e} \bar{\sigma}^*_{njk}(x(\zeta), \xi, \omega) M^c(\zeta) ds(\zeta) \tag{50}$$

$$I(q'_n) = \int_{\partial B_e} \bar{q}'_n(x(\zeta), \xi, \omega) M^c(\zeta) ds(\zeta) \tag{51}$$

are strongly singular if $\xi \in \partial B_e$.

To obtain more accurate numerical values, an indirect method can be used, based on the following relations deduced from equations (20) and (21) :

$$\frac{1}{2}\delta_{jk} + \rho\omega^2 \int_B \bar{u}^*_{jk}(x, \xi, \omega) dv(x) + \int_{\partial B} \bar{\sigma}^*_{njk}(x, \xi, \omega) ds(x) = 0 \tag{52}$$

$$\frac{1}{2} + C_\varepsilon\, i\omega \int_B \bar{T}'(x, \xi, \omega) dv(x) + i\omega T_0\, \gamma \int_{\partial B} n_j \bar{u}'_j(x, \xi, \omega) ds(x) - \int_{\partial B} \bar{q}'_n(x, \xi, \omega) ds(x) = 0 \tag{53}$$

The domain integrals in the relations (52) and (53) have been calculated by converting them into boundary integrals by using the conversion formulae (37)-(44). Other regularization procedures could be used, as for instance, proposed in [12].

6. CONCLUDING REMARKS

As it has been already noticed, the present B.E.M. formulation based on coupled fundamental solutions of the governing equations avoids an iterative procedure. The unknown functions appear only in the boundary integrals : hence the discretization for numerical implementation is confined only to the boundary of the body. Consequently, the computer running time is significantly reduced. By neglecting the coupling effects, the problem of thermal stresses can be solved by using integral representations of the displacement field, expressed by means of the classical fundamental solutions from elastostatics or elastodynamics [18] [20]. Following this, domain integrals are introduced.

For special thermal fields, the domain integrals can be converted into boundary integrals as was proposed in [24].

7. REFERENCES

[1] Rizzo, F.J., An integral equation approach to boundary value problems of classical elastostatics, Quart. Appl. Math., 25 (1967), pp.83-95.

[2] Cruse , T.A. and Rizzo, F.J., A direct formulation and numerical solution of the general transient elastodynamic problem I, J. Math. Anal. Appl. 22-1, (1968), pp.244-259.

[3] Cruse , T.A., A direct formulation and numerical solution of the general transient elastodynamic problem II, J. Math. Anal. Appl. 22-2, (1968), pp. 341-355.

[4] Rizzo, J.F., and Shippy, D.J., An application of the correspondence principle of linear viscoelasticity theory, SIAM J. Appl. Math. 21, n°2 (1971), pp.321-330.
[5] Sweldow, J.L., and Cruse, T.A., Formulation of boundary integral equations for three-dimensional elasto-plastic flow, Int. J. Solids Struct. 7 (1971), pp.1673-1684.
[6] Bui, H.D., Some remarks about the formulation of three-dimensional thermoelastoplastic problems by integral equations, Int. J. Solids Struct. 14 (1978), pp.935-939.
[7] Chaudonneret, M., Méthode des équations intégrales appliquées à la résolution de problèmes de viscoplasticité, J. de Mécanique Appliquée 1 (1977), pp.113-132.
[8] Kumar, V. and Mukherjee, S., A boundary integral equation formulation for time-dependent inelastic deformation in metals, Int. J. Mechanical Sciences 19 (1977), pp.713-724.
[9] Predeleanu, M., On a boundary solution approach for the dynamic problem of thermo viscoelasticity theory, (International Conference on Numerical Methods in Thermal Problems, Swansea, 2nd-6th July 1979) in Levis R.W., Morgan K. and Zienkiewicz O.C. (eds) Numerical Methods in Heat Transfer (John Wiley, London, 1981).
[10] Predeleanu, M. and Fleurier, J., New developments in coupled thermoelastic analysis by B.E.M., in Connors J.J. and Brebbia C.A. (eds) Betech 86 Proc. of the 2nd Boundary Element Technology, M.I.T. Cambridge (U.S.A.) (C.M.L. Publications, Southampton 1986), pp.211-221.
[11] Fleurier, J. and Predeleanu, M., On the use of coupled fundamental solutions for thermoelastic problem, Engineering Analysis (1986), in print.
[12] Sladek, V., and Sladek, J., Boundary integral equation method in thermoelasticity. Part I : General analysis, Appl. Math. Modelling, 7 (1983), pp.241-253.
[13] Tanaka, M. and Tanaka, K., A boundary element approach to dynamic problems in coupled thermoelasticity, S.M. Archives 6 (1981) 4, pp.467-491.
[14] Christensen, R.M., and Naghdi, P.M., Linear non-isothermal viscoelastic solids, Acta Mech. 3 (1967) n°1, pp.1-12.
[15] Gurtin, M.E. and Sternberg, E., On the linear theory of viscoelasticity, Arch. Rational Mech. Anal. 11 (1962), n°4, pp.391-456.
[16] Eringen, C., Mechanics of Continua, (John Wiley, London, 1967).
[17] Ionescu-Casimir, V., Problem of linear coupled thermoelasticity Theorems of reciprocity for the dynamic problem of coupled thermoelasticity, Bull Acad. Solon. Sci. Ser. Sci. Tech. 9 (1964), n°12.
[18] Maizel, V.M., Thermal problem of the elasticity theory (Kiev, 1951).
[19] Predeleanu, M., On thermal stresses in visco-elastic bodies, Bull. Math. de la Soc. Sci. Math. Phys. de la R.P.R. 3 (1959), n°2, pp.223-228.
[20] Nowacki, W., Thermal stresses due to the action of heat sources in a visco elastic space, Arch. Mech. Stos. XI (1959), pp.111-125.
[21] Nowacki, W., Dynamic problems of thermoelasticity, (Noordhoff, Leyden, 1975).
[22] Stippes, M., Rizzo, F.J., A note on the body force integral of classical elastostatic, Z.A.M.P., 28 (1977), 339.
[23] Brebbia, C.A., Telles, J.C.F., and Wrobel, L.C., Boundary Element Techniques : Theory and Applications in Engineering, (Springer-Verlag, Berlin, 1984).
[24] Rizzo, F.J. and Shippy, D.J., An advanced boundary integral equation method for three-dimensional thermoelasticity, Int. J. Numer. Methods in Engineering 11 (1977), pp.1753-1768.

Thermomechanical Couplings in Solids
H.D. Bui and Q.S. Nguyen (Editors)
Elsevier Science Publishers B.V. (North-Holland)

THE TWO-DIMENSIONAL PROBLEM OF THERMOELASTICITY FOR AN INFINITE REGION BOUNDED BY A CIRCULAR CAVITY

T. Honein, G. Herrmann

Division of Applied Mechanics, Stanford University, Stanford, CA 94305, USA

Based on an extended circle theorem, originally established by Milne-Thomson, the temperature distribution in an infinite region bounded by a circular cavity is determined in terms of the distribution existing in the same region without the cavity. The stress distribution induced by the presence of the cavity in the thermal field can be determined using the formalism of Muskhelishvili. It is found that this stress distribution is universal, i.e. the same for arbitrary original temperature fields and depends essentially only on the magnitude of the heat flux vector which existed at the origin of the circular cavity. Both insulated and perfectly conducting cavity boundaries can be treated.

INTRODUCTION

We are concerned here with the problem of determining the temperature distribution in a two-dimensional infinite region containing a circular cavity in terms of the distribution which existed in the body before the cavity was introduced. It is assumed that a temperature field in the homogeneous body has been induced by sources of arbitrary strength and location, except that no sources are located within the area of the prospective cavity.

This problem can be solved immediately invoking the circle theorem of Milne-Thomson [1]. An extension of this theorem permits to treat not only insulated cavities (heat flux vector tangent to the boundary),but also perfectly conducting cavities (heat flux vector normal to the boundary)

Further, the stress distribution induced by the presence of the cavities is determined using the formulation presented by Muskhelishvili [2]. The results indicate that this stress distribution is universal and depends essentially only on the magnitude of the heat flux vector which existed at the center of the prospective cavity in the original, homogeneous body.

TEMPERATURE DISTRIBUTION

Let the temperature distribution $T = T(x,y)$ be given with reference to a cartesian system x,y. Since it is harmonic (except at singular points or lines), it can be represented as the real part of an analytic function

$$f(z) = T(x,y) + i\,S(x,y)$$

where $z = x + iy$. The heat flux vector q_i, being proportional to the temperature gradient $T_{,i}$, is tangent to the lines S = const.

In view of Milne-Thomson's circle theorem [1] established for potential fluid flow, the modified temperature distribution in the body with a cavity of radius a will be given by the analytic function F(z) where

$$F(z) = f(z) \pm \overline{f\left(\frac{a^2}{\bar{z}}\right)}$$

The bar designates complex conjugate quantities. The plus sign applies to an insulated cavity (S = const on the rim of the cavity) and the minus sign to a conducting cavity (T = const on the rim of the cavity). Thus F(z) solves for the temperature distribution in both cases.

STRESS DISTRIBUTION

To determine the stress distribution, we make use of developments presented by Muskhelishvili [2]. In his notation, let a displacement distribution u^*, v^* be defined by

$$u^*(x,y) + iv^*(x,y) = \int F(z)dz$$

which, for a multiply-connected region can be set

$$\int F(z)dz = z\,B \log z + \gamma \log z + \text{a holomorphic function}$$

The total displacement u,v can be represented as

$$u = u' + \frac{\nu u^*}{2(\lambda+\mu)} \quad ; \quad v = v' + \frac{\nu v^*}{2(\lambda+\mu)}$$

where u', v' are two new functions which suffer the following jumps after an integration along a contour surrounding the cavity

$$[u'] + i\,[v'] = -\frac{\pi i \nu}{\lambda+\mu}(Bz+\gamma)$$

Here λ,μ are Lamé constants and ν is a positive coefficient. As pointed out in [2], the problem has been reduced to that of a dislocation.

Since f(z) is analytic in the region of the prospective cavity, it can be expanded as

$$f(z) = \alpha_o + \alpha_1 z + \alpha_2 z^2 + \ldots + \alpha_n z^n$$

It follows

$$f(\frac{a^2}{\bar{z}}) = \alpha_o + \alpha_1 \frac{a^2}{\bar{z}} + \alpha_2 \frac{a^4}{\bar{z}^2} + \ldots + \alpha_n (\frac{a^2}{\bar{z}})^n$$

and

$$\overline{f(\frac{a^2}{\bar{z}})} = \bar{\alpha}_o + \bar{\alpha}_1 \frac{a^2}{z} + \bar{\alpha}_2 \frac{a^4}{z^2} + \ldots + \bar{\alpha}_n (\frac{a^2}{z})^n$$

Thus F(z) is

$$F(z) = \alpha_o + \alpha_1 z + \ldots + \alpha_n z^n$$

$$+ \alpha_o + \alpha_1 \frac{a^2}{z} + \ldots + \alpha_n (\frac{a^2}{z})^n$$

From this we find, with validity outside the circle

$$u^* + iv^* = \bar{\alpha}_1 a^2 \log z + \text{a holomorphic function}$$

which lead to the conclusion

$$B = 0 \ , \ \gamma = a^2\overline{\alpha_1}$$

and it follows

$$[u'] + i\,[v'] = - \frac{\P i\nu}{\lambda+\mu} a^2\overline{\alpha_1} = \frac{\P\nu a^2}{\lambda+\mu} (-\mathrm{Im}\alpha_1 - i\mathrm{Re}\alpha_1)$$

This can be simplified further by a suitable choice of axes of reference, e.g. such that $\mathrm{Re}\alpha_1 = 0$. This means

$$[u'] = - \frac{\P\nu a^2}{\lambda+\mu} \mathrm{Im}\alpha_1 \quad ; \ [v'] = 0$$

The stress distribution induced by such a displacement discontinuity is well known, see e.g. Timoshenko and Goodier [3].

Before writing it down, let us note that

$$\alpha_1 = f'\,(0) = \frac{\partial T}{\partial x}(0) + i\,\frac{\partial S}{\partial x}(0) = \frac{\partial T}{\partial x}(0) - i\,(\frac{\partial T}{\partial y}(0))$$

Thus $\mathrm{Im}\alpha_1 = -\frac{\partial T}{\partial y}(0)$.

Next we recall Fourier's law

$$q_i = -kT_{,i}$$

where q_i is the heat flux and k the coefficient of thermal conductivity. Since $\mathrm{Re}\alpha_1 = 0$ it follows that $\frac{\partial T}{\partial x}(0) = 0$ and hence $q_x = 0$. Thus $q_y = q$ and

$$\mathrm{Im}\alpha_1 = -\frac{\partial T}{\partial y}(0) = q(0)/k$$

It can be noted further that

$$[u'] = -\frac{\nu a^2}{\lambda+\mu}\mathrm{Im}\alpha_1 = 2\alpha a^2 q/k$$

Here $2\alpha = \nu(\lambda+\mu)$, with α being the coefficient of thermal expansion.

With these coefficients the stress distribution may be written as follows:

$$\sigma_{rr} = \frac{E\alpha a q}{2k}\left(\frac{a}{r} - \frac{a^3}{r^3}\right)\sin\theta$$

$$\sigma_{\theta\theta} = \frac{E\alpha a q}{2k}\left(\frac{a}{r} + \frac{a^3}{r^3}\right)\sin\theta$$

$$\sigma_{r\theta} = -\frac{E\alpha a q}{2k}\left(\frac{a}{r} - \frac{a^3}{r^3}\right)\cos\theta$$

where E is Young's modulus.

It is seen that this stress distribution is universal in the sense that it is independent of the details of the temperature distributions and is simply proportional to the magnitude of the heat flux vector, existing at the center of the cavity before it was introduced, i.e. existing in the original homogeneous body. The above expressions are valid for an insulated cavity. For a perfectly conducting cavity, the sign in each formula has merely to be changed.

ACKNOWLEDGEMENTS

This work was performed under the auspices of the U.S. Department of Energy through a contract with Stanford University. This support is gratefully ackknowledged.

REFERENCES

[1] Milne-Thomson, L.M., Theoretical Hydrodynamics, (Macmillan, London, 1968)
[2] Muskhelishvili, N.I., Some Basic Problems of the Mathematical Theory of Elasticity, (Noordhoff, Groningen, 1953)
[3] Timoshenko, S. and Goodier, J.N., Theory of Elasticity (McGraw-Hill, New York, 1970).

Thermomechanical Couplings in Solids
H.D. Bui and Q.S. Nguyen (Editors)
Elsevier Science Publishers B.V. (North-Holland)

AN ANALYSIS OF THERMOMECHANICAL FIELDS NEAR FAST-RUNNING CRACK-TIPS AND IN WELDING

S.N. ATLURI
Center for the Advancement of Computational Mechanics
GEORGIA INSTITUTE OF TECHNOLOGY
School of Civil Engineering
ATLANTA, GEORGIA 30332 - U.S.A.

Abstract :

The problem of temperature distribution in a solid due to a propagating heat source is of relevance in several seemingly distinct problems such as dynamic crack propagation in a rate-sensitive (Viscoplastic) solid, metal welding, etc. The analysis of temperature rise at the tip of a fast-moving crack, due to heat generated in the plastic zone near the crack-tip, and its influence on the mechanical fields near the crack-tip, are of interest in understanding the process of fracture. Likewise, an analysis of the temperature fields in the welded materials, due to the propagating heat source in multi-pass welding for instance, is of importance in estimating the residual stress fields in the welded materials.

In the first part of the paper, a detailed computational analysis (using a moving finite element procedure) is presented of the problem of fast-crack propagation, at very high strain-rates, in viscoplastic solids, in plane-strain. Detailed results are obtained for the variations of the stress, strain, and inelastic stress-work density fields near the propagating crack-tip, for various values of the speed of crack-propagation. Thus, the rate of heat generated near the crack-tip, and its spatial distribution near the propagating crack-tip, are ascertained for various crack velocities.

In the second part of the study the transient temperature field (as seen by a moving observer) due the propagating heat source, is analyzed by a moving mesh finite element procedure. The effects of temperature-dependent material properties, and of the loss of heat to the surrounding medium through convection and radiation, are studied. Situations under which the conditions in the heat source may be labeled as 'isothermal' or 'adiabatic' are explored. Estimates of temperature rise near fast-running crack-tips in viscoplastic solids are made. It is found that at realistic speeds of crack propagation in structural steel, the front of the propagating heat source remains cold, while higher temperatures persist behind the heat source even at distances of the order of the heat source. Moreover, the temperature gradients themselves within the heat source are very intense. A coupled thermo-viscoplastic analysis is suggested for assessing the effects of high temperatures prevailing in the wake of a fast running crack-tip on the fracture process itself.

Thermomechanical Couplings in Solids
H.D. Bui and Q.S. Nguyen (Editors)
Elsevier Science Publishers B.V. (North-Holland)

THERMOMECHANICAL COUPLING IN FRACTURE MECHANICS

H.D. BUI, A. EHRLACHER, Q.S. NGUYEN

LABORATOIRE DE MECANIQUE DES SOLIDES
ECOLE POLYTECHNIQUE
91120 PALAISEAU - FRANCE

The thermal effect due to thermomechanical coupling at the tip of a moving crack is investigated. The problem of plane crack extension in dissipative continua is studied within the framework of classical thermodynamics. The consequences of the first and second principles are derived, in particular it is shown that the crack-tip behaves like a moving heat source. The temperature singularity is discussed. In elastodynamics, the asymptotic expansion of the coupled solution is given up to second order terms in linear elasticity and linear conduction. In plasticity, the mechanical and thermal response is simulated by numerical analysis for steady state motion and mode I loading. The obtained results are compared to experimental observations by infrared-thermography.

1. INTRODUCTION

Thermomechanical coupling in fracture mechanics is discussed here for a linear crack, propagating in an arbitrary dissipative medium. Several collective results, obtained in the Laboratoire de Mécanique des Solides by different theoretical, numerical and experimental investigations are presented.

At the crack-tip, the thermomechanical coupling is a priori not negligeable because of stress and strain singularities. For a propagating crack, one can observe experimentally the heat generated during crack propagation by measuring the temperature elevation. Weishert and Schönert [1] have detected 130°C., measured with a minute thermocouple at a distance of 30μm of a crack propagating at 10 m/s in steel. Fuller, Fox and Field [2] have suggested a temperature elevation of about 500°C. during fracture of polymers by the registered infrared radiation. Light emission produced by fast cracks in glass has been also interpreted as thermal radiation with a temperature higher than 1000°C. [1].

In order to understand the nature of this coupling, a detailed analysis has been performed is our Laboratory and the results obtained are presented below.

In the first part, a thermodynamical description of the running crack problem is given for arbitrary media in the framework of classical thermodynamics and continuum mechanics. Thermo-mechanical equations of evolution are derived : in particular additional terms due the presence of a crack-tip heat source are underlined.

In the second part, the temperature distribution near the crack is discussed as a function of the material characteristics and the conduction law. The nature of the theoretical singularity of the tempeature is given when a crack-tip heat source exists. In particular, the asymptotic expansion of the thermo-mechanical response has been obtained in linear elastodynamics and linear conduction. The thermomechanical coupling is however much more complex in the case of dissipative materials. In plasticity for example, the asymptotic response cannot be obtained explicitly because of the complex nature of the constitutive equations. In this case however, a numerical analysis has been performed. The mechanical and thermal responses have been simulated in the case of steady state motion and mode I loading.

In the third part, experimental results obtained by infrared-thermography are presented for small propagating velocity ($\dot{\ell} \sim 10$ mm/s).

2. THERMODYNAMICAL DESCRIPTION OF THE RUNNING CRACK PROBLEM

Fracture process is an energy consuming phenomena. A solid containing a propagating crack represents a system in irreversible process and the dissipation is due principally to different dissipative mechanisms such as internal friction, plastic deformation etc... The thermomechanical evolution of such a system can be described in the classical framework of continuum thermodynamics.

Let us recall first that for any material system the first and second principles of thermodynamics can be expressed as :

$$\left.\begin{aligned} &\dot{E} + \dot{C} = P_e + P_c && \text{(Energy balance)} \\ &\dot{S} + \int_{\partial V} \frac{q \cdot n}{T}\, da \geqslant 0 && \text{(Entropy production)} \end{aligned}\right\} \qquad (1)$$

where E, C, P_e, P_c, S, q denote respectively internal energy, kinetic energy, external mechanical power, received calorific power, entropy and heat flux.

For clarity, a bidimensional problem of linear crack propagation is considered (plane strain, plane stress or anti-plane shear), but the material is arbitrary. The system is also assumed to be *thermally isolated*. As shown in fig. 1, applied force $\underset{\sim}{F}(t)$ and displacement $\underset{\sim}{u}$ (t) are given on complementary parts $\partial\Omega_u$, $\partial\Omega_F$ of the boundary. On the crack surface Σ , an unilateral condition of non interpretation *without friction* is also assumed. Because of the presence of a possible moving singularity of thermomechanical fields near the crack-tip A, the thermodynamical description is not a classical one. Additional terms due to the singularity may play a fundamental role.

The *consequence of the energy balance* can be established in the following way :

If e, T, S denote respectively specific internal energy, temperature and specific entropy, the energy balance of the whole system can be written as :

$$\frac{d}{dt}\int_{\Omega} \rho\,(e + \frac{1}{2}\,v^2)\;d\Omega = \int_{\partial\Omega} n\,.\,\sigma\,.\,v\;da + P_c \qquad (2)\ .$$

Although the right-hand side of *(2)* is clear, the explicit expression of the left-hand side is not straightforward since the function $\rho(e + \frac{1}{2}\,v^2)^{\bullet}$ is not necessarily an integrable function because of the moving singularity. It is necessary to isolate the crack-tip with a closed curve Γ in *translation motion* with the crack which delimits a domain V_Γ. It follows that :

$$\frac{d}{dt}\int_{\Omega} \rho(e + \frac{1}{2}\,v^2)\;d\Omega = \frac{d}{dt}\int_{\Omega_\Gamma} -\,i\,d\,-\,d\Omega + \frac{d}{dt}\,-\int_{V_\Gamma} -\,i\,d\,-\,\Omega$$

In the moving reference A X Y, let us introduce the time derivative $\ast$, for example $\overset{*}{e} = e_{,t}(X,Y,t)$, which is related to the material derivative $\cdot$ by the relation :

$$\overset{*}{e} - e_{,1}\,.\,\dot{\ell} = \dot{e} \qquad . \qquad (3)$$

The *transport of singularity principle* consists in the conservation of the nature of the singularity with the crack motion. While e may be singular, $\overset{*}{e}$ is much more regular and $\dot{e} \sim -\,\dot{\ell}\,e_{,1}$. As a consequence of this regularity, one obtains :

$$\frac{d}{dt}\int_{V_\Gamma} \rho\,(e + \frac{1}{2}\,v^2)\;d\Omega = \int_{V_\Gamma} \rho\,(\overset{*}{e} + v\,\overset{*}{v})\;d\Omega$$

and thus for a vanishing curve $\Gamma \to 0$:

$$\lim_{\Gamma \to 0}\ \frac{d}{dt}\int_{V_\Gamma} \rho\,(e + \frac{1}{2}\,v^2)\;d\Omega = 0 \qquad . \qquad (4)$$

Since

$$\frac{d}{dt}\int_{\Omega_\Gamma} \rho(e + \frac{1}{2}\,v^2)\;d\Omega = \int_{\Omega_\Gamma} \rho(\dot{e} + v\,\dot{v})\;d\Omega - \int_{\Gamma} \rho(e + \frac{1}{2}\,v^2)\,\dot{\ell}\,n_1\,d\Gamma$$

one obtains finally from the energy balance of the whole systèm :

$$\lim_{\Gamma \to 0}\left\{\int_{\partial\Omega} n.\,\sigma\,.\,v\;d\Omega - \int_{\Omega_\Gamma} \rho(\dot{e} + v\,\dot{v})\;d\Omega + \int_{\Gamma} \rho(e + \frac{1}{2}\,v^2)\,\dot{\ell}\,n_1\;d\Gamma\right\} = 0 \quad (5)\ .$$

To point out the energy exchanges near the crack-tip, one may compare *(5)* to the energy balance of the system of material points occupying Ω_Γ at time t. This energy balance is :

$$\int_{\Omega_\Gamma} \rho(\dot{e} + v\,\dot{v})\, d\Omega = \int_{\partial\Omega} n.\sigma.v\, da - \int_\Gamma n \cdot \sigma \cdot v\, d\Gamma + \int_\Gamma q \cdot n\, d\Gamma \qquad (6)$$

and thus the following expression holds :

$$\lim_{\Gamma \to 0} \left\{ \int_\Gamma q \cdot n\, d\Gamma - \int_\Gamma \left[\rho(e + \tfrac{1}{2} v^2)\, \dot{\ell}\, n_1 + n \cdot \sigma \cdot v\right] d\Gamma \right\} = 0 \qquad (7)\ .$$

The non necessarily zero quantity H defined by :

$$H = \lim_{\Gamma \to 0} \int_\Gamma q \cdot n\, d\Gamma \qquad (8)$$

represents physically a concentrated *heat source* at the crack-tip. From *(7)* its expression is given by :

$$H = \lim_{\Gamma \to 0} \int_\Gamma \left[\rho(e + \tfrac{1}{2} v^2)\, \dot{\ell}\, n_1 + n \cdot \sigma \cdot v\right] d\Gamma \qquad (9)\ .$$

The *consequence of the second principle* can also be derived in the same spirit.

The entropy production of the whole system reduces to $S = \frac{d}{dt} \int_\Omega \rho\, s\, d\Omega$ and can be written as :

$$S = \lim_{\Gamma \to 0} \left\{ \int_{\Omega_\Gamma} \rho\, \dot{s}\, d\Omega - \int_\Gamma \rho\, \dot{\ell}\, n_1\, d\Gamma \right\} \geqslant 0 \qquad (10)\ .$$

To identify the entropy production at the crack-tip S_A, one may compare to the entropy production of the system of material points occupying Ω_Γ at time t :

$$S_\Gamma = \int_{\Omega_\Gamma} \rho\, \dot{s}\, d\Omega - \int_\Gamma \frac{q \cdot n}{T}\, d\Gamma \geqslant 0$$

which yields finally :

$$S_A = \lim_{\Gamma \to 0} \int_\Gamma \frac{1}{T} (q \cdot n - \rho\, T\, s\, \dot{\ell}\, n_1)\, d\Gamma \geqslant 0 \qquad (11)\ .$$

If $T \to T_A < +\infty$ when $x \to A$, it follows from *(8)*, *(9)*, *(11)* that :

$$T_A \cdot S_A = \lim_{\Gamma \to 0} \int_\Gamma \left[\rho(w + \tfrac{1}{2} v^2)\, \dot{\ell}\, n_1 + n \cdot \sigma \cdot v\right] d\Gamma \geqslant 0 \qquad (12)$$

where $w = e - Ts$ denotes the specific free energy density, $\mathcal{D} = T_A \cdot S_A$ is the crack-tip dissipation.

If $\dot{\ell} > 0$, since $v = \dot{u} \sim -\dot{\ell}\, u_{,1}$, one should remark that :

$$\left.\begin{aligned} &H = \dot{\ell} \lim_{\Gamma\to 0} \tilde{J}_\Gamma && \text{with} \quad \tilde{J}_\Gamma = \int_\Gamma (\rho(e + \tfrac{1}{2} v^2)\, n_1 - n\cdot\sigma\cdot u_{,1})d\Gamma \\ &T_A\, S_A = \dot{\ell} \lim_{\Gamma\to 0} J_\Gamma && \quad\quad\quad\; J_\Gamma = \int_\Gamma (\rho(w + \tfrac{1}{2} v^2)\, n_1 - n\cdot\sigma\cdot u_{,1})d\Gamma \end{aligned}\right\} \quad (13)\ .$$

In most applications, $T_A < +\infty$ and $s \ll r^{-1}$; the fact that $\lim_{\Gamma\to 0} \int_\Gamma \rho\, T\, s\, n_1\, d\Gamma = 0$ implies :

$$H = T_A \cdot S_A = \dot{\ell} \cdot \lim_{\Gamma \to 0} J_\Gamma = \dot{\ell}\, J_0 \geqslant 0 \quad (14)\ .$$

Thus the heat source is also the crack-tip dissipation and must be a hot one.

In fact, the assumption of continuity and regularity everywhere except at the crack-tip is tacitly admitted in the preceding analysis. If shock waves are present under the form of a surface of discontinuity S propagating in translation with the crack, classical equations of jump on discontinuity surface lead to the introduction of a surface heat source and surface entropy production on S. In particular, the contribution of a surface of discontinuity in translation with the crack is :

$$S_S = \dot{\ell} \lim_{\Gamma \to S} \int_\Gamma \frac{1}{T} \left[\rho(w + \tfrac{1}{2} v^2)\, n_1 - n\cdot\sigma\cdot u_{,1}\right] d\Gamma \quad (15)$$

in terms of entropy production.

In conclusion, the crack-tip singularity leads to the introduction of concentrated heat source H in addition to the more classical notions of volumic heat source $\sigma\,\dot{\varepsilon} - \rho\,\dot{e}$ and surface heat sources localized on surface of discontinuity.

The thermomechanical evolution of the solid is described by the mechanical equations (equilibrium, constitutive equations) and the thermal equations which express simply the different source conditions. These thermal equations are :

$$\left.\begin{aligned} &.\ \text{Local equations} &&: \forall x \in \Omega \qquad \text{Div}\, q = \sigma\,\dot{\varepsilon} - \rho\,\dot{e} \\ &.\ \text{Boundary conditions} &&: \text{Classical on } \partial\Omega \\ & && \quad \text{Non-classical at the crack-tip A} \\ & && \quad \lim_{\Gamma\to 0} \int_\Gamma q\cdot n\, d\Gamma = H \end{aligned}\right\} \quad (16)\ .$$

3 - TEMPERATURE ANALYSIS

There is thermomechanical coupling at different levels : local coupling at a regular point by energy balance div q = $\sigma \dot{\varepsilon} - \rho \dot{e}$, crack-tip coupling by *(8)*, *(9)* besides the fact that constitutive equations also involve temperature.

The most interesting question is naturally the asymptotic behaviour of the thermo-mechanical response. It is clear for most materials that the coupled equations are so complex that closed form solutions can not be explicitly obtained. However, partial results can be established in certain situations :

. Let us assume that thermal conduction obeys Fourier's law :

$$q = -k \cdot \nabla T \tag{17}$$

or in a more general manner :

$$q = -k \mid \nabla T \mid^{m-2} . \nabla T \quad , \quad m \geqslant 2 \tag{18}$$

Linear conduction *(17)* is a special case of the generalized Fourier law *(18)*, when m = 2. One can then establish the following :

. *Proposition* :

If H > 0, then the temperature field admits the asymptotic expansion :

$$\left.\begin{array}{lll} m > 2 & T = T_a - \dfrac{H}{2\pi k} r^{\alpha} \quad , & \alpha = \dfrac{m-2}{m-1} \\ m = 2 & T = -\dfrac{H}{2\pi k} \operatorname{Log} r & \end{array}\right\} \tag{19}$$

Indeed, in *(16)* the second member $\sigma \dot{\varepsilon} - \rho \dot{e}$ represents volumic heat source. Integrable conditions for volumic heat source imply that its singularity must be less than r^{-2}. The assumption H > 0 implies that $q \sim r^{-1}$ thus Div $q \sim r^{-2}$ and hence thermal equation *(16)*, asymptotically, reduces to :

$$\operatorname{div} q = 0 \tag{20}$$

Equations *(8)*, *(20)* and the boundary condition on the crack surface lead to the simple system :

$$\left.\begin{array}{l} \operatorname{Div} q = 0 \\ \dfrac{\partial T}{\partial \theta}(\pm\pi) = 0 \\ \lim\limits_{r \to 0} \displaystyle\int_{\pi}^{\pi} k \mid \Delta T \mid^{m-2} \dfrac{\partial T}{\partial r} \, 2\pi r \, d\theta = H \end{array}\right\} \tag{21}$$

If H = 0, i.e. if there is no concentrated heat source at the crack-tip, then the discussion becomes much more complex. Since $\sigma \dot{\varepsilon} - \rho \dot{e}$ may have the same singularity as Div q, the simplified equation *(20)* is not valid asymptotically.

4 - THERMOELASTICITY

This section is devoted to a detailed analysis by an analytic method of the asymptotic behaviour in the special case of linear-thermoelasticity and linear conduction [3].

If the material is elastic, the free energy density $W(\varepsilon,T)$ has the following expression :

$$\left.\begin{aligned} \rho\, W &= \frac{1}{2}\,(\lambda\, \varepsilon_{ii}\, \varepsilon_{kk} + 2\,\mu\, \varepsilon_{ij}\, \varepsilon_{ij}) - 3\,k\,\alpha(T - T_0)\, \varepsilon_{kk} \\ &\quad - c\,T\,\mathrm{Log}\left(\frac{T}{T_0}\right) - (S_0 - c)\,T + W_0 \end{aligned}\right\} \qquad (22)$$

and the constitutive equations are :

$$\left.\begin{aligned} s &= -\frac{\partial W}{\partial T} \;\Rightarrow\; \rho\, s = 3\,K\,\alpha\, \varepsilon_{kk} + c\,\mathrm{Log}\,\frac{T}{T_0} + s \\ \sigma &= \rho\,\frac{\partial W}{\partial \varepsilon} \;\Rightarrow\; \sigma_{ij} = \lambda\, \varepsilon_{ii}\, \delta_{ij} + 2\,\mu\, \varepsilon_{ij} - 3\,K\,\alpha\left(\frac{T}{T_0}\right)\delta_{ij} \end{aligned}\right\} \qquad (23)$$

where $K = \frac{1}{3}\,(3\,\lambda + 2\,\mu)$ denote the bulk modulus.

The thermomechanical equations of evolution can now be written explicitly. One obtains for example in plane strain, when linear conduction *(17)* is assumed :

$$\left.\begin{aligned} & k\,\Delta T - c\,\dot{T} - 3\,K\,\alpha\,T\,(\dot{\varepsilon}_{11} + \dot{\varepsilon}_{22}) = 0 \\ & (\lambda + \mu)\,\nabla\,(\varepsilon_{11} + \varepsilon_{22}) + \mu\,\Delta u - 3\,K\,\alpha\,\nabla T - \rho\,\ddot{u} + f = 0 \end{aligned}\right\} \qquad 24)$$

where f denotes volumic force. Initial and boundary conditions must be prescribed to complete local and crack-tip source conditions *(16)*. We can, for example, assume a traction free crack surface $\sigma \cdot n = 0$ as the mechanical condition and temperature continuity or absence of heat exchange on the crack boundary as thermal condition.

The search for the asymptotic behaviour of the thermomechanical response can be formulated in the following way : displacement u and temperature T are developed as a series of decreasing singularities :

$$\left.\begin{aligned} u &= \overset{(1)}{u} + \overset{(2)}{u} + \ldots \\ T &= \overset{(1)}{T} + \overset{(2)}{T} + \ldots \end{aligned}\right\} \qquad (25)$$

where $\overset{(n)}{u_i}$ and $\overset{(n)}{T}$ belong to the set of elementary functions $r^{\beta}\,(\mathrm{Log}\,r)^{\beta_1}\,\ldots\,(\mathrm{Log}_m\,r)^{\beta m}\,f(\theta,t)$, where β_1, β_2, ... are real numbers and $\mathrm{Log}_m\,r = \mathrm{Log}\,(\mathrm{Log}_{m-1}\,r)$; $f(\theta,t)$, representing the angular distribution, is assumed to be regular on the interval $]-\pi, \pi[$ with respect to θ.

The time derivative in *(24)* can be asymptotically evaluated since one obtains for any physical quantity g :

$$\dot{g} = - \dot{\ell}\, g_{,1} \quad + \text{ more regular terms}$$

The development *(25)* is now derived step by step from equations *(17)*, *(24)* and appropriate boundary conditions on the crack surface.

It is assumed first that $H \neq 0$ and thus from the previous discussion, one obtains :

$$\overset{(1)}{T} = - \frac{H}{2\pi k} \operatorname{Log} r \qquad (27)\ .$$

But the assumption $H \neq 0$ also implies from *(8)*, *(9)* and *(22)* that $\overset{(1)}{\varepsilon}$ is more singular than $\overset{(1)}{T}$. Under these conditions $\overset{(1)}{u}$ verifies necessarily the local equation :

$$(\lambda + \pi)\, \nabla \left(\overset{(1)}{\varepsilon}_{11} + \overset{(1)}{\varepsilon}_{22}\right) + \pi\, \Delta\overset{(1)}{u} - \rho\, \dot{\ell}^2\, \overset{(1)}{u}_{,11} = 0 \qquad (28)\ ,$$

as it follows from *(24)*. Equation *(28)* is not new since it represents exactly the classical form studied in isothermal elastodynamics by Yoffé [4] . By analytical method, it is well-known that the solution $\overset{(1)}{u}$ is a linear combination of elementary functions $K_i(t)\, g(\dot{\ell})\, r^{1/2}\, h_{ij}\, (\theta,\dot{\ell})$ which effectively leads to a value $H \neq 0$. The reader may refer to [3] for a more complete presentation of our results.

From the expression obtained for $\overset{(1)}{T}$ and $\overset{(1)}{u}$, equations *(24)* show that $\overset{(2)}{T}$ must verify :

$$k\, \Delta\overset{(2)}{T} + 3\, K\, \alpha\, \overset{(1)}{T}\, \dot{\ell}\left(\overset{(1)}{\varepsilon}_{11} + \overset{(1)}{\varepsilon}_{22}\right)_{,1} = 0 \qquad (29)$$

and thus one obtains :

$$\overset{(2)}{T} = \overline{T}(t) + r^{1/2} \operatorname{Log} r\, f(\theta,\dot{\ell}) \qquad (30)$$

with an explicit expression of $f(\theta,\dot{\ell})$ as given in the reference [3].

The angular distribution $f(\theta,\dot{\ell})$ can be easily seen from isothermal curves $\overset{(1)}{T} + \overset{(2)}{T} =$ Const.. Figure 2 represents isothermal curves $\overset{(1)}{T} + \overset{(2)}{T}$ up to a constant $\overline{T}(t)$, for a common steel (with $\lambda = \mu = 0.8.10^{11}$ N m^{-2}, $k = 42$ J.m^{-1} s^{-1}; $\alpha = 1.5.10^{-5}$ and $K_c = 10^7$N.m$^{3/2}$) under two different situations : mode I or mode II loading with perfectly isolated boundary condition as a function of the crack velocity $\dot{\ell}$.

5 - THERMO-PLASTICITY

In the context of incremental plasticity and ductile fracture, the *purely mechanical description* of the propagation of a crack has been intensively discussed. Rice has suggested, since 1968, the conjecture that a crack-tip energetic parameter does not exist when a crack propagates. This conjecture was illustrated later by the works of Chitaley and Mc Clintock [5], Slepyan [6], Rice, Drugan and Sham [7] concerning the asymptotic expansion of the mechanical solution of a growing crack in mode III and I respectively. The strain field is of course singular near the crack, but its singularity is not sufficient to contribute to energy balance as in classical elasticity.

A complete asymptotic solution of the coupled solution in perfect plasticity is still an open problem. Our objective here is to discuss by means of a numerical analysis the thermomechanical coupling under the conjecture $J_0 = 0$ in perfect plasticity. We know already that the conjecture is fully justified in mode III, because in this case the mechanical solution does not depend on the temperature distribution and thus Chitaley and Mc Clintock's result, still available, leads to the estimate $\lim_{\Gamma \to 0} J_0 = 0$. By nature, numerical analysis cannot give exactly the degree of singularity of mechanical and thermal fields. However, it illustrates the transition between the asymptotic and the far fields. We also wish to know the dimensions of different zones of interest from the thermal or mechanical point of view, and compare the results obtained with experiments performed approximately under the same condition.

The problem of the steady-state quasi-static propagation of a linear semi-infinite crack in an infinite strip has been considered [8] :

For the sake of clarity, the material is assumed to be elastic, perfectly plastic, following Mises criterion

$$f(\sigma) \equiv \frac{1}{2} \sigma'_{ij} \sigma'_{ij} - \frac{\sigma_y^2}{3} \leqslant 0 \quad .$$

It is also homogeneous with constant, isotropic thermoelastic coefficients and thus the associated energy density is :

$$\left.\begin{array}{l} \rho W = \frac{1}{2}\Big[\left(\varepsilon_{ii} - \varepsilon^p_{kk} - 3\,\alpha(T - T_0)\right)^2 + 2\,\mu\left(\varepsilon_{ij} - \varepsilon^p_{ij} - \alpha(T - T_0)\,\delta_{ij}\right) \cdot \\ \qquad \left(\varepsilon_{ij} - \varepsilon^p_{ij} - \alpha(T - T_0)\,\delta_{ij}\right)\Big] - c\,T\,\mathrm{Log}\,\frac{T}{T_0} - (s_0 - c)(T - T_0) + W_0 \end{array}\right\} \quad (31) \ .$$

The infinite strip is submitted to a mode I loading defined by a uniform pressure q applied on the portions $c \leqslant X < +\infty$ of the boundary. A X Y denotes the moving reference with the crack-tip A, in translation with constant velocity $\dot{\ell}$.

The assumption of a steady state motion enables us to express every physical quantity g as a function of relative coordinates $g = g(X,Y)$ with material time derivative $\dot{g} = -\dot{\ell} g_{,X}$.

In particular, the Prandtl-Reuss law for the plastic strain can be written as :

$$-\dot{\ell}\,\varepsilon^p_{,X} = \lambda \frac{\partial f}{\partial \sigma} \qquad \left.\begin{array}{l} \lambda \geqslant 0 \ \text{ if } \ f(\sigma) = 0 \\ \lambda = 0 \ \text{ if } \ f(\sigma) < 0 \end{array}\right\} \qquad (32) .$$

If a linear Fourier law of conduction is assumed, $q = -k\,\nabla T$, equation *(16)* leads to the associated thermal equation :

$$-k\,\Delta T + \rho T\,\dot{s} - D = 0 \qquad (33) .$$

No concentrated heat source $J_0\,\dot{\ell}$ will be considered ($J_0 = 0$) at the crack-tip A. Although surface heat sources eventually may exist on possible surfaces of discontinuity of $\dot{u}$ (i.e. $T\,\dot{s}$ and D are not necessarily volumic regular functions), the absence of a concentrated heat source at the crack-tip ensures automatically, for any solution of *(15)* :

$$\lim_{\Gamma \to 0} \int_\Gamma q \cdot n \, d\Gamma = 0 .$$

The thermomechanical response $u_i(X,Y)$, $T(X,Y)$ ... of the cracked strip is then the solution of the following boundary value problem :

Find $u_i(X,Y)$, $\sigma(X,Y)$, $T(X,Y)$ ... verifying :

$$\left.\begin{array}{lll}
\varepsilon_{ij} = \frac{1}{2}(u_{i,j} + u_{j,i}) & & \text{Compatibility} \\
\frac{\sigma}{\rho} = \frac{\partial W}{\partial \varepsilon}(\varepsilon, \varepsilon^p, T) & & \\
-\dot{\ell}\,\varepsilon^p_{,X} = \lambda \frac{\partial f}{\partial \sigma} , & \begin{array}{l} \lambda \geqslant 0 \ \text{ if } \ f(\sigma) = 0 \\ \lambda = 0 \ \text{ if } \ f(\sigma) < 0 \end{array} & \text{Elasto-plastic Constitutive Equations} \\
\operatorname{div} \sigma = 0 & & \text{Equilibrium} \\
k\,\Delta T + c\,\dot{\ell}\,T_{,X} + \alpha\,T\,\dot{\ell}(\sigma_{kk})_{,X} - \dot{\ell}\,\sigma\,\varepsilon^p_{,X} = 0 & & \text{Thermal equation}
\end{array}\right\} \qquad (34)$$

with the following additional conditions :

$$\left.\begin{array}{lll}
. \; u_x(0,0) = 0 \quad , \quad u_y(0,0 = 0 & & \\
. \text{ If } y = 0 \text{ and } X \geqslant 0 & : & u_y = 0 \; , \; \frac{\partial T}{\partial n} = 0 \quad , \quad \sigma^{xy} = 0 \\
. \text{ If } X < c \text{ and } Y = \pm h & : & \sigma^{xy} = 0 \; , \; \sigma^{yy} = 0 \; , \; \frac{\partial T}{\partial n} = 0 \\
. \text{ If } X \geqslant c \text{ and } Y = \pm h & : & \sigma^{xy} = 0 \; , \; \sigma^{yy} = q \; , \; \frac{\partial T}{\partial n} = 0 \\
. \text{ If } Y = 0 \text{ and } X < 0 & : & \sigma^{xy} = 0 \; , \; \sigma^{yy} = 0 \; , \; \frac{\partial T}{\partial n} = 0 \\
. \text{ If } X \to \pm \infty & : & \frac{\partial T}{\partial n} \to 0 \; , \; \sigma^{xx} \text{ and } \sigma^{xy} \to 0 \quad , \; \sigma^{yy} \to q \\
. \text{ If } X \to + \infty & : & \varepsilon^p_{ij} \to 0
\end{array}\right\} \quad (35) .$$

The resolution of non-linear equations *(34)* and *(35)* is obtained by iterations following two principal steps :

. If the temperature distribution is known, mechanical equations in *(34)* and *(35)* correspond to a problem of non-linear, non-local elasticity which can be resolved by internal iterations. At each internal iteration, the plastic strain distribution is re-computed from *(32)* and the infinite condition $\varepsilon^p_{ij} \to 0$ when $X \to +\infty$.

. If the mechanical distribution is known, thermal equations in *(34)* and *(35)* correspond to a diffusion-convection problem $\Delta T + \frac{c\dot{\ell}}{k} T_{,x} = f$.

The two steps are alternatively computed from a given initial temperature field until convergence of the iterative process.

The characteristics of the material considered are :

$E = 20.000$ kg/mm^2 , $\nu = 0.3$, $\sigma_y = 122.47$ kg/mm^2 , $k = 0.042$ J/mm.s.d°C ,
$\alpha = 1.5\ 10^{-5}$, $\rho = 7.8$ g/cm^3 , $c = 0.0035$ J/mm^3.d°C .

The numerical solution is presented in figure 3 in plane stress and in plane strain for $\dot{\ell} = 10$mm/s. It can be noted that again, plastic reloading of the free boundary of crack is observed as in the pure mechanical situation [5] [6], [7]. Cold temperature is due to thermo-elastic coupling beyond the crack and plastic dissipation is responsible for a maximum temperature elevation of 4°C. in plane strain, 9°C. in plane stress, in a small zone of about 1mm around the crack-tip.

For higher velocity of propagation, the influence of conduction is weaker with respect to convection at a given distance to crack. In particular one can obtain practically the thermal response with the same mesh at $\dot{\ell}$ = 100 mm/s by convection terms only. To obtain the influence of thermal conduction, a higher space resolution is needed for $\dot{\ell} > 100$ mm/s.

5 - EXPERIMENTAL OBSERVATION BY THERMOGRAPHY

It is interesting to compared these numerical results with Bui-Ehrlacher-Nguyen's thermography experiments [9]. Figure 4 represents the temperature distribution obtained by an infrared camera in a cracked thin plate at low rate of propagation $\dot{\ell}$ = 2 mm/s. In this thermogram, the temperature variation is 2 degrees per colour.

The temperature elevation is about 10°C. at the crack-tip and seems to be stronger than the computed one, at the same rate of propagation, even with the assumption of a thermically isolated system.

This difference can be attributed firstly to the fact that the propagation is not in a steady regime and the actual boundary conditions of the tested plate correspond to a stronger plastification. Secondly, one can observe a difference between the geometrically defined crack-tip and the hotest region. This difference is due to the fact that the dissipation is stronger beyond the crack. Two factors contribute to this phenomenon : a certain striction, i.e. localization, transversally observed along the crack prolongation, and a certain softening since the material is not strictly perfectly plastic near the crack-tip.

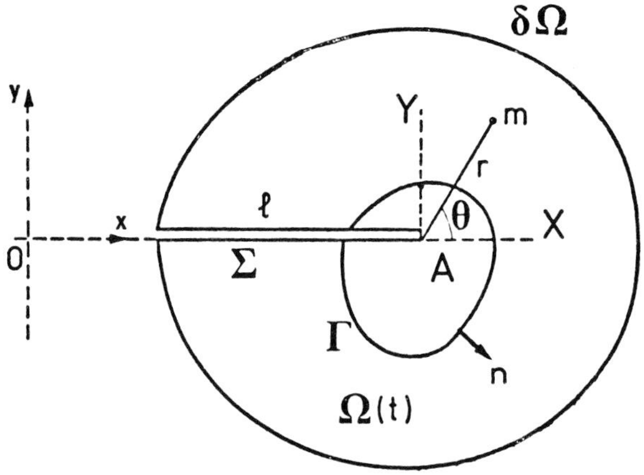

Figure 1 :
Linear crack with moving reference.

REFERENCES

[1] Weichert, R., and Schönert, K., "On the temperature rise at the tip of a fast running crack", J. Mech. Phys. Solids, 22, (1974), pp. 127-133.

[2] Fuller, K.N.G., Fox, R.G., and Field, J.E., "The temperature rise at the tip of fast moving cracks in glassy polymers", Proc. R. Soc., A-231, (1975), pp. 537-551.

[3] Bui, H.D., Ehrlacher, A., and Nguyen, Q.S., "Propagation de fissure en thermoélasticité dynamique", Journal de Mécanique, vol. 19, (1980), p.697.

[4] Yoffé, E.H., "The moving Griffith crack", Phil. Mag. 42, (1951), pp. 739-750.

[5] Chitaley, A.D. and Mc Clintock, F.A., "Elastic-plastic mechanics of steady crack growth under anti-plane shear", J. Mech. Phys. Solids, 19, (1971), pp. 147-163.

[6] Slepyan, L.I., "Growing crack during plane deformation of an elastic-plastic body", Mech. Tve. Tela, 9, (1974), pp. 57-67.

[7] Rice, J.R., Drugan, W.J. and Sham, T.L., "Elastic-plastic analysis of growing cracks", Fracture Mechanics, ASTM-STP, 700, (1980), pp. 189-219.

[8] Mézière, Y. and Nguyen, Q.S., "A numerical analysis of the thermomechanical coupling at the tip of a moving crack", Num. Meth. Fracture Mechanics, (1984), pp. 505-519.

[9] Bui, H.D., Ehrlacher, A. and Nguyen, Q.S., "Etude expérimentale de la dissipation dans la propagation de fissure par thermographie infrarouge", C.R. Acad. Sci., tome 293, (1981), pp. 71-75.

Figure 2 : Theoretical isothermal curves in thermoelasticity [3].

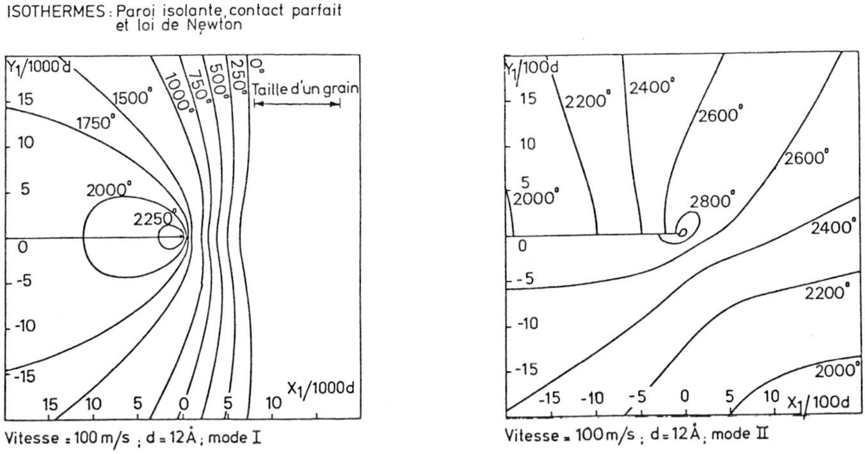

Figure 3 : Numerical isothermal and isomises curves in Thermoplasticity [8].
(Plane stress and plane strain).

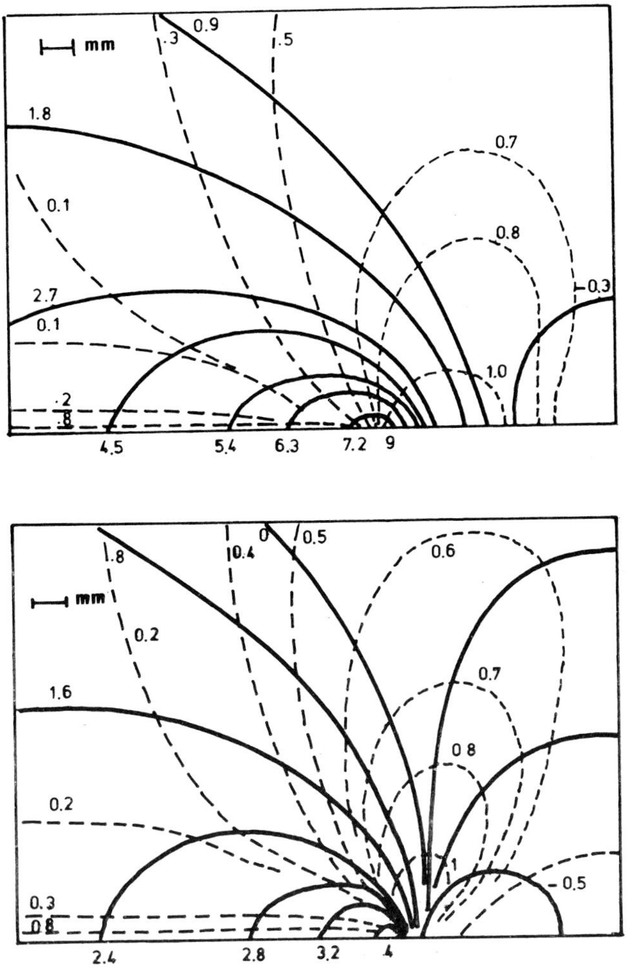

Figure 4 : Experimental observations by Infrared Thermography [9].

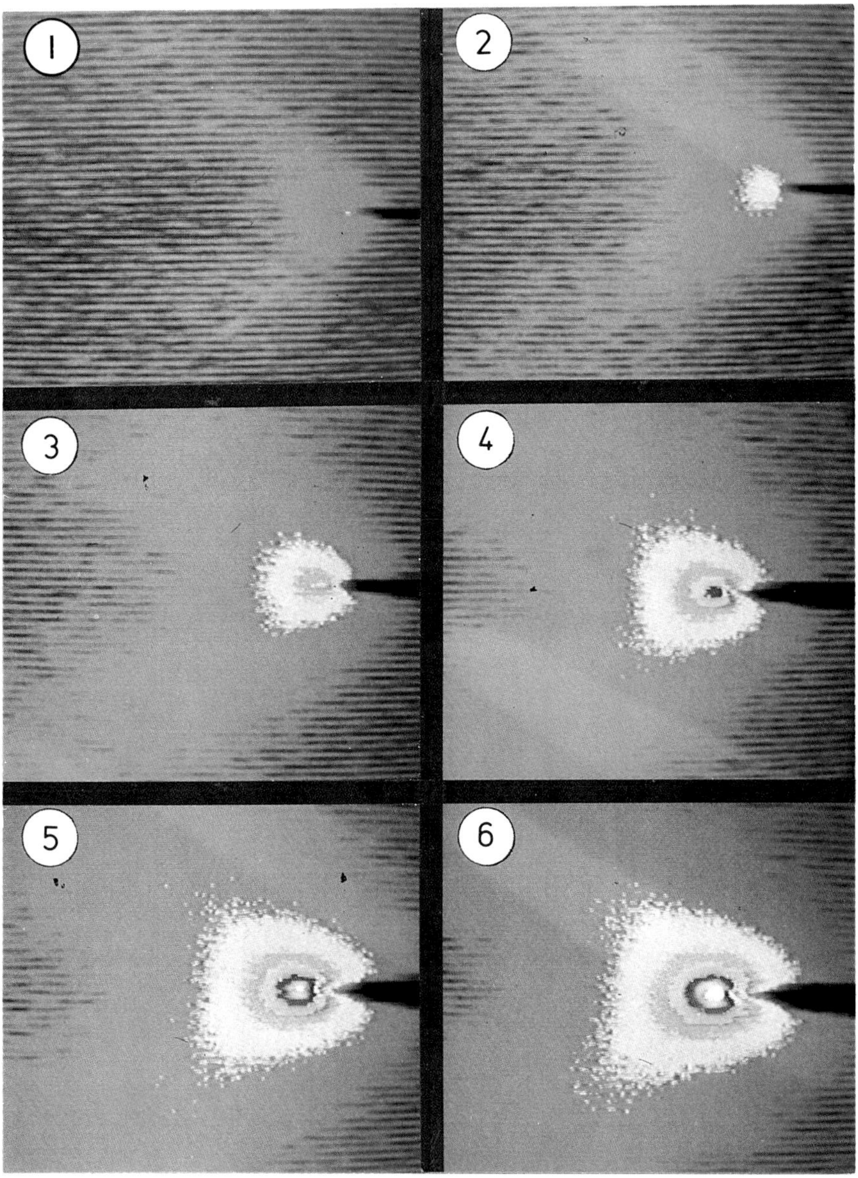

Thermomechanical Couplings in Solids
H.D. Bui and Q.S. Nguyen (Editors)
Elsevier Science Publishers B.V. (North-Holland)

THERMOMECHANICAL COUPLING DURING HOT WORKING PROCESSES

Cristian TEODOSIU

Institut National Polytechnique de Grenoble, ENSPG
Laboratoire Génie Physique et Mécanique des Matériaux
U. A. CNRS 793, Domaine Universitaire, bp 46
38402 St Martin d'Hères Cedex, France

The paper starts with the analysis of a set of viscoplastic and constitutive equations that are descriptive for the rheological behaviour of metals and alloys in the hot working range. The initial and boundary-value problems are solved by means of a 2-D finite-element code for plane and axisymmetric viscoplasticity. Both velocity and temperature fields are considered non-stationary and the formulation includes all types of mechanical and thermal boundary conditions that are relevant for hot working processes. The model is applied to the hot extrusion of axisymmetric billets of 316 stainless steel, the numerical results obtained being compared with both upper-bound estimates and experimental data.

1. INTRODUCTION

Hot working processes cover absolute temperatures larger than about half the absolute melting temperature and strain rates that range typically from 0.1 to 100 s^{-1}. Within this range materials are highly strain rate and temperature sensitive and this makes their behaviour strongly non-linear. On the other hand, the heat loss outside the deformation zones makes at least the temperature field non-stationary. Consequently, the description of the thermomechanical coupling becomes very complex and any model should realize a compromize between accuracy and manageability. With this in mind, we adopt in the present paper the following simplifying hypotheses.

(i) The elastic part of the strain rate is neglected with respect to the plastic part (rigid-plastic approximation) and the latter is assumed to proceed without volume change, hence

$$\underset{\sim}{D}^p = \underset{\sim}{D} = (\mathrm{grad}\underset{\sim}{v})^S, \qquad \mathrm{div}\underset{\sim}{v} = 0, \tag{1.1}$$

where $\underset{\sim}{D}$ and $\underset{\sim}{D}^p$ are the total strain rate tensor and its plastic part, respectively, $\underset{\sim}{v}$ denotes the velocity field, and all differential operators are taken with respect to the current configuration.

(ii) Body forces and inertial effects are neglected, and hence the Cauchy stress tensor $\underset{\sim}{T}$ must satisfy the equilibrium equation

$$\mathrm{div}\underset{\sim}{T} = \underset{\sim}{0}. \tag{1.2}$$

(iii) The evolution of the absolute temperature θ is governed by the simplified equation of heat propagation

$$\rho c\dot{\theta} = \kappa \underset{\sim}{T}:\underset{\sim}{D}^p + \text{div}(k\,\text{grad}\theta), \qquad (1.3)$$

where ρ is the mass density, c is the specific heat, k is the thermal conductivity, and κ is a coefficient that takes into account the stored elastic energy and which is very close to unity for hot deformation. Here and in the following a superposed dot denotes material time differentiation, while a colon is used to denote the scalar product of two second-order tensors, e.g., $\underset{\sim}{T}:\underset{\sim}{D}^p = T_{ij}D^p_{ij}$.

(iv) It is supposed that the material is initially isotropic and that its microstructure can be phenomenologically described by a set of scalar internal variables $\underset{\sim}{\alpha} = \{\alpha_1,\ldots,\alpha_n\}$, e.g. dislocation density, mean dislocation free path, mean grain and subgrain sizes, volume fraction recrystallized, etc. Such materials remain isotropic during the deformation and the evolution of the internal variables merely describes an isotropic hardening or softening of the material[1].

(v) The plastic strain rate $\underset{\sim}{D}^p$ may be derived from a convex viscoplastic potential $\Omega(\underset{\sim}{T},\theta,\underset{\sim}{\alpha})$ that depends on stress only through the second invariant of the stress deviator, $\underset{\sim}{T}'$, which is further proportional to the equivalent tensile stress $\bar{\sigma} = [(3/2)\underset{\sim}{T}':\underset{\sim}{T}']^{1/2}$. This leads to the viscoplastic constitutive equation

$$\underset{\sim}{D}^p = (3d^p/2\bar{\sigma})\underset{\sim}{T}', \qquad (1.4)$$

where $d^p = [(2/3)\underset{\sim}{D}^p:\underset{\sim}{D}^p]^{1/2}$ is the equivalent tensile plastic strain rate. The dependence

$$d^p = d^p(\bar{\sigma},\theta,\underset{\sim}{\alpha}), \qquad (1.5)$$

as well as the evolution equations for the internal variables,

$$\dot{\alpha}_s = \dot{\alpha}_s(\bar{\sigma},\theta,\underset{\sim}{\alpha}), \quad s = 1,\ldots,n, \qquad (1.6)$$

can be obtained from one-dimensional experiments.

Before formulating the boundary-value problems, we will discuss in the next section the range of validity of hypothesis (iv) and examine some viscoplastic constitutive and evolution equations that are suitable for the hot working range.

2. VISCOPLASTIC AND EVOLUTION EQUATIONS FOR THE HOT WORKING RANGE

Work-hardening associated with hot viscoplastic deformation may satisfactorily be described by a superposition of isotropic and kinematic hardening [2-5]. There exists, however, a large class of hot working processes, e.g. extrusion, forging or single-pass rolling, which do not depart very much from a radial loading, and for which the hardening can still be considered as isotropic. We shall concentrate, therefore, on this simplified description, our main purpose here being to emphasize the effects of the thermomechanical coupling.

For hot deformation it is generally possible to express the constitutive law (1.5) in a form which is typical of thermally activated processes, namely

$$d^p = f(\bar{\sigma},\underset{\sim}{\alpha})\exp(-Q/R\theta), \qquad (2.1)$$

where Q is an apparent activation energy and R is the universal gas constant. On the other hand, the evolution equations (1.5) assume in general a work-hardening - recovery format [2,6] :

$$\dot{\alpha}_s = h_s(\underset{\sim}{\alpha})d^p - r_s(\underset{\sim}{\alpha})\exp(-Q_s/R\theta), \quad s = 1,\ldots,n, \tag{2.2}$$

the first and second terms in the right-hand side corresponding to hardening and recovery rates, respectively. Moreover, the activation energies associated with various recovery mechanisms are approximately equal to Q, thus suggesting that glide and recovery obstacles have similar features and may even involve the same rate-controlling events [6,7].

Next, by introducing the cummulated equivalent plastic strain $\bar{\varepsilon}$, defined by the evolution equation $\dot{\bar{\varepsilon}} = d^p$ and the initial condition $\bar{\varepsilon}(0) = 0$, we can rewrite eqns (2.2) in the form

$$d\alpha_s/d\bar{\varepsilon} = h_s(\underset{\sim}{\alpha}) - r_s(\underset{\sim}{\alpha})/Z, \quad s = 1,\ldots,n, \tag{2.3}$$

where

$$Z = d^p\exp(Q/R\theta) \tag{2.4}$$

is a temperature-compensated strain rate, called the Zener-Hollomon parameter. The solution of (2.3) depends on the history of Z and, of course, on the initial conditions, say

$$\alpha_s(0) = \alpha_{so}, \quad s = 1,\ldots,n. \tag{2.5}$$

However, it could be sometimes acceptable to replace the differential system (2.3) by 'state equations' of the finite form

$$\underset{\sim}{\alpha} = \underset{\sim}{\alpha}(\bar{\varepsilon}, Z) \tag{2.6}$$

and to identify such laws by experiments performed at various *constant* values of Z, provided that both the laboratory experiment and the hot working process start from almost the same microstructure (generally a well-annealed material with the same initial grain size). Under such circumstances, introducing (2.6) into (2.1), taking into account (2.4), and solving for $\bar{\sigma}$ yields the simplified flow rule

$$\bar{\sigma} = \bar{\sigma}(\bar{\varepsilon}, Z), \tag{2.7}$$

where $\bar{\varepsilon}$ plays now the role of a unique hardening parameter, whereas the temperature and the strain rate are combined into the single variable Z.

Sah and Sellars [8] have shown that, at least for ferritic stainless steels, the evolution of the flow stress can indeed be described by using $\bar{\varepsilon}$ as unique internal state variable for various isothermal strain-rate histories simulating extrusion, one-pass rolling, and constant-strain-rate experiments.

For illustration, we indicate here also the particular form of (2.7), as obtained in [6] for a type 316 stainless steel, by torsion tests with d^p between

0.007 and 5.44 s^{-1} and θ between 1073 and 1473 K, namely

$$\bar{\sigma} = \{a_o + c_o[1 - \exp(-n_o\bar{\varepsilon})]\}\sinh^{-1}(Z/2b_o) \quad \text{for } Z \leqslant Z_1$$

and (2.8)

$$\bar{\sigma} = \bar{\sigma}_1 + c_o[1 - \exp(-n_o\bar{\varepsilon})]\sinh^{-1}(Z/2b_o) \quad \text{for } Z > Z_1,$$

where $\sinh^{-1}x = \ln[x+(1+x^2)]^{1/2}$, a_o, b_o, c_o, and n_o are material parameters, and Z_1 and $\bar{\sigma}_1$ are some limiting values defining the transition from the thermal to the athermal flow (see also Sect. 5). For constant strain rate and temperature eqn (2.8) reduces to a Voce law of the form $\bar{\sigma} = a + c[1 - \exp(-n_o\bar{\varepsilon})]$ and gives, for sufficiently large values of $\bar{\varepsilon}$, a stationary value of the flow stress, corresponding to a dynamic equilibrium between hardening and recovery. Clearly, such a law can hold only to the onset of dynamic recrystallization, while for higher values of $\bar{\varepsilon}$ a more sophisticated model would be appropriate (see e.g. [6]).

We shall see in Sect. 5 that even the simplified flow rule (2.7) already permits to obtain a fairly good modelling of the material behaviour for monotonic hot deformation. On the other hand, it will be apparent from what follows that the bulk of the model adopted remains unchanged and the amount of computation is not severely increased when passing from the simplest flow rule (2.7) to more sophisticated descriptions like (2.1), (2.2) and even when supplementary introducing a kinematic hardening.

3. THE THERMOMECHANICAL INITIAL-AND-BOUNDARY-VALUE PROBLEM

Assume that the material subjected to hot working can be divided at a current time t into a rigid region where $\underset{\sim}{D} = \underset{\sim}{0}$ and a viscoplastic region Ω, the boundary of the latter being denoted by Σ. By adopting the simplifying hypotheses listed in Sect. 1 and limiting ourselves for conciseness to the simplest flow rule (2.7) considered in the previous section, we conclude that the following *field equations* must be satisfied for any $\underset{\sim}{x} \in \Omega$ and any $t \geqslant 0$:

$$\mathrm{div}\underset{\sim}{T} = \underset{\sim}{0}, \quad \mathrm{div}\underset{\sim}{v} = 0, \tag{3.1}$$

$$\underset{\sim}{D} = (\mathrm{grad}\underset{\sim}{v})^S, \quad \underset{\sim}{T} = p\underset{\sim}{1} + 2\mu\underset{\sim}{D}, \tag{3.2}$$

$$\rho c d\theta/dt = \kappa\underset{\sim}{T}:\underset{\sim}{D} + \mathrm{div}(k\,\mathrm{grad}\theta), \tag{3.3}$$

$$d\bar{\varepsilon}/dt = d^P = [(2/3)\underset{\sim}{D}:\underset{\sim}{D}]^{1/2}, \tag{3.4}$$

$$\mu = \bar{\sigma}/(3d^P), \quad \bar{\sigma} = \bar{\sigma}(\bar{\varepsilon},Z), \quad Z = d^P\exp(Q/R\theta), \tag{3.5}$$

where $p = (1/3)\mathrm{tr}\underset{\sim}{T}$ is the mean stress, $\underset{\sim}{1}$ is the unit tensor, whereas μ plays the role of a (non-linear) viscoplastic viscosity.

By using eqns (3.2) and (3.5) to eliminate $\underset{\sim}{D}$ and $\underset{\sim}{T}$, the remaining field equations (3.1), (3.3), and (3.4) reduce to a system of six scalar differential equations for the determination of the unknown fields $\underset{\sim}{v}$, p, θ, and $\bar{\varepsilon}$. The field equations are supplemented now by adequate initial and boundary conditions.

We assume that the surface Σ can be decomposed, as regards the *mechanical boundary conditions* into three disjoint parts : the surface Σ_v, on which the velocity vector $\underset{\sim}{v}$ is prescribed, the surface Σ_σ, on which the stress vector $\underset{\sim}{t}$ is prescribed, and the surface Σ_τ, on which the normal component v_n of the velocity vector and the tangential component $\underset{\sim}{t}_\tau$ of the stress vector are prescribed. Thus

$$\begin{aligned} &\underset{\sim}{v} = \underset{\sim}{v}^* \text{ on } \Sigma_v, \quad \underset{\sim}{t} = \underset{\sim}{T}\underset{\sim}{n} = \underset{\sim}{t}^* \text{ on } \Sigma_\sigma, \\ &v_n = \underset{\sim}{v}\cdot\underset{\sim}{n} = \nu^*, \quad \underset{\sim}{t}_\tau = \underset{\sim}{t}^*_\tau \text{ on } \Sigma_\tau, \end{aligned} \tag{3.6}$$

where $\underset{\sim}{n}$ is the unit outside normal to Σ, $\underset{\sim}{v}^*$, $\underset{\sim}{t}^*$ and ν^* are prescribed functions of place and time, and $\underset{\sim}{t}^*_\tau$ is the friction force between the material and the tool. Specifically, we assume that

$$\underset{\sim}{t}^*_\tau = -\tau^*\underset{\sim}{\ell}, \qquad \tau^* = m\bar{\sigma}/\sqrt{3},$$

where $\underset{\sim}{\ell}$ is the unit vector of the relative tangential velocity of the material with respect to the tool, while m is a viscoplastic coefficient of friction, which depends on the deformed material, the tool, and the lubricant.

Next, we suppose that the boundary Σ can be decomposed, as regards the *thermal boundary conditions* into three disjoint parts : the surface Σ_θ, on which the absolute temperature θ is prescribed, the surface Σ_q, on which the heat flux $q = -k\partial\theta/\partial n$ is prescribed, and the surface Σ_e, on which the heat flux is prescribed as a linear function of the difference between the surface temperature of the material and the effective (lubricant-dependent) temperature of the surrounding medium. Thus

$$\theta = \theta^* \text{ on } \Sigma_\theta, \quad k\partial\theta/\partial n = -q^* \text{ on } \Sigma_q, \quad k\partial\theta/\partial n = -a(\theta - \bar{\theta}) \text{ on } \Sigma_e, \tag{3.7}$$

where θ^*, q^* and $\bar{\theta}$ are prescribed functions of place and time, and a is an equivalent convection-heat transfer coefficient. All material parameters are allowed to depend slowly on current temperature.

The *initial conditions* to be satisfied in the viscoplastic region are

$$\theta = \bar{\theta}_o, \qquad \bar{\varepsilon} = \bar{\varepsilon}_o \qquad \text{for } t = 0,$$

where θ_o and $\bar{\varepsilon}_o$ are prescribed functions of place. Clearly, replacing $\bar{\varepsilon}$ by a set of scalar internal variables $\underset{\sim}{\alpha}$ requires using the evolution equations (2.2) instead of (2.3) and correspondingly increasing the number of initial conditions.

Taking into account that the material elements are severely deformed during hot working processes, the Eulerian formulation is preferred to a Lagrangian one within the present finite-element context. Consequently, all material time derivatives have to be replaced by spatial time derivatives, by using the operatorial transformation $d/dt = \partial/\partial t + \underset{\sim}{v}\cdot\text{grad}$.

To obtain a weak formulation of the boundary-value problem, we first note that the velocity field has to satisfy at a given time t the field equations

(3.1), (3.2) in Ω and the boundary conditions (3.6) on Σ. We aim at replacing this 'mechanical part' of the boundary-value problem by a principle of virtual velocities. The velocity field $\underset{\sim}{v}$ given in $\bar{\Omega} = \Omega \cup \Sigma$ is said to be kinematically admissible if it satisfies the second field equation (3.1) and the first and third boundary conditions (3.6). The vector field $\delta\underset{\sim}{v}$ given in $\bar{\Omega}$ is said to be a virtual velocity field if it satisfies the homogeneous boundary conditions $\delta\underset{\sim}{v} = \underset{\sim}{0}$ on Σ_v and $v_n = \underset{\sim}{v}\cdot\underset{\sim}{n} = 0$ on Σ_τ. With these definitions it can be proved [9] that the following principle of virtual velocities holds :

A kinematically admissible velocity field $\underset{\sim}{v}$ and a scalar field p satisfy eqns (3.1), (3.2), and (3.6) iff the condition

$$\int_\Omega (2\mu\underset{\sim}{D}:\delta\underset{\sim}{D} + p\,\mathrm{div}\delta\underset{\sim}{v})\,d\Omega = \int_{\Sigma_\sigma} \underset{\sim}{t}^*\cdot\delta\underset{\sim}{v}\,d\Sigma + \int_{\Sigma_\tau} \underset{\sim}{t}^*_\tau\cdot\delta\underset{\sim}{v}_\tau\,d\Sigma \tag{3.8}$$

is satisfied for any virtual velocity field $\delta\underset{\sim}{v}$, where $\delta\underset{\sim}{v}_\tau$ is the tangential component of $\delta\underset{\sim}{v}$ on Σ_τ and $\delta\underset{\sim}{D} = (\mathrm{grad}\delta\underset{\sim}{v})^S$.

The incompressibility constraint $(3.1)_2$ is treated via a penalty-function approach [10, 11], by putting $p = \lambda\mathrm{div}v$, where λ is a very large quantity, typically $(10^7$ to $10^9)\mu_o$, where μ_o denotes some mean value of μ in Ω. This permits the elimination of p from the variational principle and hence an economy of variables, at the expense of allowing a slight compressibility.

4. FINITE-ELEMENT FORMULATION

In order to discretize eqn (3.8), we put

$$\underset{\sim}{v} = \sum_\alpha \underset{\sim}{v}_\alpha(t)\Phi_\alpha(x), \qquad \delta\underset{\sim}{v} = \sum_\alpha \delta\underset{\sim}{v}_\alpha\Phi_\alpha(x), \tag{4.1}$$

where α denotes a current global node, Φ_α are the shape functions and $\underset{\sim}{v}_\alpha$, $\delta\underset{\sim}{v}_\alpha$ denote, respectively, the velocity and virtual velocity vectors at node α. Introducing (4.1) into (3.8) and taking into account that the virtual velocities are arbitrary, we arrive at a system of non-linear algebraic equations of the form

$$\underline{K}\,\underline{V} = \underline{F}, \tag{4.2}$$

where $\underline{K}$ is a symmetric stiffness matrix, the vector $\underline{V}$ contains only the non-prescribed components of the nodal velocities, whereas $\underline{F}$ includes terms arising from the prescribed boundary values of $\underset{\sim}{t}$ and $\underset{\sim}{v}$, as well as some velocity-dependent friction terms, which are kept in the right-hand side in order to preserve the symmetry of $\underline{K}$. For the expression of these arrays in the axisymmetric case, see [9].

Since both $\underline{K}$ and $\underline{F}$ depend on $\underset{\sim}{v}$, θ, and on the internal state variables, system (4.2) has been solved iteratively with respect to $\underset{\sim}{v}$ for any fixed values of θ and $\bar{\varepsilon}$ (respectively $\underset{\sim}{\alpha}$). For many hot working processes a direct iteration involving a few steps is generally satisfactory , and the velocity field obtained is almost stationary during a significant part of the process.

The simultaneous treatment of the equation of heat propagation written in its spatial form, i.e.

$$\rho c(\partial\theta/\partial t + \underset{\sim}{v}\cdot\text{grad}\theta) = \kappa\underset{\sim}{T}:\underset{\sim}{D} + \text{div}(\text{grad}\theta), \tag{4.3}$$

requires a special consideration, because of the large convective effects. Indeed, it is well known that the coupling of lower-order time-stepping methods with a conventional Galerkin spatial discretization often leads to spurious numerical oscillations (wiggles) and/or instability. These effects become particularly unpleasant for high Péclet numbers, which are typical of rapid hot working processes.

In order to overcome this difficulty, two main ways have been applied in this paper : (i) The use of test functions that differ from the shape functions by velocity-dependent additive terms corresponding to both conduction and convection effects (so-called 'upwinded functions'). Wiggles could be eliminated by this procedure for Péclet numbers up to about 100. (ii) For higher Péclet numbers, the evolution of the temperature distribution at the entrance of the deformation zones has been determined by a separate finite-element analysis still using eqn (4.3) but with vanishing convective and heat-production terms. Next, the process has been considered as adiabatic within the deformation zone, and the temperature field has been obtained by determining the instantaneous streamlines and then integrating the material form (3.3) of the equation of heat propagation along these lines.

For conciseness, we shall limit here ourselves to sketching the former of these two methods. First, eqn (4.3) is multiplied by each of the test functions W_α (where α ranges now only over the nodes with unprescribed temperatures) and integrated over Ω. Then, the divergence-term is integrated by parts, thus obtaining

$$\int_\Omega \rho c W_\alpha(\partial\theta/\partial t + \underset{\sim}{v}\cdot\text{grad}\theta)\, d\Omega = \int_\Omega \kappa\underset{\sim}{T}:\underset{\sim}{D}\, d\Omega + \int_\Omega kW_\alpha\partial\theta/\partial n\, d\Sigma -$$
$$- \int_\Omega k(\text{grad}W_\alpha)\cdot(\text{grad}\theta)\, d\Omega. \tag{4.4}$$

Finally, $\partial\theta/\partial n$ is eliminated by means of the boundary conditions (3.7) and the temperature field is discretized by setting

$$\theta = \sum_\alpha \theta_\alpha(t)\Phi_\alpha(\underset{\sim}{x}),$$

where θ_α denotes the temperature at node α. This leads to a system of non-linear differential equations of the form

$$\underline{M}\,\dot{\underline{\Theta}} + \underline{H}\,\underline{\Theta} = \underline{U}, \tag{4.5}$$

where $\underline{\Theta}$ is the vector of the unknown nodal temperatures, while the matrices $\underline{M}$, $\underline{H}$, and the vector $\underline{U}$ depend on $\underset{\sim}{v}$, θ, and on the internal variables $\underset{\sim}{\alpha}$ (respectively on $\bar{\varepsilon}$). For the detailed expression of these arrays in the axisymmetric case we refer again to [9].

System (4.5) has been integrated by means of a semi-implicit method, by writing

$$\underline{\Theta}(t+\Delta t) - \underline{\Theta}(t) = \dot{\underline{\Theta}}(t+\beta\Delta t)\Delta t$$

and choosing $\beta\in[0.5, 1]$ in order to ensure the unconditional stability of this integration scheme. Writing system (4.5) at time $t+\beta\Delta t$ and using also the linear interpolation

$$\underline{\Theta}(t+\beta\Delta t) = (1 - \beta)\underline{\Theta}(t) + \beta\underline{\Theta}(t+\Delta t)$$

gives a system of non-linear algebraic equations of the form

$$\underline{E}\{\underline{\Theta}(t+\Delta t) - \underline{\Theta}(t)\} = \underline{Q}, \tag{4.6}$$

where $\underline{E}$ is a non-symmetric matrix (unless $W_\alpha = \Phi_\alpha$ for any α, i.e. for a conventional Galerkin scheme). Since both $\underline{E}$ and $\underline{Q}$ depend on the nodal temperatures through ρ, c, k, a, and through the plastic-power term, system (4.6) has been also solved iteratively.

A Richardson scheme has been used to monitor the time step Δt and to simultaneously improve the accuracy of the iteration scheme. Namely, the semi-implicit scheme presented in the previous section is known to give an error of the order $(\Delta t)^2$ at each step ($(\Delta t)^3$ for the special case $\beta = 0.5$). Let us denote by $\underline{\Theta}^E(t+\Delta t)$ the exact solution of (4.6) and by $\underline{\Theta}^{(1)}(t+\Delta t)$ and $\underline{\Theta}^{(2)}(t+\Delta t)$ the estimations obtained by proceeding from t to $t+\Delta t$ in a single step, respectively in two steps of magnitude $\Delta t/2$. We may write then

$$\underline{\Theta}^E(t+\Delta t) = \underline{\Theta}^{(1)}(t+\Delta t) + (\Delta t)^2\underline{C},$$

$$\underline{\Theta}^E(t+\Delta t) = \underline{\Theta}^{(2)}(t+\Delta t) + 2(\Delta t/2)^2\underline{C}.$$

Assuming that $\underline{C}$ is nearly constant within the interval $[t, t+\Delta t]$, we may solve these two equations with respect to $\underline{\Theta}^E(t+\Delta t)$ and $\underline{C}$, thus obtaining a better estimate of the nodal temperatures at time $t+\Delta t$ and also a simple way of monitoring the time step, by comparing $\max_\alpha\{(\Delta t)^2 C_\alpha\}$ with some admissible error per step.

Summarizing, the algorithm consists in repeating the following steps : (i) determination of the velocity field at a given time t by using the principle of virtual velocities, (ii) determination of the temperature field at a subsequent time $t+\Delta t$ by using the equation of heat propagation and a semi-implicit time integration scheme, and (iii) updating the internal variables and calculation of the strain rate and stress components at the Gauss points.

This algorithm has been used to generate a 2-D finite element code for plane and axisymmetric problems. The element used throughout the calculation is the nine-noded isoparametric C^o-quadrilateral with biquadratic shape functions for velocity components and temperature. Local arrays have been computed by 3x3

Gauss-Legendre numerical integration, except the penalty term in the principle of virtual velocities, for which a 2x2 reduced integration has been used.

The remaining part of the paper will be devoted to the analysis of the results obtained by applying a previous form of the code [9] to the hot extrusion of axisymmetric rods.

5. APPLICATION OF THE MODEL TO THE NUMERICAL SIMULATION OF HOT EXTRUSION

For illustration we consider the hot extrusion of an axisymmetric rod of type 316 stainless steel and compare the results obtained by the finite-element model [9] with both upper-bound estimates [12] and experimental data [13,14].

The viscoplastic properties of the material are given by eqn (2.8), where $Q = 499$ kJ/mol, $R = 8.314$ J/molK, $n_o = 3.86$, $a_o = 10.07$ MPa, $b_o = 5.43\times10^{15}\ s^{-1}$, $c_o = 11.13$ MPa, $Z_1 = 7.6\times10^{20}\ s^{-1}$, and $\bar{\sigma}_1 = 121.24$ MPa. According to [15], the dependence of the density ρ, specific heat c, and thermal conductivity k on absolute temperature θ is given by

$$\rho(\mathrm{kg/m^3}) = 8200 - 0.5\theta, \quad c(\mathrm{J/kgK}) = 587.8 + 0.0683\theta, \quad k(\mathrm{W/mK}) = 17.1 + 0.011\theta.$$

The same authors have determined an empirical law for the air-cooling heat flow during the transfer of the billet from the furnace to the container. This law may be written in the equivalent form

$$q = a_A(\theta_S - \bar{\theta}_A),$$

where $\bar{\theta}_A = 476$ K is an effective temperature of the surrounding air which takes into account the insulating effect of the lubricant, θ_S is the surface temperature of the billet, whereas

$$a_A(\mathrm{W/m^2K}) = 21.2 + 4.763\times10^{-8}(\theta_S + \bar{\theta}_A)(\theta_S^2 + \bar{\theta}_A^2).$$

Finally, as shown in [16], the heat flux from the billet to the container after upsetting depends almost linearly on the surface temperature of the billet and may be expressed as

$$q = a_C(\theta_S - \bar{\theta}_C),$$

where $\bar{\theta}_C = 783$ K is an effective temperature of the container and $a_C = 11$ kW/m^{2}K is an effective heat transfer coefficient, both of which taking empirically into account the presence of the lubricant.

In a previous investigation [12], the process has been analysed by an upper-bound method, as follows. The region inside the container and the die has been divided into a rigid part Ω_r and a deformation zone Ω. Based on experimental evidence, it was assumed that the separation surface Σ_i between Ω_r and Ω has a spherical shape and that the streamlines in the longitudinal section of the billet have a cosine form (fig. 1). The temperature field has been calculated by starting from the temperature distribution at the entrance of the deformation

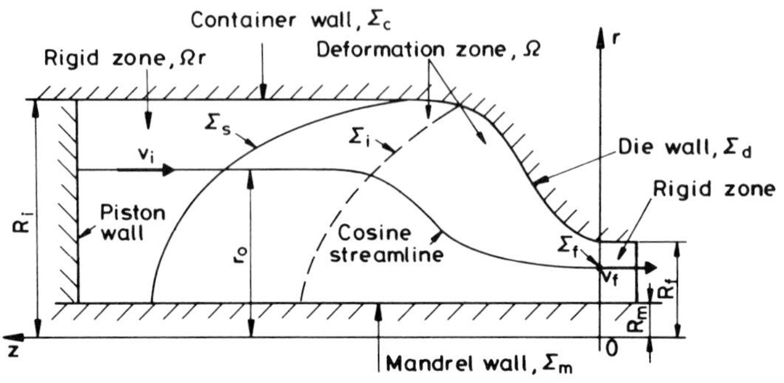

FIGURE 1
Rigid and deformation zones assumed in the simulation of hot extrusion of an axisymmetric product.

zone, as determined experimentally in [16], and then integrating the equation of heat propagation along the streamlines, under the assumption that the deformation proceeds adiabatically in Ω. The results obtained have been used in the present finite-element calculation both to rationally generate the mesh with the aid of the cosine streamline pattern (fig. 2) and to initialize the velocity and temperature fields.

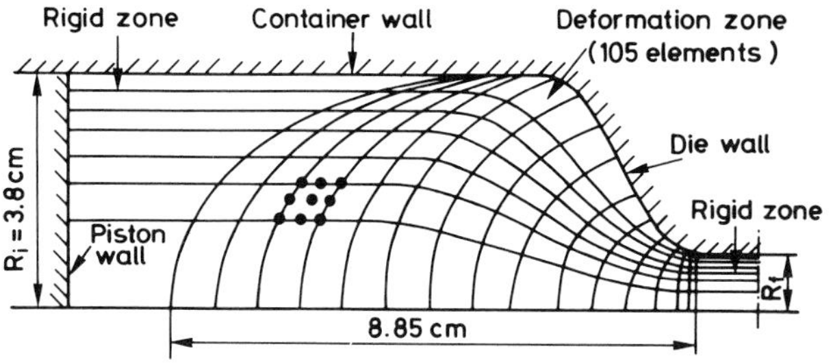

FIGURE 2
Cosine streamlines and finite-element mesh.

In order to permit a direct comparison of the finite-element and experimental results, the basic combination of technological variables considered in the numerical simulation has been taken to coincide with one used in [13,14], namely : initial radius of the billet after upset 0.038 m, initial length of the billet 0.220 m, extrusion speed 0.19 m/s, and semi-cone die angle 85° (flat die). Friction effects between the material and the container-and-die walls at the boundary of the viscoplastic region have been taken into account by adopting a

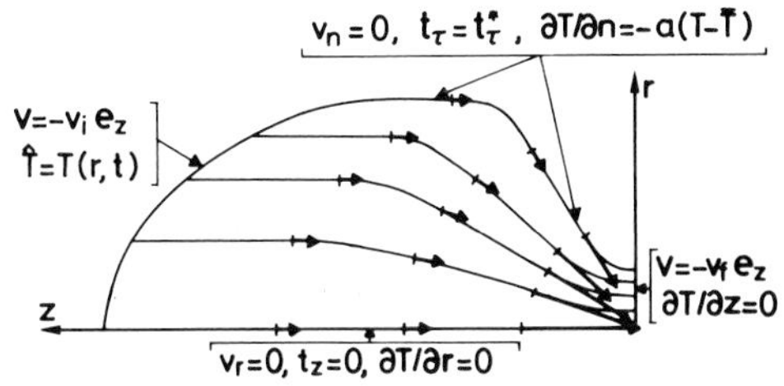

FIGURE 3
Boundary conditions used in the finite-element analysis. Arrows show some nodal velocity vectors, calculated by the finite-element method at 0.45 s after upsetting.

viscoplastic friction coefficient m = 0.8. The deformation zone has been extended with respect to the upper-bound analysis up to the surface Σ_s shown in fig. 1, in order to derive a more realistic velocity field.

Fig. 3 shows the velocity vectors at some nodal point, as obtained by three iterations at 0.45 s after upsetting, with 1% admissible relative error per step. Clearly, the finite-element solution is in rather good agreement with the form of the streamlines chosen in the upper-bound calculation. On the contrary, as shown by fig. 4, the cosine pattern of the streamlines and the adiabatic calculation of the temperature evolution along these streamlines may lead to overestimations of the exit temperature and cummulated equivalent plastic strain, especially at the surface of the extruded bar.

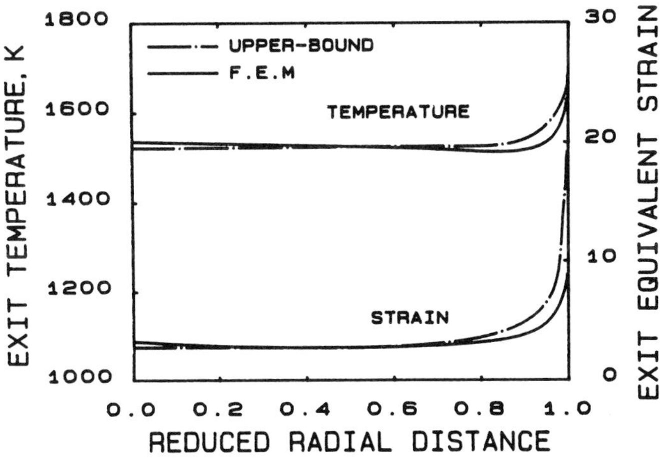

FIGURE 4
Radial profiles of the temperature and cummulated equivalent plastic strain at the exit of the die at 0.45 s after upset, calculated by the finite-element method [9] and estimated by the upper-bound method [12].

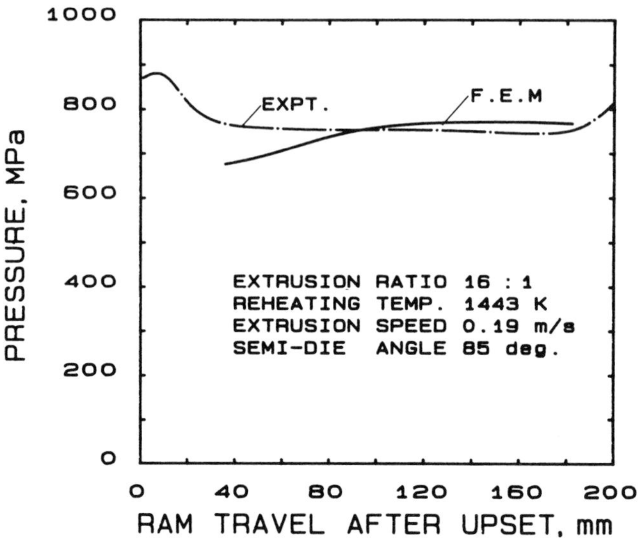

FIGURE 5
Extrusion pressure - ram travel curves for type 316 stainless steel, calculated by the finite-element method [9] and measured by Hughes et al.[13].

Fig. 5 shows the evolution of the extrusion pressure with ram travel after upset. The agreement between the finite-element calculation and the experiment is remarkably good, especially for the stationary part of the running pressure, where the calculated pressure is merely 4% higher than the measured one. On the contrary, the calculated value is about 10% less than the experimental one for the initial running pressure. This difference originates mainly from the neglect in the finite-element model of the friction between the rigid part of the billet and the container. Indeed, according to the evaluation in [13], the total billet/container friction can contribute in our case some 7% of the initial running pressure, but has little effect on the final pressure for a discard of about 20 to 30 mm. It should be also noticed that both the upper-bound analysis and the finite-element calculation apply only after the material has filled the die, in our case some 28 mm stroke after upset. Consequently, at least the same stroke is necessary in order to eliminate the effect of possible errors in the initial evaluation of the thermokinematical fields. That is why, the further comparison between the finite-element results and experiment has been done only on the final running pressures, at about 190 mm stroke after upset.

Fig. 6 shows in a semi-logarithmic plot the dependence of the final running pressure on the extrusion ratio, calculated by the finite-element model. Clearly, the extrusion pressure follows approximately a linear dependence on $\ln\psi$,

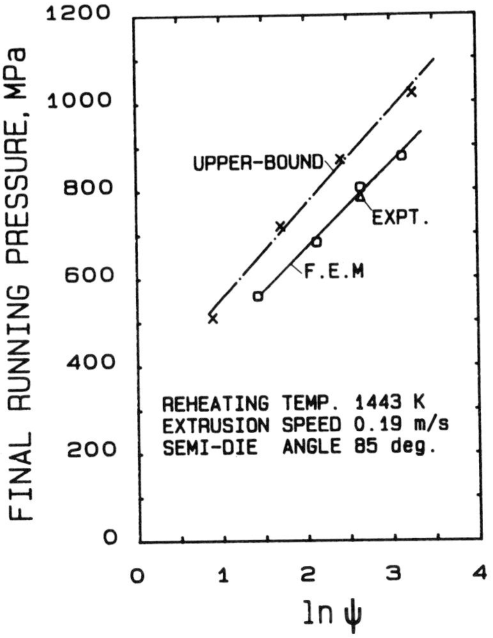

FIGURE 6
Dependence of the final running pressure on extrusion ratio ψ for type 316 stainless steel, calculated by the finite-element method [9] and estimated by the upper-bound method [12]. The experimental point is taken from Hughes et al.[13].

where ψ is the extrusion ratio, in agreement with the prediction of the upper-bound analysis [12]. Finally, fig. 7 indicates the dependence of the final running pressure on reheating temperature. It can be seen once more that the agreement is quite satisfactory, the absolute difference in pressure being less than 6% of the measured value.

6. CONCLUSIONS

The comparison of the calculated and measured values of the extrusion pressure shows that the finite-element model provides a precise tool for the numerical simulation of hot working processes. However, a successful numerical simulation requires an accurate evaluation of the material properties and of the heat transfer conditions, by experiments performed in the hot working range of temperature, strain and strain rate, and for the materials, tools and lubricants involved in the process.

Unlike upper-bound methods, which allow to obtain only a global information about the thermokinematical fields, the finite-element model permits the determination of the velocity and temperature fields with the desired accuracy, thus

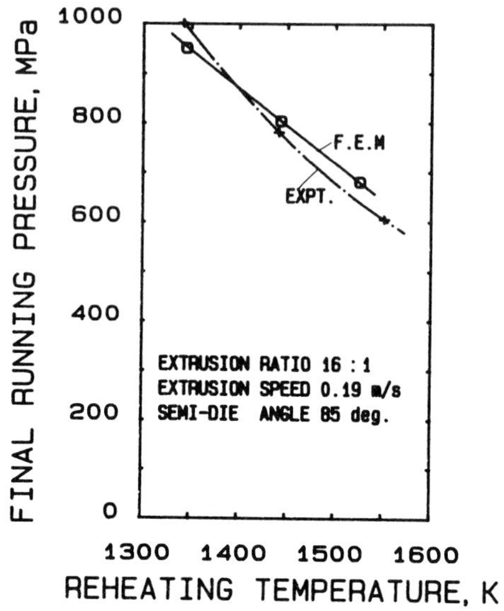

FIGURE 7
Dependence of the final running pressure on reheating temperature for type 316 stainless steel, calculated by the finite-element model [9] and measured by Gupta et al.[14].

avoiding possible rough estimations of the local values of strain, strain rate, and temperature. On the other hand, the knowledge of the real temperature and deformation conditions permits the study of the microstructure and damage evolution along the streamlines, either by laboratory simulation of the thermomechanical history, or by integrating the evolution equations of the microstructural quantities along the streamlines.

With the present model, a complete finite-element simulation of a hot working process, corresponding to a given set of technological variables, requires typically a few hundred seconds CPU time on a CDC-CYBER 720 computer. This computational effort is certainly justified for selected application and/or for optimizing mass production processes, where the alternative of large-scale industrial experiments could be indeed much more expensive.

REFERENCES

[1] Mandel, J., Plasticité classique et viscoplasticité (Int. Centre for Mech. Sci., Udine, 1971, Springer, 1972).
[2] Miller, A.K., ASME J. Eng. Mat. Techn. 96 (1976) 97, 108.
[3] Miller, A.K. and Sherby, O.D., Acta Metall. 26 (1978) 289.
[4] Chaboche, J.-L., Description thermodynamique et phénoménologique de la viscoplasticité cyclique avec endommagement (Thèse de Doctorat d'Etat, Univ. P. et M. Curie, Paris, 1978).

[5] Lemaitre, J. and Chaboche, J.-L., Mécanique des matériaux solides (Dunod, Paris, 1985).
[6] Teodosiu, C., Nicolae, V., Soós, E., and Radu, C.G., Rev. Roum. Sci. Techn.-Méc. Appl. 24 (1979) 13, 225.
[7] Mc Queen, H.J. and Jonas, J.J., Recovery and Recrystallization during High Temperature Deformation, in : Arsenault, R.J., (ed.), Plastic Deformation of Materials (Academic Press, New York, 1976).
[8] Sah, J.P. and Sellars, C.M., Effect of Deformation History on Static Recrystallization and Restoration in Ferritic Stainless Steel, in: Sellars, C.M. and Davies, G.J., (eds.), Hot Working and Forming Processes (Metal Soc., London, 1980), pp. 62-66.
[9] Teodosiu, C., Soós, E., and Roşu, I., Rev. Roum. Sci. Techn.- Méc. Appl. 28 (1983) 575, 30 (1985) 49, 32 (1987) in print.
[10] Zienkiewicz, O.C., Oñate, E., and Heinrich, J.C., Int. J. Num. Meth. Engng., 17 (1981) 1497.
[11] Zienkiewicz, O.C., The Finite Element Method (Mc Graw Hill, London-New York -St Louis, 1977).
[12] Teodosiu, C., Roşu, I., and Dumitrescu, T., Rev. Roum. Sci. Techn. - Méc. Appl. 28 (1983) 317, 395, 495.
[13] Hughes, K.E., Nair, K.D., and Sellars C.M., Metals Technology 1 (1974) 161.
[14] Gupta, A.K., Hughes, K.E., and Sellars, C.M., Metals Technology 7(1980)323.
[15] Sellars, C.M. and Whiteman, J.A., Metals Technology 8 (1981) 10.
[16] Hughes, K.E. and Sellars, C.M., J. Iron Steel Inst. 210 (1972) 661.

Thermomechanical Couplings in Solids
H.D. Bui and Q.S. Nguyen (Editors)
Elsevier Science Publishers B.V. (North-Holland)

DEFECT MODELS IN ANISOTROPIC THERMOELASTIC MATERIALS

R.K.T. Hsieh
Department of Mechanics
Royal Institute of Technology
S-100 44 Stockholm, Sweden.

Defects are always present in solids and their presence changes considerably many of the material properties. A quantitative treatment of these properties can be given by the continuum model. However each type of defect has to be treated as a separate problem. Using the concept of material multipoles, we present a unified approach to describe the quantitative physical properties of all defects in anisotropic thermoelastic materials. It is shown that all defects can be represented by a distribution of thermoelastic multipoles dP^F_{ij} and dP^H_i. Particular emphasis is laid on the study of the properties of lattice defects, the influence of which on the thermo-mechanical behaviour of materials is considerable.

1. INTRODUCTION

Defects are always present in solids and their presence changes considerably many of the material properties. A quantitative treatment of these properties can be given by the continuum model. However each type of defect has to be treated as a separate problem. Using the concept of material multipoles, a unified approach is presented to describe the quantitative physical properties of all defects in anisotropic thermoelastic materials. Particular emphasis is laid on the study of the properties of lattice defects, the influence of which on the thermo-mechanical behaviour of materials is considerable. The results are applicable for the study of the effects of temperature in solids with defects. Such problems are of importance for aeronautical, metallurgical materials, ...

The thermoelastic solid is assumed to be homogeneous and anisotropic. Its mechanical and thermal behaviours are governed respectively by the linear anisotropic constitutive relationship and the anisotropic Fourier's law. The fields of the mechanical displacements and temperatures can be obtained by solving the coupled system consisting of the force equation, the heat conduction, the material behaviour relationships together with the boundary and initial conditions. In particular, analytical solutions can be found with the help of the Green functions and the material multipole representations.

During these last decades, the importance of the inelasticity of solids has been widely recognized. Eshelby [1] refers to it as stress free transformation strains. Kröner [2] refers to it as elastic polarization. Mura [3] referred to it as eigenstress, the concept of which was generalized by Minagawa [4]. Kröner [2] introduced the concept of elastic force dipole to model a point defect in linear elasticity. Kovàcs [9] used the concept of elastic force monopole to model also a dislocation. Hsieh, Kovàcs and Vörös [8] extended the concept of elastic multipoles to micropolar solids. Hsieh and his co-workers also used the concept of material multipoles and successfully applied the model to describe the mechanical properties of point defects (vacancy and interstitial atoms) and line defects (dislocation) in nonlocal micropolar elasticity [5], in statistical elasticity [7], in uncoupled thermoelasticity [6]. (Monographs on lattice defects can be found in e.g. Kovàcs and Zsoldos [10] and Steeds [11], on thermoelasticity in e.g. Nowinski [12]). Barr and Cleary [13], Zhou and Hsieh [6], Wu and Hui [14] also apply the model of internal stress to crack problems. Engineers usually employ the term residual stresses for the self-equilibrated internal stresses when they remain in materials after fabrication or plastic deformation. The present material defects are viewed as sources of internal thermo-elastic stresses and are defined as following. A cut is first made in the body along the surface A_L bounded by the curve L. The two sides of the cut surface are then slipped relative to each other with a displacement $\underline{a}$ and a temperature T across the cut surface. Finally, the two sides of the cut are welded together (perhaps by adding or removing material). A defect is said to be created along the curve L. Analytically, the mechanical displacement fields and the temperature field are thus both suffering a discontinuity when crossing the slip surface. These continuum descriptions of the material microstructure can be described in terms of the material multipoles and are shown to be equally applicable for any type of defects in an anisotropic thermoelastic medium. For a certain type of defects, it is however necessary to give the physical surface A_L and its transformations $\underline{a}$ and T. In particular, the line defects (dislocations and disclinations) can be created by a rigid body transformation. The Volterra dislocations are seen to be a particular case of this approach. Results in the case of cracks, volume defects, vacancies, etc... have also been obtained. A common property of all these defects is that they have been uniformly modelled as distributions of thermoelastic monopoles. It is shown that the problem of finding the self-stress in a thermoelastic body can be reduced to the problem of determining the deformation field and the thermal field of a thermoelastic body with fictitious (but always present) body force loadings and fictitious body heat source loadings. These loadings are given in terms of the thermoelastic multipoles. These results are used in particular to study the influence of temperature on the mechanical properties of the materials with structural defects.

Explicit results are shown with the assumption that both inertia and dynamical coupling terms may be neglected. This is the case of a large collection of problems and in our case simplifies the algebra. The general thermoelastic problem then decomposes into two separate field equations that must be solved consecutively but independently. Using the Duhamel Neumann analogy theorem and the Green function representation, closed form analytical solutions for the mechanical fields can be found in terms of the (defectious) material monopoles. The obtained results are in good agreement with experimental data.

2. GREEN FUNCTION REPRESENTATION OF THERMOELASTICITY

The thermoelastic solid is assumed to be anisotropic. Its linear mechanical constitutive equation reads

$$t_{ij} = C_{ijk\ell}\, u_{k,\ell} - \beta_{ij}\theta \tag{1}$$

where t_{ij} is the stress, u_k is the displacement, θ denotes small changes in temperature from the reference, stress-free state, constant temperature θ_o, $C_{ijk\ell}$ are the elastic coefficients and β_{ij} are the thermal coefficients. The linear constitutive relationship for the specific entropy η per unit volume is given by

$$\eta = \beta_{ij}u_{i,j} + \alpha\theta \tag{2}$$

where the coefficient α is defined as $\alpha = \rho \dfrac{c_v}{\theta_o}$, ρ being the density of the material, c_v is the specific heat.
The thermal law is the anisotropic Fourier's law

$$q_i = -k_{ij}\, \theta,_j \tag{3}$$

where q_i is the heat flux and k_{ij} are the heat conduction coefficients.
The coefficients $C_{ijk\ell}$ and β_{ij} verify the relationships

$$C_{ijk\ell} = C_{k\ell ij} = C_{jik\ell} = C_{ij\ell k}$$

$$\beta_{ij} = \beta_{ji}$$

We also adopt Onsager's principle and write for the coefficients k_{ij}

$$k_{ij} = k_{ji} \tag{4}$$

The balance equations are (small deformations, dynamic case)

$$t_{ij,j} - X_i - \rho\ddot{u}_i = 0 \tag{5}$$

$$q_{i,i} + \theta_o\dot{\eta} - r' = 0 \tag{6}$$

where r' is the external heat source per unit volume, X_i is the force density per unit volume.

Assuming in addition that the thermoelastic solid is homogeneous, eliminating eqs (1) - (3) into (5) - (6), we obtain

$$C_{ijk\ell}\, u_{k,\ell j} - \beta_{ij}\, \theta_{,j} - X_i - \rho\ddot{u}_i = 0 \tag{7}$$

$$\theta_o\beta_{ij}\, \dot{u}_{i,j} - k_{ij}\, \theta_{,ij} + \alpha\theta_o\dot{\theta} - r' = 0 \tag{8a}$$

$$\beta_{ij}\, \dot{u}_{i,j} - \xi_{ij}\, \Theta_{,ij} + \alpha\dot{\theta} - r = 0 \tag{8b}$$

where we defined $\xi_{ij} = \frac{k_{ij}}{\theta_o}$ and $r = \frac{r'}{\theta_o}$

For an infinite medium, the solutions of eq (7), (8b) can be got using the Green function representations. Such solutions can be written as

$$u_i(\underline{x},t) = -\int[G^F_{iq}(\underline{x}-\underline{x}',t-t')X_q(\underline{x}',t')+G^H_i(\underline{x}-\underline{x}',t-t')r(\underline{x}',t')]dV'dt' \quad (9)$$

$$\theta(\underline{x},t) = -\int[S^F_i(\underline{x}-\underline{x}',t-t')X_i(\underline{x}',t')+S^H(\underline{x}-\underline{x}',t-t')r(\underline{x}',t')]dV'dt' \quad (10)$$

where the Green's functions G^F_{iq}, G^H_i, S^F_i and S^H are defined by

$$C_{ijk\ell}\, G^F_{kq,\ell j} - \rho\ddot{G}^F_{iq} - \beta_{ij}\, S^F_{q,j} + \delta_{iq}\, \delta(\underline{x}-\underline{x}')\, \delta(t-t') = 0$$

$$\beta_{ij}\, \dot{G}^F_{iq,j} - \xi_{ij}\, S^F_{q,ij} + \alpha\dot{S}^F_q = 0$$

$$C_{ijk\ell}\, G^H_{k,\ell j} - \rho\ddot{G}^H_i - \beta_{ij}\, S^H_{,j} = 0$$

$$\beta_{ij}\, \dot{G}^H_{i,j} - \xi_{ij}\, S^H_{,ij} + \alpha\dot{S}^H + \delta(\underline{x}-\underline{x}')\, \delta(t-t') = 0 \quad (11)$$

For isotropic infinite media, the Green's functions have been derived [15].

3. THERMOELASTIC MONOPOLES

The displacement fields and the temperature field due to a point force and a point heat source can be written as

$$u_i(\underline{x},t) = -[G^F_{iq}(\underline{x}-\underline{x}',t-t')\, X_q(\underline{x}',t') + G^H_i(\underline{x}-\underline{x}',t-t')\, r(\underline{x}',t')] \quad (12)$$

$$\theta(\underline{x},t) = -[S^F_i(\underline{x}-\underline{x}',t-t')\, X_i(\underline{x}',t') + S^H(\underline{x}-\underline{x}',t-t')\, r(\underline{x}',t')] \quad (13)$$

where the Green's functions G^F_{iq}, G^H_i, S^F_i and S^H were defined by eq (11). For an array of point forces and point heat sources, we shall have

$$u_i(\underline{x},t) = -[\sum X^\alpha_q\, G^F_{iq}(\underline{x}-(\underline{x}'+\underline{d}^\alpha),t-t') + + \sum r^\alpha\, G^H_i(\underline{x}-(\underline{x}'+\underline{d}^\alpha),t-t')]$$

where $\underline{d}^\alpha$ is the position vector of the point force X^α_q and the point heat r^α.

An expansion around $(\underline{x}',t')$ will give

$$u_i(\underline{x},t) = -[\sum X^\alpha_q\, G^F_{iq}(\underline{x}-\underline{x}',t-t') + \sum X^\alpha_q d_m\, G^F_{iq,m'}(\underline{x}-\underline{x}',t-t') + \ldots + + \sum r^\alpha\, G^H_i(\underline{x}-\underline{x}',t-t') + \sum r^\alpha d_m\, G^H_{i,m'}(\underline{x}-\underline{x}',t-t') + \ldots]$$

We now introduce the notations

$$P^F_{qm\ldots} = \sum X^\alpha_q d_m \ldots$$

$$P^H_{m\ldots} = \sum r^\alpha d_{m\ldots}$$

To represent a physically possible defect, a thermal elastic array is required to fulfil the conditions

$$P_q^F = 0 \tag{14a}$$

$$P^H = \gamma_{k\ell}\dot{P}_{k\ell}^F \tag{14b}$$

where $\gamma_{k\ell}$ is the expansion tensor and we have the relationship

$$\beta_{ij} = -C_{ijk\ell}\gamma_{k\ell}.$$

The precedent expression for the displacement reduces to

$$u_i(\underline{x},t) = -(P_{qm}^F\ G_{iq,m'}^F + P_m^H\ G_{i,m'}^H + \gamma_{k\ell}\ \dot{P}_{k\ell}^F G_i^H) \tag{15}$$

In a similar way one can derive the expression for the temperature

$$\theta(\underline{x},t) = -(P_{qm}^F\ S_{q,m'}^F + P_m^H\ S_{,m'}^H + \gamma_{k\ell}\dot{P}_{k\ell}^F S^H) \tag{16}$$

Expressions (15) and (16) are respectively the monopole approximation of the thermoelastic displacement fields and the temperature field.

4. DEFECTS AS THERMOELASTIC MONOPOLES

4.1 General equations

In this section, it will be shown that any defect in a thermoelastic body can be described by a continuous distribution of thermoelastic monopoles. The relationships (15) and (16) then read

$$u_i(\underline{x},t) = -\int(G_{iq,m'}^F\ dP_{qm}^F + G_{i,m'}^H\ dP_m^H + G_i^H\ \gamma_{qm}d\dot{P}_{qm}^F)dt' \tag{17}$$

$$\theta(\underline{x},t) = -\int(S_{q,m'}^F\ dP_{qm}^F + S_{,m'}^H\ dP_m^H + S^H\gamma_{qm}d\dot{P}_{qm})dt' \tag{18}$$

For this purpose, let a surface A be chosen in an arbitrary way in the medium and suffer a transformation such that the element dA_m' at a position $\underline{x}'$ has a displacement a_ℓ and a temperature T. As a result of these transformations, the surface becomes a source of thermoelastic singularities with strengths which we define by the following expressions:

$$dP_{qm}^F = C_{qmk\ell}\ a_k\ dA_\ell' \tag{19}$$

$$dP_m^H = -\xi_{\ell m}T\ dA_\ell' \tag{20}$$

Equations (17)-(20) are equally applicable to any type of defect. To specify these equations for a certain type of defect, it is however necessary to give the surface A and its transformations. As far as the modelling of the defect is concerned, we can see from eqs. (17) and (18) that they are respectively the total displacement and the total temperature at any point of the thermoelastic body with defects. The initial conditions will be assumed to be homogeneous, $u_k = \dot{u}_k = 0$ and $\Theta = 0$ at $t'=\pm\infty$.

From the constitutive standpoint, the thermoelastic body with defects will be described by equations (1)-(3) with however always present fictitious body forces X_i and fictitious body heat sources r, see eqs. (7) and (8b). The expressions for these fictitious body loadings can be obtained by inserting the relationships (17) and (18) into (7) and (8b).

It has to be noted that if one is only interested in calculating the displacement, the temperature, the stress and the heat flux at any point of the thermoelastic body it is not necessary to introduce the defect-material multipole body loadings X_i and r. Instead of that, one can calculate them directly by inserting the defect-material multipole displacement u_i, eq. (17) and temperature, eq. (18) into the relationships (1) and (3) for respectively the stress and the heat flux.

It has also to be noted that though eqs. (17) and (18) give in principle the displacement and temperature fields due to any defect in a thermoelastic body, in practice due to the anisotropic material properties and to the mathematical difficuties in the determination of the Green's functions, closed form solutions of (17) and (18) can only be found for special configurations of the defects and of the anisotropy of the body.

Finally, the knowledge of the transformations of both a_ℓ and T lead to the macroscopic modelling of a microscopic thermoelastic defect. Should one be interested in the effect of elastic defects only, one has then to put T = 0 in relationships (19) - (20). In what follows, the definitions of a_ℓ and T for some basic defects will be given.

4.2 Defects defined by cut surfaces (line defects)

Let us define a defect in the following way. Make a cut in the material over the slip surface $A_L(t)$ bounded by a curve L(t). The positive direction of the normal to the surface is related to the curve L by the right-hand rule. If the originally coincident points of the positive and negative surfaces at the point $\underline{x}'$ are moved relative to each other so that we have $a_\ell^L(\underline{x}')$ and $T^L(\underline{x}')$, then a line defect is created. If $a_\ell^L(\underline{x}')$ and $T^L(\underline{x}')$ are given by

$$[u_\ell]_{A_{L(t')}} = a_\ell^L(\underline{x}') = b_\ell + \varepsilon_{\ell ps}\Omega_p(x_s'-x_s^o) \tag{21a}$$

$$[\theta]_{A_{L(t')}} = T^L(\underline{x}') = T \tag{21b}$$

where b_ℓ, Ω_ℓ are constant vectors, T is a constant scalar and $\underline{x}' \varepsilon A_{L(t)}$, then a general dislocation type is produced. For isotropic thermoelastic solids, we have that T = 0, eq. (21a) models a general mechanical dislocation in thermoelastic solids. We note that this expression is the same as the expression for general dislocation in elastic solids.

With the help of the expressions (17) and (18), the displacement and the temperature fields due to rigid body produced line defects write

$$\begin{aligned} u_i(\underline{x},t) = &-C_{k\ell qm}\int G^F_{iq,m'}(b_k+\varepsilon_{kps}\Omega_p(x_s'-x_s^o))dA_\ell'dt' \\ &+ \xi_{m\ell}T\int G^H_{i,m}dA_\ell'dt' \\ &- C_{k\ell qm}\gamma_{qm}\int G^H_i \frac{d}{dt}(b_k+\varepsilon_{kps}\Omega_p(x_s'-x_s^o))dA_\ell'dt' \end{aligned} \tag{22a}$$

$$\theta(\underline{x},t) = -C_{qmk\ell}\int S^F_{q,m'}(b_k+\varepsilon_{kps}\Omega_p(x'_s-x^o_s))dA'_\ell dt' + \xi_{m\ell}T\int S^H_{,m'}dA'_\ell dt' - C_{k\ell qm}\gamma_{qm}\int S^H \frac{d}{dt}(b_k+\varepsilon_{kps}\Omega_p(x'_s-x^o_s))dA'_\ell dt' \tag{22b}$$

The stress and the heat flux can now be found by inserting eqs. (22) into the constitutive relationships (1) and (3). The fictitious body forces and fictitious heat sources can be obtained by substituting these stresses and heat fluxes into the balance equations (7) and (8b).

4.3 Defects defined by closed surfaces (volume defects)

Let A now be a surface enclosing a volume $V \neq 0$ and define the thermoelastic monopoles along this surface by the displacement and temperature fields

$$a_\ell = \begin{cases} u^\tau_\ell(\underline{x}',t'), & \text{if } \underline{x}' \in V \\ 0 & \text{otherwise} \end{cases} \tag{23}$$

$$T = \begin{cases} T^\tau(\underline{x}',t'), & \text{if } \underline{x}' \in V \\ 0 & \text{otherwise} \end{cases} \tag{24}$$

This defect is due to a homogeneous volume defect i.e. a defect in which its volume differs from that of the hole in the matrix. For isotropic thermoelastic solids, we have that $T^\tau=0$ for permanent multipoles (homogeneous volume defects,...) while relation (24) applies for induced multipoles. Considering, for example, thermoelastic inhomogeneities, these effects, due to the bulk moduli difference between the defect and its matrix, will appear.

With the the use of eqs. (17), (18) and applying Gauss's theorem, after some algebra one obtains for the displacement and temperature fields

$$u_i(\underline{x},t) = \begin{cases} -\int_V [t^\tau_{q\ell}G^F_{iq,\ell'} + q^\tau_\ell G^H_{i,\ell'}/\theta_o + C_{qmk\ell}\gamma_{qm}\dot{u}^\tau_{k,\ell'}G^H_i]\, dV'dt' + u^\tau_i & \text{if } \underline{x}' \in V \\ -\int_V [t^\tau_{q\ell}G^F_{iq,\ell'} + q^\tau_\ell G^H_{i,\ell'}/\theta_o + C_{qmk\ell}\gamma_{qm}\dot{u}^\tau_{k,\ell'}G^H_i]\, dV'dt & \text{if } \underline{x}' \not\in V \end{cases} \tag{25}$$

$$\theta(\underline{x},t) = \begin{cases} -\int_V [t^\tau_{q\ell}S^F_{q,\ell'} + q^\tau_\ell S^H_{,\ell'}/\theta_o + C_{qmk\ell}\gamma_{qm}S^H\dot{u}^\tau_{k,\ell'}]\, dV'dt' - T^\tau & \text{if } \underline{x}' \in V \\ -\int_V [t^\tau_{q\ell}S^F_{q,\ell'} + q^\tau_\ell S^H_{,\ell'}/\theta_o + C_{qmk\ell}\gamma_{qm}S^H\dot{u}^\tau_{k,\ell'}]\, dV'dt' & \text{if } \underline{x}' \not\in V \end{cases} \tag{26}$$

5. THERMOELASTIC-ELASTIC CORRESPONDENCE PRINCIPLE

The thermoelastic solid in this section is also assumed to be linear homogeneous and anisotropic. Its constitutive equations then read

$$t_{ij} = C_{ijk\ell}\, u_{k,\ell} - \beta_{ij}\, \theta \tag{27}$$

$$q_i = -k_{ij}\, \theta_{,j} \tag{28}$$

In addition, it is assumed in the following that both inertia and dynamical coupling terms may be neglected which is the case of a large collection of problems and also simplifies the algebra. For these cases, the general thermoelastic problem decomposes into two separate field equations which must be solved consecutively but independently ([6]).

The equation of heat conduction then reduces to

$$-k_{ij}\, \theta_{,ij} + \rho c_v \dot{\theta} - r' = 0 \tag{29}$$

where ρ is the density and c_v is the specific heat.
The condition of static equilibrium reduces to

$$C_{ijk\ell}\, u_{k,\ell j} - \beta_{ij}\, \theta_{,j} - X_i = 0 \tag{30}$$

where X_i are body forces.

The solution of eq. (29) is known while the solution of eq. (30) can be found using the Duhamel-Neumann analogy theorem. This thermoelastic-elastic correspondence principle states that instead of considering a thermal elastic solid with thermoelastic stresses (27) and body forces X_i, an equivalent elastic solid can be considered with elastic stresses σ_{ij} and body forces f_i given by

$$\sigma_{ij} = C_{ijk\ell}\, u_{k,\ell} \text{ and } f_i = X_i + \beta_{ij}\, \theta_{,j} \tag{31}$$

the displacements u_i being unchanged and the thermal changes θ being identically equal to zero.

The analytical solution of equation (30) can now be found in terms of elastic material multipoles only, defined as

$$P^F_{jsm_1\ldots m_{q-1}} = \Sigma\; f^\alpha_j\, d^\alpha_s d^\alpha_{m_1} \ldots d^\alpha_{m_{q-1}} \tag{32}$$

where d^r_s is the relative position vector of the point body force f^r_i.

Up to the monopole approximation, the displacement writes

$$u_i = \int G^F_{ij} dP_j + \int G^F_{ij,m'} dP_{jm} \tag{33}$$

where G^F_{ij} is the elastic Green function defined by the equation

$$C_{ij\ell m} G_{\ell n,mj}(\underline{x},\underline{x}') + \delta_{in}\delta(\underline{x}-\underline{x}') = 0 \tag{34}$$

In order that the material multipoles should represent material defects viewed here as sources of internal elastic stresses, it is required that

$$dP_j = -\beta_{jk}\Theta_{,k}dV' \text{ and } dP_{jm} = C_{jmnp}a_n dA'_p \quad (35)$$

The relation (35a) is the mechanical equilibrium condition to be satified by all physically possible defects.
The relation (35b) is the continuum mathematical description of the modelling of a defect through a transformation of the surface element dA'_p by a displacement $a_n(\underline{x}')$, the surface element then creating a source of elastic singularities. This point of view is compatible with a generalization of the Volterra dislocation if a cut is first made in the body along the surface A_L bounded by the curve L, then the two sides of the cut surface are slipped relative to each other by a displacement $a_k(\underline{x}')$

$$a_k(\underline{x}') = u_k \big|_{A_L^+} - u_k \big|_{A_L^-} \qquad \underline{x}' \varepsilon A_L \quad (36)$$

and finally the two sides are welded (perhaps by adding or removing material). The results (33), (35) are shown to be equally applicable for any type of defects in an anisotropic thermoelastic medium. For a certain type of defect, it is however necessary to give the physical surface A_L and its transformation $a_k(\underline{x}')$.
In particular, line defects (dislocation and disclination) can be created by a rigid body transformation defined by

$$a_k(\underline{x}') = b_k + \varepsilon_{kpq}\,\Omega_p(x'_q - x^o_q) \quad (37)$$

where b_k is the Burgers vector, Ω_p is the Frank vector. Modellings for the cases of cracks, volume defects, vacancies, etc ... can also be obtained.

We have therefore shown that using a thermoelastic-elastic correspondence principle, thermoelastic material defects in uncoupled thermoelastic solids can be modelled uniformly as continuum distributions of elastic monopoles. These results are being used to study the influence of the temperature on the mechanical properties of the materials with structural defects.

6. CONCLUSIONS

It has been shown that all defects in thermoelastic solids can be uniformly modelled by using the concept of distribution of thermoelastic surface multipoles dP^F_{ij} and dP^H_i. It was also shown that the problem of finding the self-stress in a thermoelastic body can be reduced to the problem of determining the deformation field and the thermal field of a thermoelastic body with fictitious body force loadings and fictitious body heat source loading. These latter can be derived by using the concept of thermoelastic multipoles and the Green's functions.

REFERENCES

[1] Eshelby, J.D., Progress in Solid Mechanics 11, Sneddon, I.N. and Hill, R., (eds.), (North Holland, Amsterdam, 1961).

[2] Kröner, E., Kontinuumstheorie der Versetzungen und Eigenspannungen (Springer-Verlag, Berlin, 1958).

[3] Mura, T., Micromechanics of Defects in Solids (Martinus Nijhoff Publishers, The Hague, 1982).

[4] Minagawa, S., Phys. Stat. Sol. (B) 124 (1984) 565.

[5] Hsieh, R.K.T., Int. J. Engng.Sci. 20, N:o 2 (1982) 261.

[6] Zhou, S.A. and Hsieh, R.K.T., Int.J.Engng.Sci. 23, N:o 11 (1985) 1197.

[7] Zhou, S.A. and Hsieh, R.K.T., A Statistical Theory of Elastic Materials with Micro-Defects, Int. J. Engng. Sci. 1986, in print.

[8] Hsieh, R.K.T., Vörös, G. and Kovàcs, I., Physica 101B (1980) 201.

[9] Kovàcs, I., Physica 94B (1978) 177.

[10] Kovàcs, I. and Zsoldos, L., Dislocations and Plastic Deformation (Pergamon Press, London, 1973).

[11] Steeds, J.W., Anisotropic Elasticity Theory of Dislocations (Clarendon Press, Oxford, 1972).

[12] Nowinski, J.L., Theory of Thermoelasticity with Applications (Sijthoff & Noordhoff International Publishers, Alpen Aan den Rijn, 1978).

[13] Barr, D.T. and Cleary, M.P., Int. J. Solids Structures 19 (1983) 73.

[14] Wu, K.C. and Hui, C.Y., A Complex-Variable Method for Two-Dimensional Internal Stress Problems and its Applications to Crack Growths in Nonelastic Materials, submitted to J. Applied Mechanics.

[15] Nowacki, W., Bull.Acad.Pol.Sci.,Ser.Sci.Techn. 12 (1964) No. 6 and 9; vol. 13 (1965) N:o 4.

Thermomechanical Couplings in Solids
H.D. Bui and Q.S. Nguyen (Editors)
Elsevier Science Publishers B.V. (North-Holland)

THE RELATIONSHIP BETWEEN BOUNDING THEORY AND THE THERMODYNAMICS OF METALS WITH SPECIAL EMPHASIS ON HIGH TEMPERATURE BEHAVIOUR

A.R.S. Ponter and J.A. Scaife
Department of Engineering, University of Leicester, Leicester, U.K.

The paper explores properties of a class of constitutive equations which model the high temperature creep of metals. The class is characterised by a quadratic free energy in terms of a set of internal state variables and the existance of a potential relating the thermodynamic fluxes and affinities. A local stability condition gives rise to a range of general properties of the continuum problem for cyclic load and temperature. The theory is related to Druckers Stability Postulates for time independent behaviour and to simplified methods of creep analysis.

1. INTRODUCTION

In recent years the modelling of the micromechanisms of deformation and degeneration in solids has advanced considerably, with a thermodynamic framework to the constitutive equation forming an essential element in the argument. For the more complex aspects of material behaviour, such as high temperature creep, there remains, however, the problem of providing sufficient material data to ensure that the resulting constitutive equations adequately predict behaviour over complex cycles and the considerable computational effort involved in constructing step-by-step numerical solutions. There is some advantage, therefore, in gaining insight into extreme modes of behaviour and the construction of simplified methods of analysis which make a reduced demand upon material data and computational effort.

This paper is concerned with theoretical results which assist in this endeavour for a class of constitutive relationships which model high temperature behaviour where the material is capable of exhibiting steady state creep strain rates. The constitutive equation is constructed from the assumptions that the free energy is quadratic in the state variables and that a potential exists for the rate equations. This class of materials is identical to the class discussed by Ponter [1] and the results in this paper may be regarded as extensions of these earlier results.

Section (2) discusses, from first principles, the thermodynamic assumptions which underly the chosen class of constitutive models. In section (3) two distinct modes of behaviour are then discussed. In the first there exist regions of stress space at a given state where the inelastic strain rate $\underline{\dot{\varepsilon}}'$ is zero. Processes which commence and terminate at this condition may then be regarded as plastic processes i.e. the time independent mode of behaviour of the solid. This view accords with the experimental view that plasticity is a particular mode of behaviour associated with a certain range of strain rate and temperature. In this way it is possible to demonstrate the relationship between restrictions placed upon the time dependent constitutive equations and the Drucker Stability Postulates for plastic processes. The second mode of behaviour is when there exists no region of stress space where $\dot{\underline{\varepsilon}}' = 0$, corresponding to high temperature creep behaviour. In this case we look for extreme modes which bound the cyclic state set up in a body subjected to cyclic load and temperature. We show that there exists two extreme modes which correspond to the actual behaviour when the cycle time is very small and very large compared with characteristic material time scales, giving rise to rapid and slow cycle solutions. As well as giving an understanding of the dependency of the state histories on the cycle time, the extreme solutions can be evaluated rather more easily that the full cyclic solutions and therefore form the basis for simplified methods of analysis. Finally, the relevance of these solutions to creep analysis is briefly discussed.

2. THE CONSTITUTIVE RELATIONSHIP

Classical thermodynamics ([2], for example) assumes that the internal free energy Ψ is a function of a set of extensive parameters which are observable, in principle, and that a reversible path can be conceived, if not realised, from a ground state where $\Psi = 0$ to any given state. In metals the internal state involves a complex dislocation structures and the free energy is the elastic strain energy associated with this structure. The conceptual reversible path involves the formation of dislocations by an external agency [3]. The characterisation of the set of such structures which occur during a deformation process seems to require a large number of scalar and tensor parameters, but attempts to model metallic creep have usually assumed a single scalar and a single second order tensor, representing the isotropic and kinematic observed behaviour.

As we are concerned with metallic behaviour at high temperature, there is a need to emphasise that an important characteristic of creep behaviour under constant stress is the existance of a stationary state of creep where the inelastic strain increases with time with no change in the mean properties of the dislocation structure. In this process the inelastic strain $\underline{\varepsilon}'$ increases at constant Ψ . In our construction of Ψ we must, therefore, accept that there exists reversible paths where $\underline{\varepsilon}'$ changes but Ψ remains constant. There will, of course, be particular processes where a functional relationship exists between $\underline{\varepsilon}'$ and the internal state (e.g. kinematic hardening plasticity) but this will arise from integration of the rate equations over the state history and not from the definition of Ψ .

With this preamble in mind we can now define the reversible path from the ground state to final state, characterised by the total strain $\underline{\varepsilon}$, the inelastic strain $\underline{\varepsilon}'$, temperature change $\Theta - \Theta_o$ and internal state variable $\underline{x}$ which characterise the mean state of the dislocation structure, as follows:

a) The formation of the dislocation structure so that

$$\rho \Psi = f(\underline{x})$$

where f, the latent energy of the dislocation structure, is quadratic in $\underline{x}$. This process will cause inelastic strain of $\underline{\varepsilon}' = \underline{\varepsilon}^a$.

b) A change in the inelastic strain from $\underline{\varepsilon}^a$ to $\underline{\varepsilon}'$ with no change in Ψ .

c) A change in temperature from Θ_o to Θ , producing a change in inelastic strain to $\underline{\varepsilon}' + \underline{\delta}\ \alpha\ (\Theta - \Theta_o)$ where δ is the Kroneker delta tensor and α the coefficient of linear expansion.

d) A change in $\underline{\varepsilon}$ to its final value with Θ and $\underline{x}$ maintained constant.

The resulting free energy per unit mass is given by

$$\rho\ \Psi = \frac{1}{2}\ \underset{\sim}{C}\ \underline{e} \cdot \underline{e} + f\ (\underline{x}) + \rho\ g\ (\Theta), \tag{1}$$

where

$$\underline{e} = \underline{\varepsilon} - \underline{\varepsilon}' - \underline{\delta}\ \alpha\ (\Theta - \Theta_o), \tag{2}$$

$$f\ (\underline{x}) = \frac{1}{2}\ \underset{\sim}{H}\ \underline{x}\ \underline{x} \tag{3}$$

Here ρ is the material density which is assumed to be constant, $\underset{\sim}{C}$ is the elastic compliance tensor, $\underline{e}$ the elastic strain and $\underset{\sim}{H}$ a symmetry constant tensor.

The Gibbs equation and the thermodynamic forces are then given by

$$d\ \Psi = \underline{\sigma}\ d\ \underline{\varepsilon} - \underline{\sigma}\ d\ \underline{\varepsilon}' - \zeta\ d\ \Theta + \underline{X}\ d\ \underline{x} \tag{4}$$

where

$$\left.\begin{aligned} \rho\ \frac{\partial \Psi}{\partial \underline{\varepsilon}} &= \underset{\sim}{C}\ [\ \underline{\varepsilon} - \underline{\varepsilon}' - \underline{\delta}\ \alpha\ (\Theta - \Theta_o)\ ] = \underline{\sigma} \\ \rho\ \frac{\partial \Psi}{\partial \underline{\varepsilon}'} &= -\ \underline{\sigma} \\ -\rho\ \frac{\partial \Psi}{\partial \Theta} &= \underline{\sigma}\ \underline{\delta}\ (\ \alpha\ (\Theta - \Theta_o)\ ) - \rho\ g\ (\ \Theta\ ) = \zeta \\ \rho\ \frac{\partial \Psi}{\partial \underline{x}} &= \underset{\sim}{H}\ \underline{x} = \underline{X} \end{aligned}\right\} \tag{5}$$

The internal energy $U\ (\underline{\varepsilon}, \underline{\varepsilon}', \underline{x}, \zeta)$ is defined by

$$U = \Psi + \zeta\ \Theta \tag{6}$$

and, by the first law,

$$\dot{U} = \dot{W} + \dot{q}\ , \tag{7}$$

where $\dot{W} = \underline{\sigma}\ \underline{\dot{\varepsilon}}$ is the rate of work done on the material element and $\dot{q}$ is the rate of heat transfer to the element. The internal dissipation D per unit volume is now given by the difference between the rate of heat production $\rho\ \Theta\ \dot{\zeta}$ due to entropy change and heat transfer $\rho\ \dot{q}$. Hence, from (4) (5) (6) and (7)

$$D = \rho\ (\Theta\dot{\zeta} - \dot{q}) = \{\underline{\sigma}\,\dot{\underline{\varepsilon}}' - \underline{X}\ \dot{\underline{x}}\} = \underline{\Sigma}\ \dot{\underline{\eta}} \tag{8}$$

where the fluxes $\dot{\underline{\eta}}$ and affinities $\underline{\Sigma}$ are given by

$$\dot{\underline{\eta}} = (\dot{\underline{\varepsilon}}', \dot{\underline{x}})\ , \underline{\Sigma} = (\underline{\sigma}, -\underline{X}) \tag{9}$$

The dissipation D is composed of the difference between the work done on the element associated with inelastic strain $\underline{\sigma}\ \dot{\underline{\varepsilon}}'$ and the rate of increase in the stored or latent energy $\dot{f} = \underline{X}\ \dot{\underline{x}}$. The second law requires D to be positive. In experimental studies Taylor and Quinney [4],and many others, have found that the latent energy in a uniaxial plastic test is approximately 10 - 15% of the work associated with inelastic strain.

Finally we assume the existance of a flow potential $\Omega(\underline{\Sigma})$, [5]

$$\dot{\underline{\eta}} = \frac{\partial\Omega}{\partial\underline{\Sigma}} \tag{10}$$

or

$$\dot{\underline{\varepsilon}}' = \frac{\partial\Omega}{\partial\underline{\sigma}} \quad \text{and} \quad \dot{\underline{x}} = -\frac{\partial\Omega}{\partial\underline{X}}\ . \tag{11}$$

For an elastically isotropic material these relationships may be put in a more specific form if we assume that the state may be adequately represented by a scalar parameter β and a second order tensor α_{ij} with zero trace. As f is both quadratic and a function of the invariants of α_{ij}

$$f = L\ \frac{\beta^2}{2} + \frac{M}{2}\ \alpha_{ij}\ \alpha_{ij}\ , \tag{12}$$

where L and M are material constants. The rate equation (11) then takes the form

$$\left.\begin{aligned} \dot{\beta} &= -\frac{\partial\Omega}{\partial B}\ , \quad B = L\,\beta \\ \dot{\alpha}_{ij} &= -\frac{\partial\Omega}{\partial A_{ij}}\ , \quad A_{ij} = M\ \alpha_{ij} \end{aligned}\right\} \tag{13}$$

This form of constitutive equation was adopted by Ponter [1] to establish a number of results related to stability of behaviour in creep. In the general case when f is given by (3) as f is a positive definite quadratic form there exists a linear transformation which changes f to the canonical form

$$f = \frac{1}{2} H_i \; x_i^2 \quad \text{and hence} \quad \dot{x}_i = - \frac{\partial \Omega}{\partial X_i} \quad \text{and} \quad X_i = H_i \; x_i \tag{14}$$

the rate equations then also have the canonical form adopted by Ponter [1] where the individual state variable and affinities are proportional to each other.

3. THE CONVEXITY OF Ω AND RELATED INEQUALITY CONSTRAINTS

The primary concern of this paper is the relationship between local instantaneous properties of the constitutive equation and the behaviour of both the material element and a continuum when subject to a process in the form of history of load and temperature. The existance of Ω and the positive definiteness of the free energy provides certain local properties. For a prescribed $\underline{\sigma}$, there exists certain states with associated affinities $\underline{X} = \underline{X}^{ss} (\underline{\sigma})$, when $\dot{\underline{x}} = 0$. These states define the steady state properties of the materials where the inelastic strain rate $\dot{\underline{\varepsilon}}' = \dot{\underline{\varepsilon}}'^{ss}$ may be expressed in terms of a potential $\Omega^{ss} (\underline{\sigma}) = \Omega (\underline{\sigma} , \underline{X}^{ss})$, as

$$\frac{\partial \Omega^{ss}}{\partial \underline{\sigma}} = \frac{\partial \Omega}{\partial \underline{\sigma}} - \frac{\partial \Omega}{\partial \underline{X}} \frac{\partial \underline{X}}{\partial \underline{\sigma}} = \dot{\underline{\varepsilon}}' \tag{15}$$

When, for a range of values of $\underline{\sigma}$, the steady state rate $\dot{\underline{\varepsilon}}^{ss} = 0$, and $\Omega = 0$ the material may be considered to have time independent properties as a change in stress can initiate and terminate in such states. The instantaneous yield surface is then the boundary of the region in stress space where $\dot{\underline{\varepsilon}}^{ss} = 0$.

For a small but finite change $\Delta \underline{\sigma}$ so that

$$\underline{\sigma} (t) = \underline{\sigma} (t_o) + \Delta \underline{\sigma} \; (t-t_o) / \Delta t \quad \text{for } t_o \leq t \leq t_o + \Delta t = t_1$$

then $$\int_{t_o}^{t_1} \dot{\Omega} \, dt = \int_{t_o}^{t_1} \{ \dot{\underline{\sigma}} \, \dot{\underline{\varepsilon}}' - \dot{\underline{x}} \, \dot{\underline{X}} \} \, dt = \Omega (t_1) - \Omega (t_o) = 0$$

as $$\dot{\underline{x}} \, \dot{\underline{X}} = 2 \, f \, (\dot{\underline{x}}) \geq 0 \quad \text{then}$$

$$\int_{t_o}^{t_1} \dot{\underline{\sigma}} \, \dot{\underline{\varepsilon}}' \, dt \geq 0 \quad \text{and hence} \quad \Delta \underline{\sigma} \, \Delta \underline{\varepsilon}' \geq 0 , \tag{16}$$

where $\Delta \underline{\varepsilon}'$ is the inelastic strain produced during Δt.

Inequality (16) may be recognised as Druckers stability postulate in the small [6] .

Further process properties may only be derived by placing further restrictions concerned with a definition of stable material behaviour. At constant state an increase in stress $d\,\underline{\sigma}$ produces an increase in strain rate $d\,\dot{\underline{\varepsilon}}'$ or, more generally, an increase in affinity $d\,\underline{\Sigma}$ produces an increase in flux $d\,\dot{\underline{\eta}}$, i.e.

$$d\,\underline{\Sigma}\;d\,\dot{\underline{\eta}} = d\,\underline{\sigma}\;d\,\dot{\underline{\varepsilon}}' - d\,\underline{X}\;d\,\dot{\underline{x}} \geq 0 \tag{17}$$

Inequality (17) is a necessary and sufficient condition [1] for the convexity of Ω

$$\Omega\,(\underline{\Sigma}_1) - \Omega\,(\underline{\Sigma}_2) - (\underline{\Sigma}_1 - \underline{\Sigma}_2)\,\frac{\partial\Omega}{\partial\underline{\Sigma}_2} \geq 0 \tag{18}$$

When $\underline{x}$ arises from dislocation structure inequality (17) implies that an increase in slip involves an increase in back stress which resists the motion. It should be noted, however, that state variables which describe the development of internal damage [7] would not satisfy this inequality.

Before proceeding to a discussion of time dependent properties it is worth considering the consequences for a finite plastic process conducted at a sufficiently slow rate that Ω remains close to zero during an entire process from t_o to t. Identifying $\underline{\Sigma}_1$ as the affinities at t_o and $\underline{\Sigma}_2$ as those at a subsequent time t', then (18) leads to

$$\int_{t_o}^{t} \{\underline{\sigma}(t') - \underline{\sigma}(t_o)\}\,\dot{\underline{\varepsilon}}'\,dt' \geq \int_{t_o}^{t} \{\underline{X}(t') - \underline{X}(t_o)\}\,\dot{\underline{x}}\,dt' - \int_{t_o}^{t} \{\Omega(t') - \Omega(t_o)\}\,dt'$$

now $$\int_{t_o}^{t} \{\underline{X}(t') - \underline{X}(t_o)\,\dot{\underline{x}}\,dt = \int_{t_o}^{t} \dot{f}\,(\underline{x}(t') - \underline{x}(t_o))\,dt = f(\underline{x}(t) - \underline{x}(t_o)) \geq 0$$

For a sufficiently slow process lying at the boundary of the region where $\Omega = 0$ the value of $\Omega\,(t)$ may be made arbitrarily small whereas $f\,(\underline{x}\,(t) - x\,(t_o))$ is finite for a finite change in state. Hence

$$\int_{t_o}^{t} \{\underline{\sigma}(t') - \sigma(t_o)\}\,\dot{\underline{\varepsilon}}'\,dt \geq 0 \tag{19}$$

which may be recognised as Druckers Stability Postulate in the large [6] which can be seen as a property of a particular type of process associated with time independent behaviour.

4. EXTREME MODES OF BEHAVIOUR OF A CONTINUUM FOR TIME DEPENDENT BEHAVIOUR

Our primary concern is not with the time independent processes but time dependent processes which take place at high temperatures and over long periods of time, where there exists no finite regions of stress space in which $\Omega = 0$. Equality in (18) may then occur only when $\underline{\Sigma}_1 = \underline{\Sigma}_2$. In such circumstances the convexity condition gives rise to a rang of inequalities which form the bases of simplified methods of analysis for bodies subjected to cyclic loading.

Consider a body with volume V and surface S subjected to a history of surface traction $\underline{P}$ (t) over part of the surface S_T and zero displacements over the remainder of S, S_u. The volume is subjected to a history of temperature Θ and it is assumed that the strains are small. Initially, at $t = 0$, the body has a residual stress field $\underline{\rho}(0) = \underline{\sigma}(0) - \hat{\underline{\sigma}}(0)$ where $\hat{\underline{\sigma}}$ is the elastic stress distribution corresponding to $\underline{\varepsilon}' = 0$, and initial state $\underline{x}(0)$.

In [1] a number of results were derived which demonstrated that a material which satisfies the inequality (17) and hence the convexity condition (18) exhibit certain stable properties. Two identical bodies with differing initial states $\underline{\rho}(0)$ and $\underline{X}(0)$ but identical histories of loading will approach a common history of $\underline{\sigma}$ and $\underline{X}$ with increasing time although they will differ in their total strain. For cyclic loading the common asymptotic state is also cyclic. For constant $\underline{P}$ and Θ this defines a unique asymptotic state, the steady state, characterised by a constitutive relationship for the total strain rate in terms of the steady state potential Ω^{ss} ;

$$\dot{\underline{\varepsilon}} = \left[\frac{\partial \Omega^{ss}}{\partial \underline{\sigma}} \right]_{\underline{\sigma}^{ss}} \tag{20}$$

Further Ω^{ss} is convex, and $\int_V \Omega^{ss}(\underline{\sigma}^{ss})\, dV$ is an absolute minimum amongst all stress fields in equilibrium with $\underline{P}$ on S_T. Hence the classical steady state creep analysis [13] is seen as the definition of a limit state which requires knowledge of a limited range of material behaviour, the observed performance of the material when it approaches a constant strain rate some time after initial loading.

This situation is shown schematically in Figure (1) where contours of constant Ω are shown for uniaxial stress σ and a single state variable x. A contour C D exists along which $\dot{x} = 0$ and $X = X^{ss}(\sigma)$. If at time $t = 0$ the material is at point A and σ remains constant at σ_o then X will increase in time until it reaches the steady state at B.

In the more important problems when $\underline{P}$ (t) and Θ (x ,t) are cyclic with cycle time Δt, the extreme modes are somewhat more complex in form as the value of Δt compared with characteristic material time scale becomes important. At each instant t we may define a corresponding steady state solution where $\dot{\underline{x}} = 0$, $\underline{X} = \underline{X}^{ss}$ and $\dot{\underline{\varepsilon}}$ is given by (20) . From the convexity condition (18) integrated over the volume it follows that the actual value of Ω at any time is bounded from below by this steady state value.

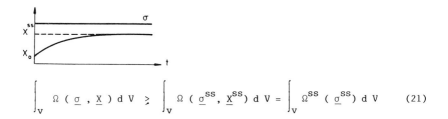

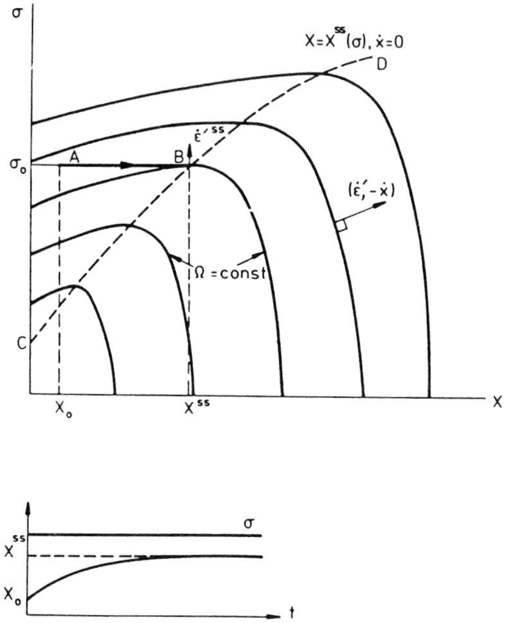

Figure 1: Schematic representation of the behaviour of the model for constant uniaxial stress σ for a single state variable x .

$$\int_V \Omega(\underline{\sigma}, \underline{X})\, dV \geq \int_V \Omega(\underline{\sigma}^{ss}, \underline{X}^{ss})\, dV = \int_V \Omega^{ss}(\underline{\sigma}^{ss})\, dV \qquad (21)$$

The mean value of Ω over the cycle may also be bounded from above. For this we need to define a cyclic steady state solution $\underline{\sigma}$ as the limiting case as $\Delta t \to 0$. This "rapid cycle" solution is discussed in [8] and has the form

$$\underline{\sigma} = \underline{\sigma}^* = \hat{\underline{\sigma}} + \bar{\underline{\rho}}\ , \quad \underline{X} = \underline{X}^* = \text{constant} \qquad (22)$$

where $\hat{\underline{\sigma}}$ is the thermoelastic solution (i.e. $\Omega = 0$) and $\bar{\underline{\rho}}$ is a constant residual stress field. The stationary affinities $\underline{X}^*$ and $\bar{\underline{\rho}}$ are given by the following condition on the average rate of change of the flux taking $t = 0$ as the beginning of a cycle,

$$\dot{\tilde{\underline{\varepsilon}}} = \frac{\Delta \underline{\varepsilon}'}{\Delta t} = \lim_{\Delta t \to 0} \frac{1}{\Delta t} \int_0^{\Delta t} \frac{\partial \Omega}{\partial \underline{\sigma}}\, dt \qquad (23)$$

$$\dot{\tilde{\underline{x}}} = \frac{\Delta x}{\Delta t} = - \lim_{\Delta t \to 0} \frac{1}{\Delta t} \int_0^{\Delta t} \frac{\partial \Omega}{\partial \underline{X}} \, dt = 0 \tag{24}$$

where the average strain rate $\dot{\tilde{\underline{\varepsilon}}}$ satisfies the compatibility equations. Although this solution is defined for $\Delta t \to 0$, it provides a bounding solution for finite values of Δt . If we take $(\underline{\sigma}_1, \underline{X}_1) = (\underline{\sigma}^*, \underline{X}^*)$ and $(\underline{\sigma}_2, \underline{X}_2) = (\underline{\sigma}, \underline{X})$, the actual cyclic history, in the convexity condition (18) we obtain

$$\Omega (\underline{\sigma}^*, \underline{X}^*) - \Omega (\underline{\sigma}, \underline{X}) - \dot{\underline{\varepsilon}}' (\underline{\sigma}^* - \underline{\sigma}) + \dot{\underline{x}} (\underline{X}^* - \underline{X}) \geq 0 \tag{25}$$

Noting that, if E is the complimentary elastic strain energy

$$E (\underline{\sigma}) = \frac{1}{2} \underset{\sim}{C}^{-1} \underline{\sigma} \, \underline{\sigma} \geq 0 ,$$

then
$$\int_V \frac{d}{dt} E (\underline{\sigma}^* - \underline{\sigma}) \, dV = - \int_V \dot{\underline{e}} (\underline{\sigma}^* - \underline{\sigma}) \, dV$$

and also that $\int_V \dot{\underline{\varepsilon}} (\underline{\sigma} - \underline{\sigma}^*) \, dV = 0$ as $(\underline{\sigma} - \underline{\sigma}^*)$ is a residual stress field, then the convexity condition (25) becomes, on integrating over the volume V and cycle Δt

$$\int_0^{\Delta t} \int_V \Omega (\underline{\sigma}^*, \underline{X}^*) \, dV \, dt \geq \int_0^{\Delta t} \int_V \Omega (\underline{\sigma}, \underline{X}) \, dV \, dt$$

$$+ \int_0^{\Delta t} \int_V \frac{d}{dt} \left\{ E (\underline{\sigma} - \underline{\sigma}^*) + f (\underline{x} - \underline{x}^*) \right\} \underline{dV} \, \underline{dt} \tag{26}$$

When $\underline{\sigma}$ and $\underline{X}$ have reached the asymptotic cyclic state so that $\underline{\sigma} (0) = \underline{\sigma} (\Delta t)$ and $\underline{X} (0) = \underline{X} (\Delta t)$ then inequality (26) shows that the average of Ω over both the volume and the cycle is bounded from above by the value associated with the rapid cycle solution $\underline{\sigma}^*$ and $\underline{X}^*$,

$$\int_0^{\Delta t} \int_V \Omega (\underline{\sigma}^*, \underline{X}^*) \, dV \, dt \geq \int_0^{\Delta t} \int_V \Omega (\underline{\sigma}, \underline{X}) \, dV \, dt \tag{27}$$

We see, therefore, that the two solutions $(\underline{\sigma}^*, \underline{X}^*)$ and $(\underline{\sigma}^{ss}, \underline{X}^{ss})$ bound, in an average sense, the actual behaviour for any cycle time. The actual histories $(\underline{\sigma}, \underline{X})$ approaches these two extremes in the limits as Δt

becomes small and long respectively compared with characteristic material times.

At the level of a material element the situation is demonstrated schematically in Figure (2) for stress varying between σ_1 and σ_2. For slow cycling the (σ, X) point follows the $\dot{x} = 0$ steady state line whereas for rapid cycling $X = X^*$ remains constant and the total change in x over the cycle is zero.

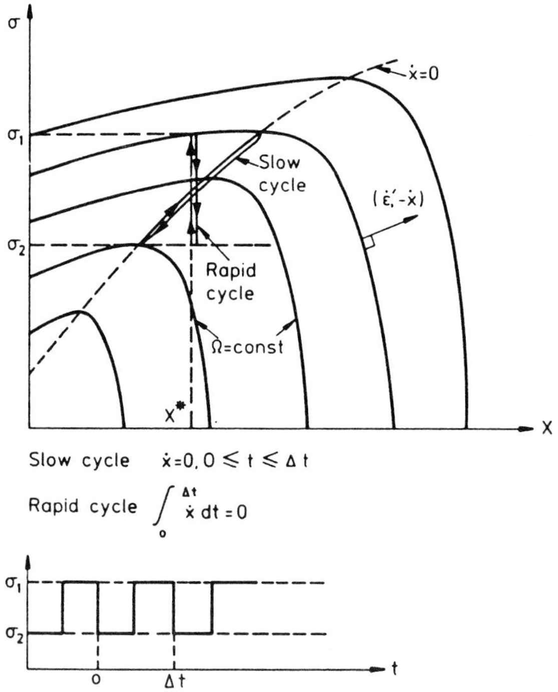

Figure 2: Schematic representation of the behaviour of the model for variable unaxial stress $\sigma(t)$. The two extreme modes of response, the rapid and slow cycles, are shown.

In reality metals and alloys exhibit a range of material time scales which stretch for a fraction of a second, i.e. the phenomena shown in dynamical plasticity tests, to times associated with thermally activated processes in long term creep tests. Constitutive equations, however, which have been developed to model material behaviour taken from a specific range of material tests usually involved only a limited range of time scales. For creep constitutive equations there are usually a small number, associated with steady state creep behaviour, primary creep strains and recovery processes.

For practical calculations results involving directly observable quantities are of greater interest. For the particular case when Ω is homogenous of degree $(n + 1)$ in both $\underline{\sigma}$ and $\underline{X}$ then Ω and the dissipation D are proportioned to each other,

as $$D = \underline{\sigma}\,\dot{\underline{\varepsilon}}' - \underline{X}\,\dot{\underline{x}} = \underline{\Sigma}\,\frac{\partial\Omega}{\partial\underline{\Sigma}} = (n+1)\,\Omega$$

In a cyclic state where $\underline{\sigma}(0) = \underline{\sigma}(\Delta t)$ and $\underline{X}(0) = \underline{X}(\Delta t)$ the average value of Ω then becomes

$$\int_0^{\Delta t}\int_V \Omega \, dV\, dt = \frac{1}{n+1}\int_0^{\Delta t}\int_V \underline{\sigma}\,\dot{\underline{\varepsilon}}'\, dt\, dV$$

as $$\int_0^{\Delta t} \underline{X}\,\dot{\underline{x}}\, dt = f(\Delta t) - f(0) = 0$$

When the temperature remains constant

$$\int_V\int_0^{\Delta t} \underline{\sigma}\,\dot{\underline{\varepsilon}}'\, dt\, dV = \int_V\int_0^{\Delta t} \underline{\sigma}\,\dot{\underline{\varepsilon}}\, dt\, dV = \int_0^{\Delta t}\int_S \underline{P}\,\dot{\underline{u}}\, dS\, dt$$

as $\int_0^{\Delta t} \underline{\sigma}\,\dot{\underline{e}}\, dt = 0$. The inequalities (21) and (27) now become bounds on the work done by the applied load, implying that the average velocity fields $\dot{\underline{u}}$ over a cycle given by the two extreme solutions bound the actual velocity field, so that the rapid cycle solution would tend to give an overestimate and the slow cycle solution an underestimate of the steady state creep rate.

This point of view has been investigated experimentally using commercially pure copper and ASTM 316 austenitic stainless steel for cyclic thermal loading problems for a two bar structure [9,10] and for a thermally shocked plate [11] . The constitutive equation adopted was a simple form of the Bailey-Orowan recovery model, involving a single scalar affinity S ,

$$\Omega(\underline{\sigma}, S) = F\left[\phi(\underline{\sigma}) - S\right] + g(S)$$

where $\phi(\underline{\sigma})$ is the Von-Mises effective stress, F a function whose derivative F^1 is given by

$$F^1 \geq 0\,,\quad \phi = S$$

$$F^1 = 0\,,\quad \phi < S$$

and $$g(S) = k\,S^{n+1}/(n+1)$$

For uniaxial stress $\phi = \sigma$ contours of constant Ω are shown schematically in Figure (3). The rapid cycle behaviour for stress ranging between σ_1 and σ_2 is shown as well as the slow cycle. The evaluation of the rapid cycle solution is particularly easy for this material as the average strain rate is given in terms of the steady state strain rates at the extremes of the stress cycle.

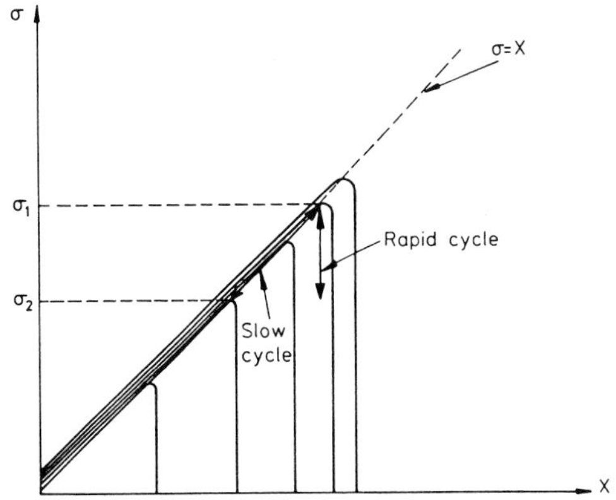

Figure 3: For the Bailey-Orowan model the inelastic strain occurs at the extreme of the cycle when $\sigma = X = S$.

All the experiments [9,10,11] demonstrated that the two solutions provided bounds to the experimentally observed average velocity of deformation when a steady cyclic state had been reached with a tendancy for the experimental results to approach the rapid cycle solutions when the applied loads are less than the high temperature shakedown limit.

The results of this paper may be extended to bounds on a wider range of deformation properties by evaluating bounds on the functional

$$W_o^T (\underline{\sigma}, \underline{\sigma}^*) = \int_o^T (\underline{\sigma}^* - \underline{\sigma}) \, \dot{\underline{\varepsilon}}' \, dt$$

where $\underline{\sigma}^*$ is a prescribed history of stress and $\underline{\sigma}$ is the stress history corresponding to $\dot{\underline{\varepsilon}}'$. An upper bound on W_o^T for arbitrary $\underline{\sigma}$ (t) but prescribed initial state forms the basis for bounds on the inelastic work and displacements [12] . These methods have been explored by Scaife [14] for the particular case where

$$\Omega = \Omega_1 + \Omega_2$$

where Ω_i is homogenous of degree n_i in its arguments. The resulting bounds are insensitive to the details of both ψ and Ω and gives results which are very similar to the bounds for the Bailey-Orowan model, provided the steady state creep behaviour under constant load and temperature are identical. This implies that the rapid cycle solution for the Bailey-Orowan model may be accepted as a safe bound for a range of constitutive relationships. The implication of this work, therefore, is that safe estimates of long term deformation rates in creep can be calculated from knowledge of constant stress data without the need for a detailed material description.

REFERENCES

[1] Ponter, A R S., "Convexity and Associated Continuum Properties of a Class of Constitutive Relationships", Jn. de Mecanique, Vol. 15, (1976) pp527-542.
[2] Callen, H B., "Thermodynamics", (John Wiley, 1960).
[3] Ponter, A R S., Bataille, J., and Kestin, J., "A Thermodynamic Model for the Time Independent Plastic Deformation of Solids", Jn. de Mecanique, Vol. 18, (1979), pp511-539.
[4] Taylor, G I., and Quinney, H., "The Latent Energy Remaining in a Metal After Cold Working", Proc. Roy. Soc., Vol. 143, (1984), p307.
[5] Rice, J R., Jn. Mech. Physics Solids, Vol. 19, (1971), pp433-455.
[6] Drucker, D C., "A More Fundamental Approach to Plastic Stress-Strain Relations", Proc. 1st U.S. Nat. Cong. Appl. Mech. (1950), p487.
[7] Cocks, A C F., and Leckie, F A., In this volume.
[8] Ponter, A R S., "The Analysis of Cyclically Loaded Creeping Structures for Short Cycle Times", Int. J. Solids Structures, (1976), Vol. 12, pp809-825.
[9] Ponter, A R S., and Megahed, M M., "Creep and Plastic Ratchetting in Cyclically Thermally Loaded Structures", Proc. IUTAM Symposium "Physical, Non-linearities in Structural Analysis", (Ed J Hult and J Lemaitre), Springer-Verlag, (1981), pp220-227.
[10] Megahed, M M., Ponter, A R S., and Morrison, C J., "An Experimental and Theoretical Investigation into the Creep Properties of a Simple Structure of 316 Stainless Steel", Int. J. Mech. Sci., (1984), Vol. 26, pp149-154.
[11] Ponter, A R S., Jakeman, R R., and Morrison, C J., "An Experimental Study of Simplified Methods for the Prediction of the Deformation of Structures Subject to Severe Cyclic Thermal Loading", Jn. of Strain Analysis, Vol. 20, (1985), pp225-240.
[12] Ponter, A R S., "General Displacement and Work Bounds for Dynamically Loaded Bodies", J. Mech. Phys. Solids, Vol. 23, (1975), pp151-163.
[13] Odquist, F K G., "Mathematical Theory of Creep and Creep Rupture", Oxford University Press, 2nd Ed. (1973).
[14] Scaife, J A., PhD Thesis, University of Leicester, (1985).

Thermomechanical Couplings in Solids
H.D. Bui and Q.S. Nguyen (Editors)
Elsevier Science Publishers B.V. (North-Holland)

THEORETICAL SCHEMES OF THERMOMECHANICAL COUPLING

D. FAVIER GPM2,UA CNRS, INPG, BP.46-38402 St. Martin d'Heres-France
P. GUELIN I.M.G.,UA CNRS, BP.68 - 38402 St. Martin d'Heres- France
W.K. NOWACKI I.P.P.T., Swietokrzyska, 21,00-49 Warsaw, Poland
P. PEGON C.C.E, C.C.R., ISPRA, 21020 Ispra, Italy

The objective of this paper is two-fold. First, we shall present an analysis of the links between the discrete memory concept, the constitutive schemes of finite elastoplasticity and the thermal effects arising from deformation: the case of variable temperature is introduced through the definition of a provisional scheme of superelasticity and shape memory effects. Secondly, some results of quasi-static cyclic tests are given in order to underline the validity of the use of the discrete memory notion.

1.- INTRODUCTION

The purpose of this paper is to show the ability of a constitutive scheme of discrete memory type to describe qualitatively some aspects of metals behaviour: thermomechanical hysteresis, combined effects of hysteresis type and reversible type such as superelasticity and shape memory, second order effects such as ratchet. However, the rate independent hardening and the viscous hardening are not taken into consideration. In so far as the study remains in the field of continuum idealization , it may be considered as a three-fold proposal : first, it gives a scheme of finite perfect elastoplasticity for constant temperature, isotropic, homogeneous continuum, under arbitrary non rotational or rotational cyclic loading path ; secondly, the new definition of an hyperelastic stress contribution is introduced through a method well suited for the analysis of coupling effects and second order effects ; thirdly, the combined schemes of plasticity and hyperelasticity are used to introduce a theory of thermoplasticity. Both the problem of heat rate effects arising from deformation and the various problems of relative orientation arising from rotational kinematics are underlined throughout the presentation. The outstanding rôle of the discrete memory notion expressed through material descriptions in dragged along coordinate systems is also pointed out.

2.- FROM ONE-DIMENSIONAL PURE HYSTERESIS BEHAVIOUR TO THE INTRODUCTION OF DIFFERENTIAL GEOMETRY THROUGH THE DISCRETE MEMORY CONCEPT.

2.1- Previous studies have shown |1 to 12| that the heuristic consideration of symbolic models with springs and friction sliders points out the outstanding rôle that the discrete memory concept plays in the development of thermomechanical schemes describing hysteresis. In particular, it has been shown |3| that the discrete memory forms of the intrinsic dissipation and of the reversible power can be, at least partly, derived from the complete thermomechanical study of such symbolic models: such a possibility is of importance |13|. It must be pointed out that well known experimental facts are at the origin of the analysis based on the consideration of symbolic models |14 to 18|. The only readily suggested patterns are those for which mechanical properties are periodically restored under periodic loading. Some of these schemes are, for example, the symbolic models consisting of an infinite parallel succession of pairs springs-friction sliders associated in serie .Using such models, one has to take care to make the distinction between the stress contributions of pairs whose friction sliders move (σ_2), and the others (σ_1). Then, the power of external forces is: $P_e = \sigma_1 D + \sigma_2 D$,where D denotes the strain rate.

The model enables one not to use the notion of boundary between an elastic and a plastic domain.It allows also a simple and straight-forward introduction of the discrete memory notion.Moreover,the model enables one to point out the rôle of the similarity functional ω expressing the similarity of each branch, with respect to the first loading branch; for a given deformation ε the stress supported by the model is for all the branches :

(1) $\sigma - {}^t_R\sigma = \omega S((\varepsilon - {}^t_R\varepsilon)/\omega)$; $\omega = 1$ or 2 ,where ${}^t_R\sigma$ and ${}^t_R\varepsilon$ denote the stress and the strain at the origin of the current branch.

The Masing coefficient ω equals one for the first loading branch and two for the other branches. Consequently, ω is a piecewise constant functional of the loading history. Furthermore the two fundamental and non-classical thermomechanical properties of the model are that, first, each possible path is irreversible and that, secondly, it is always possible to return the model to its original state; this can be obtained by a great number of almost symmetrical cycles, whose amplitudes decrease slowly (fundamental cyclic path). Such non classical properties underline the interest which must be devoted in search of the well founded expressions of the reversible power π and of the intrinsic dissipation ϕ. One defines π as:

(2) $\pi = {}^t_R\sigma \; D$

Consequently π is, like ω and ${}^t_R\sigma$ a piecewise constant functional of the specified loading history.The obtained form of π suggests a three dimensional extension, leading also to the expression of the intrinsic dissipation ϕ:

$$\phi(t) = (\sigma\,(t) - {}^t_R\sigma)D = \Delta^t_R\sigma\; D$$

and to the definition of an inversion criterion (or "loading-unloading" criterion). In fact,let one introduce a "help function" W defined by :

$$dW = \phi\; dt \; ; \; W\,(t_{I+}) = 0 \; ; \; W\,(0_+) = 0$$

where t_I is the inversion time. Then the inversion criterion is as follows : the Cauchy stress $\sigma(t_I)$ is an inversion state σ_I if the help function virtual variation δW is negative on the time interval $]t_I, t_I+dt]$:

(3) $$\sigma(t_I) = \sigma_I \text{ if } \delta W < 0 \ \forall t \in]t_I, t_I+dt] \ ; \ dW = \phi dt$$

For the evolution immediately following this inversion state, the reference stress ${}^t_R\sigma$ is equal to the inversion stress σ_I. It is worthwhile to emphasize that the criterion expresses almost everywhere the second law of thermodynamics and that there is no inversion if the evolution path is such that ϕ and the help function virtual or real variation is nul.

Consequently, the above analysis leads, first, to the introduction of three piecewise constant functionals of the loading history, ω, ${}^t_R\varepsilon$, ${}^t_R\sigma$ and, secondly, to a pure hysteresis scheme which is constituted by :

a) a discrete memory existence condition;

b) "constitutive relations" such as $\partial\sigma/\partial t = h(D, \Delta^t_R\sigma, \omega)$ instead of (1), relations to which the set of thermomechanical rates must be added ;

c) the inversion criterion (3) which defines each inversion state ;

d) an algorithm $\mathbf{A}$ |9|, which defines at each time the set of inversion stresses σ_I still memorized, the value of the reference stress ${}^t_R\sigma$ and the Masing coefficient ω for that time;

e) a non-restrictive initial neutral condition related both to the restoration properties associated with the fundamental cyclic path and to the condition $W(t_{I+}) = 0$ |3 and 9|.

2.2- Regarding the thermomechanical rates which must be included in point b) of the above scheme, the analysis is simple in the case of the first loading branch ($\omega = 1$). One notices that by definition of σ_1 and σ_2 the internal intrinsic heat rate supply Q_{ii} and the rate of internal energy are :

(4) $$-\dot{Q}_{ii} = \sigma_2 D = \sigma_2\dot{\varepsilon} = \sigma(\varepsilon)\dot{\varepsilon} - \varepsilon\dot{\sigma}(\varepsilon)$$
$$\dot{E} = \sigma_1 D = \sigma_1\dot{\varepsilon} = \varepsilon\dot{\sigma}(\varepsilon)$$

The symbolic model is strongly heuristic for the analysis of cyclic situations. Being entirely well defined, it enables one, without introduction of internal variables, to derive the expression of the rate $\dot{E}$ of internal energy supplied by the springs and of the rate $\dot{Q}_{ii}$ of internal intrinsic heat supplied by the friction of the sliders. One obtains :

(5) $$-\omega\dot{Q}_{ii} = \Delta^t_R\sigma\,\dot{\varepsilon} - \dot{\sigma}\,\Delta^t_R\varepsilon = \phi - C$$

The result strongly suggests a three-dimensional extension. Moreover, it points out the rôle of the discrete memory concept in the definition of the thermomechanical coupling and the interest which must be devoted to the three-dimensional extension of the quantities ${}^t_R\varepsilon$ and ${}^t_R\sigma$.

2.3-Regarding ${}^t_R\varepsilon$, the extension to three dimensions is well known to be obtained with the aid of a dragged along coordinates description and to be physically meaning-full under the same assumption as for the one-dimensional case ; that is to say "in so far as the configuration at such some time may be supposed to have a permanent significance" |19|. In the dragged along coordinate system x^k (k=1,2,3) of associated convected frames $(M(x^k), \vec{g}_i(x^k,t), \vec{g}_j(x^k,t))$ at the material point M (x^k) the Almansi tensor is :

$$(6) \qquad \overset{t}{\Delta}_R \varepsilon = 1/2\ G(t) - 1/2\ {}^t_R G$$

where G(t) is the current metric and ${}^t_R G$ is the Cauchy tensor such as:

$$(7) \qquad {}^t_R G = g_{ij}(x^k, t_R)\ \vec{g}^i(x^k,t) \otimes \vec{g}^j(x^k,t) = {}^t_R g_{ij}\ \vec{g}^i \otimes \vec{g}^j$$

for which holds the invariance requirement of the discrete memory condition $L_{V\circ\circ}\, {}^t_R G = 0$. The vanishing of the Lie derivative expresses the fact that ${}^t_R G$ is the covariant transport from t_R to t of the metric tensor $G(t_R)$. As soon as the implication of the discrete memory concept at the root of continuum mechanics is recognized, the well-known formalism of differential geometry appears natural, simple and straightforward.

3.- FIRST STEP TOWARD THE EXTENSION TO THREE DIMENSIONS : FROM DIFFERENTIAL GEOMETRY TO THE VAN DANTZIG METHOD OF DEFINITION.

3.1- To perform the transport of a tensor field ${}^t_R\sigma$ associated with the Cauchy stress field σ, four different specific manners exist, depending on the kind of the implemented stress tensor. The dragged along process can be applied using the derivative $L_{V\circ\circ}$ associated with σ of type (0,2) and with :

$$(8) \qquad {}^t_{R\circ\circ}\sigma = \sigma_{ij}(t_R)\ \vec{g}^i(t) \otimes \vec{g}^j(t)\ ;\ \overset{t}{\Delta}_{R\circ\circ}\sigma = \sigma - {}^t_{R\circ\circ}\sigma$$

or conversely using $L_V^{\circ\circ}$ associated with σ of type (2,0), ${}^{t\circ\circ}_R\sigma$ and $\overset{t\circ\circ}{\Delta}_R\sigma$. One can also make use of $L_V{}^{\circ}{}_{\circ}$ and $L_{V\circ}{}^{\circ}$ associated with σ of type (1,1) and ${}^{t\,\circ}_{R\ \circ}\sigma$ and ${}^t_{R\circ}{}^{\circ}\sigma$ but this last possibility is open in so far as the symmetry requirement regarding σ is fulfilled. Consequently $L_V{}^{\circ}{}_{\circ}$ and $L_{V\circ}{}^{\circ}$ must be used at once leading to a third type of extensions to three dimensions |11,20,21|.

3.2- The properties of the three types of schemes can be investigated with the help of a simple form of "constitutive relations" (cf.paragraph 2.1). More precisely a simple use of the transport can be implemented using a simple tensorial transposition of the symbolic models :

$$L\,\overset{t}{\Delta}_R\sigma + \overset{t}{\Delta}_R\sigma \ \underset{====}{\text{symbolic}}\ \overset{t}{\Delta}_R\varepsilon + L\,\overset{t}{\Delta}_R\varepsilon$$

For example, a contravariant Zaremba scheme (ZCN) is specified in $(M, \vec{g}_i, \vec{g}_j)$ by:

$$\text{a)}\ \frac{\partial}{\partial t}\, {}^{t\circ\circ}_R g^{ij} = 0\ ;\ \frac{\partial}{\partial t}\, {}^{t\circ\circ}_R \sigma^{ij} = 0$$

$$\text{b)}\ \frac{\partial}{\partial t}\sigma^{ij} = a_o\, g^{ij} + a_1\, D^{ij} + a_2\, \overset{t}{\Delta}_R\sigma^{ij}$$

$$(9) \qquad \begin{bmatrix} \dot{\phi} \\ \dot{Q}_{ii} \\ \dot{I} \\ \dot{E} \end{bmatrix} = \begin{bmatrix} -1 & -1 & 0 \\ 1/\omega & 1/\omega & 1/\omega \\ (1-\omega)/\omega & (1-\omega)/\omega & 1/\omega \\ (1-\omega)/\omega & 1/\omega & 1/\omega \end{bmatrix} \qquad \begin{bmatrix} P_i \\ \pi \\ C \end{bmatrix}$$

$$\pi = {}^{t}_{R}\sigma^{ij} D_{ji} \quad ; \quad C = (L_V \Delta^{too}_{R}\, \sigma)^{ij} (\Delta^{too}_{R}\, \varepsilon)_{ji}$$

$$a_o = \sqrt{g}\, \overset{\infty}{\lambda}\, D^i_{\ i} \; ; \; a_1 = 2\sqrt{g}\, \overset{\infty}{\mu} \; ; \; a_2 = -\sqrt{g}\, \overset{\infty}{\mu}/(\omega\sqrt{g}\, S_o)^2$$

and by the conditions c), d) and e) already presented in paragraph 2.1. Among the thermomechanical densities of rate, one notices that $\dot{I}$ is the rate of loss of order that is to say the density of intrinsic dissipation under non calorific form, for ϕ is a sum of two rates $-\dot{Q}_{ii}$ and $\dot{I}$. The case of the covariant Zaremba schemes (ZCV) is similar. On the contrary the case of the Zaremba schemes of mixed type (ZM) leads to the introduction of $\overset{1}{\sigma}$ and $\overset{2}{\sigma}$ the components of which satisfy the formal symmetry expressed through the equality of $\overset{1}{\sigma}{}^{i}_{\ j}$ and $\overset{2}{\sigma}{}_{j}^{\ i}$. Consequently the ZM schemes are :

$$(10)\quad \left|\begin{array}{l} L_{V^o\,o} G = L_{V^o}{}^{o}\, G = 0 \quad ; \quad \dfrac{\partial}{\partial t}\, {}^{t^o}_{R\,o}\overset{1}{\sigma} = \dfrac{\partial}{\partial t}\, {}^{t\,o}_{R^o}\overset{2}{\sigma} = 0 \\[2ex] \left.\begin{array}{l} \dfrac{\partial\, \overset{1}{\sigma}{}^{i}_{\ j}}{\partial t} = a_o\, \delta^i_{\ j} + a_1\, D^i_{\ j} + a_2\, \Delta^{t^o}_{R\,o}\, \overset{1}{\sigma}{}^{i}_{\ j} \\[2ex] \dfrac{\partial\, \overset{2}{\sigma}{}_{i}^{\ j}}{\partial t} = a_o\, \delta_i^{\ j} + a_1\, D_i^{\ j} + a_2\, \Delta^{t\,o}_{R^o}\, \overset{2}{\sigma}{}_{i}^{\ j} \end{array}\right| \quad 2\,\sigma = \overset{1}{\sigma} + \overset{2}{\sigma} \end{array}\right.$$

The ZM schemes are in accordance with the Van Dantzig general method of definition of constitutive schemes |22|.

Homogeneous problems of simple push-pull type, isochoric push-pull type or simple shear type can be studied as particular cases of the transformation:

$$(11)\quad z^i = (1+ K_i(t))\, Z^i = J_i\, Z^i \; ; \; i = 1 \text{ and } 3 \quad ; \; z^2 = J_2\, Z^2 + 2\tau\, Z^3$$

Such a study has been done recently |11| performing the integrations under the assumptions : $\overset{\infty}{\lambda}/\overset{\infty}{\mu} = 2$; $S_o/\overset{\infty}{\mu} = 8/3000$; owing to the influence of the results on the subsequent developments, it is worthwhile to give a self-consistent recalling of the main conclusions and associated illustrations.

3.3- For the simple push-pull tests such as $|\mathrm{Ln}\, J_3| < 0.02$, the evolutions, with respect to Ln J_3 (LNJ3), of the reduced intensity p_3/S_o of the absolute stress vector acting on the surface $x^3 = 1$, are shown on figure 1 (A and C)for the ZCN and the ZM schemes, respectively. For the isochoric test and in the same loading conditions, the evolutions of p_3/S_o (P3), I_σ/S_o (1S), $\sqrt{II_{\bar{\sigma}}}/S_o$ (2SB) ,$-\dot{Q}_{ii}/S_o\sqrt{II_{\bar{D}}}$ (-QP) and $\dot{E}/S_o\sqrt{II_{\bar{D}}}$ (EP) are shown on figure 1 for the two schemes.For the simple shear test, the evolution is defined with respect to τ (TO) and the same quantities are shown on figure 2 with, in addition, the intensity T_3 (TO3) of the tangential component of the absolute stress vector acting on the surface $x^3 = 1$. Furthermore one gives the results regarding the ZCV scheme in column B and the results obtained by the method introduced by one of the authors to take into account the rotational situation (cf paragraph 4.2) in columns D of figures 1 and 2.

By definition of the schemes, the Cauchy stress is obtained from the relation : $\sigma = \Delta^t_R\sigma + {}^t_R\sigma$ which points out the easiest way to recall the conclusions of |11|. Regarding $\Delta^t_R\sigma$; the properties of each branch are similar to those of the first loading branch where $\Delta^t_R\sigma = \sigma$; the stress rate choice only holds for secondary effects (fig.2 Ab, Bb, Cb) ; the obtainment of a plastic limit is not achieved. This last point is not linked with the simplicity of the constitutive coefficients. It focuses the attention on the effects of the transport process which is implemented in the cyclic situation introducing the discrete memory functional ${}^t_R\sigma$. Regarding this field, the conclusions are as follows :

a) the geometrical effects are closely dependent on the loading conditions;

b) the mixed transport does not perturbate the behaviour of the invariants of σ in the isochoric non rotational case ;

c) the results obtained in the rotational case do not allow the selection of a type of transport but underline both the importance of second order effects and the increasing divergence of the behaviour with respect to the usual plastic limit assumption.

Consequently, the subsequent analysis will, in a first step, concentrate on the notion of plasticity understood as denoting not only time independance but also existence of a limit value in the relevant space of invariants. This work, remaining mainly in the field of continuum idealization , will be done using the schemes of mixed type which are in accordance with the general Van Dantzig method of definition of constitutive schemes.

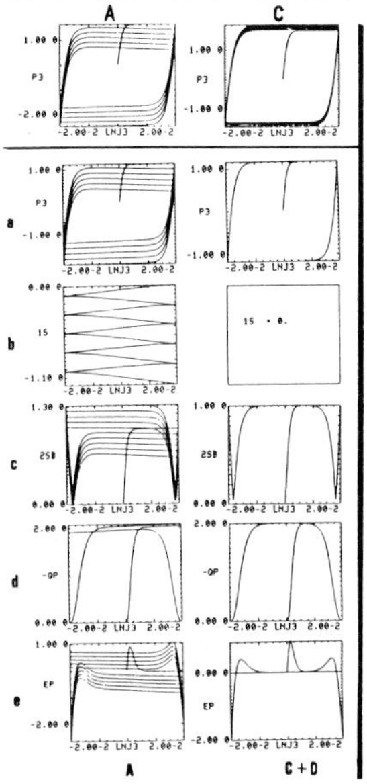

Figure 1

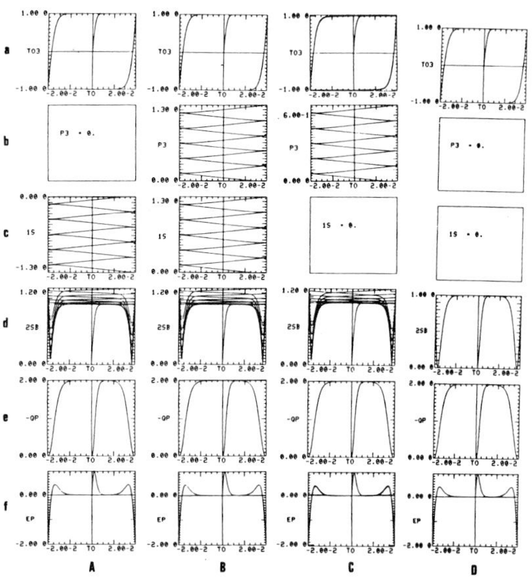

Figure 2

4.- SECOND STEP TOWARD THE EXTENSION TO THREE DIMENSIONS : FROM MIXED TRANSPORT TO PLASTICITY.

4.1- In the non-rotational case the study of the ZM schemes is strongly simplified both by the fact that distinction between σ, $\overset{1}{\sigma}$ and $\overset{2}{\sigma}$ is not necessary for : $\sigma = \overset{1}{\sigma} = \overset{2}{\sigma}$, and by the fact that the analysis can be performed making use of the invariant form |12|:

$$\dot{I}_{\Delta\sigma} = 3\,a_o + a_1\,\zeta + a_2\,I_{\Delta\sigma}\ ;\ I_{\Delta\sigma} = \Delta\sigma^i_{\ i}\ ;\ \zeta = D^i_{\ i}$$

$$(12)\quad \dot{\overline{II}}_{\Delta\bar\sigma} = a_1\,\bar\phi + a_2\,2\overline{II}_{\Delta\bar\sigma}\ ;\ Q^2 = 2\,\overline{II}_{\Delta\bar\sigma} = \Delta\bar\sigma^i_{\ j}\,\Delta\bar\sigma^j_{\ i}\ ;\ \bar\phi = \Delta\bar\sigma^i_{\ j}\,\bar D^j_{\ i}$$

$$\dot{\overline{III}}_{\Delta\bar\sigma} = a_1\,\bar\Psi + a_2\,3\overline{III}_{\Delta\sigma}\ ;\ 3\overline{III}_{\Delta\sigma} = \Delta\bar\sigma^i_{\ j}\,\Delta\bar\sigma^j_{\ k}\,\Delta\bar\sigma^k_{\ i}\ ;\ \bar\Psi = \Delta\bar\sigma^i_{\ j}\,\Delta\bar\sigma^j_{\ k}\,\bar D^k_{\ i}$$

where $(\dot{\ })$ denotes $\partial/\partial t$ and where $\bar a^i_{\ j} = a^i_{\ j} - 1/3\,I_a\,\delta^i_{\ j}$

One notices that the quasi-linearity leads to :

$$(13)\qquad \frac{\dot{\overline{II}}_{\Delta\sigma}}{2\overline{II}_{\Delta\bar\sigma}} - \frac{\dot{\overline{III}}_{\Delta\sigma}}{3\overline{III}_{\Delta\bar\sigma}} = a_1\,(\bar\phi/2\overline{II}_{\Delta\bar\sigma} - \bar\Psi/3\overline{III}_{\Delta\bar\sigma}) = \mathrm{tg}\,(3\,\varphi_\Delta)\quad \dot\varphi_\Delta$$

where φ_Δ is the orientation of RR_2 in the deviatoric plane of the principal stresses space (Figure 3).The time independence is obtained through :

$$(14)\quad a_o = \alpha_o\,\zeta + \alpha_3\,\bar\phi + \alpha_5\,\bar\Psi\ ;\ a_1 = \alpha_1\ ;\ a_2 = \alpha_2\,\zeta + \alpha_4\,\bar\phi + \alpha_6\,\bar\Psi$$

where the α_i are scalar functions of ω and of the invariants of $\Delta^t_R\sigma$ and $^t_R\sigma$. To these assumptions one adds first : $\alpha_o = \sqrt{g}\,\overset{\infty}{\lambda}$; $\alpha_1 = 2\sqrt{g}\,\overset{\infty}{\mu}$, in order to express the quasi-reversibility and the restoration properties which are typical of pure hysteresis behaviour and secondly : $\alpha_2 = 3\,\alpha_3 + \alpha_4\,I_{\Delta\sigma} = 3\alpha_5 + \alpha_6\,I_{\Delta\sigma} = 0$ in order to prescribe an isotropic-deviatoric uncoupling. This second assumption, which is physically strongly restrictive, leads to the easiest analysis : first the isotropic part of the behaviour is then reversible; secondly, the problem can be stated in a deviatoric plane as the investigations of the forms :

$$(15)\qquad \Delta\dot{\bar\sigma} = a_1\,\bar D + a_2\,\Delta\bar\sigma\ ;\ a_2 = \alpha_4\,\bar\phi + \alpha_6\,\bar\Psi = b_4\,\bar\phi_m = b_4\,(\bar\phi + \bar b\,)$$

which are compatible with a specified plastic yield locus (P) of the deviatoric plane (cf. fig.3) where (P) is the circle associated with the Von Mises yield surface of radius $Q_o = \sqrt{2}\,S_o$, if S_o denotes the yield value in simple shear.

In order to warrant the compatibility of the deviatoric properties of the ZM schemes with respect to the specified plastic limit, the assumptions are three-fold in the non rotational case. First, the thermomechanical rates are supposed to be defined by a set of assumptions :

$$(16)\qquad \begin{bmatrix} \pi_m \\ -\dot Q_{ii} \\ \dot E \\ \dot I \end{bmatrix} = \begin{bmatrix} -1 & -1 & 0 \\ 0 & -1/\omega & 1/\omega \\ -1 & -1/\omega & 1/\omega \\ 0 & (\omega-1)/\omega & 1/\omega \end{bmatrix} \begin{bmatrix} P_i \\ \bar\phi_m \\ \bar C \end{bmatrix}$$

$$\bar\phi_m = \Delta^t_R\bar\sigma^i_{\ j}\,\bar D^j_{\ i} + \bar b\ ;\qquad \bar C = \Delta\bar\varepsilon^i_{\ j}\,\Delta\bar\sigma^j_{\ i}\,(\dot{\overline{II}}_{\Delta\bar\sigma} + a_1\,(\bar\phi_m - \bar\phi))/2\,\overline{II}_{\Delta\bar\sigma}$$

giving back the results of the one-dimensional analysis as a limit case owing to the subsequent assumptions. Secondly, the two dimensional character (in the deviatoric plane) of the required deviatoric schemes is associated with the rôle of the similarity functional ω and of the functional $\vec{\phi}$ introducing the distinction between radial paths (such as OR, fig.3) and neutral paths (such as RN fig.3) defined as follows ; along a neutral path $\bar{\phi}_m$ is nul and one has :

$$\partial/\partial t\, \Delta\bar{\sigma}^i_j = a_1 \bar{D}^i_j \quad ; \quad \dot{\bar{II}}_{\Delta\bar{\sigma}} = a_1 \vec{\phi}$$

so that a closed locus in the deviatoric stress plane (Q_σ, φ_σ) is associated with a closed locus in the plane (Q_ε, φ_ε) ; along a radial path, $\dot{\varphi}_\Delta$ is nul, one has $\vec{\phi}_m = \bar{\phi}$ and the properties are deduced from one another by similarity with a limit behaviour depending of φ if the yield locus is not a circle (Coulomb like case). Consequently, one has :

$$\alpha_4 = b_4 \quad ; \quad \alpha_6 = 0 \quad ; \quad \alpha_4 \vec{\phi} = b_4 \vec{\phi}_m$$

along the radial paths. One notices that along neutral paths, the rates $\bar{C}$, $\dot{Q}_{ii}$ and $\dot{I}$ are nul. Consequently π_m , $\dot{E}$ and $-P_i$ are equals and the definition leads to a strongly "reversible" type of paths. Moreover the thermomechanical rate forms obtained in the one-dimensional case are given back along the radial paths. Thirdly, the one parameter family of similar neutral lines are labelled by a modified help function W_m, acting as a pseudo-potential during the evolution along a branch of cycle, and defined by :

$$\delta\, W_m = -\sqrt{g}\, b_4 \vec{\phi}_m\, \delta t$$

At this point of the statement, it remains to complete the second assumption pointing out the outstanding rôle of the discrete memory notion. In the same way as the one-dimensional scheme is discontinuous in character through π, ${}^t_R\sigma$ and updating at the origin of each branch, the one parameter family of neutral lines, deduced from the yield locus P (of radius $Q_o = \sqrt{2} S_o$ in the Mises case) by similarity, follows a discontinuous evolution updated at each inversion (fig.3). During the first loading branch ($\omega = 1$), the family of neutral curves is fixed and centered on the origin (as well as $\pi = 0$ when $\omega = 1$ in the one-dimensional scheme). At right ($t > t_R$) of the first inversion point R (fig.3) the neutral locus moves instantaneously and warrants the quasi-reversibility in the vicinity of R and the restoration of the properties, as indicated on figure 3. One notices that such a discontinuous pattern is both entirely similar to the one-dimension thermomechanics previously recalled and quite different from the usual viscoplastic material models with kinematic "hardening" where are implemented continuously moving loading locus and continuously varying internal variables.

The description of the method is achieved but it is useful to give some details at least in a simple particular case. Let us, for example, make use, as previously, of the simple choice for which b_4 and α_4 are equal to $2\sqrt{g}\, \overset{\infty}{\mu} /(\omega\, S \sqrt{g})^2$ in order to describe a radial path along which one has :

$$\vec{\phi}_m = (\alpha_4 \vec{\phi} + \alpha_6 \bar{\psi})/b_4 = \vec{\phi} + c\, \dot{\varphi}_\Delta = \vec{\phi} + c\, \mathrm{cotg}3\varphi_\Delta\, (\dot{\bar{II}}_\Delta/2\, \bar{II}_\Delta - \dot{\bar{III}}_\Delta/3\, \bar{III}_\Delta)$$

It remains to give the explicit forms of ωS and of the scalar c associated with a specified yield locus. Regarding the first question the similarity rule and the discontinuous process described on figure 3 give the solution. The scalar c is derived from the equation of the neutral locus and from the neutral condition :

$$\dot{\overline{II}}_\Delta = A\,\overline{II}_\Delta\,\dot{\varphi}_\Delta \;;\quad 2\,Q_\Delta\dot{Q}_\Delta = A\,Q_\Delta^2\,\dot{\varphi}_\Delta ;\quad \dot{\bar{\phi}}_m = \dot{\bar{\phi}} + c\,\dot{\varphi}_\Delta = \dot{\overline{II}}/\alpha_1 + c\,\dot{\varphi}_\Delta = 0$$

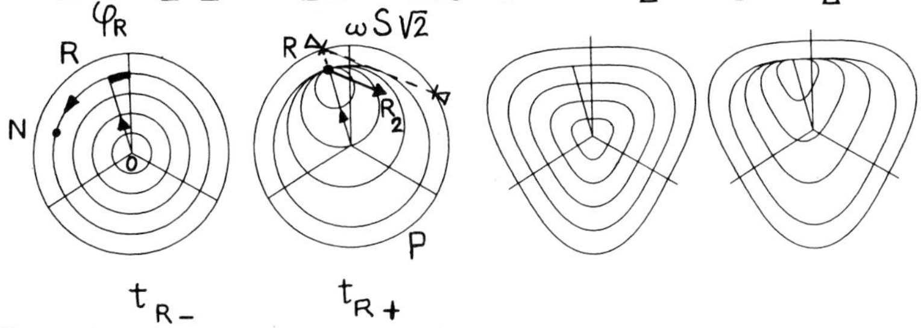

Figure 3: the discontinuous sliding process and the $\omega S\sqrt{2}$ definition

Using the usual generalized Coulomb yield locus such as :

$$Q\,(1 + \gamma\cos 3\varphi)^n = Q_o$$

one has : $A = -\,3n\,\gamma\sin 3\varphi/(1+\gamma\cos 3\varphi)$; $c = -\,A\,Q_\Delta^2/2\,a_1 = -\,A\,\overline{II}_\Delta/a_1$

The differential-difference equation giving the Zaremba stress rate field of mixed type is then defined under the form :

$$(17)\qquad \dot{\bar{\sigma}}^i{}_j - b_4\,c\,\mathrm{cotg}3\varphi_\Delta\,(\dot{\overline{II}}/2\overline{II} - \dot{\overline{III}}/3\overline{III})\,\Delta\bar{\sigma}^i{}_j = a_1\,\bar{D}^i{}_j + b_4\,\dot{\bar{\phi}}\,\Delta\bar{\sigma}^i{}_j$$

As an illustration of the correspondance between the stress and the strain paths,one gives on figure 4 the results of the integrations of a Mises case and of a "Coulomb" case along various similar loading paths. The locus of ω=2 is emphasized by dashed lines. One example of heat rate (QP)evolution is also given for a spiral loading in the deviatoric plane of the stress space (fig.5), where Q_σ and Q_ε denote Q_σ/Q_o and $2\mu Q_\varepsilon/S_o$.

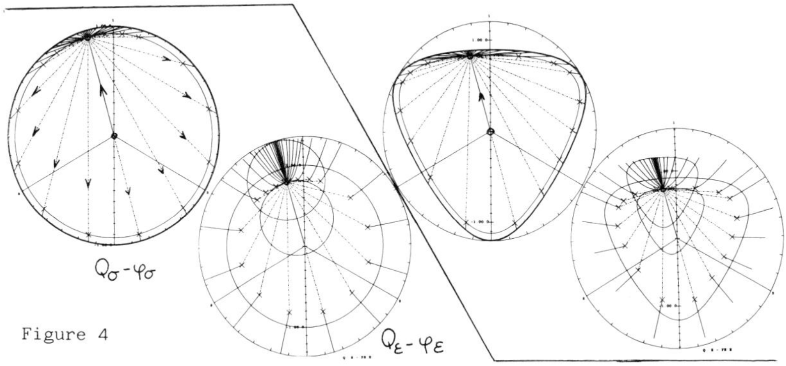

Figure 4

Before dealing with the rotational situations, it is useful to make several remarks.

a) In the constant temperature, isotropic, without hardening case, the method of definition of continuum idealization of pure hysteresis behaviour has been achieved in the best conditions with respect to a classical three-fold criterion (general invariances, relevant stability and well founded thermodynamics).

b) The implemented tools are also simple in the sense that the quasi linearity appears as an important ingredient. Moreover, a complementary study leads to the conclusion that the use of more general tensorial isotropic function is irrelevant to take into account the plastic limit behaviour in a simple and straightforward manner.

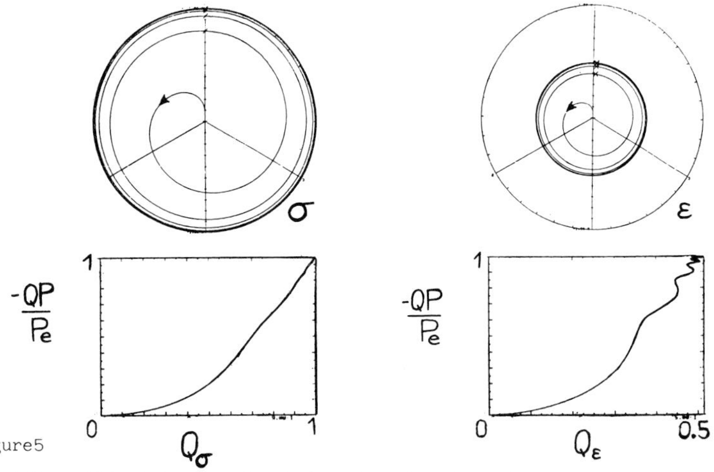

Figure5

c) However, the analysis of rotational situations (paragraph 3) has already pointed out a discrepancy of the Van Dantzig method with respect to the plastic limit requirements. More generally, it can be said that, at this stage of the analysis, one is able to describe a first loading large deformation followed by a cyclic evolution which is restricted to be constituted by "small" amplitude branches if rotational kinematics come into play. In fact, both the Van Dantzig method and the non-rotational case focus the attention on the mixed case but not on some prefered frames reflecting specific processes at some microscale levels : the plastic limit requirement remains taken into account in a phenomenological way in the non rotational case where a closed correspondance between the continuum idealization and the prefered directions of microstructural processes can be assumed . Such a correspondance cannot be assumed for the study of rotational situations where the analysis of Mandel and Drucker type acquire a particular interest |23,24|.

4.2- It is possible to take into account the rotational situations with the aid of a method which leads once more to what may be called a simple idealization in the sense defined above (remark b). Let one consider the homogeneous kinematics(11) defined previously. Owing to the remarks given above, let one suppose that the frame of definition of the stress derivative remains very close to the initial cartesian frame which has been introduced with the simple choice $x^i = Z^i$. It is clear that such an assumption is strongly connected with the fact that, in the case under consideration, it exists a constant loading direction : when the K_i are nul, the illustration is well known (fig.6). Let M_1 and M_2 be two material points located at $(0,-\tau,1)$ and $(0,0,1)$ at time 0 and $(0,0,1)$ and $(0,\tau,1)$ at the current time. Owing to the prefered sliding direction of a rigid plastic process, the "internal forces" on the material points M_1 and M_2 are the same and M_1, located at $(0,0,1)$ may be substituted for M_2 to define these forces in an eulerian way, in spite of the fact that M_2, located at $(0,\tau,1)$ and associated with x^3 and $\vec{g}_3$, remains used to define the geometry and the kinematics in a Lagrangian way.

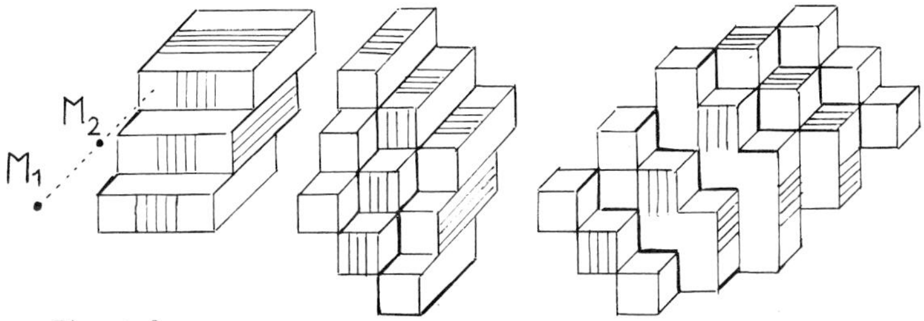

Figure 6

In-so-far-as the rigid plastic process is only an approximation, the analysis may be taken as provisional. However, the subsequent scheme gives, as well as in the non-rotational case, relevant results regarding the limit behaviour of pure hysteresis type. The stress rate being defined in a frame (the initial frame), the direction of which is fixed with respect to a constant loading direction, one assumes once more the isotropic deviatoric uncoupling and one considers the deviatoric part :

$$\bar{\sigma} = \sum_{i=1}^{3} \bar{\sigma}_i \, \vec{e}^{\,i} \otimes \vec{e}_i + \bar{\sigma}_4 \, (\vec{e}^{\,2} \otimes \vec{e}_3 + \vec{e}^{\,3} \otimes \vec{e}_2)$$

of the stress tensor. It is useful to notice at first, that if $\bar{\sigma}_1$ is nul the constant stress phase ($\varphi_\sigma = \pi/6$) is no more an orientation parameter but that a new orientation parameter must be introduced to take into account the rotation of the principal directions of σ. The use of the proper vectors or of the associated Euler angles leading to cumbersome forms, a simple method has been introduced by one of the authors. Let two orientations parameters φ and θ be defined by :

$$(18) \quad \bar{\sigma}_1 = q\sqrt{2/3}\cos\varphi \;;\quad \bar{\sigma}_2 = q\sqrt{2/3}\cos(\varphi - 2\pi/3) \;;\quad \bar{\sigma}_3 = q\sqrt{2/3}\cos(\varphi + 2\pi/3)$$

$$q^2 = \sum_{i=1}^{3} \bar{\sigma}_i^{\,2} \;;\quad \sqrt{2}\,\bar{\sigma}_4 = Q\cos\theta \;;\quad q = Q\sin\theta \;;\quad Q^2 = 2\,II_{\bar{\sigma}} = q^2 + 2\,\bar{\sigma}_4^{\,2}$$

In these relations, the choice of the notations φ and θ is only suggested by formal analogy with respect to the non rotational case previously studied.

The constitutive differential-difference equations are similar to (17) and the ωS form is once more obtained following the method suggested on fig.3, in which the limit circle is now replaced by a limit surface. However the terms similar to those introduced by $c\dot{\varphi}$ are now more complicated, even in situations of Mises type. This difficulty may be suggested as follows. Let OR be a rectilinear first loading defined by Q_R, φ_R, θ_R and let RI be a first rectilinear unloading making the angle α with OR. The direction cosines of RO and RI being $\sin\theta_R \cos\varphi_R$, $\sin\theta_R \sin\varphi_R$, $\cos\theta_R$ and $\sin\theta_\Delta \cos\varphi_\Delta$, $\sin\theta_\Delta \sin\varphi_\Delta$, $\cos\theta_\Delta$ respectively, one has :

$$\cos\alpha = \sin(\pi - \theta_R)\cos(\pi + \varphi_R)\sin\theta_\Delta\cos\varphi_\Delta + \sin(\pi - \theta_R)\sin(\pi + \varphi_R)\sin\theta_\Delta\sin\varphi_\Delta + \cos(\pi - \theta_R)\cos\theta_\Delta$$

Consequently $\dot{\alpha}$ can be expressed (through $-\sin\alpha\,\dot{\alpha}$) as a function of $\dot{\varphi}_\Delta$ and $\dot{\theta}_\Delta$ Moreover, in a situation of Mises type, one has :

$$\bar{b} = Q^2 \operatorname{tg}\alpha\,\dot{\alpha}/a_1 = \left(-\frac{Q^2_\Delta}{2\mu\cos\alpha}\right)(-\sin\alpha\,\dot{\alpha})$$

In the non rotational case the relation between $\dot{\varphi}_\Delta$, $\dot{\overline{II}}_{\Delta\bar{\sigma}}$ and $\dot{\overline{III}}_{\Delta\bar{\sigma}}$ was obvious. In the rotational case, the derivation of the equivalent relation is based on (18) and is rather cumbersome. In spite of there interests, detailed analysis of orientation problems are rarely given. Therefore it seems useful to note briefly several step of the derivation. The choice of orientation parameters gives :

$$\dot{\bar{\sigma}}_1 = \dot{q}\sqrt{2/3}\cos\varphi - q\sqrt{2/3}\sin\varphi\,\dot{\varphi}\ ;\quad \dot{\bar{\sigma}}_1 - \dot{\bar{\sigma}}_3 = \dot{\sigma}_1 - \dot{\sigma}_3\ ;\ \ldots\ldots.$$

$$\sqrt{2}\,\dot{\sigma}_4 = \dot{Q}\cos\theta - Q\sin\theta\,\dot{\theta}\ ;\quad \dot{q} = \dot{Q}\sin\theta + Q\cos\theta\,\dot{\theta}\ ;\quad \dot{\overline{II}}_\Delta = q\,\dot{q} + 2\,\bar{\sigma}_4\,\dot{\bar{\sigma}}_4$$

Consequently, the left member of the stress rate equations are more complicated than in the non rotational case ,for one has ,for example :.

$$\dot{\varphi}_\Delta = \operatorname{cotg} 3\varphi_\Delta \sum_{i=1}^{3}\left(\frac{\bar{\sigma}_i}{2T_2} - \frac{2\bar{\sigma}_j\bar{\sigma}_k + \bar{\sigma}_i^2}{9T_3}\right)\dot{\bar{\sigma}}_i\ ;\quad T_2 = \sum_{i=1}^{3}\bar{\sigma}_i\bar{\sigma}_i\ ;\quad T_3 = \sum_{i=1}^{3}\bar{\sigma}_i\bar{\sigma}_i\bar{\sigma}_i$$

Some illustrations of the properties of the proposed scheme are given above (cf. point 4.3). However it is useful to make once more several remarks regarding the assumptions lying at the origin of the provisional analysis :

a) The discrete memory form of the scheme leads to the introduction of the two tensors ${}^{t}_{R}\Delta\sigma$ and ${}^{t}_{R}\sigma$. Consequently the problem of relative orientation arises: this happens whatever the state of affairs regarding the problem of the prefered frames for the definition of these tensors. The relative orientation problem is, by itself, a difficulty clearly pointed out through the use of the discrete memory concept.

b) The relative orientation problem is the first way by which appears the unsolved problems of the complicated links between discontinuous small scale processes and continuum idealization . Despite the apparently inexhaustible riches of the simple and intuitive pattern of the plastic limit surface, the problem of the relative orientation of $\Delta^t_R\sigma$ and ${}^t_R\sigma$ underlines the relevance of the Friedel warning |25| regarding the ouststanding rôle of solid friction in crystal (a rôle of which fig.6 tries to give an illustration).

c) This state of affairs comes out again through the problem of the preferred frames where $\Delta^t_R\sigma$ and ${}^t_R\sigma$ would be defined. What may be called the discrete memory form of the Mandel problem is introduced as follows. Let $(M, \vec{G}_i, \vec{G}_j)$ be the initial field of frames and $(M, \vec{g}_i, \vec{g}_j)$ the associated actual dragged along field of frames. The quasi-dragged along preferred field of frames $(M, \vec{h}_i, \vec{h}_j)$ would be defined as nearly "similar" to $(M, \vec{G}_i, \vec{G}_j)$ in the sense of α_{ij} nearly equal to $\overset{o}{\alpha}_{ij}$ (fig.7), and however nearly "dragged along" like $(M, \vec{g}_i, \vec{g}_j)$.

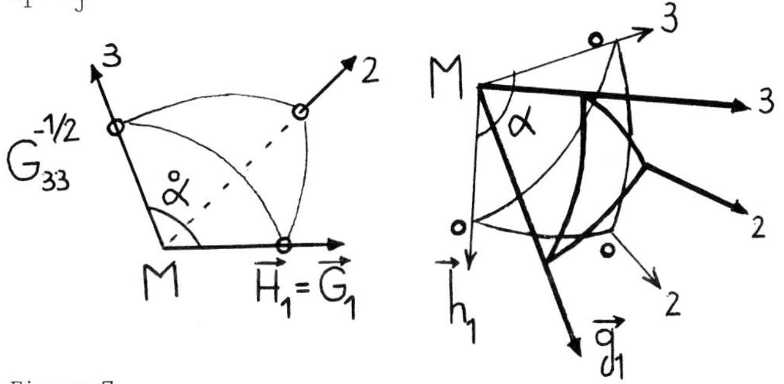

Figure 7

Therefore, a second relative orientation problems appears through the last requirement. It is constitutive in character whereas the first relative orientation problem regarding $\Delta^t_R\sigma$ and ${}^t_R\sigma$ is quite mathematical. The proposed scheme of pure hysteresis appears to be defined through a method which allows a clear distinction of its provisional character. Due to second orientation problem the evolution of the state of affairs is linked with the appearance of new accurate experimental results regarding cylic finite strains.

4.3 The aim of this point is three fold : first to extend the remark given above on the experimental results (Fig 8); secondly to underline the fact that the obtained idealized schemes are relevant to provide a description of the limit behaviours of real materials ; thirdly, to illustrate once more the fact that the intrinsically discontinuous character of the differential-difference scheme |26 and 27| based on a discrete memory thermomechanics is compatible with the classical requirements regarding "stability" and causality (Fig 9) |28,29 and 30| .

The homogeneous kinematics (11) are used once more in the particular case where σ_1 is nul and the normal component p_3 of the absolute stress vector acting on the surface $x^3 = 1$ is constant; the associated tangential component is specified like for cyclic shear test. Consequently the problem under consideration is similar to that of cyclic torsion under constant axial load (path 1). The torsion load (path 2) is followed by two torsion cycles (path 3) and the last torsion branch is extended further (path 4). The ratchet obtained along the paths 2 and 4 vanishes along the fixed cyclic path 3, making conspicuous the limit behaviour of the scheme owing to the experimental results.

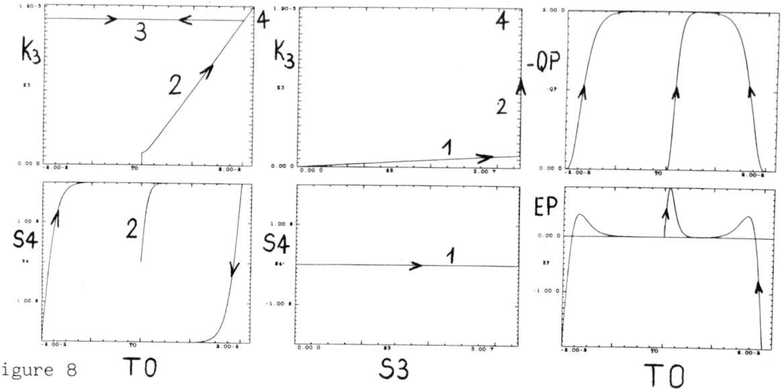

Figure 8

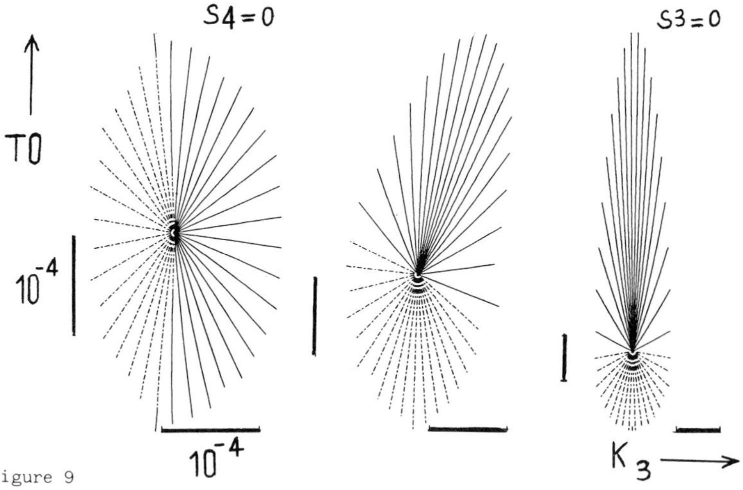

Figure 9

The second application deals once more (cf |9,10,12|) with non-proportional loading obtained through both a "large" first loading OR in the stress space and several linear "small" stress paths of equal lengths ($S_o/8$ for fig.9). To the set of all these "small" paths is associated, in the response space, a set of kinematical paths. The result of three local studies are given on figure 9. The response shown at left is associated with a purely oedometric "large" loading. The transition between loading and unloading in the response space (K_3-K_{3R} ,$\tau-\tau_R$, where K_{3R} and τ_R are related to the end R of the "large" first loading) is underlined by full and dotted lines. With the proposed scheme there is no unattainable point in the response space but one notices once more that the discontinuity always exists, tends to disappear rapidly with decreasing loading and is especially drastic in the non-rotational situations (Figure 9). These results are in contradictions with the conclusions given in |31|.

5.- DEFINITION OF A REVERSIBLE STRESS CONTRIBUTION.

5.1- Let E be the density of elastic energy ($E = \sqrt{g}\,\bar{E} = \rho\,\sqrt{g}\,e$ if $\bar{E}$ and e are the absolute scalar per unit volume and mass, respectively). The reversible stress contribution is defined in ($M, \vec{g}_i, \vec{g}_j$) by the rate form:

(18) $$\delta E/\partial t = (\sigma_{rev})^i{}_j\; D^j{}_i$$

where E is taken as a function of strain invariants, but not through $\partial/\partial\varepsilon$ |32|.

5.2- If these invariants are those regarding the dragged along Cauchy strain tensor ${}^t_o g = G_{ij}\,(x^k,o)\;\vec{g}^i(x^k,t)\;\otimes\;\vec{g}^j(x^k,t)$ one obtains, at the current time t, of metric tensor g_{ij} :

$$\Delta^t_o\sigma^{ij} = 2\,[-\,G^{ir}\,G_r{}^j\,\partial E/\partial II_G + G^{ij}\,(\partial E/\partial I_G + I_G\,\partial E/\partial II_G) + III_G\,g^{ij}\,\partial E/\partial III_G]$$

if one uses the principal invariants of ${}^t_o g$, denoted by I_G, II_G, III_G. If one uses the set $I_G, \overline{II}_G, \overline{III}_G$ or the set $I_G, \overline{II}_{\bar{G}}, \overline{III}_{\bar{G}}$ other expressions are obtained |12|.

5.3- In fact the current strain may be defined in two ways : using classically ${}^t_o g$ (as above) or using the Almansi strain tensor $\Delta^t_o\varepsilon$ defined at the current time t of metric tensor $g_{ij}\;\vec{g}^i\otimes\vec{g}^j$. The use of $\Delta^t_o\varepsilon$ is therefore relevant to define $\Delta^t_o\sigma$ at the current time t in ($M, \vec{g}^i, \vec{g}^j$). Moreover, instead of the physically irrelevant sets I_ε, II_ε, III_ε or I_ε, $\overline{II}_\varepsilon$, $\overline{III}_\varepsilon$ or I_ε, $\overline{II}_{\bar{\varepsilon}}$, $\overline{III}_{\bar{\varepsilon}}$, it is useful to make use of the set : $V = \sqrt{g}/\sqrt{G}$, $\overline{II}_{\bar{\varepsilon}}$, $\overline{III}_{\bar{\varepsilon}}$ or of the associated polar set V, Q, φ where $Q = \sqrt{2\,\overline{II}_{\bar{\varepsilon}}}$; $\cos 3\varphi = 3\sqrt{6}\,\overline{III}_{\bar{\varepsilon}}/Q^3$. One obtains first :

(19) $$\Delta^t_o\sigma^i{}_j = \alpha_o\,\delta^i{}_j + \alpha_1\,\Delta^t_o\varepsilon^i{}_j + \alpha_2\,\Delta^t_o\varepsilon^i{}_r\,\Delta^t_o\varepsilon^r{}_j$$
$$\dot{I}_\varepsilon = \dot{V}/V - 2\,\Delta^t_o\varepsilon^i{}_r\,D^r{}_i\;(= D^i{}_r\,G^r{}_i)$$

$\alpha_o = V\,\partial E/\partial V + b_Q\,\partial E/\partial Q + b_\varphi\partial E/\partial\varphi;\ \alpha_1 = c_Q\,\partial E/\partial Q + c_\varphi\partial E/\partial\varphi;\ \alpha_2 = d_Q\,\partial E/\partial Q + d_\varphi\,\partial E/\partial\varphi$

and $b_Q = -I/3Q$; $c_Q = (1 + 2I/3)/Q$; $d_Q = -2/Q$

$b_\varphi\,Q^3\sin 3\varphi = -I_\varepsilon\,Q\cos 3\varphi/3 + 2Q^3\cos 3\varphi/3 + 2\sqrt{6}\,I^3/27 - \sqrt{6}\,Q^2 I/3 - \sqrt{6}\,I^2/9 + \sqrt{6}\,Q^2/3$

$c_\varphi\,Q^3\sin 3\varphi = Q\cos 3\varphi\,(1 + 2I/3) + \sqrt{6}\,Q^2/3 - 4\sqrt{6}\,I^2/9 + 2\sqrt{6}\,I/3$

$d_\varphi\,Q^3\sin 3\varphi = -\,2Q\cos 3\varphi - \sqrt{6} + 2\sqrt{6}\,I/3$

Secondly the coupling effects, second order effects and correspondances of strain and stress paths can be studied with the associated invariant form :

$$I_{\Delta^t_o\sigma} = 3\,V\,\partial E/\partial V - 2\,Q\,\partial E/\partial Q$$
$$Q^2_{\Delta^t_o\sigma} = e_o\,(\partial E/\partial Q)^2 + (e_1/Q)\,(\partial E/\partial Q)\,(\partial E/\partial\varphi_\varepsilon) + (e_2/Q^2)\,(\partial E/\partial\varphi_\varepsilon)^2$$

$$Q^3_{\Delta^t_o\sigma} \cos 3\varphi_{\Delta^t_o\sigma} /\sqrt{6} = f_o\,(\partial E/\partial Q)^3 - 3(f_1/Q)\,(\partial E/\partial Q)^2\,(\partial E/\partial \varphi_\varepsilon)$$
$$- 3(f_2/Q^2)\,(\partial E/\partial Q)\,(\partial E/\partial \varphi_\varepsilon)^2 + (f_3/Q^3)\,(\partial E/\partial \varphi_\varepsilon)^3$$

with $J = 1 - 2I/3$; $e_o = J^2 + 2\,Q^2/3 - 4\,JQ\cos 3\varphi_\varepsilon/\sqrt{6}$

$$e_1 = 8\,JQ\sin 3\varphi_\varepsilon/\sqrt{6}\ ;\ e_2 = J^2 + 2Q^2/3 + 4\,JQ\cos 3\varphi_\varepsilon/\sqrt{6}$$

and:

$$f_o = J^3\cos 3\varphi_\varepsilon/\sqrt{6} - Q\,J^2 + 2\,Q^2\,J\cos 3\varphi_\varepsilon/\sqrt{6} - 2\,Q^3\cos 6\varphi_\varepsilon/9$$
$$f_1 = J^3\sin 3\varphi_\varepsilon/\sqrt{6} + 2\,Q^2\,J\sin 3\varphi_\varepsilon/3\sqrt{6} - 2\,Q^3\sin 6\varphi_\varepsilon/9$$
$$f_2 = J^3\cos 3\varphi_\varepsilon/\sqrt{6} + Q\,J^2/3 - 2\,Q^2\,J\cos 3\varphi_\varepsilon/3\sqrt{6} - 2\,Q^3\cos 6\varphi_\varepsilon/9$$
$$f_3 = J^3\sin 3\varphi_\varepsilon/\sqrt{6} - 2\,Q^2\,J\sin 3\varphi_\varepsilon/\sqrt{6} - 2\,Q^3\sin 6\varphi_\varepsilon/9$$

For non rotational kinematics defined by (11), one obtains :

$$I = \sum_{i=1}^{3} (J_i^2-1)/2J_i^2\ ;\quad Q_\varepsilon^2 = \sum_{i=1}^{3} (J_i^2-1)^2/4J_i^4) - I^2/3$$

In the simple shear case one has :

$I = -2\,\tau^2$; $Q^2 = 2\,\tau^2\,(1 - 2I/3)$; $\cos 3\varphi_\varepsilon = 2\sqrt{6}\,\tau^4\,(4I/9 - 1)/Q^3$

Illustrations are given above (dotted lines on figure 10,paragraph 6), using:

$$\partial E/\partial Q_\varepsilon = Q_{or}\,(1-\alpha_r\,\exp(-Q_\varepsilon/Q_{1r}))\ \mathrm{th}(Q_\varepsilon\mu_r/Q_{or});\ \partial E/\partial V = \partial E/\partial\varphi_\varepsilon = 0$$

with Q_{or}= 400 MPa, μ_r= 25000 MPa, α_r= 0.9 and Q_{1r}= 10%.

6.- INTRODUCTION TO THE SCHEMES OF SUPERELASTICITY.

It has been underlined previously that experimental evidence does not oblige to distinguish an elastic part in the strain rate and one will come back at this point later (paragraph 8) giving some results regarding shape memory alloys and references to equivalent results regarding others alloys.

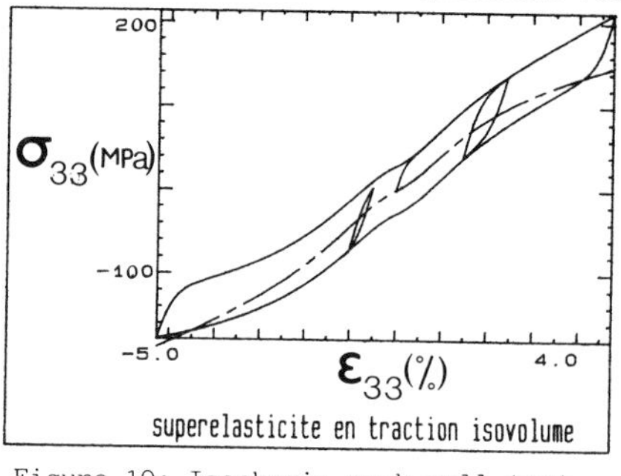

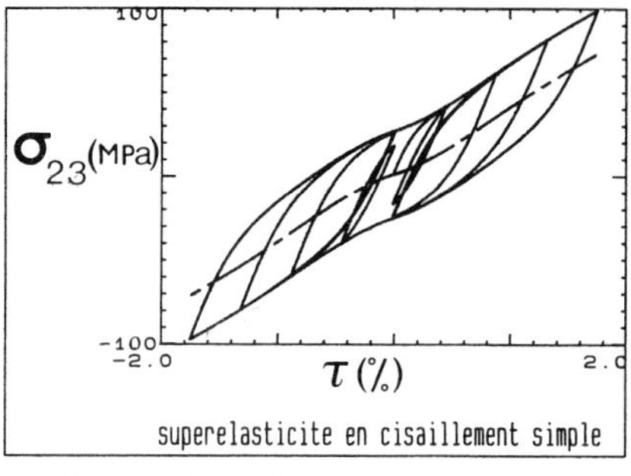

Figure 10: Isochoric push-pull test ; Simple shear test.

To take into account different types of microstructural processes, it seems therefore preferable to split the Cauchy stress tensor : the intrinsic dissipation ϕ of the real material is then considered as the sum of three types of terms reflecting the addition (and interaction) of dissipative effects or reversible effects. If the pure hysteresis, the hardening and the reversible stress contributions are denoted σ_a, σ_e and σ_r respectively, one has $\sigma = \sigma_a + \sigma_e + \sigma_r$ and if π_e is supposed to be nul :

$$\phi = - P_i - \pi = (\sigma_a + \sigma_e + \sigma_r)\, D - ({}^t_R\sigma_a\, D + 0 + \sigma_r\, D) = \phi_a + \phi_e \;;\; \phi_r = 0$$

In this study the hardening effects (of viscous and rate independent types) are neglected and the scheme is based on the definition $\sigma = \sigma_a + \sigma_r$;the hysteresis contribution coefficients are taken as $\overset{\infty}{\lambda} = 15000$ MPa ; $\overset{\infty}{\mu} = 7500$ MPa and $S_o = 25$ MPa; one simulates one isochoric push-pull test and one simple shear test (Figure 10).

7.- INTRODUCTION OF THE CASE OF VARIABLE TEMPERATURE : TOWARD THE DEFINITION OF THE SCHEME OF SHAPE MEMORY EFFECTS.

The previously defined scheme of superelasticity is taken as a point of departure. In this work, the reversible stress contribution σ_r is the only temperature dependent |4|. In spite of this restrictive assumption the obtained scheme allows a qualitative description of the "low" temperature aspects of the stress-induced martensite effects. For example one considers the results of push tests given in figure 4 of |33|.These results are recalled on figure 11;

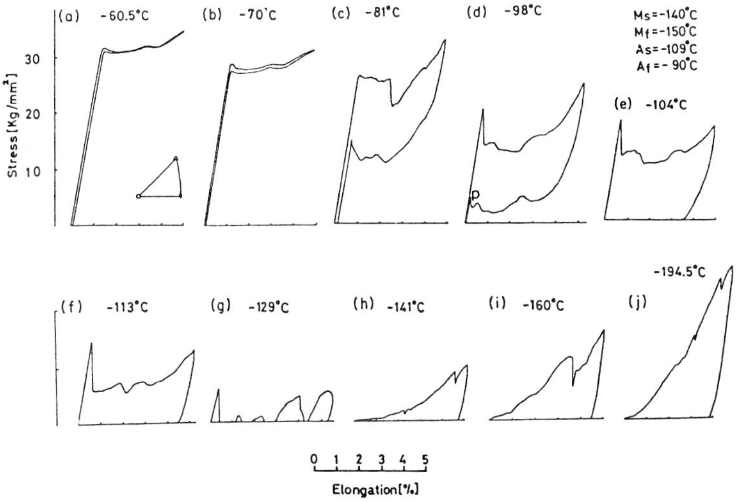

Figure 11

The scheme is obtained for $T < -80°C$ with:

$$\alpha_r = \alpha_{Tr} + \alpha_T \,\mathrm{th}\big[(T - A_f)\, \mu_\alpha / \alpha_T\big] \;;\; \alpha_{Tr} = \alpha_T = 0.8;\; \mu_\alpha = 0.04°C^{-1}$$

$$Q_{1r} = Q_{Tr} + Q_T \,\mathrm{th}\big[(M_s - T)\, \mu_Q / Q_T\big] \;;\; Q_{Tr} = 0.075;\; Q_T = 0.025 \;;\; \mu_Q = 0.001°C^{-1}$$

$$Q_{or} = 350 \text{ MPa};\; \mu_r = 20000 \text{ MPa};\; \overset{\infty}{\lambda} = 2\overset{\infty}{\mu} = 15000 \text{ MPa};\; S_o = 60 \text{ MPa}$$

In addition to the loading-unloading process studied in |33| the behaviour under symmetric cyclic loading is shown on figure 12 in the cases c, e ,h and j of the figure 11. This provisional study is sufficient to underline the important rôle played by the notion of discrete memory (or "erasable micromemory" |34|) and justifies looking for a simple scheme of shape memory effects |35|.

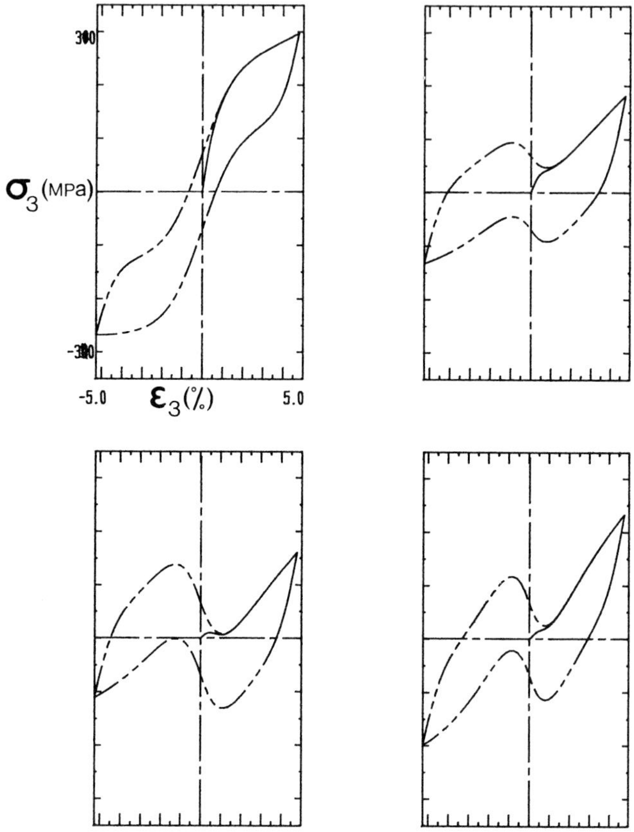

Figure 12

8 - DISCRETE MEMORY NOTION AND QUASI-STATIC CYCLIC TESTS.

8.1- It has been pointed out previously (cf paragraph 2) that well known experimental results are at the origin of the analysis based on the consideration of symbolic models. Results of such type has been obtained recently for various alloys through high accuracy cylic tests (of combined traction-torsion type) allowing the analysis of both first and second order effects (cf |14 and 17|, for example). Obviously other interesting results are already available regarding both a wide range of materials (such as granular medium, grease, molten polymers) and a wide range of phenomena in the field of physics.

8.2- In this paragraph, one gives only some results regarding a particular shape memory alloy. These results have been chosen for two reasons: first, hardening effects are not large and secondly the symmetric centred cyclic torsion tests have been performed at ordinary temperature on an alloy specially prepared to be in the transition zone at this temperature. Consequently, the results given here seem to be not often encountered in the available references.

8.3- The typical small cycle path which may be used to introduce the discrete memory notion is obtained at various stages of a centred cyclic test (fig.13). Both facts that the evolution is always irreversible and that the properties are restored just after an inversion point |15| are underlined with the aid of the accurate Han-Wack method for the study of the tangential behaviour |18| (fig.14). The elastic behaviour appears as a limit behaviour at the origin of each loading branch|15 to 18|. Finally, the fundamental cyclic tests which are performed before and after a cyclic test are quite identical (fig.15).

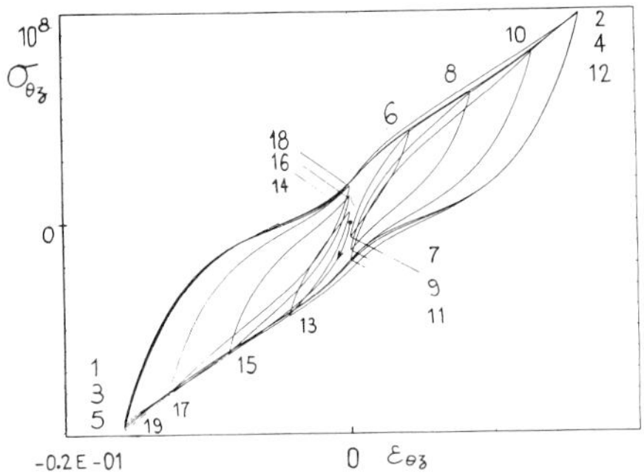

Figure 13.

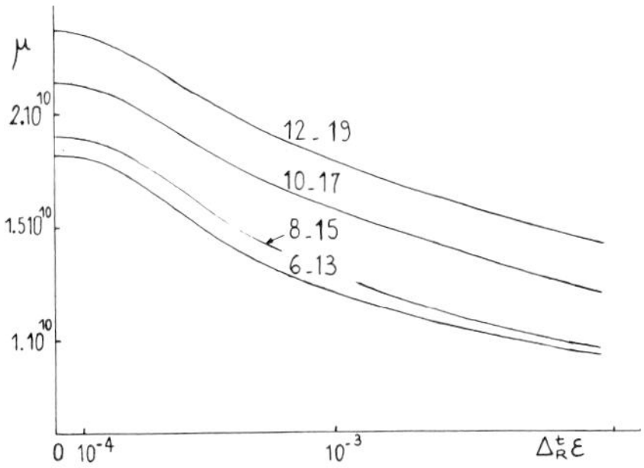

Figure 14.

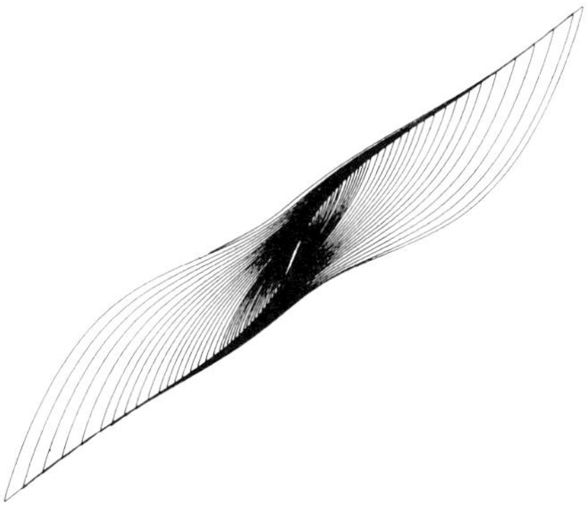

Figure 15.

ACKNOWLEDGEMENTS

This work was supported in part by the Commission of the European Communities by a grant to the fourth author, by the Centre National de la Recherche Scientifique (under contract G.R.E.C.O "G.D.E.") and by the Ministere de la Recherche et de l'Industrie (under contract G.I.S. "Rupture a chaud).

REFERENCES

/1/ BOISSERIE, J.M. and GUELIN, P., Proc. 4th SMIRT, L1/9, 1977
/2/ BOISSERIE, J.M. and GUELIN, P., Proc. 5th SMIRT, L1/6, 1979
/3/ GUELIN, P., J. de Mecanique Theorique, 19, 2, 217-247, 1980
/4/ BOISSERIE, J.M. and GUELIN, P., Proc. 6th S.MIRT, L3/5, 1981
/5/ GUELIN, P. and BOISSERIE, J.M., Res. Mech. Letters, 2, 13-17, 1982
/6/ WACK, B., TERRIEZ, J.M. and GUELIN P., Acta Mechanica, 50, 9-37, 1983
/7/ BOISSERIE, GUELIN, TERRIEZ, and WACK, Trans. ASME, JEMT, 105, 1983
/8/ GUELIN, P. and NOWACKI, W.K, Arch. Mech., 36, 1, 1984
/9/ GUELIN, P., NOWACKI,W.K. and PEGON, P., Arch. Mech., 37, 4-5, 1985
/10/ GUELIN, P. and PEGON, P., Proc. 8th SMIRT, L5/7, 1985
/11/ PEGON, P. and GUELIN, P., Res. Mechanica, 1986, in print
/12/ FAVIER, D., GUELIN, P. and PEGON, P., Cahiers du Groupe Francais de Rheologie, numero special, 1986
/13/ GILES,R.,Mathematical fundations of thermodynamics, Pergamon, 1964
/14/ FAVIER, D., these, Grenoble 1981
/15/ HAN, S., these, Grenoble 1985
/16/ HAN, S., WACK, B., Cahiers du Groupe Francais de Rheologie, 1986
/17/ HAN, S.,and WACK, B., Res. Mechanica, 1986, in print
/18/ HAN, S.,and WACK, B., Arch. Mech., 1986, in print
/19/ OLDROYD, J.G., Proc. R... Soc. London, A200, 1950
/20/ MASUR, E.F., Quart. Appl. Math., 19, 160-163, 1961
/21/ GUELIN, P., Note on the Cauchy tensors $\vec{g}^{i} \otimes \vec{g}^{j}\, {}^{t}_{R}g_{ij}$ and $\vec{g}_{i} \otimes \vec{g}^{j}\, {}^{t}_{R}\sigma_{j}$ expressing the discrete memory concept, "The Summer School on Two Phase Medium Mechanics", 57-132, ed. E. Dembicki, Gdansk 1983.
/22/ VAN DANTZIG, D., Proc. Cambridge, Phil. Soc., 30, 1934
/23/ MANDEL, J., Int. J. Solids Structures, 9, 725, 1973
/24/ PALGEN, L. and DRUCKER, D.C., Int. J. Solids Structures, 19, 519, 1983
/25/ FRIEDEL, J.,Dislocations et deformations plastiques, Yravals, 1979
/26/ BELLMAN, R. and COOKE, K.L. (Academic Press, 1963), Vol. 6 of the series on Math. in Sc. and Eng.
/27/ MINORSKY, N., Theorie des oscillations, Gauthiers Villars 1963
/28/ FER, F., Thermodynamique, Gordon and Breach, 1970
/29/ VOGEL, T., Theorie des systemes evolutifs, Gauthiers Villars 1963
/30/ POPPER, K., Sir., l'Univers irresolu, Hermann, Paris, 1984
/31/ DIBENEDETTO, H. and DARVE, F., J. de Meca. Theo. Appl., 2, 5, 767; 1983
/32/ COHEN, H. and WANG, C.C., Arch.Rat.Mech.Analysis, 85, 213, 1984
/33/ OTSUKA, K., WAYMAN, C.M., NAKAI, K., SAKAMOTO, H., SHIMIZU, K., Acta metallurgica,24,pp 207-226 ,1976
/34/ VERGUTS, H., DELAEY, L., AERNOUDT, E., VERMEERSCH, W., Euromech, colloquium 71, 1983
/35/ VERGUTS, H., AERNOUDT, E., Proceedings on the 7th Int. Conf. on the Strength of Metals and Alloys, pp 563-568,Montreal, August 12-16 1985.

Thermomechanical Couplings in Solids
H.D. Bui and Q.S. Nguyen (Editors)
Elsevier Science Publishers B.V. (North-Holland)

MODELLING OF ANISOTHERMAL EFFECTS IN ELASTO-VISCOPLASTICITY

A. BENALLAL - A. BEN CHEIKH

Laboratoire de Mécanique et Technologie
E.N.S. de CACHAN/C.N.R.S./Université PARIS 6
61, Avenue du Président Wilson - 94230 CACHAN (France)

ABSTRACT
A phenomenological constitutive three-dimensional model of metal behavior in elasto-viscoplasticity with variable temperature is proposed in order to predict the response of components subjected to both mechanical and thermal loadings. It covers several phenomena such as thermo-elasticity, plasticity (isotropic and kinematic hardening), viscous effects (creep, relaxation, strain rate effects), and temperature effects with large temperature variations. Data from an experimental program consisting of various tests including thermal and mechanical cyclings show a good agreement with the predicted results.

INTRODUCTION

The study of the mechanical behavior of metals is often restricted to the case of isothermal conditions. Because of this convenient simplifying assumption, the behavior of metals under isothermal loadings is better understood than under thermo-mechanical loadings. But in some cases, such as aeronautics and nuclear engineering, thermo-mechanical fatigue problems are encountered in structures subjected to thermal stresses cycling, anisothermal creep, resulting from start-up, shutdown, ground-air-ground flight,...

This work is an attempt to predict the response of metal structures suffering such conditions. To this end, a unified anisothermal elasto-viscoplastic three-dimensional model is presented. Using the formalism of thermodynamics of irreversible processes with internal variables, temperature effects are described by the introduction of temperature in the free energy and in the potential of viscoplasticity. The basic viscoplastic model is that of J.L. CHABOCHE, formulated for isothermal cases [1], [2].

CLASSICAL ISOTHERMAL FORMULATION

In the framework of thermodynamics of irreversible processes, the thermomechanical state of any particle of a system can be defined by the knowledge of a thermodynamical potential and a potential of dissipation [3]. These potentials are functions of two kinds of state variables : observable variables and for dissipative mechanisms : internal variables. Let us write the viscoplastic model of J.L. CHABOCHE in its simplest form, which considers two internal variables $\underline{\alpha}$ and p. $\underline{\alpha}$ is a tensor associated with the kinematic hardening, p is a scalar associated with the isotropic hardening/softening. Assuming isotropy of the material, partition of the total strain and HOOKE's law for the elastic part, the thermodynamical potential (free energy) W and the viscoplastic potential Ω are [1] :

$$W = \frac{1}{2}\,\underline{C}:\underline{\varepsilon}^e:\underline{\varepsilon}^e + h_1(\underline{\alpha}) + h_2(p) \;; \qquad \Omega = (\mu/n+1)\,\langle J-(R+k)\rangle^{n+1} \qquad (1)$$

where : $J = [\frac{3}{2}(\underline{S}-\underline{X}):(\underline{S}-\underline{X})]^{1/2}$ $\underline{S} = \underline{\sigma} - \frac{1}{3}\ \mathrm{Tr}\ [\underline{\sigma}]\ \underline{1}$ (2)

$h_1(\underline{\alpha}) = \frac{2}{3}\ c\ a\ \underline{\alpha}:\underline{\alpha}$ $h_2(p) = h\ [\ p + (\exp(-\gamma p))/\gamma\ ]$ (3)

where $\underline{\varepsilon}^e$ is the elastic strain tensor, $\underline{C}$ the HOOKE's tensor, $\underline{\sigma}$ the stress tensor and $\underline{1}$ the unit tensor. a, c, γ, h, n, μ are material constants.
$\underline{X}$ is the conjugate variable of $\underline{\alpha}$, representing in the stress space the translation of the elastic domain surface. R is the conjugate variable of p, representing the increase/decrease of the elastic domain size, k its initial value (initial yield point).

The state laws : $\underline{\sigma} = \partial W/\partial\underline{\varepsilon}^e$; $\underline{X} = \partial W/\partial\underline{\alpha}$; $R = \partial W/\partial p$ (4)

and the evolution laws : $\dot{\underline{\varepsilon}}^p = \frac{\partial\Omega}{\partial\underline{\sigma}}$; $\dot{\underline{\alpha}} = -\frac{\partial\Omega}{\partial\underline{X}} - \frac{3}{2}\frac{\underline{X}\dot{p}}{a}$; $\dot{p} = -\frac{\partial\Omega}{\partial R}$ (5)

(where $\dot{\underline{\varepsilon}}^p$ is the inelastic strain rate tensor) give the constitutive equations.
In uniaxial loading, the viscoplastic model is then (see figure 1.b) :

$$\sigma = X + \lambda(R + k) + \lambda\sigma_v \qquad (6)$$

where $\sigma_v = |\dot{\varepsilon}^p/\mu|^{1/n}$ is the viscous over-stress, $\lambda = \pm 1$ gives the flow direction.
Building an anisothermal viscoplastic model consists in a first time to know how the temperature is involved in the evolution of X, R + k, σ_v.

TEMPERATURE DEPENDENCE OF THE HARDENING VARIABLES AND THE VISCOUS OVER-STRESS IN THE CASE OF THE INCONEL 718 SUPER-ALLOY

In this purpose, an experimental program consisting of isothermal uniaxial tension-compression strain controlled tests with increasing strain ranges has been achieved (figure 1.a). Each test is ended by a relaxation period. Several tests have been carried out between 20°C and 800°C on INCONEL 718 alloy [4].

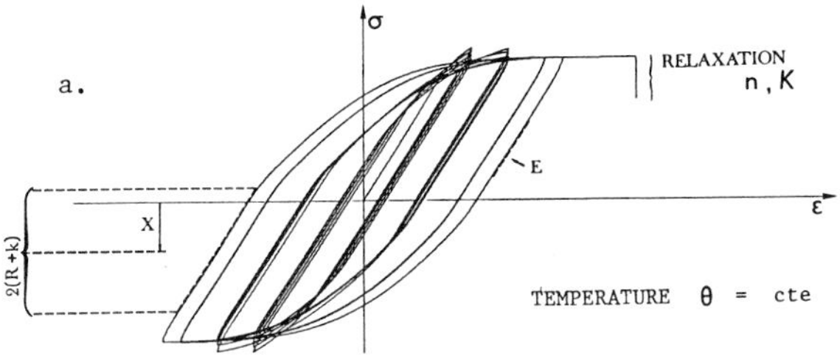

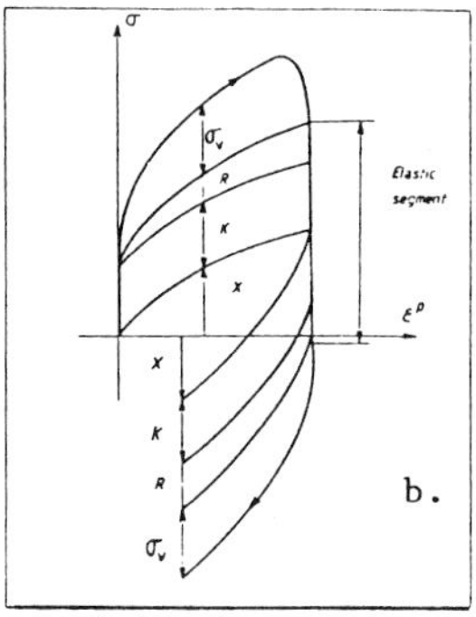

Figure 1 : a) Typical isothermal test for material identification
b) Partition of the stress according to formula (6)

• The identification of R + k is done by measuring the elastic domain each half hysteresis loop from the first one up to the steady state cycle. In order to determine the evolution law of R + k, the values of R + k are plotted versus the accumulated plastic strain p for different temperatures (figure 2). The identification leads to a temperature independent value of γ and to an exponential temperature dependence of h and k.

$$h(T) = a_1(1 - \exp(-a_2T)) + a_3 \quad ; \quad k(T) = k_1(1 - \exp(k_2T)) + k_3 \qquad (7)$$

• As shown in figure 1, X is measured by the middle of the elastic range for the steady state corresponding to each strain range level. Under these conditions, it can easily proved that :

$$(\Delta X/2) = a \text{ th } (c \, \Delta\varepsilon^p/2) \qquad (8)$$

where $\Delta\varepsilon^p$ denotes the steady state plastic strain range, ΔX the kinematic hardening range. This procedure allows to obtain a and c.

The results are summarized in figure 3 where it can be seen that for the studied material, X is insensitive to temperature.

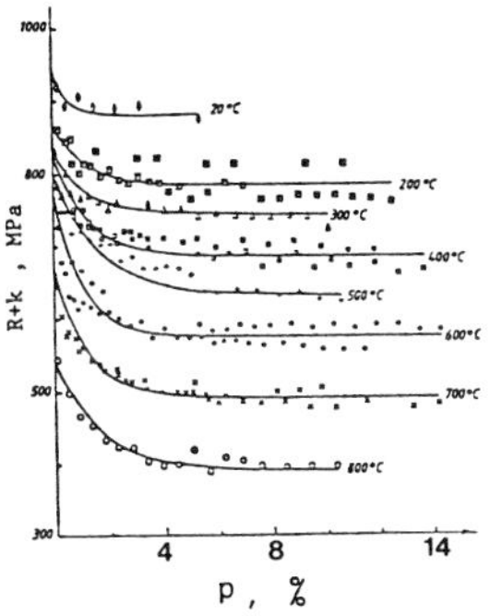

Figure 2 : Identification of R+k at various temperatures
.,X,O : experimental results
—— : identification

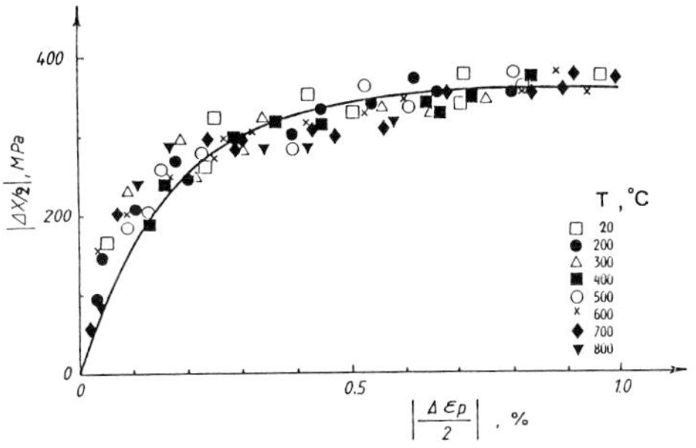

Figure 3 : Identification of X at various temperatures
.,X,O : experimental results
—— : identification

• Viscous effects (strain rates effects, creep, relaxation) are related to n and μ. Calculations of these terms are performed from relaxation tests at stabilized conditions (R = h), and at a quasi-satured value of X (ε^p > 1 % ⟶ X ∼ a). Combining the constitutive equations in this case, we obtain :

$$\sigma = [\mu \, E(n-1)t - (\sigma_0 - \sigma_\infty)^{1-n}]^{1/1-n} + \sigma_\infty \quad (9)$$

where t = time, $\sigma(t=0) = \sigma_0$, $\sigma(t=\infty) = \sigma_\infty$

The identification from experimental results gives a temperature independent value of n and a strongly dependent value of μ. Furthermore, experimental data show very small viscous effects at low temperature (under 500°C). These

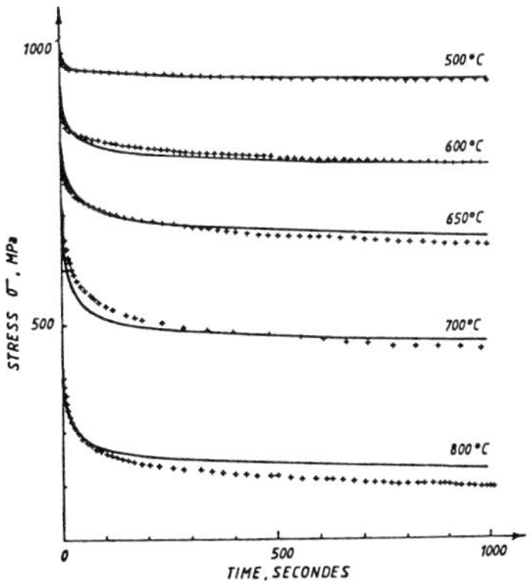

Figure 4 : Relaxation tests at ε =1.4% at various temperatures ; + : experimental results ; —— : identification

phenomenological remarks lead to take μ in the form :

$$\mu(T) = A \exp(-BT) \tag{10}$$

• The identification of the coefficients related to the thermoelastic behavior leads to the following evolution of YOUNG's modulus, POISSON's ratio and the thermal expansion coefficient :

$$E(T) = e_1(1-\exp(e_2 T)) + e_3 \quad ; \quad \nu(T) = \text{constant} \quad ; \quad \alpha(T) = \alpha_1 T + \alpha_2$$

CONSTITUTIVE EQUATIONS FOR ANISOTHERMAL ELASTO-VISCOPLASTICITY

Gathering the above results, we are now ready to write the thermodynamical potentials :

$$W = m(T) + \underline{M}(T):\underline{\varepsilon}^e + \frac{1}{2}\,\underline{C}(T):\underline{\varepsilon}^e:\underline{\varepsilon}^e + h_1(\underline{\alpha}) + H(T,p) \tag{11}$$

$$\Omega = (\mu(T)/n+1)\ \langle J-(R+k)\rangle^{n+1} \tag{12}$$

with :

$$H(T,p) = [\ a_1(1-\exp(-a_2 T) + a_3\ ]\ .\ [\ p + \exp(-\gamma p)/\gamma\ ]$$

where $m(T)$ is related to the thermal dissipation and $\underline{M}(T)$ a tensor related to the thermal expansion. The constitutive equations are derived from these potentials by combining the state laws (4) and the evolution laws (5).

$$\underline{\varepsilon} = \underline{\varepsilon}^e + \underline{\varepsilon}^p \quad ; \quad \underline{\varepsilon}^e = \frac{1+\nu}{E(T)}\,\underline{\sigma} - \frac{\nu}{E(T)}\,\mathrm{Tr}[\underline{\sigma}]\ \underline{1} + \alpha(T).(T-T_0)\underline{1}$$

$$\dot{\underline{\varepsilon}}^p = \frac{3}{2}\,\mu(T)\ \langle J-(R+k)\rangle^n \frac{\underline{S}-\underline{X}}{J}\ ;\ \dot{\underline{X}} = c\ [\ \frac{2}{3}\,a\ \dot{\underline{\varepsilon}}^p - \underline{X}\dot{p}\ ]\ ;\ R = h(T)\ (1-\exp(-\gamma p))$$

For the studied material, the identification procedure has led to the following set of material constants (between 20°C and 800°C).

T_0 = 20°C

e_1 = 30000 MPa	e_2 = 1.45 10^{-3} °C^{-1}	e_3 = 206000 MPa
ν = 0.3	α_1 = 4.475 10^{-9} °C^{-2}	α_2 = 1.307 10^{-5} °C^{-1}
a = 340 MPa	c = 500	γ = 60
a_1 = -180 MPa	a_2 = 3.10^{-3} °C^{-1}	a_3 = -40 MPa
k_1 = 50 MPa	k_2 = 3.4 10^{-3} °C^{-1}	k_3 = 920 MPa
A = 4.10^{-7} MPa^{-n} s^{-1}	B = 0.02 °C^{-1}	n = 4

The free energy W and the viscoplastic potential being identified, we can also, for a complete thermo-mechanical description, write the heat equation :

$$\varrho\beta\dot{T} - \mathrm{div}\ [\ K(T)\ \mathrm{grad}T\ ] = D_1 + r + TF \tag{13}$$

where K(T) is the conduction coefficient taken function of temperature

$$\beta = -T\ (\partial^2 W/\partial T^2) \quad ; \quad D_1 = \underline{\sigma}:\dot{\underline{\varepsilon}}^p - \underline{X}:\dot{\underline{\alpha}} - R\dot{p}$$

$$F = (\partial^2 W/\partial\underline{\varepsilon}^e\partial T)\ \dot{\underline{\varepsilon}}^e + (\partial^2 W/\partial\underline{\alpha}\partial T)\ \dot{\underline{\alpha}} + (\partial^2 W/\partial p\partial T)\ \dot{p}$$

where D_1 is the mechanical dissipation, r the internal heat supply, ϱ the material density.

THERMO-MECHANICAL TESTS AND SIMULATION

Various thermo-mechanical tests have been performed in order to verify the accuracy of the model. The different strain (or stress) and temperature histories programmmed are given in the following table which shows that the proposed formulation describe qualitatively and quantitatively all the presented tests. Only test VI and VII are not reproduced qualitatively : indeed, experimental results of tests VI and VII are unexpected. In all previous tests and in isothermal tests (at each temperature), the INCONEL 718 presents an isotropic softening ($h(T)<0$). In test VI and especially in test VII (which start with a cooling), the tendency is reversed. This phenomena cannot, in any case, be predicted by the model. The cycling hardening occuring in this case of loading is under investigation.

CONCLUSION

An easily identifiable model of metal behavior has been developed in order to describe complex thermo-mechanical processes. The predictions of anisothermal mechanical tests are shown to be in good agreement with experiments.

For the prediction of the response of metal components, the model has been implemented in an in-house finite element code. Furthermore, the heat equation (13) is also implemented which allows the evaluation of the thermo-mechanical coupling and the temperature distribution when the adequate thermal boundary conditions are known. In order to test the accuracy of this anisothermal model, an experimental program designed to reproduce thermal and mechanical conditions encountered in an aircraft engine turbine disc is set up. The specimen is a notched plate representing an alveole of blade fixation, instrumented locally in its most stressed part with mechanical and optical extensometers. Space-time dependent temperature field is applied and recorded through several thermocouples providing the nodal temperature inputs for finite element analysis. Comparisons will be made between local-global experimental measurements and finite element results.

REFERENCES

[1] Chaboche, J.L., Thèse d'Etat ès Sciences, Université PARIS 6, 1978.
[2] Lemaitre, J., Chaboche, J.L., Mécanique des matériaux solides, DUNOD, 1986.
[3] Germain, P., Nguyen, Q.S., and Suquet, P., J. of Appl. Mech., pp.1010-1020, vol.50, 1983.
[4] Benallal, A. and Ben Cheikh, A., Constitutive equations for anisothermal elasto-vicsoplasticity, Proceedings of the second international conference and short course on constitutive laws for engineering materials, TUCSON (U.S.A.), 1987 (to appear).

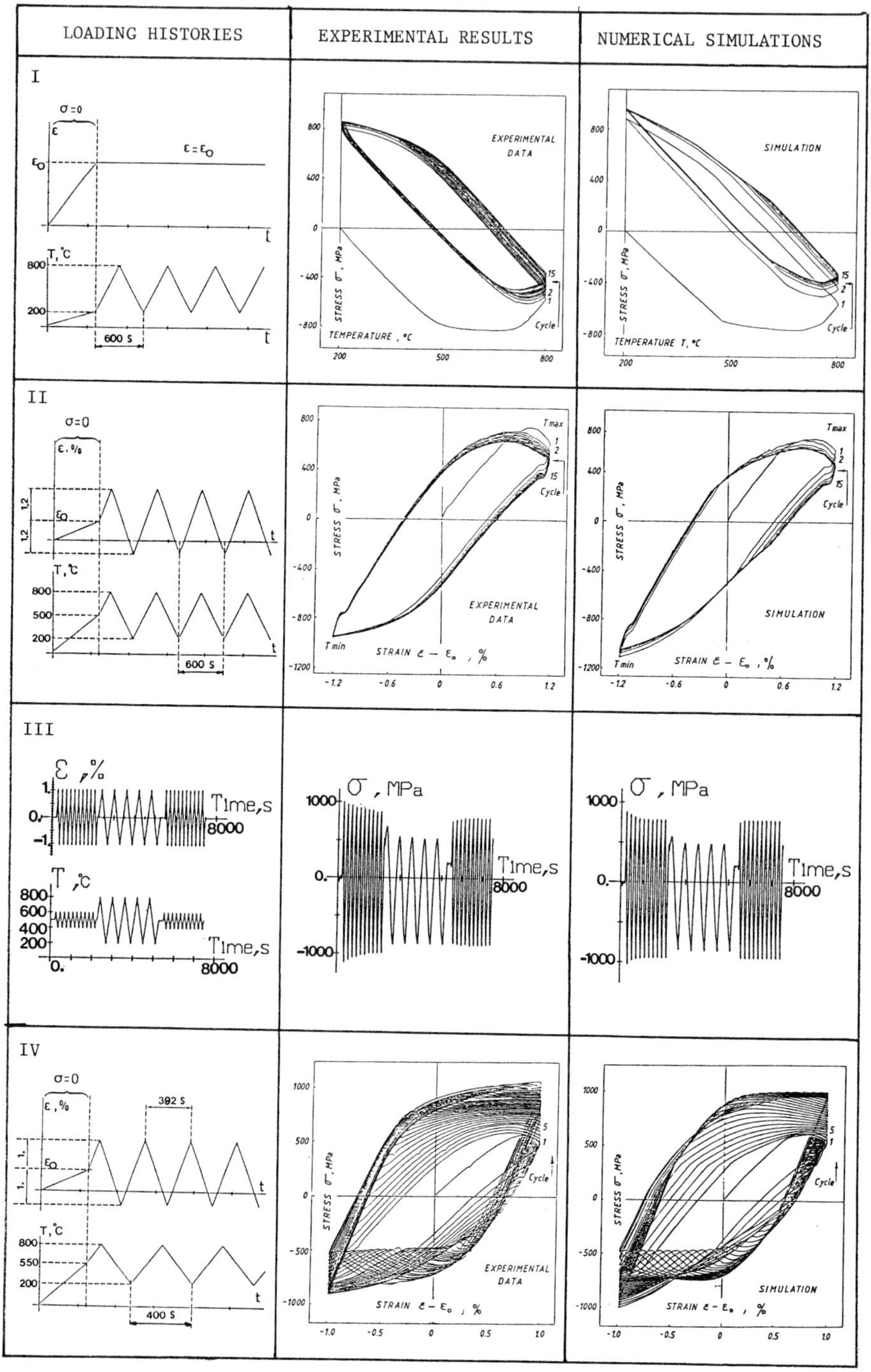

LOADING HISTORIES
EXPERIMENTAL RESULTS
NUMERICAL SIMULATIONS
I
σ=0
ε
ε=ε0
ε0
t
T, °C
800
200
600 s
EXPERIMENTAL DATA
SIMULATION
STRESS σ, MPa
TEMPERATURE, °C
TEMPERATURE T, °C
Cycle
15
2
1
II
ε, %
1.2
T max
T min
STRAIN ε − ε0, %
500
III
σ, MPa
Time,s
8000
IV
392 s
400 s
550

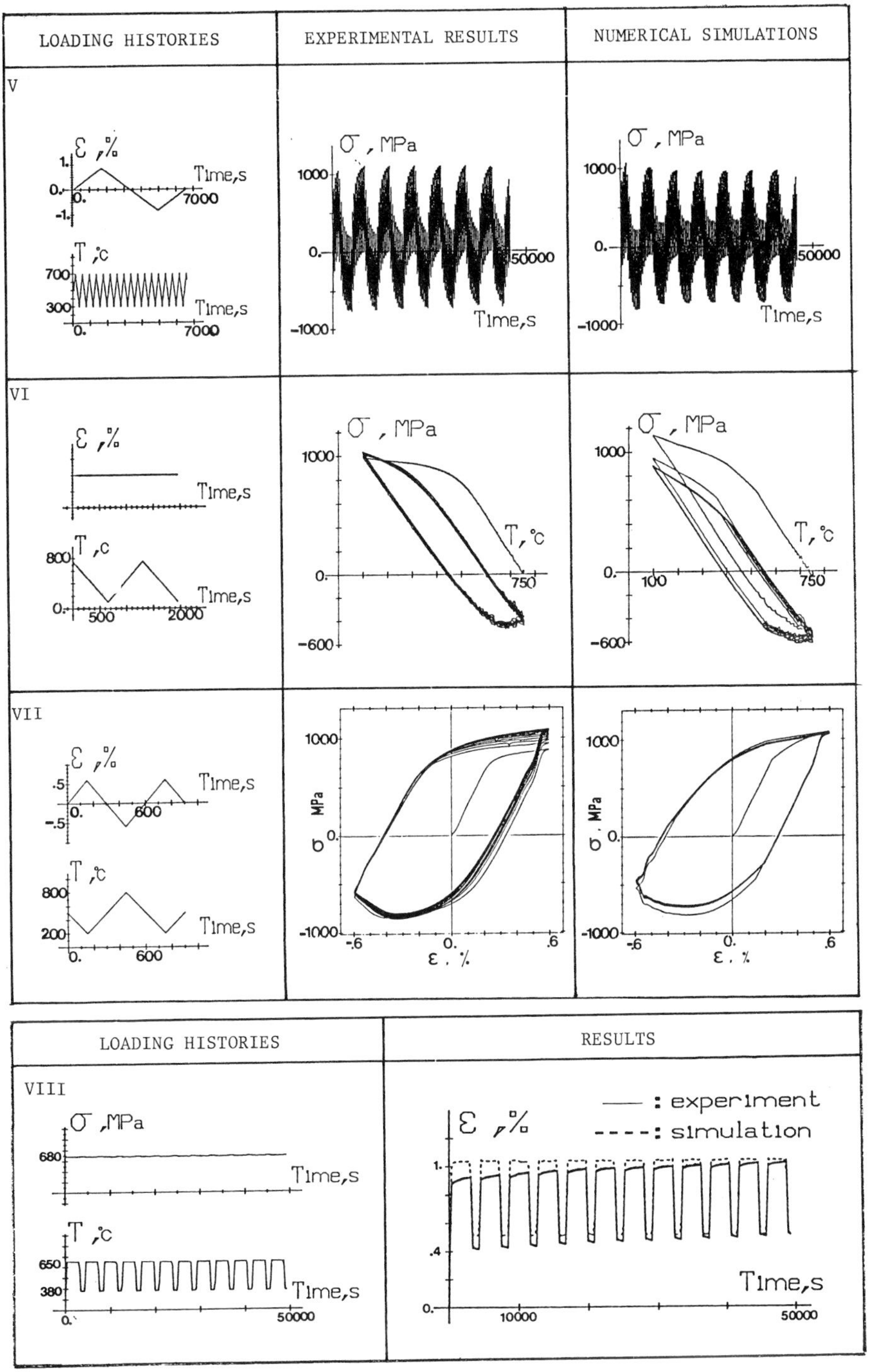

LOADING HISTORIES
EXPERIMENTAL RESULTS
NUMERICAL SIMULATIONS
V
ε ,%
Time,s
7000
T ,°c
700
300
σ , MPa
1000
0.
50000
-1000
Time,s
VI
800
500
2000
T, °c
750
100
-600
VII
.5
-.5
600
200
-1000
-.6
.6
σ, MPa
ε , %
LOADING HISTORIES
RESULTS
VIII
680
650
380
50000
—— : experiment
- - - - : simulation
1.
.4
10000

Thermomechanical Couplings in Solids
H.D. Bui and Q.S. Nguyen (Editors)
Elsevier Science Publishers B.V. (North-Holland)

PHASE CHANGE WITH DISSIPATION

Dominique BLANCHARD, Michel FREMOND

Laboratoire Central des Ponts et Chaussées
58 boulevard Lefebvre
75732 PARIS Cedex 15, France

Augusto VISINTIN

Istituto di Analisi Numerica del C.N.R.
Corso Carlo Alberto, 5
27100 PAVIA, Italy

1. INTRODUCTION

Continuum mechanics thermodynamics give models for dissipative phenomenons occuring during liquid-solid phase-changes. Two examples are given in the sequel.

Freezing of water and ice thawing are reversible phenomenons. Other materials can present irreversible solid-liquid phase-changes. For instance, during cooling, a cooked egg remains solid and does not liquefize ! Thermal hardening materials, like glues, are also examples of materials with irreversible phase-changes.

Freezing of water saturated soils will be our second example. Two phenomenons must be emphasis during soil freezing

- the porosity of the soil increases,
- liquid water does exist at negative temperatures.

Those two phenomenons result in large water movements producing important soil heavings during freezing [2]. They can be modelized by using irreversible thermodynamics.

2. PHASE CHANGE WITH DISSIPATION

2.1. Free energy

Let there be a material whose temperature T evolves in the neighbourhood of the absolute phase change temperature T_0. Let us denote by χ, the volumetric proportion of the phase which exists at temperatures greater than T_0. We intend to describe the evolution of the temperature and of the composition of a structure Ω made of this material. We assume that the free energy Ψ is a function of T and χ

$$\psi(T,\chi) = -\mu(T-T_0)\chi + Tg(\chi) - CT \operatorname{Log} T$$

where C is the heat capacity and μ a positive constant whose physical meaning is given. The function $g(\chi)$ obliges the proportion χ to take values between 0 and 1. For instance g can be the indicator function I of the segment [0,1]

$\left(I(x) = 0 \text{ if } x \in [0,1],\ I(x) = +\infty \text{ if } x \notin [0,1]\right)$ or the sum of I and a smooth convex function $\hat{g}$. With those assumptions Ψ is a convex function with respect to χ and a concave function with respect to T.

Let us compute the specific energy

$$(0) \qquad e = \Psi + Ts = \Psi - T\frac{\partial \Psi}{\partial T} = \mu T_0 \chi + CT$$

where $s = -\frac{\partial \Psi}{\partial T}$ is the entropy. This expression shows that μT_0 is equal to the specific phase-change latent heat L.

2.2. Second principle of thermodynamics

Let us assume that all effects, except the thermal effects are neglectable. The second principle of thermodynamics is in this case [3]

$$(1) \qquad -\frac{\partial \Psi}{\partial \chi}\dot{\chi} - \frac{\vec{q}\,\mathrm{grad}\,T}{T^2} \geqslant 0, \text{ where } \dot{\chi} = \frac{\partial \Psi}{\partial t},$$

for all heat flux vectors $\vec{q}$ and velocities $\frac{\partial \chi}{\partial t}$. It is classical to replace inequality (1) by two inequalities

$$(2) \qquad -\frac{\partial \Psi}{\partial \chi}\dot{\chi} \geqslant 0, \quad -\frac{\vec{q}\,\mathrm{grad}\,T}{T^2} \geqslant 0.$$

The last inequality is satisfied by assuming the Fourier's constitutive law :

$$(3) \qquad \vec{q} = -k\,\mathrm{grad}\,T,$$

where k is the thermal conductivity.

In order to satisfy the first inequality we assume that there exists a pseudo-potential of dissipation $\Phi(\dot{\chi})$ which is a positive and convex function with respect to $\dot{\chi}$ and such that $\Phi(0) = 0$. By assuming the constitutive law

$$(4) \qquad -\frac{\partial \Psi}{\partial \chi}(T,\chi) = \frac{\partial \Phi}{\partial \dot{\chi}}(\dot{\chi}),$$

the first inequality of (2) is satisfied.

Note. The derivative of Ψ and Φ in formulas (2) and (4) are subgradient in the sense of convex analysis [4].

2.3. Energy conservation law

The energy conservation law, assuming there is no mechanical effects and no heat sources, is

$$(5) \qquad \rho\frac{\partial e}{\partial t} + \mathrm{div}\,\vec{q} = 0$$

which gives with (0) and (3)

$$(6) \qquad \rho C\frac{\partial T}{\partial t} + \rho L\frac{\partial \chi}{\partial t} - k\Delta T = 0,$$

where ρ is the density.

To find the evolution of the proportion $\chi(x,t)$ and the temperature $T(x,t)$ $(x \in \Omega)$ of the structure during the time interval $[0,\hat{T}]$ we have to solve the equations (4) and (5) where ρ, C, k, L, $\Phi(\dot{\chi})$ and convenient initial and boundary conditions are known.

Let us give examples of pseudo-potential Φ.

2.4. Non dissipative constitutive law ($\Phi \equiv 0$)

The relation (4) gives

$$\mu(T-T_o) \in T\,\frac{\partial g}{\partial \chi}(\chi). \tag{7}$$

If $g = I$, it results that

$$\begin{aligned} &\text{if } \theta = T - T_o > 0, && \chi = 1 \\ &\theta = 0 && , \; 0 \leqslant \chi \leqslant 1, \\ &\theta < 0 && , \; \chi = 0. \end{aligned}$$

The whole phase-change occurs at $\theta = 0$. Consequently this is exactly the classical Stefan problem which appears to be a non dissipative problem.

If $g = I + \hat{g}$, the relation (7) means that a part of the phase-change occurs at $T = T_o$, the remaining part occurs at temperatures around T_o. A practical example in soil freezing is described later.

2.5. Dissipative constitutive laws ($\Phi \not\equiv 0$)

Let us choose

$$\Phi(\dot{\chi}, \theta) = \frac{1}{2}\,\eta(\theta)\,(\dot{\chi})^2$$

where $\eta(\theta)$ is a strickly positive function of θ. The relation (4) gives

$$\eta(\theta)\dot{\chi} + T\,\frac{\partial g}{\partial \chi}(\chi) \ni \mu(T-T_o). \tag{8}$$

This equation modelizes delays to phase-change or classical hysteresis loops observed on the $\{\theta(t), \chi(t)\}$ graph.

Let us assume that $g = I$ and that the viscosity parameter $\eta(\theta) = \eta_\varepsilon(\theta)$ depends on a parameter ε, with the following conditions

$$\begin{aligned} &\text{if } \theta \geqslant 0, \; \mu_\varepsilon(\theta) \rightarrow \bar{\mu} > 0 \text{ as } \varepsilon \rightarrow 0 \,; \\ &\text{if } \theta < 0, \; \mu_\varepsilon(\theta) \rightarrow +\infty \quad \text{as } \varepsilon \rightarrow 0 \,. \end{aligned}$$

Relation (8) is equivalent to

$$\dot{\chi} + \frac{\partial I}{\partial \chi}(\chi) \ni \frac{\mu(\theta)}{\eta_\varepsilon(\theta)}, \tag{9}$$

wich gives as $\varepsilon \rightarrow 0$,

$$\dot{\chi} + \partial I(\chi) \ni \frac{\mu\theta^+}{\bar{\eta}}, \tag{10}$$

with ($\theta^+ = \sup\{0, \theta\}$).

The equations (6) and (10) modelize the cooking of an egg. The proportion χ of cooked egg can only increase even as the temperature decreases (relation (10) proves that $\dot{\chi}$ is always positive or equal to 0). This is a completely irreversible phase-change.

Note. When $\Phi(\dot{\chi})$ is not differentiable but only subdifferentiable [4], relation (8) is replaced by

$$\frac{\partial \Phi}{\partial \dot{\chi}}(\dot{\chi}) + T\,\frac{\partial g}{\partial \chi}(\chi) \ni \mu\theta. \tag{11}$$

2.6. Mathematical results

Let Ω be a smooth, bounded domain of $\mathbb{R}^3$ and the time interval $(0,\hat{T})$. We define

$$Q = \Omega \times]0,\hat{T}[,$$

and

$$\forall u,v \in H^1(\Omega), \quad \langle Au,v \rangle = k\int_\Omega \text{grad}u \cdot \text{grad}v \, d\Omega.$$

Let be

(12) $M : \mathbb{R} \to \mathbb{R} \cup \{+\infty\}$, a convex, lower semi-continuous function whose domain is $[0,1]$,

(13)
$$\beta \in C^0(\mathbb{R}), \quad f \in L^2\big(0,\hat{T};(H^1(\Omega))'\big), \quad \theta^0 \in L^2(\Omega)$$
$$\chi^0 \in L^\infty(\Omega)\ ;\ 0 \leq \chi_0 \leq 1, \text{ a.e. in } \Omega,$$

where $\big(H^1(\Omega)\big)'$ is the dual space of $H^1(\Omega)$.

We want to solve problem P :

Find $\theta \in L^2\big(0,\hat{T},H^1(\Omega)\big)$ and $\chi \in H^1\big(0,\hat{T};L^2(\Omega)\big)$ such that

(14) $$\frac{\partial}{\partial t}(\rho C\theta+\rho L\chi) + A\theta = f \text{ in } \big(H^1(\Omega)\big)' \text{ and a.e. in }]0,\hat{T}[,$$

(15) $$\frac{\partial \chi}{\partial t} + \frac{\partial M}{\partial \chi}(\chi) \ni \beta(\theta), \text{ a.e. in } Q,$$

(16) $$\theta(0) = \theta_0 \text{ in } \big(H^1(\Omega)\big)', \quad \chi(0) = \chi_0, \text{ a.e. in } \Omega.$$

Note. The equation (14) is equivalent to the equation (6) with a classical Neumann boundary condition (the heat flux is given on the boundary of Ω).

The equation (15) of which particular cases are the equations (9) and (10) is equivalent to the variational inequality

for every v such that $0 \leq v \leq 1$,

$$\Big(\frac{\partial \chi}{\partial t} - \beta(\theta)\Big)(v-\chi) + M(v) - M(\chi) \geq 0, \quad \text{a.e. in } Q,$$

$$0 \leq \chi \leq 1, \text{ a.e. in } Q.$$

We have the following theorem which generalizes the theorem 2 of [6] :

THEOREM. *If we assume* (12), (13) *and* β *is a non decreasing function,*

$$|\beta(\theta)| \leq a|\theta|^p + b, \quad (a,b \in \mathbb{R}^+,\ 1 \leq p < 10),$$

$$f \in L^2(Q), \quad \theta^0 \in H^1(\Omega),$$

then the problem P *has a solution such that*

$$\theta \in H^1\big(0,\hat{T};L^2(\Omega)\big) \cap L^2\big((0,\hat{T};H^2(\Omega)\big).$$

If β *is a Lipschitz function, the problem* P *has a unique solution which depends continuously on the data* θ^0, χ^0 *and* f [6].

The theorem is proved by regularization and a limit process, by using monotony and compacity methods and results of [5] and [6].

The theorem applies when we choose for (15) the equations (9) and (10). Similar theorems can be proved for the equations (8) and (11).

3. FREEZING OF WATER SATURATED SOILS

In this model the free energy of the wet saturated soil depends on the temperature T, the volumetric unfrozen water content χ and the porosity ε,

$$\Psi(T,\chi,\varepsilon) = - CT \operatorname{Log} T - \frac{L\varepsilon\chi}{T_o}(T-T_o) + \bar{\lambda}(\varepsilon-\varepsilon_{ng})(T-T_o) + \frac{k}{2}T(\varepsilon-\varepsilon_{ng})^2 + T\varepsilon g(\chi) + Th(\varepsilon)$$

where C is the heat capacity of the soil, L the volumetric latent heat of the water, T_o the phase change temperature of the water (T_o = 273°K), ε_{ng} is the unfrozen porosity of the soil, $\bar{\lambda}$ and k are two positive parameters. The function g and h oblige χ and ε to take values between 0 and 1.

The specific energy e is

$$e = \Psi - T\frac{\partial\Psi}{\partial T} = CT + L\varepsilon\chi - T_o\bar{\lambda}(\varepsilon-\varepsilon_{ng}). \tag{17}$$

The last term of e, $T_o\bar{\lambda}(\varepsilon-\varepsilon_{ng})$, can be neglected, assuming that its variations with respect to the time are small.

The second principle of thermodynamics gives

$$-\frac{\partial\Psi}{\partial\chi}\dot{\chi} - \frac{\partial\Psi}{\partial\varepsilon}\dot{\varepsilon} - \frac{\vec{q}\ \mathrm{grad}\ T}{T^2} \geqslant 0 \tag{18}$$

for every $\dot{\chi}$, $\dot{\varepsilon}$ and grad T. We replace (18) by the classical relations

$$-\frac{\partial\Psi}{\partial\chi}\dot{\chi} - \frac{\partial\Psi}{\partial\varepsilon}\dot{\varepsilon} \geqslant 0 \quad \text{and} \quad -\frac{q\ \mathrm{grad}\ T}{T^2} \geqslant 0. \tag{19}$$

The last relation of (19) is satisfied assuming the Fourier's law.

3.1. Dissipation with respect to unfrozen water content χ

Following the experimental results mentionned in the introduction, we assume that there is no dissipation with respect to χ. We have then

$$0 \in \frac{\partial\Psi}{\partial\chi} = -\frac{L\varepsilon\theta}{T_o} + \varepsilon T\frac{\partial g}{\partial\chi}(\chi)$$

or

$$\frac{\theta}{\theta+T_o}\frac{L}{T_o} \in \frac{\partial g}{\partial\chi}(\chi)$$

or

$$(\theta,\chi) \in G \tag{20}$$

where G is the graph shown on figure 1 which can be obtained from experiments. It gives the unfrozen water content versus the Celsius temperature. This result is one of the more important in soil freezing : it does exist liquid water at temperature lower than 0°C in soils.

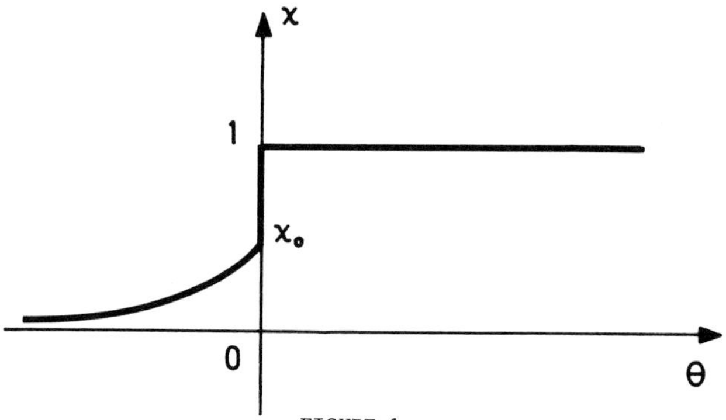

FIGURE 1
The (θ,χ) graph G. It shows the unfrozen water content χ as a function of θ. It is obtained from experiments. In a soil a part of the water remains unfrozen at negative temperatures.

3.2. Dissipation with respect to the porosity ε

Experiments show that the porosity depends on the temperature.

The first choice is to suppose that there is no dissipation with respect to ε. We have then

$$0 \in \frac{\partial \Psi}{\partial \varepsilon}$$

or

$$0 \in \frac{\partial \Psi}{\partial \varepsilon} = -\frac{L\chi\theta}{T_0} + \overline{\lambda}\theta + kT(\varepsilon-\varepsilon_{ng}) + Tg(\chi) + T\frac{\partial h}{\partial \varepsilon}.$$

This relation means that ε is a function of the temperature θ. Assuming that $h = I$, we have for $0 < \varepsilon < 1$,

$$k(\varepsilon-\varepsilon_{ng}) = \frac{1}{T_0+\theta}\left\{\left(\frac{L\chi}{T_0} - \overline{\lambda}\right)\theta - g(\chi)\right\} = f(\theta).$$

It can be seen that this function f of θ is in practical situations a decreasing function of θ, which is in accordance with experiments. For instance, this is the case if $g = I$ and $\overline{\lambda} - \frac{L}{T_0} > 0$.

Experiments show that ε is not a one to one function of θ. The results depend on the experiment. This proves that the first choice is not convenient. We are led to assume that there is a dissipation (or a viscosity) with respect to ε.

As a second choice we assume that the pseudo-potential of dissipation is

$$\Phi(\dot{\varepsilon},\theta) = \frac{C(\theta)}{2}\dot{\varepsilon}^2$$

where $C(\theta)$ is a strictly positive function of θ. So

$$C(\theta)\dot{\varepsilon} + \frac{\partial \Psi}{\partial \varepsilon} \ni 0.$$

From experiments we know that the porosity varies mostly for temperatures negative and close to 0°C. According to this result we choose

$$C(\theta) = a\, e^{C_1|\theta|}$$

where a and C_1 are positive constants. We assume also that $\overline{\lambda}$ is such that

$\bar{\lambda} - \frac{L\chi}{T_o}$ is large for θ negative ($\chi \simeq 0$) and small for θ positive ($\chi \simeq 1$).

In the neighbourhood of $\theta = 0$, $-\frac{\partial \Psi}{\partial \varepsilon}$ is equivalent to

$$(\frac{L}{T_o}\chi - \bar{\lambda})\theta - k(\theta+T_o)(\varepsilon-\varepsilon_{ng}) - (\theta+T_o)g(\chi).$$

To simplify we assume that the last two terms are small and neglectable with respect to $\bar{\lambda}$.

With all these assumptions, we have

for the negative temperatures ($\theta < 0$),

$$\dot{\varepsilon} \simeq -\frac{\bar{\lambda}\theta}{a} e^{-C_1|\theta|}$$

and we can choose as a practical law,

$$\dot{\varepsilon} = - C_2\theta e^{-C_1|\theta|} ; \tag{21}$$

for the positive temperatures ($\theta > 0$),

$$\dot{\varepsilon} \simeq \left\{(\frac{L}{T_o} - \bar{\lambda})\theta - kT(\varepsilon-\varepsilon_{ng})\right\}\frac{1}{ae^{C_1|\theta|}} .$$

Several hypotheses are possible. For instance it is possible to choose as a practical law

$$\dot{\varepsilon} = \frac{- C_3(\varepsilon-\varepsilon_{ng})}{e^{C_1|\theta|}} . \tag{22}$$

The three constants C_1, C_2 and C_3 must be obtained from experiments or from knowledge derived from experiments. The equations to obtain $\theta(x,t)$, $\chi(x,t)$ and $\varepsilon(x,t)$ are (3), (5), (17), (20), (21) and (22). They are completed by classical boundary and initial conditions.

Notes. The two main facts of the model are

1/ the large dissipation with respect to ε for θ far away from 0°C,

2/ the important variation of χ in the neighbourhood of 0°C, which results in different behaviour of ε for θ negative or positive.

When θ is negative the choice of the relation $\dot{\varepsilon}$, ε, θ is unique. When θ is positive different choices are possible. The computer program of the Laboratoire Central des Ponts & Chaussées for soil freezing has choosen the relation $\dot{\varepsilon} = - \tilde{C}_3\theta(\varepsilon-\varepsilon_{ng})$ [1].

If one whishes to be more precise when the dissipation is considered, one must not forget that the material we deal with is a soil. The results we have obtained with this model describe correctly the qualitative behaviour of soils submitted to cold temperatures.

Conclusion. Rather simple choices of free energies together with pseudo-potential of dissipation allow to describe phase-changes with dissipative or irreversible phenomenons.

REFERENCES

[1] D. Blanchard, A. Dupas, M. Frémond, M. Lévy. Soil Freezing and Thawing. Modeling and Applications. Symposium on Soil Water Problems in Cold Regions. American Society of Civil Engineers, Detroit, 1985.
[2] Gel des Sols et des Chaussées, Ecole Nationale des Ponts et Chaussées, M. Frémond, P. Williams Editeurs, Paris, 1979.
[3] P. Germain. Mécanique des milieux continus, Masson, Paris, 1973.
[4] J.J. Moreau. Fonctionnelles Convexes, Collège de France, Paris, 1967.
[5] J.-L. Lions et E. Magenes. Non homogeneous boundary value problems and applications, II, Springer-Verlag, Berlin, 1972.
[6] A. Visintin. Stefan problem with phase relaxation, J. Applied Math., I.M.A., 34, 1985, p. 225-246.

Thermomechanical Couplings in Solids
H.D. Bui and Q.S. Nguyen (Editors)
Elsevier Science Publishers B.V. (North-Holland)

THERMOVISCOELASTIC BEHAVIOUR OF HIGH POLYMERS: AN INFRARED RADIOMETRY STUDY

R.H. BLANC, Laboratoire de Mécanique et d'Acoustique du CNRS
BP 71, 13402 Marseille cedex 9, France

E. GIACOMETTI, Aérospatiale, 13725 Marignane, France

1. INTRODUCTION

The thermomechanical behaviour of high polymers is investigated. When these media are subjected to mechanical vibration, dissipation of the energy added causes a rise in temperature, which is accentuated by the fact that these media are poor heat conductors. The rise in temperature then reduces the stiffness moduli. In the case of structures, this can lead to considerable changes in the distribution of stresses and strains. This thermo-viscoelastic coupling has been studied in the context of solid propellants ; most authors have taken the heat released over a period to be approximately equal to the work added (for review see[1]),which implies that the transformation in question is both total and instantaneous. These two assumptions are shown to be inaccurate, and are replaced by more realistic formulations based on the following experiments. The energy balance between the mechanical energy added and the heat released by the material has already been investigated [2-3]; here, it is proposed to extend this research as follows. Calorimetric experiments have shown that viscoelastic dissipation involves delayed phenomena [3-4]. With infrared radiometry it is possible, due to the very fast responses obtained with this method, to investigate this problem in the frequency domain and to propose a simple law for the phase shift of the temperature which characterizes the medium thermomechanically.

2. THEORY

2.1. Statement of the problem

A thin specimen of the medium under investigation was subjected to periodical loading under simple uniaxial traction. Providing the time is sufficiently short, the initial phase of the phenomenon can be said to be adiabatic and the temperature at instant t to be uniform over the sample. The equation accounting for the phenomenon [5] can hence be reduced to

$$(1) \qquad \dot{\theta} = -\alpha\dot{\sigma} + \frac{D}{\rho c}$$

with $\alpha = \frac{\lambda}{\rho c}T$ and $T = T_0 + \theta$

where θ is the temperature variation of the sample from its initial absolute value T_0 , σ the stress, ρ the mass per unit volume, c the specific heat, λ the coefficient of linear thermal expansion and D the energy dissipation rate. In α, the coefficient $\lambda/\rho c$ of the temperature is called the thermoelastic constant. The term $-\alpha\dot{\sigma}$ is commonly known as the thermodynamic coupling term in the differential equation (1), which becomes:

$$(2) \qquad \rho c\,\dot{\theta} + \lambda\dot{\sigma}\theta = -\lambda T_0\dot{\sigma} + D.$$

The hypothesis that all the work resulting from the mechanical power added $\sigma\dot{\varepsilon}$ is transformed into heat leads to $D = \sigma\dot{\varepsilon}$, where ε is the strain. But previous experiments [2], [6] as well as the following ones show that actually only a fraction F of this work is recovered in the form of heat, and we propose the more real hypothesis :

(3) $$D = F\sigma\varepsilon \qquad \text{with} \qquad 0<F<1$$

Consider the case of an excitation having the form

$$\sigma = \sigma_0 + \sigma_1 \sin \omega t$$

In the steady state, this gives the strain

$$\varepsilon = \varepsilon_0 + \varepsilon_1 \sin(\omega t - \delta)$$

δ being the mechanical loss angle of the medium ; with

(4) $$\sigma_1 = |E^*(i\omega)|\, \varepsilon_1 \qquad \sigma_0 = E^*(0)\, \varepsilon_0$$

where $E^*(i\omega)$ is the complex modulus of the material.

The work added at instant t per unit volume of the sample is

(5) $$W = \int_0^t \sigma\dot{\varepsilon}dt = \sigma_0\varepsilon_1 [\sin(\omega t - \delta) + \sin \delta] - \frac{1}{4}\sigma_1\varepsilon_1 [\cos(2\omega t - \delta) - \cos \delta] + \frac{1}{2}\omega\sigma_1\varepsilon_1 \sin \delta . t.$$

The work added per period $2\pi/\omega$ is

$$\Delta W = \int_0^{2\pi/\omega} \sigma\dot{\varepsilon}dt = \pi\sigma_1\varepsilon_1 \sin \delta$$

2.2. **Temperature Variation**

Integration of the differential equation in θ (1) involves the factor exp $[\lambda\sigma/\rho c]$. The exponent $\lambda\sigma/\rho c$ is generally small compared to one : in the following experiments, it ranges from 10^{-4} to 10^{-3} . Taking exp $[\lambda\sigma/\rho c]$ to be equal to one and substituting for D from (3), we obtain

$$\theta(t) = -\alpha_0\sigma + \frac{FW}{\rho c}$$

with

(6)

$$\alpha_0 = \frac{\lambda T_0}{\rho c}$$

$\theta(t)$ can be expressed as the sum of two terms [7-8], a "fast" term $\hat{\theta}$ and a "slow" term $\bar{\theta}$,

$$\theta = \hat{\theta} + \bar{\theta},$$

where, taking (5) into account :

$$\hat{\theta} = -\alpha_0\sigma_1 \sin\omega t + \frac{F\varepsilon_1}{\rho c}[\sigma_0 \sin(\omega t - \delta) - \frac{1}{4}\sigma_1 \cos(2\omega t - \delta)] ,$$

a periodic term with null mean value in which the first term predominates, and

$$\bar{\theta} = -\alpha_0\sigma_0 + \frac{F\varepsilon_1}{\rho c}[\sigma_0 \sin\delta + \frac{1}{4}\sigma_1 \cos\delta + \frac{1}{2}\omega\sigma_1 \sin\delta . t] ,$$

a linear time function accounting for the mean heating.

The fundamental harmonic $\hat{\theta}_1$ of $\hat{\theta}$ is :
(7)

$$\hat{\theta}_1 = -\alpha_0\sigma_1 \sin \omega t + \frac{F\sigma_0\varepsilon_1}{\rho c} \sin(\omega t - \delta)$$

This is the term we compare with the mesured $\hat{\theta}_1$.

3. EXPERIMENTAL SET-UP

The set-up is shown in figure 1. It can be used to obtain and process the four following quantities : the force and displacement from which σ and ε are deduced, the particle velocity on which phase measurements are based, and the temperature θ. The latter is measured by means of an infrared radiometer equipped with an InSb detector, after calibration of amplitude and phase versus frequency. A frequency response analyser yields the amplitude and the phase of the signal under consideration, or those of its fundamental harmonic, in relation to the chosen origin . A more detailed description of the set-up is given in [6], "Annexe I".

Six high polymers were tested : polyvinyl chloride (PVC), polypropylene (PPH), Nylon 6, acetal, Rilsan (polyamide), and Teflon (tetrafluorethylene). The following results were obtained.

4. ENERGY BALANCES

With the first three media, we measured the fraction F of the work added that was released in the form of heat :

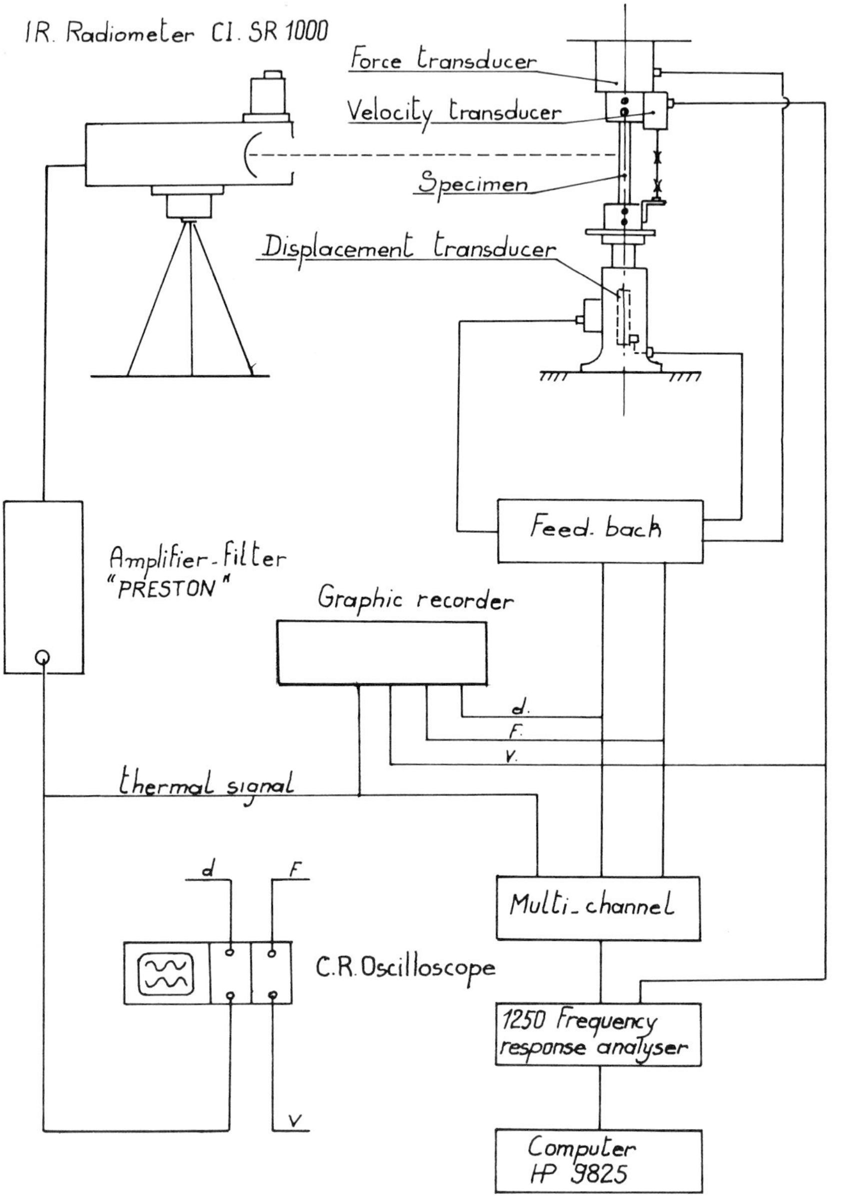

Fig. 1 - Diagram of the experimental set-up

$$F = \frac{\Delta Q}{\Delta W} = \frac{\rho c \Delta\theta \,/\, \text{cycle}}{\pi \sigma_1 \varepsilon_1 \sin\delta}$$

At 12 Hz and 24 °C, we obtained :

	PVC	PPH	Nylon
F	54%	55%	28%

5. TEMPERATURE SIGNAL PHASE

5.1. **A frequency-invariant**

In each of the six media, the phase angle φ between $\hat{\theta}_1$ and ε was measured versus the frequency f. It was proposed to compare φ with the measured value of the loss angle δ. $\varphi_\pi = \varphi \pm 180^\circ$ and δ were both plotted on the same graph, as shown in figure 2. In all the media tested, the difference $\delta - \varphi_\pi$ was found to be independent of the frequency :

(8) $\delta - \varphi = \text{constant}$

The following values were obtained :

	PVC	PPH	Nylon	Acetal	Rilsan	Teflon
$\delta - \varphi_\pi (^\circ)$	1.3	2.5	2.0	1.7	0.5	2.1

5.2 - **Effects of loading**

According to (7), the phase φ depends explicitly on the loading. Now, the above relation (8) accounts for the behaviour of the material only if φ results not only from the loading, but also from the medium.

As we have just seen, the measured φ_π is always smaller than δ, whereas, on constructing the vector $\hat{\theta}_1$, it can be observed that the calculated phase φ_π is always greater than δ . This confirms that the temperature signal phase cannot result from the loading alone.

5.3. - **Intrinsic phase shift**

We propose the hypothesis that the thermomechanical behaviour of the material causes an intrinsic phase lag, which it is now proposed to confirm by establishing the respective contributions of the two phenomena, namely the intrinsic phase lag and the loading. Let us therefore examine the variation of φ with the stress. The loading is taken to have an undulating shape as was the case in our experiments, so that $\sigma_0 = \sigma_1 = \sigma$. To simplify the formulation, let us take $|E^*(i\omega)| = E$; substituting from (4), relation (7) then becomes :

$$(9)\quad \hat{\theta}_1 = -\alpha_0 \sigma \sin \omega t + \frac{F\sigma^2}{\rho c E} \sin(\omega t - \delta)$$

Let us write $\Delta\varphi(\sigma)$ for the phase shift introduced here by the loading. Let φ_0 be the phase shift introduced between ε and θ by the medium according to our hypothesis,such that :

$$(10)\qquad \varphi(\sigma) = \varphi_0 + \Delta\varphi(\sigma)$$

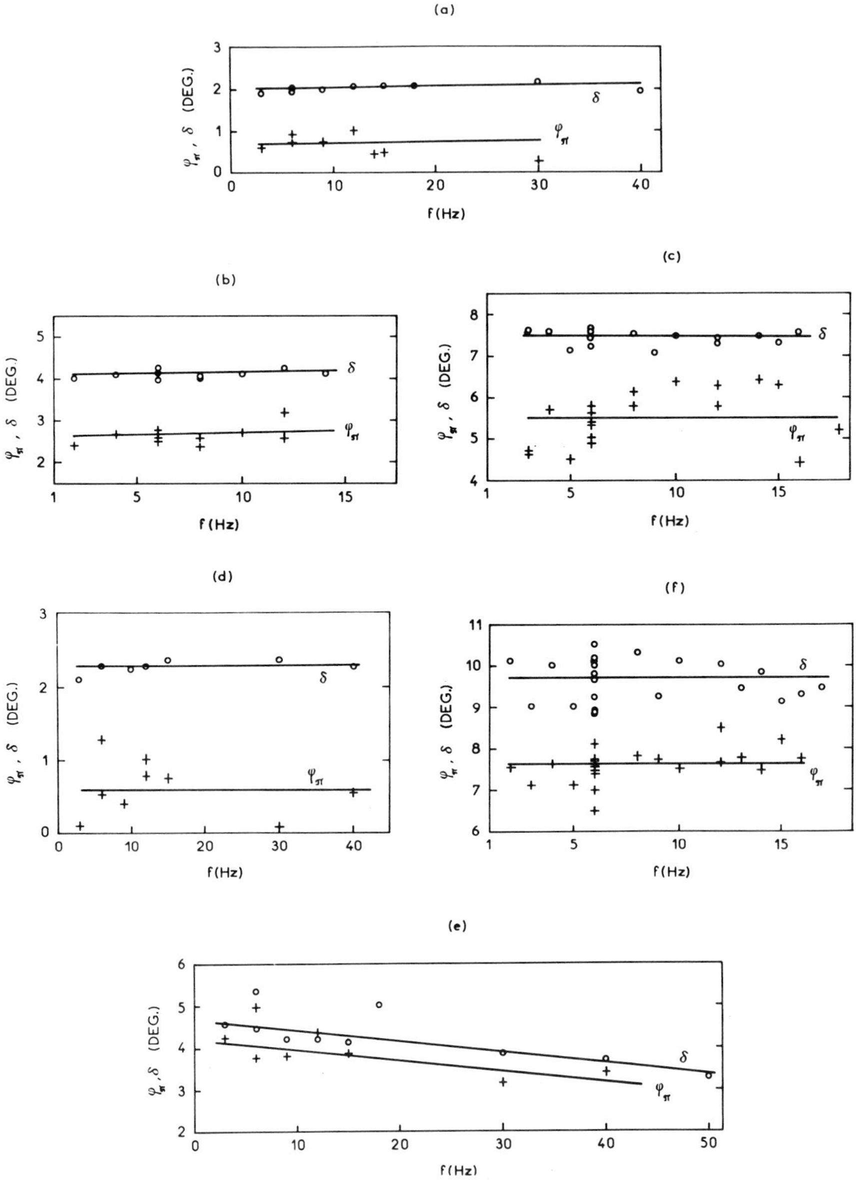

Fig. 2 - Temperature phase φ_π in relation to the strain $\pm$ 180° and loss angle δ in six high polymers under cyclic loading, as a function of frequency f. (a) : PVC, σ = 7,8 MPa ; (b) : PPH, σ = 4,9 MPa ; (c) : Nylon, σ = 6,9 MPa ; (d) : acetal, σ = 15 MPa ; (e) : Rilsan, σ = 3,9 MPa ; (f) : Teflon, σ = 3,5 MPa.

From (9), substituting from (6), we obtain to within the 2nd. order :

$$\Delta\varphi(\sigma) = \frac{F\delta\sigma}{E\lambda T_0}\left(1 + \frac{F\sigma}{E\lambda T_0}\right) \qquad (11)$$

It can be seen that $\Delta\varphi(\sigma) \to 0$ when $\sigma \to 0$, which shows that φ_0 is the intrinsic temperature phase of the material under null stress undulating loading.

The phase φ was measured as a function of the stress σ and plotted on a diagram for each of the following media : PVC, PPH, and Nylon (these graphs are presented in [6]). On the other hand, $\Delta\varphi(\sigma)$ was calculated using relation (11), and the result was substituted into (10). By identification with the experimental results, we obtain φ_0 . This gives

	PVC	PPH	Nylon
$\varphi_0(^\circ)$	0.3	2.7	5.2

In [6], we carried out the same calculation, taking F = 1 ; we obtained slightly lower values for φ_0 , but they always differed from zero.

6. CONCLUSION AND SUMMARY

Infrared radiometry, thanks to the highly rapid responses obtained, was found to be a most appropriate technique for studying fast thermomechanical phenomena. In particular, it provides a means of measuring the temperature phase, which was not previously possible. In the present study, the fraction of the work added that is dissipated in the form of heat is determined in high polymers subjected to forced vibration. Far from being 100%, this fraction is found to range between 28% and 55%. Our most noteworthy finding is the fact that between the strain and the temperature there exists a phase angle, which is an intrinsic property of the viscoelastic medium under investigation. In all the polymers tested, the difference between the temperature phase and the loss angle is found to be independent of the frequency and shown to be a thermomechanical constant of the medium.

REFERENCES

[1] Lee, E.H., Thermo-viscoleasticity, in = Hult, J. (ed.), Mechanics of visco-elastic media and bodies, Proc. IUTAM Symp., Gothenburg, 1974 (Springer-Verlag, Berlin, 1975), pp.339-357.
[2]Tauchert, T.R., Thermomechanical Coupling in Viscoelastic Solids, in : Boley, B.A. (ed.), Thermoinelasticity, Proc. IUTAM Symp., East Kilbride, 1968, (Springer-Verlag, 1970)-, pp. 316-326.
[3]Müller F.H., Thermodynamics of Deformation processes, in : Eirich, F.R., (ed.), Rheology, Vol. 5 (Acad. Press, N.Y., 1969), pp. 417-489.
[4] Persoz, B. et Rosso, J.C. Chaleur de déformation des polymères, in: Thermodynamique des comportements rhéologiques, Colloque 1977 (Cahiers Groupe Fr. Rheol., n° spéc., janv. 1980) pp. 11-35.
[5] Bouc, R. et Nayroles B., J. Méc. Théor. Appl. (1985), 4 (1), pp. 27-58.
[6] Blanc, R.H., et Giacometti E., Etude du comportement thermomécanique des matériaux viscoélastiques par radiométrie infrarouge, in : Aérospatiale, Etudes et contrôle de la structure interne et du comportement thermomécanique des matériaux non métalliques, C.R. fin d'étude, M.R.T., n° 79-7-1566, (févr 1985), 3ème partie.
[7]Blanc, R.H. and Giacometti E., Internat. J. Solids Structures (1981), 17 (5), pp. 531-540.
[8]Nayroles B., Bouc, R., Caumon, H., Chezeaux, J.C. et Giacometti, E., Internat. J. Engrng Sci. (1981), 19, pp. 929-947.